THE AURELIAN LEGACY

British Butterflies and their Collectors

'The Aurelians'
This fine painting, dated 1909, is by John Cooke,
who exhibited regularly at the Royal Academy at the turn of the twentieth century.
It hangs in the hallway of the Royal Entomological Society of London
and depicts two distinguished Fellows of the Society,
G. B. Longstaff and Professor Selwyn Image.

THE AURELIAN LEGACY

BRITISH BUTTERFLIES AND THEIR COLLECTORS

Michael A. Salmon

with additional material by

Peter Marren and Basil Harley

'I would go through fire and water for insects.'

J. C. Dale (in 1864)

2000

Published by Harley Books
(B.H. & A. Harley Ltd)
Martins, Great Horkesley,
Colchester, Essex CO6 4AH, England

Text set in New Baskerville, and designed by Alison and Peter Guy
Index prepared by Susan Williams
Printed by Midas Printing Ltd, Hong Kong, China

British Library Cataloguing-in-Publication Data.
A catalogue record for this book is available from the British Library

ISBN 0 946589 40 2

FOR SUSIE, JESSICA, AND MADELEINE

'When having shaken off all pursuers, I took the rough, red road that ran from our house towards field and forest, the animation and lustre of the day seemed like a tremor of sympathy around me. Black *Erebia* butterflies ('Ringlets' as the old English Aurelians used to call them), with a special gentle awkwardness peculiar to their kind, danced among the firs. From a flower head two male Coppers rose to a tremendous height, fighting all the way up – and then, after a while came the downward flash of one of them returning to his thistle. These were familiar insects, but at any moment something better might cause me to stop with a quick intake of breath. I remember one day when I warily brought my net nearer and nearer to a little *Thecla* that had daintily settled on a sprig. I could clearly see the white *W* on its chocolate-brown underside. Its wings were closed and the inferior ones were rubbing against each other in a curious circular motion – possibly producing some small, blithe crepitation pitched too high for a human ear to catch. I had long wanted that particular species and when near enough, I struck.'

Vladimir Nabokov,
Speak, Memory: an Autobiography revisited,
1951

'Back came the sun, dazzlingly.

It fell like an eye upon the stirrups, and then suddenly and yet very gently rested upon the bed, upon the alarum clock, and upon the butterfly box stood open. The pale clouded yellows had pelted over the moor; they had zigzagged across the purple clover. The fritillaries flaunted along the hedgerows. The blues settled on little bones lying on the turf with the sun beating on them, and the painted ladies and the peacocks feasted upon bloody entrails dropped by a hawk. Miles away from home, in a hollow among teasles beneath a ruin, he had found the commas. He had seen a white admiral circling higher and higher round an oak tree, but he had never caught it. An old cottage woman living alone, high up, had told him of a purple butterfly which came every summer to her garden. The fox cubs played in the gorse in the early morning, she told him. And if you looked out at dawn you could always see two badgers. Sometimes they knocked each other over like two boys fighting, she said.'

Virginia Woolf,
Jacob's Room,
1922

TABLE OF CONTENTS

LIST OF ILLUSTRATIONS 11

FOREWORD
The Hon. Miriam Rothschild, D.B.E., F.R.S. 15

PREFACE AND ACKNOWLEDGEMENTS
Michael Salmon 17

PICTURE AND TEXT CREDITS 21

CHAPTER 1
'The Grand Panacea' 25
A short history of butterfly collecting in Britain

CHAPTER 2
'Weapons of the Chase' 55
Pins, nets and collecting boxes

CHAPTER 3
'A Company of Serious, Thoughtful Men' 93
The butterfly collectors

1. Thomas Moffet (1553–1604) 95
2. Christopher Merrett (1614–1695) 98
3. John Ray (1627–1705) 99
4. Leonard Plukenet (1642–1706) 101
5. Charles duBois (*c.*1656–1740) 101
6. James Petiver (1663–1718) 103
7. Joseph Dandridge (1664–1746) 105
8. Eleanor Glanville (*c.*1654–1709) 106
9. William Vernon (*fl.*1660–*c.*1735) 108
10. Eleazar Albin (*fl.*1690–1742) 109
11. Benjamin Wilkes (*fl.*1690–1749) 110
12. James Dutfield (*fl.*1716?–1757?) 112
13. George Edwards (1694–1773) 112
14. Dru Drury (1725–1803) 114
15. Moses Harris (1730–*c.*1788) 115
16. Johann (John) Reinhold Forster (1729–1798) 117
17. William Jones (1750–1818) 120
18. John Berkenhout (*c.*1730–1791) 121
19. William Lewin (?–1795) 122
20. John Abbot (1751–1840) 123
21. Charles Abbot (1761–1817) 124
22. William Kirby (1759–1850) 124

23. Adrian Hardy Haworth (1767–1833) 127
24. Edward Donovan (1768–1837) 129
25. Joseph Grimaldi (1779–1837) 131
26. William Spence (1783–1859) 132
27. Richard Weaver (*fl.* 1790–1860) 133
28. Lætitia Jermyn (1788–1848) 136
29. George Samouelle (*c.* 1790–1846) 137
30. John Curtis (1791–1862) 138
31. James Charles Dale (1792–1872) 139
32. James Francis Stephens (1792–1852) 141
33. James Rennie (1787–1867) 142
34. Abel Ingpen (1796–1854) 143
35. Frederick William Hope (1797–1862) 144
36. Edward Newman (1801–1876) 145
37. James Duncan (1804–1861) 147
38. John Obadiah Westwood (1805–1893) 148
39. Henry Doubleday (1808–1875) and 150
40. Edward Doubleday (1811–1849) 150
41. John Arthur Power (1810–1886) 153
42. Francis Orpen Morris (1810–1893) 154
43. Frederick Bond (1811–1889) 155
44. John William Douglas (1814–1905) 156
45. William Buckler (1814–1884) 157
46. Charles Stuart Gregson (1817–1899) 159
47. Emma Sarah Hutchinson (1820–1906) 160
48. Thomas Vernon Wollaston (1822–1878) 161
49. Henry Tibbats Stainton (1822–1892) 162
50. John Jenner Weir (1822–1894) 164
51. Joseph Greene (1824–1906) 165
52. George Carter Bignell (1826–1910) 166
53. Henry Guard Knaggs (1832–1908) 167
54. Joseph William Dunning (1833–1897) 168
55. Octavius Pickard-Cambridge (1828–1917) 169
56. Charles Golding Barrett (1836–1904) 171
57. William West (1836–1920) 172
58. James Platt Barrett (1838–1917) 173
59. William Henry Harwood (1840–1917) 174
60. Albert Brydges Farn (1841–1921) 175
61. Thomas Algernon Chapman (1842–1921) 176
62. Thomas de Grey, 6th Baron Walsingham (1843–1919) 177
63. Francis Buchanan White (1842–1894) 178
64. William Forsell Kirby (1844–1912) 179
65. William Holland (1845–1930) 180
66. Henry John Elwes (1846–1922) 181
67. Henry Harpur Crewe (1828–1883) and 181
68. Sir Vauncey Harpur Crewe (1846–1924) 181
69. Richard South (1846–1932) 184
70. Robert Adkin (1849–1935) 186
71. Sir Edward Bagnall Poulton (1856–1943) 187

72. James William Tutt (1858–1911) 189
73. Edward Meyrick (1854–1938) 191
74. Frederick William Frohawk (1861–1946) 193
75. Heinrich Ernst Karl Jordan (1861–1959) 197
76. Margaret Elizabeth Fountaine (1862–1940) 199
77. Percy May Bright (1863–1941) 202
78. Sidney George Castle Russell (1866–1955) 202
79. Lionel Walter Rothschild, 2nd Baron Rothschild of Tring (1868–1937) 205
80. Edward Bagwell Purefoy (1868–1960) 207
81. Leonard Woods Newman (1873–1949) 209
82. Henry Attfield Leeds (1873–1958) 211
83. William Rait-Smith (1875–1958) 211
84. Hon. Nathaniel Charles Rothschild (1877–1923) 213
85. Edward Alfred Cockayne (1880–1956) 214
86. John William Heslop Harrison (1881–1967) 216
87. Philip Bertram Murray Allan (1884–1973) 218
88. Norman Denbigh Riley (1890–1979) 219
89. Lionel George Higgins (1891–1985) 220
90. Henry Charles Huggins (1891–1977) 221
91. Arthur Francis Hemming (1893–1964) 222
92. Sir Robert Henry Magnus Spencer Saundby (1896–1971) 223
93. Edmund Brisco Ford (1901–1988) 224
94. Baron Charles George Maurice de Worms (1904–1980) 226
95. Russell Frederick Bretherton (1906–1991) 227
96. Henry Bernard Davis Kettlewell (1907–1979) 227
97. Robert Watson (1916–1984) 229
98. Edward Charles Pelham-Clinton, 10th Duke of Newcastle (1920–1988) 230
99. Richard Lawrence Edward Ford (1913–1996) 231
100. Peter William Cribb (1920–1993) 232
101. John Heath (1922–1987) 233

CHAPTER 4
'Immense Swarmes of Butterflies as e'en to Darken the Skyes' 240
Some species of historical interest

1. The Chequered Skipper (*Carterocephalus palaemon*) 241
2. The Essex Skipper (*Thymelicus lineola*) 244
3. The Lulworth Skipper (*Thymelicus acteon*) 245
4. The Mallow Skipper (*Carcharodus alceae*) 249
5. The Apollo (*Parnassius apollo*) 249
6. The Swallowtail (*Papilio machaon*) 252
7. The Scarce Swallowtail (*Iphiclides podalirius*) 254
8–10. The Clouded Yellows (*Colias* spp.) 257
8. The Pale Clouded Yellow (*Colias hyale*) 258
9. Berger's Clouded Yellow (*Colias alfacariensis*) 260
10. The Clouded Yellow (*Colias croceus*) 263
11. The Cleopatra (*Gonepteryx cleopatra*) 266
12. The Black-veined White (*Aporia crataegi*) 268

13. The Bath White (*Pontia daplidice*) 271
14. The Black Hairstreak (*Satyrium pruni*) 274
15–17. The Coppers (*Lycaena* spp.) 277
15. The Large Copper (*Lycaena dispar*) 278
16. The Scarce or Middle Copper (*Lycaena virgaureae*) 285
17. The Purple-edged Copper (*Lycaena hippothoe*) 290
18. The Long-tailed Blue (*Lampides boeticus*) 291
19. The Short-tailed or Bloxworth Blue (*Everes argiades*) 293
20. The Northern Brown Argus (*Aricia artaxerxes*) 295
21. The Mazarine Blue (*Cyaniris semiargus*) 297
22. The Large Blue (*Maculinea arion*) 300
23. Albin's Hampstead Eye (*Junonia villida*) 309
24. The Painted Lady (*Cynthia cardui*) 311
25. The American Painted Lady (*Cynthia virginiensis*) 316
26. The Large Tortoiseshell (*Nymphalis polychloros*) 318
27. The Camberwell Beauty (*Nymphalis antiopa*) 321
28. The European Map Butterfly (*Araschnia levana*) 327
29. Weaver's Fritillary (*Boloria dia*) 329
30. The Queen of Spain Fritillary (*Argynnis lathonia*) 331
31. The Niobe Fritillary (*Argynnis niobe*) 335
32. The Mediterranean Fritillary or Cardinal (*Argynnis pandora*) 338
33. The Glanville Fritillary (*Melitaea cinxia*) 339
34. The Arran Brown (*Erebia ligea*) 343
35. The Monarch or Milkweed Butterfly (*Danaus plexippus*) 346

CHAPTER 5
'To the Worthy Successors of the Aurelian Society' 364
Conservation and Collecting

APPENDIX 1
List of the British and Irish butterflies,
together with their past and present common names 377

APPENDIX 2
Entomological societies, publications and significant events 397

BIBLIOGRAPHY AND FURTHER READING 409

INDEX 425

LIST OF ILLUSTRATIONS

COLOUR PLATES

Frontispiece, 'The Aurelians', by John Cooke (1909) *facing title page*

Plates

1 Dedication to the worthy members of the Aurelian Society in Benjamin Wilkes's *Twelve new designs of English butterflies* (1742) 30
2 An early London Coffee House, *c.*1705 31
3 The Brown Hairstreak from Petiver's collection, dated 1702 58
4 The Revd Adam Buddle's pressed butterflies, *c.*1700 59
5 Leonard Plukenet's pressed butterflies, *c.*1695 62
6 Charlton's Brimstone butterfly – a notorious 18th-century forgery 66
7 The Small Mountain Ringlet from James Stephens's *Illustrations of British Entomology* (1827) 135
8 The Manor House, Glanville's Wootton, Dorset 139
9 The remains of Sir Vauncey Harpur Crewe's butterfly collection at Calke Abbey 184
10 Valezina Bolingbroke (née Frohawk) in the New Forest with f. *valesina* of the Silver-washed Fritillary 196
11 Some of H. A. Leeds's Chalkhill Blue butterfly aberrations 212
12 The Chequered Skipper from Edward Donovan's *Natural History of British Insects* (1799) 243
13 The Lulworth Skipper in John Curtis's *British Entomology* (1833) 247
14 The Mallow Skipper – an historic specimen captured in Surrey, 1950 249
15 W. S. Coleman's plate of 'Reputed British species' from *British Butterflies* (1860), including The Apollo 251
16 John Curtis's illustration (1836) of a Scarce Swallowtail taken in 1822 255
17 Pale Clouded Yellow male and female, illustrated by William Lewin as two different species in his *Papilios of Great Britain* (1795) 259
18 Historic specimens of Berger's Clouded Yellow, taken in 1900 and 1947 prior to its recognition as a species 261
19 Historic specimens of the Clouded Yellow 265
20 Brimstone and Cleopatra butterflies compared 267
21 British specimens of the Bath White 273
22 Black and Brown Hairstreak butterflies from H. N. Humphreys' plate in *British Butterflies and their Transformations* (1841) 275
23 The first Large Copper recorded in England, from the minute book of the Spalding Gentlemen's Society (1749) 280
24 Historic English specimens of the Large Copper taken around the mid-19th century 283
25 Henry Seymer's skilful additions in *c.*1776 of the Scarce and Large Coppers to his copy of Moses Harris's *Aurelian* of 1766 287
26 The Northern Brown Argus with related species, first illustrated by William Lewin in his *Papilios of Great Britain* (1795) 296
27 The Holly, Small and Mazarine Blue butterflies from H. N. Humphreys' plate in *British Butterflies and their Transformations* (1841) 299
28 The Large Blue, added by Henry Seymer to his copy of Harris's *Aurelian* 301
29 Historic specimens of the extinct Large Blue from three English localities 303
30 Albin's Hampstead Eye from F. O. Morris's *British Butterflies* (1853) (after James Petiver's engraving) 311
31 The Painted Lady and Marbled White butterflies from Harris's *Aurelian* 313
32 An American Painted Lady captured in south-west England in 1970 316
33 The Large Tortoiseshell, drawn by Eleazar Albin for his *Natural History of English Insects* (1720) 319
34 The Camberwell Beauty, first recorded and illustrated by Benjamin Wilkes in *English Moths and Butterflies* (1749) 323
35 Historic English specimens of the Camberwell Beauty 326

36 The Map butterfly with other 'doubtful British species' in W. Wood's *Index Entomologicus* (1854) 327
37 Historic specimens of the Queen of Spain Fritillary taken in 1882 333
38 The Glanville Fritillary, first illustrated by James Dutfield in his exceedingly rare work of 1748, with the Orange-tip butterfly 341
39 Historic specimens of the Arran Brown, taken in the 1860s 345
40 F. W. Frohawk's plate of the Monarch butterfly's life history, from his *Natural History of British Butterflies* (1924) 347
41 Historic British specimens of the Monarch butterfly 349

BLACK AND WHITE ILLUSTRATIONS

Figures

1 Title page of Linnaeus's *Systema Naturae* of 1758 28
2 Moses Harris's frontispiece for *The Aurelian*, an early self-portrait 32
3 The Aurelian Macaroni, a foppish collector, in 1773 33
4 Moses Harris's vignette of collecting equipment from *The Aurelian* 34
5 Portrait of J. C. Fabricius 34
6 Terms for membership of the third Aurelian Society 35
7 The Thatched House Tavern, venue of the first general meeting of the Entomological Society of London 37
8 An example of H. T. Stainton's 'At Home' invitations 39
9 (a, b) Pill boxes for collecting insects 43
10 A 'South London' field trip in 1905 46
11 41 Queen's Gate, home of the Royal Entomological Society of London 48
12 'Releasing the Lepidoptera – a great moment at the Annual Dinner of the Entomological Society', a *Punch* cartoon of 1930 49
13 W. Farren's advertisement of Mazarine Blues, 1861 50
14 'The Chase' by Edward Hopley (*c.*1860) 56
15 'Bagged' by Edward Hopley (*c.*1860) 56
16 Petiver's 'Brief Directions' issued to collectors (*c.*1700) 57
17 (a, b) Two details from a 14th-century manuscript, *The Romance of Alexander* 68
18 Detail from *The Aurelian* showing a bat-fowling net in action 70
19 A bat-fowling net, still in use *c.*1900 71
20 Watkins & Doncaster advertisement of 1892 71
21 Perry's contraption 'for catching all kinds of flying insects', 1831 73
22 Nets and equipment, illustrated in Morris's *British Butterflies* (1853) 73
23 A high net in use in 1955 74
24 Setting boards and setting methods, *c.*1900 75
25 Sir Hans Sloane 77
26 Interior of Watkins & Doncaster in the 1950s 79
27 Watkins & Doncaster in The Strand with its famous Swallowtail sign 79
28 Dr David Sharp with 'water net' 81
29 Henry Gulliver, a New Forest guide, with his wife, *c.*1930 88
30 The Bull Inn near Darenth, Kent, in the 1830s 88
31 Thomas Moffet 95
32 The unpublished frontispiece for Moffet's *Insectorum Theatrum*, *c.*1600 96
33 Edward Wotton 97
34 Thomas Penny 97
35 Sir Theodore de Mayerne 98
36 John Ray, father of English botany 99
37 Francis Willughby 100
38 An extract from Charles duBois' notebook on the Comma butterfly 102
39 Petiver's 'Hogs', as he called the Large and Small Skipper butterflies 103
40 Eleazar Albin on horseback 109
41 George Edwards 113
42 Dru Drury 114
43 Moses Harris 116
44 Johann Reinhold Forster 118
45 Two pages of Forster's *Catalogue of British Insects* (1770) 119
46 William Jones's silhouette portrait 120
47 John Berkenhout 121

48 John Abbot, a self-portrait 123
49 The Revd William Kirby 125
50 A. H. Haworth, from a drawing of a lost bust 127
51 Joseph Grimaldi 131
52 William Spence 132
53 Pages from Lætitia Jermyn's *A Butterfly Collector's Vade Mecum* (1824) 136
54 John Curtis 138
55 J. C. Dale 140
56 J. F. Stephens, an early daguerreotype photograph 141
57 The Revd F. W. Hope 144
58 Edward Newman 145
59 Professor J. O. Westwood 148
60 Henry Doubleday 151
61 Edward Doubleday 151
62 Dr J. A. Power 153
63 The Revd F. O. Morris 155
64 Frederick Bond 156
65 J. W. Douglas 157
66 William Buckler 158
67 The Revd John Hellins 158
68 C. S. Gregson 159
69 Mrs Emma Hutchinson 160
70 T. V. Wollaston 161
71 H. T. Stainton 163
72 J. Jenner Weir 165
73 The Revd Joseph Greene 166
74 G. C. Bignell 167
75 Dr Henry Guard Knaggs 168
76 J. W. Dunning 169
77 The Revd Octavius Pickard-Cambridge 170
78 C. G. Barrett 171
79 William West 173
80 J. Platt Barrett 173
81 A. B. Farn 175
82 T. A. Chapman 176
83 Thomas de Grey, Lord Walsingham 177
84 Francis Buchanan White 178
85 W. F. Kirby 179
86 H. J. Elwes 181
87 The Revd Henry Harpur Crewe 182
88 Sir Vauncey Harpur Crewe 182
89 Richard South 185
90 Robert Adkin 187
91 Professor Sir Edward Poulton 188
92 J. W. Tutt 189
93 Edward Meyrick 192
94 F. W. Frohawk 193
95 F. W. Frohawk with his second wife, Mabel, and their daughter Valezina 196
96 Karl Jordan 197
97 Margaret Fountaine 199
98 Margaret Fountaine in the field 201
99 Percy M. Bright 202
100 S. G. Castle Russell 203
101 Walter, Lord Rothschild 205
102 Lord Rothschild driving his zebras 206
103 Captain E. Bagwell Purefoy 208
104 E. B. Purefoy's butterfly garden at East Farleigh 208
105 Frohawk's famous drawing of an ant carrying a Large Blue larva 208
106 L. W. Newman 209
107 Newman's Butterfly Farm (sleeves) 210
108 Newman's Butterfly Farm (cages) 210
109 William Rait-Smith 211

110 Nathaniel Charles Rothschild with his wife 213
111 Dr E. A. Cockayne 215
112 Professor J. W. Heslop Harrison 216
113 P. B. M. Allan 218
114 Norman Riley 219
115 Dr L. G. Higgins 221
116 Francis Hemming 222
117 Air Marshal Sir Robert Saundby 223
118 Professor E. B. Ford 224
119 Baron Charles de Worms 226
120 R. F. Bretherton 227
121 Dr Bernard Kettlewell 228
122 Robert Watson 229
123 E. C. Pelham-Clinton (later 10th Duke of Newcastle) 230
124 R. L. E. Ford 231
125 Peter Cribb 232
126 John Heath 233
127 The announcement in 1890 of the discovery of the Essex Skipper 245
128 The Swallowtail, a woodcut from Moffet's *Insectorum Theatrum* (1634) 253
129 The Scarce Swallowtail, a woodcut from Moffet's *Insectorum Theatrum* 256
130 The text announcing the discovery of a new species of *Colias* – Berger's Clouded Yellow 260
131 Woodcut of the Clouded Yellow from Moffet's *Insectorum Theatrum* 263
132 Petiver's 'White butterflies' including the Black-veined White (1717) 269
133 Petiver's 'Yellow, White and ... mixt Butterflies' including the Bath White 272
134 (a) Portrait of Henry Seymer 278
(b) Seymer's notes on the Scarce and Large Coppers 279
135 The report in 1885 of the Short-tailed Blue's discovery in Britain 293
136 First description of the Large Blue in England by Henry Seymer, *c.*1776 300
137 Albin's Hampstead Eye in Petiver's engraving of 'Brittish eye-winged butterflies' (1717) 310
138 Charles duBois' notebook entry for the Painted Lady (*c.*1695) 312
139 Moses Harris's description of the discovery of the Camberwell Beauty 321
140 The first report of Weaver's Fritillary in 1832 from *Loudon's Magazine of Natural History* 329
141 The Niobe Fritillary, from Newman's *Illustrated Natural History of British Butterflies* (1871) 336
142 Dr John Walker, the discoverer in Bute (*c.*1760) of the Scotch Argus 343
143 A shocking advertisement of 1827 for 20 Gross of Northern Brown Argus 365
144 'After an Entomological Sale. *Beati possidentes*' by Edward Armitage R.A. (1878) 367
145 Stevens's Auction House in King Street, Covent Garden, in the 1830s 368
146–158 Woodcuts from Moffet's *Insectorum Theatrum* (1634):
146 The Apollo 379
147 The Brimstone 380
148 The Small White 381
149 The Orange-tip 382
150 The Common Blue 386
151 The Red Admiral 387
152 The Painted Lady 388
153 The Small Tortoiseshell 388
154 The Peacock 389
155 The Comma 389
156 The Dark Green Fritillary 391
157 The Speckled Wood 392
158 The Wall 393
159 Moses Harris's dedication to members of the second Aurelian Society 398
160 William Watkins, with his butterfly cages 403
161 William Watkins' advertisement, *c.*1892 404
162 Eric Classey, entomological bookseller and publisher 406

FOREWORD

A volume of this kind has been awaited with impatience for decades, for we have all longed to know the history of the bug hunters, their nets and their lamps, their eccentricities and motivation as well as the chronicle of the ephemeral butterflies themselves. And here it is at last. For none of us, heretofore, had the patience and the skill to chase up and arrange all the scattered records and obscure references and lay before the delighted reader a clear and concise account of the British Buttterfly collectors linked to their alluring quarry.

The author has given us a wealth of contemporary information as a background to this study. It came as a shock to learn that 22,000 people died of cholera in England only 165 years ago – and among them one of the most noted collectors of his day. It was also news to me that in about 1700 Skippers were known as Hogs. Nor did I dream that the vermilion in Albin's matchless portraits of insects was mixed with boy's urine and Brandy Wine. It was also of great interest to learn that Westwood, despite winning and accepting the Royal Society Medal, refused nomination as a Fellow. Was this, because like Groucho Marx, he would not join a society which would accept him as a member? Or like N. C. Rothschild (who likewise declined the honour) because he felt passionately that someone else deserved it (in this case Karl Jordan) ahead of himself?

Every entomologist must find something of personal interest in this history. I myself was delighted to discover that one of my childhood heroes, F. O. Morris, who collected such a plethora of rare species at Ashton Wold, was strongly opposed to vivisection. There is something here for every one of us.

Quite apart from the facts and anecdotes about the collectors themselves, the information about their equipment is full of interest, from setting pins to clap nets and bag nets. 'Shrimping?' enquired ex-Prime Minister Wilson when he met one of the world's most distinguished scientists, E. B. Ford, butterfly hunting in the Isles of Scilly. There will be no excuse for such confusion – even on the part of Prime Ministers – once they have read Chapter 2.

What is it that has set naturalists apart from the rest of mankind? The author has turned a searchlight of patient investigation on to the British subspecies. It is said that the love of natural history lies at the root of the ineptitude of many of our politicians. Wasn't Sir Edward Grey, the bungling Foreign Secretary at the outbreak of World War I, a fanatical and very knowledgeable bird lover? Wasn't the youthful Neville Chamberlain – an equally bungling Prime Minister at the outbreak of World War II – an ardent butterfly collector? That arch procrastinator, Lord Simon, a member of the War Cabinet, was a keen conchologist and had (so he himself proclaimed) described a new species of snail from the sandhills of Wales. Malcolm MacDonald, a splendid ornithologist, lost us the use of the Irish ports and might, as a result, have lost us World War II into the bargain. As for Francis Hemming, who unquestionably knew more about butterfly nomenclature than anyone before or since, he made an inconceivable mess of non-intervention in the Spanish Civil War.

I fancy that Dr Salmon indirectly gives us a perspicacious answer to this question. For just those qualities which they demonstrate are needed to equip the wonderfully gifted naturalist and butterfly lover spell disaster for the politician. How glorious are the trees, how entrancing are the subtly different greens of the foliage, how breath-taking and delicate are the shape of the leaves and the tracery of their venation, and how majestic the growth of branches vanishing in the canopy! But where is the wood? It is simply overlooked.

This book is written in a simple and direct style, but gives the reader much room for thought. Recently the medicos have announced that we live longer and have fewer heart attacks if we keep a dog. Judging from the octogenarians and nonagenarians who skip through these pages, net in hand, there is something to be said for the study of entomology. Reading about the lives of the collectors we sense that they are people who, however inept at dealing with matters of state, have accidentally found the secret of happiness – concentrating with astonishing tenacity to the details of another parallel world – rising above the ills to which human flesh is heir, on the wings of the angelic butterfly.

Miriam Rothschild,
Ashton Wold,
June 1999

PREFACE

'To distinguish between eccentricity and genius may be difficult, but it is surely better to bear with singularity than to crush originality.'
Anon. (1861)

Eccentricity or genius? The two are often intertwined. The question, which was originally asked of Henry Doulton, manufacturer of fine porcelain, might apply equally to many of the great butterfly collectors of days gone by. At one time the collector was an outcast, the object of incredulous stares and the jeers of small boys. The scientific and intellectual reasons for chasing butterflies were barely understood: it was seen as no fit occupation for a gentleman, even in his leisure hours. Nor did his appearance help matters: the formal garb of a country parson or physician were ill-matched by a baggy clap-net and a clanking satchel full of collecting tins and other paraphernalia. Yet, even as early as the mid-eighteenth century – a quarter of a millenium ago – entomology was by no means only the pursuit of the solitary eccentric. Butterfly collectors, or 'Aurelians' as they called themselves (from the Latin '*aureolus*' – said to refer to the highly gilded and decorated chrysalids of certain butterflies) used to meet regularly at the Swan Tavern in London, where they housed their collections, library, and regalia. One of their number, Moses Harris, was to write the most popular book about butterflies published up to that time. Suitably enough, it was called *The Aurelian.*

A century later, collecting and the study of natural history had become respectable, even fashionable activities. Butterfly collectors now had the potential to attain a more dignified position in society, as men of learning, and sometimes as popular lecturers. For a few, butterfly collecting became a full-time activity, supplying rarities and livestock to wealthy clients, curating the national and provincial collections, or editing an entomological journal. For the many, however, natural history collecting had become a passion. Victorian households often boasted a cabinet of shells, butterflies or fossils, together with a series of George Routledge's shilling volumes on natural history, or a book of newspaper cuttings from the weekly 'nature' column. For some it was an all-consuming passion. It is recalled that Mrs Philip Gosse once interrupted family prayers to point out a small brown moth fluttering near the curtains. 'Oh Philip!, do you think that can be *Boletobia*?' 'No dear, it is only the common Vapourer Moth, *Orgyia antiqua*', he replied, returning to his Bible. This was the period when much of what we know about British butterflies was discovered: the 'localities' where rare species flourished; the food-plants of their caterpillars and the details of their life-cycles; and the curious varieties or 'aberrations' – Peacock butterflies without 'eyes', 'blues' with dashes instead of dots, yellow Brimstones with splashes of flaring vermillion – all dignified by Latin names, though their genetic cause and meaning were as yet unknown.

We need to remind ourselves of the extraordinary profusion of butterflies in the nineteenth century. There was mile after mile of short-cropped downland sheep-walks alive with Adonis and Chalkhill Blues; every parish contained ancient woods managed as coppice with a regular supply of warm, sheltered, flower-filled glades for Fritillaries, White Admirals and Hairstreaks. There were no insecticides, no artificial fertilizers, no motorways, filling stations or out-of-town supermarkets. Major roads and highways only infrequently criss-crossed the landscape. The railway system, well developed by the 1860s, did not really intrude – after all, it often provided a good habitat for wildlife along its railway banks. Pollution was confined to the larger cities. The patchwork countryside, which since time immemorial had supplied hay and pasture for horses and farm animals, timber and fuel-wood, corn and orchard fruit, was also a paradise for insects. At the turn of the century much of the land was still permanent pasture, full of flowers and bounded by thick hedges, with ponds in the corners. J. F. Stephens was moved to write ecstatically of a collecting trip, in the 1820s, in Surrey. 'The boundless profusion with which the hedgerows for miles, in the vicinity of Ripley, was enlivened by myriads [of White-letter Hairstreak butterflies] that hovered over every flower and bramble blossom, last July, exceeded anything of the kind that I have ever witnessed.' He added that, 'some notion of their numbers may be formed when I mention that I captured, without moving from the spot, nearly 200 specimens in less than half an hour'. F. W. Frohawk remembered the New Forest in the latter years of the century, when the Silver-washed Fritillary was so extraordinarily abundant 'that it was common to see forty or more assembled on the blossoms of a large bramble bush, in company with many White Admirals, Meadow Browns, Ringlets, and here and there among the swarm one or two of the ab. *valesina.* When the congregation was disturbed, they would rise in a fluttering mass and the majority would again settle to continue their feast on the sweet blossoms of the bramble'.

Much of this countryside was made accessible for the first time by the growth of the railways; and indeed the collector followed the rail routes, often with his bicycle in the guard's van, as surely as the hound followed the hare. Some localities swarmed with collectors on a few weekends in the summer, each working his own patch of 'territory' – and woe betide anyone who trespassed on to it. The New Forest was the greatest magnet of all, but other places had their individual fame: Royston Heath for blue 'varieties', Blean Woods, near Canterbury, for the Heath Fritillary, the Cornish coast for Large Blues, Castle Eden Dene, and Arthur's Seat, near Edinburgh, for the Northern Brown Argus. There is little doubt that butterfly collecting as a pastime reached its heyday in the late Victorian era, and for a few decades everyone seemed to be dashing off to the countryside, collecting specimens for the family cabinet, or listening to lantern slide lectures on the wonders of nature.

Did the collector help to exterminate our native butterflies? He was certainly accused of doing so, and there were many examples of collecting for collecting's sake, hundreds of Large Blues and Purple Emperors captured and killed simply to fill a cabinet drawer. But collecting did not cause the long-term decline of our butterflies; at most the collector might have wiped out a few local colonies of heavily collected species, like the Large Blue. The real cause lies in more fundamental changes in the countryside from one that favoured butterflies to one that very largely does not. While some butterflies continue to thrive in gardens, field corners and lanes, and on other land not under intensive cultivation, others are more or less confined to nature reserves. Around the mid-nineteenth century, we lost perhaps the most beautiful species of all, the Large Copper, when the last of the primaeval fens was drained. At the turn of the century we lost two more, the Black-veined White and the Mazarine Blue, for reasons which remain obscure though overall climatic change might have played a part. More recently we have lost the Large Blue and, apparently, the Large Tortoiseshell. We have learned by how slender a thread these beautiful but fragile insects survive. The difference of a degree or so in average temperature; a few centimetres in the height of the grass, or an unnoticed increase in the shade may be enough to propel a declining isolated colony to the finality of extinction.

The natural history and conservation of butterflies have been described many times. The stars of this book are not so much the butterflies themselves as their pursuers and greatest admirers: the butterfly collectors. If the ghosts of those great collectors, James Petiver, Eleazar Albin, Moses Harris, Edward Newman, and Henry Stainton, should smile at what they may read between these covers, I would be more than a little pleased. These gentlemen were a remarkable lot. Eccentric? Yes! Obsessional? To a degree! But, above all, they were romantics in pursuit of a hobby to which their enthusiasm and devotion knew no bounds. Today the pendulum has swung so far away from collecting that it is all too easy to belittle or disregard their achievement. But we can only conserve butterflies today because of what collectors discovered about them. There is also something in the culture of butterfly collecting, that peculiarly intense enthusiasm which is about discovery as much as about capture, as well as in the personalities of the collectors themselves, that deserves to be celebrated. Many were characters as extraordinary, as colourful as any that appear in the pages of Trollope or Dickens. The pursuit of butterflies has produced in Britain Dru Drury the silversmith, the gentle, artistic Moses Harris in his tricorne hat, Henry Doubleday the quietly spoken Epping grocer, the irascible schoolmaster J. W. Tutt, Margaret Fountaine the intrepid traveller, and E. B. ('Henry') Ford, whose youthful craze for butterflies led to his demonstration of how genetics works in the field. What binds them together is a romantic devotion to the most beautiful of all insects, to which they dedicated much of their leisure and faced adventures and trials, jealous quarrels and convivial companionships. Their quest, part acquisitiveness, part curiosity, is exemplarized by Tutt, whose honeymoon (claimed some wag) consisted of collecting butterflies all day, and collecting moths all night.

This book is an anatomy of that passion. The history of butterfly collecting in Britain is described as it unfurls from the end of the seventeenth century to the present time. It includes the 'brief lives'of many of the most famous and eminent collectors. Behind these giants of this once popular and – leaving views about its morality to one side – undoubtedly healthy activity can be counted thousands of English, Scots and Welsh who have collected quietly for their own private pleasure. Eric Classey tells the story of a dear old clergyman who once purchased a butterfly net from him. 'I only collect in the vicarage garden', confided the old man, apologetically. Then, looking over his shoulder, and dropping his voice, he added, 'but I do occasionally allow myself to lean over the fence'.

Michael A. Salmon
Woodgreen, New Forest
April, 2000

ACKNOWLEDGEMENTS

It is always a pleasure to acknowledge the help of others and this book, in its final form, owes a very great deal to Peter Marren and Basil Harley. I am grateful to both. Peter read through the whole of the text and introduced a number of important changes, especially in Chapter 3 – the brief lives of one-hundred-and-one famous butterfly collectors. He did much to ensure that these individual biographies were lively, and at the same time brought to my attention details of a number of ancient and relatively unknown lepidopterists. From the outset we agreed to include only historical characters and omit living lepidopterists thus avoiding controversy. His conviction that the history of butterfly collecting occupied a respectable niche in the social history of previous generations proved most reassuring.

Basil Harley has devoted an enormous amount of time on my behalf over the past few years, researching the history of butterfly collecting in Britain. He has checked and rechecked dates, references and quotations throughout the text – always insisting on the utmost accuracy when compiling a text so replete with historical facts. His enthusiasm for the subject and his confidence that such a book on the emotive subject of butterfly collecting was needed, has proved most infectious. It was entirely due to him that my attention was directed towards the Revd Adam Buddle's ancient herbarium which he told me contained historic butterflies as well as plants; one of the earliest collections in existence. Basil also completely revised my original chapter on butterflies of historical interest, adding to it a number of new species and enlarging the text concerning others, and has greatly augmented the Appendixes and Bibliography. To him I owe a great debt of gratitude.

The marvellous help given by Brenda Leonard, Jacqueline Ruffel, Katherine Watkins and Berit Pedersen, successive Librarians to the Royal Entomological Society, is gratefully acknowledged, as is that of Pamela Gilbert and Julie Harvey, former and present Entomology Librarians at the Natural History Museum in London. Paul Cooper, assistant librarian in the General Library, and Lorna Mitchell, Vicki Veness and Zöe Gerrard, assistant librarians in the Entomology Library, devoted a great deal of time to tracing obscure references, and I am extremely grateful to them too. Mr Greg Bentley, lately Registrar to the Royal Entomological Society, very kindly allowed me to copy photographs of famous old lepidopterists from the Janson Collection and suggested that I might include a reproduction of '*The Aurelians*', a large portrait of two distinguished entomologists that hangs in the hallway of the Society's home in South Kensington. His assistance and that of his successor, Mr W. H. F. Blakemore, is much appreciated.

For encouragement and for the loan of some remarkable material, I am delighted to thank Dame Miriam Rothschild, John Redshaw of the Spalding Gentlemen's Society, Terry Dillon, and Michael J. Perceval whose copy of Henry Seymer's *Aurelian* is truly historic.

I would like to thank my friends Peter Edwards, Tudor Morgan-Jones and Eric Classey who have all encouraged me from the start, as well as David Elliston Allen, Karl Bailey, David Carter, Michael Chalmers-Hunt, June Chatfield, Allan Davies, Martin Evans, Joe Firmin, Mike Fitton, Robin Ford,

Brian Gardiner, Robert and Rosemary Goodden, John Gulliver, Pat Hall, Tony Harman, Alec Harmer, Jacques Hecq, Martin Honey, Tony Irwin, John Ounsted, Christopher Palmer, Elizabeth Platts, Claude Rivers, Dorothy Sharp, Jonathan Spencer, Christine Taylor, Andrew Wakeham-Dawson, Terry Wickett and Roy Vickery, as well as the late Viscountess Bolingbroke (Valezina Frohawk), Lionel Higgins, Humphrey Mackworth-Praed, Ernest Neal and John Purefoy, all of whom helped in so many different ways. To Annette Harley I owe more than I can say. Her word processor has seen so many changes of text and format that she must have felt at times that she was working on several books at once. She has sub-edited the whole book with great skill and determination and it has been a pleasure to receive her advice and acknowledge her corrections on so many occasions.

The illustrations are an important part of this work and I must acknowledge Barry Rickman, Richard Weaver, John Webb, and the staff of Messrs Rodney Todd-White & Son, all of whose photographic skills are so apparent. Richard, in particular, devoted hours to the seemingly endless task of copying sheaves of faded sepia-toned portraits of Victorian and early twentieth-century lepidopterists, enabling these worthies to appear once again in print. Most of the photographs reproduced in this book were taken by him.

Last but by no means least I must thank my wife Susie for her patience during the decade or more that this volume has required to take shape. There is little doubt that her continued interest means that, by now, she must be almost as familiar as I am with the lives of many of the lepidopterists of the past 300 years. I am sure that she will agree that this brief history of butterfly collecting would never have been written if the lives of many of the Aurelians had been anything less than remarkable.

PICTURE AND TEXT CREDITS

PICTURE CREDITS

The illustrations in this book are reproduced with the kind permission of the following institutions and individuals in whom the copyright is vested or who have supplied photographs from material out of copyright. Others have been reproduced from material generously lent by private collectors.

Colour Plates are reproduced courtesy of the following:
The Royal Entomological Society of London: Frontispiece & Plate 1; British Museum, London, UK/Bridgeman Art Library: Plate 2; The Natural History Museum, London: Plates 3 & 5 (Dept. of Entomology), Plate 4 (Botany Library), Plates 38 & 40 (Entomology Library); The Linnean Society of London: Plates 6, 8, 13, 16, 17, 22, 26 & 27; The National Trust: Plate 9 (photograph: Michael Freeman); John Gulliver: Plate 10; The Spalding Gentlemen's Society: Plate 23; The Editor, *The Entomologist's Record and Journal of Variation*: Plate 39.
Plates 7, 11, 12, 14, 15, 18–21, 24, 25, 28–37 & 41 are all from private sources.

Black and white illustrations are reproduced courtesy of the following:
The Natural History Museum, London: Figs 1, 5, 16, 38, 42, 45, 48, 50, 63, 112, 114 & 138; The Royal Entomological Society of London: Figs 6, 7, 11, 14, 15, 21, 30, 36, 41, 46, 54, 55, 57, 59–62, 64–74, 76–78, 83–87, 89–91, 93, 96, 100, 103, 109, 116, 117, 127, 135, 140 & 142; British Entomological and Natural History Society: Figs 10, 79 & 80; Punch Ltd.: Fig. 12; The Bodleian Library, University of Oxford: Figs 17a,b (MS. Bodl. 264, fol. 44r detail of lower border; and fol. 135r whole lower border); Robin Ford: Figs 19, 26, 27, 124 & 160; Newsquest (Sussex) Ltd.: Figs 23 & 29; The British Museum: Fig 25; The British Library: Figs 31–34 (*Insectorum Theatrum*, 434f.10); Gem Publishing Co. (*The Entomologist's Monthly Magazine*): Figs 28, 75, 82 & 121; C. MacKechnie-Jarvis: Fig. 56; The Editor, *The Entomologist's Record and Journal of Variation*: Figs 81 & 92; The National Trust: Fig. 88; Dr June Chatfield: Figs 94 & 95; W. F. Cater/News Corporation: Fig. 97; Norfolk Museums Service: Fig. 98; Dame Miriam Rothschild: Figs 101, 102, 105 & 110; Mrs Rosemary Goodden: Fig. 104; the late L. Hugh Newman: Figs 106–108; E. W. Classey: Figs 113 & 162; the late Professor E. B. Ford: Fig. 118; The London Natural History Society: Fig. 119; A. J. Pickles: Fig. 122; The National Museums of Scotland: Fig. 123; The Amateur Entomologist's Society: Fig. 125; Henry R. Arnold: Fig. 126; The Editor, *Lambillionea*: Fig. 130; The Linnean Society of London: Fig. 134a; The Wellcome Institute Library, London: Fig. 144; Philip Wilson Publishers Ltd.: Fig. 145 (from *Natural History Auctions 1700–1972*, compiled by J. M. Chalmers-Hunt, published by Sotheby Parke Bernet Publications, 1976).
Figs 2–4, 8, 9, 13, 18, 20, 22, 24, 35, 37, 39, 40, 43, 44, 47, 49, 51–53, 58, 99, 111, 115, 120, 128, 129, 131–133, 134b, 136, 137, 139, 141, 143, 146–59 & 161 are all from private sources.

TEXT CREDITS

Passages from journals and other sources quoted in the text are included courtesy of the following copyright holders:
The Editor of *Amateur Entomologist's Bulletin*; the Editor of *The Journal of the British Entomological and Natural History Society*; the proprietors of Gem Publishing Co. for *Entomologist's Gazette* and *Entomologist's Monthly Magazine*; The Royal Entomological Society of London for extracts from several of its publications; the Editor of *British Birds* for extracts from *The Zoologist*; the Editor of *Nature* for the passage on the Painted Lady by S. B. J. Skertchley; Punch Ltd for the anonymous poem about J. W. Tutt; Dr June Chatfield for passages by F. W. Frohawk; E. W. Classey for extracts from P. B. M. Allan's *Leaves from a moth-hunter's notebook*; Joe Firmin for letter about W. H. Harwood; Robin Ford for material concerning the history of Messrs. Watkins and Doncaster, and for extracts from P. B. M. Allan's *A moth-hunter's gossip* and *Moths and memories*; David Higham Associates for extract from *Jacob's Room* by Virginia Woolf; Professor William R. Mead for part of his translation of Pehr Kalm's account of his visit to Sir Hans Sloane's Museum; The National Trust for excerpt from Sir Howard Colvin's *Calke Abbey*; the late Dr Ernest Neal for extracts from *The Badger Man, memoirs of a biologist*; the late L. Hugh Newman for extracts from *Living with Butterflies*; The Director of the Oxford University Museum of Natural History for extracts from A. Z. Smith's *A History of the Hope Entomological Collections*; Dame Miriam Rothschild for extracts from her appreciations of Karl Jordan and E. B. Ford, and from *Dear Lord Rothschild*; Smith, Skolnik Literary Management, by arrangement with the estate of Vladimir Nabokov, for UK edition, and Vintage Books, a division of Random House, for US edition, for extract from *Speak, Memory; an autobiography revisited* (all rights reserved.); extract from *The Old Century and Seven More Years* copyright Siegfried Sassoon by kind permission of George Sassoon; The Spalding Gentlemen's Society for the extract from their Minute Book.

THE AURELIAN LEGACY

British Butterflies and their Collectors

CHAPTER 1

'The Grand Panacea'

A SHORT HISTORY OF BUTTERFLY COLLECTING IN BRITAIN

'I pity unlearned gentlemen on a rainy day,'
Lucius Cary, Viscount Falkland, 1634

'Things that are unusual are too often esteemed ridiculous.'
W. Kirby and W. Spence, 1826

'I AM ON THE LOOK-OUT FOR AN ENTOMOLOGICAL WIFE', confided Samuel Stevens, a member of the recently founded Entomological Society of London, to his friend J. C. Dale, in 1845. 'I want one that will be useful as well as ornamental.' Dale would have understood, and indeed probably shared, Stevens's singlemindedness. He at least had been fortunate enough to have found a wife who was tolerant of entomology, and who accompanied him on some excursions. Yet, Dale was prepared to 'go through fire and water for insects'; he once rode forty miles in a day, and discovered three insects new to science at his destination.

Clearly insects, especially Lepidoptera, and most of all, butterflies, were never far from the thoughts of Stevens and Dale. Their passion for entomology has been shared by countless others during the past three hundred years. For some, like the poet Thomas Gray, or the Prime Ministers Neville Chamberlain and Winston Churchill, butterflies were a leisure activity, affording moments of relaxation and pleasure in a crowded life.[1] Indeed, the Revd William Kirby claimed that collecting was 'a grand panacea for the *tædium vitae*',[2] while for many small boys butterfly collecting was a passion that like other passions – marbles, birds' eggs, comics – came and went, as an episode in the pageant of youth. For Lord Walsingham, it meant asking Henry Stainton, the leading lepidopterist of his day, if he knew 'of a clergyman entomologist wanting a living of about £100 a year in Norfolk'. For others, however, boyhood butterfly chasing grew into more serious forms of quest. Some travelled far to collect unusual forms or 'aberrations': some bred gorgeous tropical butterflies on potted plants and in sleeves of muslin; others painted them in fine detail; still others spent hours each day writing correspondence, editing journals and hosting social gatherings, in the evangelical spirit of Victorian naturalists. Half of the history of butterflies can be summed up as a passion for beauty and a lust for 'curiosities'. Another, the more enduring half, was a quest for understanding, which has its highest expression in the works of Gilbert White and Charles Darwin. Many naturalists bred, watched and described butterflies in all their stages, and in the minutest detail, for their own sake. Some of the Victorians saw in entomology a route for moral as well as intellectual improvement. Later on, some scientists began to study butterflies for the more universal truths they revealed on subjects like heredity, physiology and evolution.

Butterfly enthusiasts came from all walks of life, from factory workers and handloom weavers to duchesses and prime ministers, and even a King, (for George III was quite interested in butterflies). A large proportion of the leading collectors were clergymen, or from the medical or legal professions. Some few were artists and illustrators. What united them all was a passion – a sense of wonder, an admiration for beauty, a quest for understanding – which we can readily share today, for a love of butterflies provides a sense of kinship down the ages. Those of us who collect, photograph, breed or merely admire butterflies today share the same excitement, the same intense curiosity, that gripped Henry Stainton and James Tutt, in the last century, or James Petiver and Moses Harris, in the century before that, or perhaps even Thomas Moffet during the reign of the first Elizabeth.

How does one capture in words that sense of excitement on first catching sight of a rarity? Some collectors enjoyed many such moments, for in collecting luck always seems to play a part. For a few, it has taken the form of an epiphany so extreme that it shook the senses. Take Alfred Russel Wallace for example, on recalling his first vision of the magnificent birdwing butterfly, *Ornithoptera croesus*: 'My heart began to beat violently, the blood rushed to my head, and I felt much more like fainting than I have done when in apprehension of immediate death. I had a headache for the rest of the day.'[3] Siegfried Sassoon experienced a quieter moment of profound satisfaction when he at last saw the butterfly of his dreams, a Camberwell Beauty (see page 325). Two centuries earlier, Benjamin Wilkes, the first person to see this butterfly in Britain, must have felt something similar, and Moses Harris too for he named it his Grand Surprize.[4]

Charles Darwin, who collected beetles as an undergraduate, wrote memorably of the forgetfulness that attends the entomologist afflicted by The Passion. 'No pursuit at Cambridge was followed with nearly so much eagerness, or gave me so much pleasure', he wrote in his autobiography. 'I give a proof of my zeal: one day, on tearing off some old bark, I saw two rare beetles, and seized one in each hand; then I saw a third and new kind, which I could not bear to lose, so that I popped the one which I held in my right hand into my mouth. Alas! it ejected some intensely acrid fluid, which burnt my tongue so that I was forced to spit the beetle out, which was lost, as was the third one.'[5] So intense was his experience of collecting that he could still recall the exact appearance of certain posts, old trees and banks where he had captured rare beetles forty years earlier. Darwin also understood the value of collections. He once expressed to John Lubbock his 'surprise that he had never met with any one who collected odd-shaped biscuits'. Lubbock commented that, 'Though the idea seems at first sight quite ludicrous, yet a collection of the biscuits of different nations would possess many more points of interest than can be found in postage stamps'.[6]

This book is about the people who have collected butterflies in Britain. I have attempted to provide a sense of their world – the field trip and the meetings, the journals and the cabinets, the often strange paraphernalia they used, and the passions and eccentricities of the collectors themselves. I have called it the Aurelian Legacy to remind readers of the name that the early collectors called themselves. The Aurelians were enthusiasts who held their winged subjects in such awe that it seems they desired to borrow their golden glory and bestow it upon themselves. These were the people who discovered the basic facts about our native butterflies. But, more than that, they were emblematic of an age of natural history that enthused all levels of society (it does so still, but in more passive forms – conservation charities, wildlife programmes on television, and nature holidays). Although collecting in general is the most private of pleasures, collecting butterflies was also a social pursuit. Those Aurelians whose names are remembered today were dedicated, to an extent driven, men and women, but they have also been journalists, tour leaders, lecturers (and preachers), and genial hosts, fond of food, wine and gossip. They lived life to the full.

THE AURELIANS

British butterflies have a history as well as a science. Naturalists have collected and studied butterflies for over three hundred years; the earliest surviving specimens date from the late seventeenth century. Few of the earlier naturalists, at least, confined themselves to butterflies alone. And many of the great collectors of the nineteenth century had an equal enthusiasm for the much more numerous moths, and some also for beetles, bees, ants and other insects. The earliest collectors, like James Petiver, stacked boxes and stuffed cabinets with fossils, shells, skins, bones and every imaginable natural object. What started this sudden craze for natural history, which began, as far as we can tell, in the 1680s? In part, no doubt, it was the product of a more stable and wealthy society, with town houses and country mansions filled with beautiful furniture, paintings and decorations. Such people owned curio-cabinets, and the more unusual or brightly coloured insects were, without doubt, 'curiosities'. These conditions might have created an ardour for collecting, at least among the wealthier classes.

More fundamentally, the late seventeenth century was a time when science – that is, the study of the workings and relationships of the natural world – was exciting. Fear and superstition were on the wane, and people were growing more curious about their environment. The new age was epitomized by the founding of the Royal Society in 1662. Very slowly at first, perceptions of the natural world were changing. Francis Bacon had insisted that man is the interpreter of Nature, and that knowledge can be acquired from exact measurement, observation and experiment. Later in the century, Isaac Newton demonstrated the natural laws of light, physics and universal motion. What the Royal Society, of which he was President, did was to establish a forum in which new discoveries in physics, chemistry, medicine and nature could be shared through discussion and publication in the Society's Transactions. It in turn spawned a number of local learned societies, of which one of the earliest was the Spalding Gentlemen's Society, founded in 1710 to 'include all arts and sciences and exclude nothing from our conversation but politics, which would throw us all into confusion'. Members of this Society (which is still extant) took a keen interest in entomology (see Chapter 4: 15).

Among the first generation of Fellows of the Royal Society were John Ray and James Petiver. Ray, who was primarily a botanist, used Bacon's process of inductive reasoning to establish the first natural classification of plants and animals, based on their structural affinities. It was he that brought forth order out of the seeming chaos of the natural world, long before Carl von Linné, better known as Linnaeus.[7] Ray was able to do so because, like all the best British naturalists, he was a field man, familiar with the wild flowers, birds and insects of his native country. And like all the best naturalists of his generation he was a collector – although his success in collecting butterflies was limited by his inability to set specimens after *rigor mortis* had set in. Linnaeus built on Ray's work and, in the tenth edition of *Systema Naturae*, published in 1758 (Fig. 1), broke new ground in providing binominal names for all known animal species, a scheme which has been universally regarded as the starting point for all modern zoological nomenclature.

CAROLI LINNÆI
EQUITIS DE STELLA POLARI,
ARCHIATRI REGII, MED. & BOTAN. PROFESS. UPSAL.;
ACAD. UPSAL. HOLMENS. PETROPOL. BEROL. IMPER.
LOND. MONSPEL. TOLOS. FLORENT. SOC.

SYSTEMA
NATURÆ
PER
REGNA TRIA NATURÆ,
SECUNDUM
CLASSES, ORDINES,
GENERA, SPECIES,
CUM
CHARACTERIBUS, DIFFERENTIIS,
SYNONYMIS, LOCIS.

TOMUS I.

EDITIO DECIMA, REFORMATA.

Cum Privilegio S:æ R:æ M:tis Sveciæ.

HOLMIÆ,
IMPENSIS DIRECT. LAURENTII SALVII,
1758.

Fig. 1. The title page of the first volume of Linnaeus's *Systema Naturae* (Edn 10) of 1758, which covers the Animal Kingdom. He had no illusions about the magnitude of the task of classifying all living things. 'Those we know', he wrote, 'form only a fraction of those of which we know nothing'. He divided the Class Insecta into seven Orders, among which were the Lepidoptera. The butterflies (Papiliones) were in turn divided into six 'phalanges'.

The real expert on insects was not Ray but his friend, the apothecary Petiver, who has rightly been called the father of British entomology. Petiver collected specimens from Britain and from all over the world, somewhat indiscriminately, like stamps, and described each one briefly in a long series of pamphlets. It was Petiver who first gave English names to our butterflies – though not, usually, the familiar ones of today. Remarkably, the majority of British butterflies were known by around 1700, including rarities like the Mazarine Blue, Bath White and Queen of Spain Fritillary, and elusive, retiring ones like the Brown and White-letter Hairstreaks. Some may never have been 'discovered'. The common Red Admirals, Peacocks, Brimstones and 'Cabbage' Whites were as familiar to the sixteenth-century Thomas Moffet – our first entomological author – as they no doubt were to every country parson, squire or shepherd. But it is only in the brief, rather arid catalogues of Petiver and his contemporaries, and the preserved specimens on which they were based, that the documented history of butterflies really begins.

We know so little about that enviable period in British natural history when undescribed species were as common as cockroaches in a contemporary tavern, and when any reasonably observant naturalist might soon expect to throw his hat in the air as he found his 'Grand Surprize'. We can only imagine the moment when William Vernon, for example, chanced on the first Bath White (though he called it a Half-mourner), or when Eleanor Glanville found her eponymous fritillary in, of all places, Lincolnshire. A large collection of these butterflies was amassed by that prince of collectors, Sir Hans Sloane (1660–1753), physician to the Royal Household and Secretary of the Royal Society. He was once described as collecting everything and anything. Sloane spent enormous sums on butterflies and other specimens, and was well able to afford it (wrote Alexander Pope) as he received equally enormous fees for his medical opinions.[8] His museum included at least 5,500 insects; his library, 50,000 volumes and 3,560 original manuscripts. The books survived; most of the butterflies, unfortunately, did not. Some fell to bits; others were purloined, or even burnt. George Shaw (1751–1813), Keeper of Natural History and Modern Curiosities at the British Museum, made annual 'cremations' of Sloane specimens. 'When I came to the Museum most of these objects were in an advanced state of decomposition, and they were buried or committed to the flames one after another.'[9] Fortunately some of the earliest specimens survived the bonfires and are still in the Natural History Museum, London. The collection of Leonard Plukenet

(1642–1706), in a volume of about 140 pages which was acquired by Sloane, may be the oldest insect collection in existence, though those butterflies collected by the botanist Adam Buddle (*c.*1660–1715) and pressed among his pages of dried grasses are almost as old. Plukenet's collection is now in the Department of Entomology, whereas Buddle's is to be found in the Botany Department (see Chapter 2). Some of Petiver's own specimens, which he preserved between mica slips, have also survived, and a few pinned and mounted (but alas dataless) specimens, like the famous Bath White in the Hope Collection, Oxford, are the other mute monuments of the first few decades of butterfly collecting in Britain.

Entomology had become a social, even a fashionable, pursuit by the early 1700s; it was popular enough to be satirized by the poet Alexander Pope in 1712 in *The Rape of the Lock.* Among the first known collectors were women of high social standing, like Lady Margaret Cavendish Bentinck, only daughter of Edward Harley, 2nd Earl of Oxford, and later Duchess of Portland, whose collection formed part of a large private museum and menagerie at Bulstrode in Buckinghamshire. Another was Mary Somerset, widow of the first Duke of Beaufort, who reared butterflies and moths, probably with the aid of her large collection of potted plants, and acted as patron to the entomological artist, Eleazar Albin. This interest by society women in Lepidoptera continued well into the eighteenth century. In 1750, a quarter of the wealthy subscribers to *English Moths and Butterflies,* Benjamin Wilkes's sequel to *Twelve New Designs of English Butterflies* of 1742 (Plate 1), were women. The list of members of the first Society of Aurelians suggests, however, that by then male domination of organized societies was already firmly established, for all were men.

In the time of Ray and Petiver, collecting seems to have been a means to an end – a stocktaking of the natural world. Their method of mounting butterflies between slips of mica, like pressed flowers, scarcely suggests much interest in purely aesthetic considerations. The best-known entomologists of the eighteenth-century enlightenment were, however, men of a different stamp. Joseph Dandridge, Eleazar Albin, Benjamin Wilkes and Moses Harris were artists and illustrators, clearly enthralled by the brilliant wings and beautiful patterns of butterflies, as well as by the contrast between that 'crawling worm', the caterpillar, and the unconfined soaring imago. A poem by Henry Baker, in the preface to Wilkes's book, captures this eighteenth-century sense of wonder at nature's glories.

> See, to the Sun the Butterfly displays
> His glistering wings and wantons in his Rays:
> In Life exulting, o'er the Meadow flies,
> Sips from each Flow'r, and breathes the vernal Skies.
> His splendid Plumes, in graceful order show
> The various glories of the painted Bow.
> Where Love directs, a Libertine it roves,
> And courts the fair ones through the verdant Groves.
> How glorious now! How chang'd since Yesterday
> When on the ground a crawling Worm it lay,
> Where ev'ry foot might tread its Soul away!
> Who rais'd it thence and bid it range the Skies?
> Gave its rich plumage and its brilliant Dyes?
> 'Twas God: its God and thine, O Man, and He

Plate 1
Benjamin Wilkes's elaborate dedication in his *Twelve New Designs of English Butterflies* of 1742 provides the earliest known reference to the Society of Aurelians, the world's first learned society devoted specifically to entomology.

In this thy fellow Creature lets thee see
The wond'rous Change that is ordained for thee.
Thou too shalt leave thy reptile form behind,
And mount the Skies, a pure ethereal Mind,
There range among the Stars, all pure and unconfin'd.'[10]

Plate 2
'An early London Coffee House', *c.*1705, by an anonymous artist ('A. S.'). 'You have a good fire, which you may sit by as long as you please; you have a dish of coffee; you meet your friends for the transack of Business, and all for a Penny'.

These men had advantages denied to the first generation of lepidopterists – their own society. In the early 1700s in the City of London it was the fashion to meet for learned discussions in Coffee Houses (Plate 2), or even taverns, if they were respectable. Botanists used the Rainbow Coffee House in Watling Street. The entomologists, who may have formed an offshoot of botanical society, began to meet at the Swan Tavern in Exchange Alley, probably from the 1720s onwards. At some stage, after 1720 but before 1742, they decided to place their meetings on a more formal basis. They founded the first entomological society in history, and chose a name for it in keeping with the allusive classical temper of the time. They called themselves the Society of Aurelians. The name derives from the Latin, *aureolus*, meaning golden, and referring to the gilt-decorated chrysalides of certain nymphalid butterflies. To be an Aurelian was to borrow some of the glories of their admired insects – for an Aurelian was both a butterfly *and* a collector. At the same time, the name has a rather charming, boyish, mock-heroic flavour of which these artistically inclined dilettantes were no doubt aware.

Unfortunately we know little about the Society of Aurelians, and would know nothing at all were it not for the passing remarks by Wilkes, Dutfield and Moses Harris in their respective books. Very probably their founder and leading light was the elderly Joseph Dandridge, the great forgotten father-figure of British entomology.[11] Dandridge was a noted collector. He had lent many specimens to Ray, and was an inspiration to the younger naturalists of the mid-century, not least because of his talent as a water-colourist. We know some of the other Aurelians. Moses Harris was at that time considered too young for membership – he was only twelve – and so, in his words, was 'deprived of that pleasure' until with greater maturity he 'might become fitting for the Company of that ingenious and curious Body of People' (Fig. 2). But his uncle, Moses Harris senior, was a member as was Benjamin Wilkes. Others were Peter Collinson, F.R.S. (1694–1768), a wealthy Quaker and accomplished naturalist who later corresponded with Linnaeus; Thomas Knowlton, the horticulturalist; James Leman [?Lemon], the weaver and silk-fabric designer who owned a 'curious cabinett' of insects; and the above-mentioned Henry Baker, F.R.S. (1698–1774), a distinguished microscopist, man of letters and an amateur poet. The other known members are little more than names: Stephen Austin and Samuel Lee were dignified by an 'esquire'. The others, presumably humbler men, were Samuel Hartley, Elias Brownsword, Walter Blackett, Philip Constable junior, Thomas Grace, Daniel Marshal, Edmund Overall and William Wells.

Fig. 2. The frontispiece to Moses Harris's *Aurelian* of 1766 is generally believed to be an early self-portrait. He was evidently a well-equipped as well as a well-dressed young collector.

The Society's records and collections were permanently held in a back room of the Swan Tavern, on Exchange Alley, and this was to prove its undoing. On 25 March, 1748, a great fire swept through 'Change Alley, gutting every house on the street, including the Swan. It very nearly succeeded in roasting the Aurelians too. Moses Harris described the conflagration:

> '...*the great Fire happened in Cornhill*, in which the *Swan Tavern* was burnt down, together with the Society's valuable Collection of Insects, Books, *&c.* and all their Regalia: The Society was then sitting, yet so sudden and rapid was the impetuous Course of the Fire, that the Flames beat against the Windows, before they could well get out of the Room, many of them leaving their Hats and Canes; their Loss so much disheartened them, that altho' they several Times met for that Purpose, they never could collect so many together, as would be sufficient to form a Society, so that for fourteen Years, and upward, there was no Meeting of that Sort, till Phoenix-like our present Society arose out of Ashes of the Old.'[12]

It is a pity that we do not know more about these Books and Regalia. The only known illustrated books on British butterflies of that time are Albin's and Wilkes's, although the Society might have owned more general works by Ray,

Martin Lister and Moffet. Harris may have been referring to their manuscript journals, records and accounts – if so, to a unique body of historic scientific evidence, lost at a stroke and forever. As for regalia, these gilded lepidopterists may have adopted a pomp and ceremony reminiscent of the Linnean Society in 1829, when the President 'wore a three-cornered hat of ample dimensions, and sat in a crimson armchair in great state'. New Fellows 'were marched one by one to the President, who rose, and taking them by the hand, admitted them'.[13] A caricature of an 'Aurelian Macaroni' [*Maccaroni*: a dandy who affected foreign manners and style], said to be Moses Harris himself, also hints at a self-conscious interest in ostentatious plumage (Fig. 3).

Fig. 3. 'An Aurelian Macaroni'. An appropriate costume for an entomological dandy of the Aurelian Age (though his butterfly net is pathetic) – but was it entirely fanciful, or might the Aurelians in fact have dressed up as butterflies?

The Great Cornhill Fire swept through part of the City of London and a number of contemporary reports told of the devastation caused. *The Daily Advertiser* informed its readers that:

> 'Yesterday Morning, about One O'Clock [i.e. 1.00 a.m., 25th March], a Fire broke out at Mr Eldridge's, a Peruke-Maker, in Exchange-Alley, which consum'd several Houses in the said Alley, Birchin Lane and Cornhill; but the wind being South-South-West, all the Bankers Houses in Lombard Street, and their Effects are safe. No Publick Office has been burnt, except the London-Assurance ... Garraway's, the Jerusalem and Jonathan's Coffee Houses, the Swan Tavern, with the rest of the Houses in Change-Alley, are destroyed, except Baker's and Sam's Coffee-Houses, which are greatly damaged. The Flames extended themselves into Cornhill ... It's said, by People well acquainted in the Neighbourhood, that upwards of 160 Houses were burnt down.'

The London Magazine added that 'Mr Eldridge the wig-maker, his wife, two children, and a journeyman all perished in the flames'.

What were the Aurelians doing there at one o'clock in the morning? Ronald Wilkinson reminds us that March 25th was New Year's day by the Old Style Julian calendar† – the Society had probably been enjoying port and speeches after an excellent dinner. If so they were no doubt in a dangerously tipsy state as the flames engulfed the building.[14]

The second Aurelian Society 'arose out of the Ashes of the Old' in 1762,

† In the sixteenth century, the Julian calendar, introduced throughout the Roman Empire, was adjusted in most of Europe in favour of the Gregorian calendar, but the British Isles and their dependencies did not follow suit. In 1751, following the passing of the Calendar Act, the Gregorian calendar was finally adopted here, and the 3rd September that year became 15th September, suppressing the eleven intervening days completely. In addition, from 1752, New Year's Day, which until 1751 had been celebrated on 25th March, was changed to 1st January. Thus the so-called 'Old Style' calendar's date of 25th March ceased to be New Year' Day and became 5th April in the 'New Style', that date remaining to this day the start of the British fiscal year.

this time with Moses Harris as its Secretary (Fig. 4). By then, Harris had matured into an experienced entomologist. In his famous book, *The Aurelian*, published in 1766, he tells us how he had taken 'all opportunities to get

Fig. 4. Moses Harris designed this vignette which appears on the title page of *The Aurelian.* It is based on entomological apparatus of the day, and includes a clap net, racket net, collecting box and rearing cage, together with a weighted stick and sheet for beating larvae from overhanging boughs.

Knowledge in the Times, Seasons and Manner of Breeding' of Lepidoptera, and had found a wealthy patron in Dru Drury, the silversmith. Unfortunately this second Society was even shorter-lived than the first and a Society of Entomologists which followed it in 1780 lasted less than two years. We know that, like the first, the second Aurelian Society had premises and collections, for these as well as many private collections were visited and examined by the great Danish entomologist, Johann Christian Fabricius (Fig. 5), in 1767. They may not have impressed him much since, on his later visit to the capital in 1780, by which time the Society had ceased to exist, he nevertheless found 'the different collections had been considerably enriched'.[15] According to Drury, the Society disintegrated from internal dissension. We do not know what caused the quarrel, but it seems to have been a matter of incompatible personalities, unlike the cause of the collapse of the third and last Aurelian Society in 1803. This was the Society's claim of priority in requiring each member to surrender specimens for its collections

Fig. 5. A contemporary portrait of J. C. Fabricius (1745–1808), the great Danish entomologist and systematist. He visited Britain on many occasions to study the collections of British 'Aurelians' and sometimes to collect insects in their company.

(Fig. 6). Adrian Haworth, who had founded the third Society in 1801, insisted that the rule be followed to the letter, since its main object was 'to form a complete and standard Cabinet of the Entomological productions of Great Britain'. In other words, science before selfishness. He claimed that each member – there were only 10 of them, all keen collectors – had 'every one hitherto contributed articles to the *Cabinet*, which could not have been procured from any other source whatever. All these gentlemen have given up, with unexampled zeal, from their respective Collections, to the *Aurelian Cabinet*, every *British insect* which that did not contain; thus assembling together the most extensive collection in British Entomology hitherto made'.[16] Not everyone was as generous and high-minded as this, and it seems there was no great rush from others to join the Society. Then, as later, hard-sought rarities are to the collector the most precious of possessions, with a sentimental importance wholly transcending their scientific value. The third and last Aurelian Society was consequently disbanded in the spring of 1806.

TERMS

OF BECOMING

A MEMBER OF THE AURELIAN SOCIETY.

ANY person desirous of becoming a Member of this Society, must be approved of by every one of the Members for the time being; and if he possesses a collection of British Insects, he must give up from it, to the *Aurelian Cabinet*, at least one specimen of every species and variety, which the latter does not possess: for which he will immediately receive from the Curator of the Cabinet for the time being, the fullest value in rare insects; or money if he choose to accept it; and afterwards occasional duplicates of scarce Insects, which will continue to be collected, both by purchase and personal industry; to answer the claims of such *Aurelians* as may hereafter wish to have their names enrolled in the annals of the society.

The above sacrifice of rare or unique species, considering the truly advanced state of the (*a*) *Aurelian Collection* (the most extensive in British Entomology hitherto made) can seldom be great, and will evidently become less and less.

By these means the Aurelian Cabinet must ultimately arrive at the standard of perfection; and the separate collection of every Member of the Society will gradually increase both in number and value.

The Members will have a right of inspecting the Cabinet as often as they please, either for the purposes of pleasure or instruction.

(*a*) This Cabinet, and all the British Insects it contains, I here pledge myself to give up to the *Aurelian Society*, bonâ fide, and without fee or reward, as soon as the Society shall amount to 20 living Members, in such manner as I shall then explain. *Vide page* xiii. *of the Preface.*

Fig. 6. Membership terms of the third Aurelian Society, including the pledge extorted from members by A. H. Haworth to sacrifice specimens 'not yet possessed by the Society' to the 'Aurelian Cabinet'. It became a subject of dissension, which brought the Society to a premature end in 1806.

Despite their chequered history, the three Aurelian Societies had helped to keep alight the flame of Ray and his contemporaries during a rather depressed period in the history of the study of natural history. The series of books, which reached its apogee in *The Aurelian*, indicates that the study of Lepidoptera, at least, maintained some of its early popularity. However, the reduction in influence of the Royal Society after the death of Isaac Newton left a kind of scientific black hole. Newton's successor as President, Martin Folkes, was apt to fall asleep in his great chair during meetings. Matters started to improve from the mid-eighteenth century onwards, due to the genius of Linnaeus and his revolutionary method of naming and classifying plants and animals. Linnaeus was no expert on the Lepidoptera, and he acknowledged the work of British entomologists, especially Moffet, Petiver, Ray, Albin and Wilkes (though his debt to Ray was greater than he admitted). That the Swede Linnaeus was able, in his *Systema Naturae*, to describe and classify nearly all the butterflies known to occur in the Britain of 1758 may be due to the extensive collections and records documented by the Aurelians. However, it was left to his favourite pupil and disciple, Fabricius to provide a more accurate method of classifying insects based on the structure of their mouthparts.

In 1807, shortly after the demise of the third Aurelian Society, The Entomological Society (later in the same year renamed the Entomological Society of London) was established. It is uncertain exactly how long this Society was in existence, for the sources conflict. Between 1807 and 1812 it published three volumes of *Transactions* – the earliest of any entomological society. Among other things, they included the announcement of the discovery of a new British butterfly, the Small Mountain Ringlet, at Ambleside in 1808.[17] After 1812, the year which saw the issue of the third and last part of the Society's *Transactions*, Adrian Haworth lamented that 'owing to the death of some of the members and the resignation, neglect or departure from London of others, it only held quarterly, yearly or occasional meetings until the year 1822'. The half dozen members then formed yet another society – the Entomological Society of Great Britain – without dissolving the Entomological Society of London but only adjourning it for one year, and there is in fact no proof that it ever formally ceased to exist.[18] The new society seems to have been very inactive, and was wound up in 1824.

Meanwhile, with the adoption of the binominal system of nomenclature came another of those periodic surges in the progress of the study of natural history. In 1788, the Linnean Society was formed, with the purpose of 'cultivating Natural History in all its branches'. The collections, library and manuscripts of Linnaeus had been purchased in 1784 from Fru von Linné (his widow) for £1,088. 5s. od. by Sir James Smith (who became the first President of the Society), and were brought to Britain and stored in premises at 14 Paradise Row (now part of Royal Hospital Road, Chelsea), rented by Smith for the purpose. In 1787 the collection was moved to Smith's new home in Great Marlborough Street, and after several further moves, to Norwich, where Sir James died in 1828. Negotiations for the purchase of the Linnaean and Smith Collections by the Linnean Society were soon begun, completion being finally accomplished in 1835. In the meantime, the collections were transferred in 1829 to Burlington House, Piccadilly, but it was not until 1856 that the Society obtained permanent premises there, though initially in that part currently occupied by the Royal Academy. In 1873 the collections were finally moved to the current home of the Linnean Society in the western half of the Piccadilly frontage of Burlington House.[19]

Following the demise of the Entomological Society of London in 1822, entomologist members of the Linnean Society felt keenly the lack of a society devoted solely to entomology. However, William Kirby was opposed to the establishment of yet another body and proposed instead separate committees of the Linnean Society devoted to individual branches of Natural History. As a result, with the support of Haworth and other entomological fellows, 'The Zoological Club of the Linnean Society of London' was formed in 1823. However, there was no vehicle for publication of papers or proceedings other than the *Transactions* of the parent society and the lack of such a journal caused growing dissatisfaction. At the same time other zoologist members began to think of other outlets for coordinating their activities and promoting their discoveries and so, in 1826, the Zoological Society of London was founded by Sir Stamford Raffles with the objectives of forming a collection of living as well as preserved animals, and also a zoology library. There was as yet no suggestion of starting any rival publication and many of the Linnean Society fellows were happy to become fellows of the Zoological Society too. However it grew so strongly that it was able to obtain its own charter in 1829 and, from 1830, to publish its own *Proceedings* and, from 1833, its *Transactions*. Spurred on by these developments, entomologist members of the Linnean Society gathered in 1833 with others who were not members to found a new Entomological Society of London with the intention of publishing its *Proceedings* and *Transactions*. The first general meeting of this new society was held on 22nd May 1833 at The Thatched House Tavern in St James's Street (Fig. 7), with J. F. Stephens in the Chair. The Revd William Kirby was elected Honorary Life President, with instructions that his portrait should hang henceforth above the President's chair, and J. G. Children of the British Museum became the Society's first President.

The Linnean Society proved an inspiration for one natural history society after another, first in Scotland, then in different parts of Britain. Hitherto, natural history had been a London-based activity. This was soon to grow into a network of local and regional clubs and societies that introduced men and women in every part of the country to nature study, aided by improvements in technology – the microscope, which, by the 1830s, could be purchased for two or three guineas,[20] the aquarium, the fernery, and, later, the lantern slide. If the seventeenth century had been an age of entomological pioneering, and the eighteenth of rationalization as a result of the activities of the Aurelians, the nineteenth was to be the great age of the burgeoning of field clubs and entomological and natural history journals – and the golden age of natural history collecting in Britain.

Fig. 7. The Thatched House Tavern, St James's Street, where the first general meeting of the Entomological Society of London was held in May 1833.

'AT HOME' WITH HENRY STAINTON

Ralph Waldo Emerson has said that 'there is properly no history; only biography',[21] and sometimes history does seem like a pageant of famous names. Entomology is no exception. In Britain, a succession of giant figures bestrode the nineteenth century: A. H. Haworth, J. F. Stephens, Edward Newman, J. O. Westwood, Henry Doubleday, H. T. Stainton and J. W. Tutt. All of them wrote great works, edited journals, entered vigorously into every issue of the day and promoted entomology as a social and educational pastime. Stephens, Newman and Stainton even threw open their homes for this purpose. Their purpose was as much social and moral as well as scientific. These Victorians were the opposite of the 'ivory tower scientists'. Henry Stainton in particular felt a duty to share his collections, his knowledge and above all his enthusiasm. These men were missionaries for butterflies (and moths), and were leaders of a febrile entomological world, gossiping, arguing, writing, dining together and above all sharing the pleasures of the field and the excitement of the chase.

A good example of a Victorian evangelist was the Revd J. G. Wood (1827–89), an all-round naturalist in the Christian spirit of Charles Kingsley and Philip Henry Gosse. His popular, shilling handbooks, like *Common Objects of the Country* (1857), were among the first natural history books that the poor could afford – unlike the sumptuous volumes by Wilkes, Harris and Westwood. He lectured throughout Britain and went on a tour of the United States, illustrating his talks with sketches in coloured chalk on canvas. He had a large family to maintain, and so threw himself into a prodigious daily routine. 'His power of work was simply astonishing ... He was always at his desk by half-past four or five o'clock in the morning at all seasons of the year, lighting his own fire in the winter, and then writing steadily until eight. Then, in all weathers, he would start off for a sharp run of three miles over a stretch of particularly hilly country, winding up with a tolerably steep ascent of nearly a quarter of a mile, and priding himself on completing the distance from start to finish without stopping, or even slackening his pace. Then came a cold bath, followed by breakfast.'[22] Edward Newman shared this ethic to some extent; he not uncommonly arose long before dawn to snatch an extra hour or two for entomology.

This dedication was matched by a sense of confidence – that armour-plated self-assurance of the Victorian middle classes. Writing of eccentricity among the British, Edith Sitwell remarked that it was 'because of that peculiar and satisfactory knowledge of infallibility that is the hallmark and birthright of the British nation'. To some extent, field entomologists did require a thick skin. A butterfly net was to some almost a badge of frivolity and oddness, especially when it was accompanied by a stove-pipe hat (lined with cork), collecting tins, a few yards of sheeting and a pincushion bristling with pins, worn like a medallion. One of the first and most enjoyable and influential works aimed at the field collector, written by William Kirby and William Spence, warned the novice to expect stares and catcalls from the ignorant. But, it went on, hopefully, 'they will soon become reconciled to you, and regard you no more than your brethren of the angle and the gun. Things that are unusual are too often esteemed ridiculous'.[23] Charles Kingsley – the famous author of *The Water Babies* and *Hereward the Wake*, who possessed a lifelong enthusiasm for natural history – recalled that during his childhood the naturalist had always been regarded as a harmless enthusiast. He went bug-hunting simply because 'he had not the spirit to follow the fox'. Edward Newman was more resigned to a low-rating in 'the opinion of the unlettered rustic'. The opinion of 'ninety-nine persons out of a hundred', he wrote, in ironic vein, was that anyone 'who could take an interest in pursuing a butterfly is a madman. The collector of insects must, therefore, make up his mind to sink in the opinion of his friends, to be the subject of undisguised pity and ridicule of the mass of mankind, from the moment he commences so insignificant a pursuit'.[24]

Henry Guard Knaggs was another who kept his tongue firmly in his cheek. He advised lepidopterists seeking foodplants for their larvae, that a 'botanist's collecting-box would enable him to bring home a plentiful supply of fresh food, though the ordinary chimney-pot hat of daily wear answers very well for the purpose and saves the extra burden'. Collectors of a 'solitary, retiring disposition', were recommended to wear 'rifle green, or some such mournful tint, and, above all, shun the use of a shiny cap, brass fittings, and such glittering articles as may be seen for miles even by the naked eye'.[25]

Confidence of another sort was required by evangelicals like Henry Tibbats Stainton. Stainton was a remarkable man: a philanthropist and would-be politician as well as an entomologist, he was worried by the broadening inequalities in Victorian society, and strove to bring different classes together through entomology. He also addressed himself to the young by writing pieces in his *Entomologist's Annual* in the spirit of Dr Arnold of Rugby, who would 'romp and play in the garden, or plunge with a boy's delight into the Thames, entering into his pupils' amusement with scarcely less glee than themselves'. Stainton insisted that he too, was no 'sedate elderly person with no fellow feeling for a mischievous school boy'. On the contrary, he continued (with perhaps a slightly forced gaity), 'I have no sedateness about me, and am as full of fun as any one'. He insisted that he should not be addressed as 'Sir' even by a perfect stranger but 'Dear Sir', and was 'quite ready to participate in the delight of the youngest Entomologist, on adding some species to his collection, or some new fact to his knowledge'.[26]

For Stainton, as for Charles Kingsley, entomology was not only a science, but a morally correct pursuit. In *The Entomologist's Annual*, he reassured the young novice that 'if the Entomologist learns to be cheerful, and learns not to be conceited, no one can tell him that his pursuit is a useless one'. He added, almost as an afterthought, that he would not recommend *everyone* to become an entomologist nor would he recommend anyone to devote his whole time to entomology. 'Take a pleasure in your business', he advised, 'and make a business of your pleasure'.

From 1856, Stainton advertised weekly 'At Homes' in *The Entomologist's Weekly Intelligencer*, a chatty entomological newspaper he had helped to found in that year. Any entomologist above the age of fourteen who happened to be passing his house in Lewisham was welcome to drop in (Fig. 8). So popular did these occasions become, that Stainton even advertised when he was *not* 'At Home', as for example when he and Mrs Stainton would be 'sailing down the Rhine'. Stainton explained the origin and purpose of these regular socials in characteristic fashion:

Mr. Stainton will be at home, as usual, on Wednesday next, at 6 p. m. Trains by the North Kent Railway leave London Bridge at 5.30, 6.0, and 6.45, returning from the Lewisham Station at 9.13, 10.23, and 10.43, p. m.

On these occasions Mr. Stainton is happy to see any entomologist above 14 years of age (whether previously known to him or not), who may wish to look at his collection or consult him on any entomological matter.

Fig. 8. An example of H. T. Stainton's 'At Home' invitations, published in *The Entomologist's Weekly Intelligencer* of 10th May 1856.

> 'It is now nearly fourteen years ago since we were calling one evening at the house of an entomological friend, and we happened to ask him the names of some two or three moths, which he was not prepared to give with certainty off-hand, and he proposed that, as it was *Wednesday evening*, and Mr. [J. F.] Stephens was "at home" on Wednesday, we should consult him on the knotty points. Mr Stephens had then for many years devoted his Wednesday evenings to the service of entomologists. Anyone, incipient or professor, who wished to obtain information on any branch of Entomology, or to see Mr. Stephens' valuable and interesting collection, had only to call at Eltham Cottage some Wednesday evening, without any previous appointment or intimation that he was coming, and there he would be sure to find Mr. Stephens, with imperturbable good humour, happy to place his collections and books at the service of the stranger, and also to assist him personally in his investigations.'

Stainton then explained how during Stephens's lifetime he himself had held regular 'at homes' on Thursdays during winter months. Following Stephens's death, he invited visitors to come on Wednesdays too to consult Stephens's catalogue but, after one season, the numbers soon dropped off. 'It then occurred to us' he wrote, 'that if we diminished the frequency of our "at homes" we should ensure a better supply of visitors, and it would be a relief to everyone; incipients were often frightened when they found they were in for a *tête-à-tête* of three hours. We accordingly promulgated a new regulation, that we should be "at home" the first Wednesday in each month, and to this we still endeavour to adhere. Last Wednesday, for instance we were sailing down the Rhine, but next Wednesday we hope to be "at home".'[27]

The beaten track to Stainton's house, Mountsfield, became so well known that an anonymous poet, perhaps J. W. Douglas, even gave the directions in verse. This amusing ditty captures something of the convivial spirit of *The Intelligencer*.

'The Way to Mountsfield (A guide for strangers)

Would you seek the way to Mountsfield –
Would you know it – would you find it,
You must kindly learn these verses;
If they're rude, you must not mind it.

When you leave the Lewisham Station,
You must take the left-hand turning,
Keep the road straight through the turnpike,
The sun upon your right cheek burning.

When the turnpike lies behind you,
A road upon your left you see;
That's the road you must not follow,
For if you do you'll go to Lee.

Keep the main road through the village,
Until you've gone full half a mile;
The gravell'd footpath's rough with pebbles,
And p'raps torments your corns the while.

When a wood-yard you are passing,
And glimpse of the church tower just catch,
Take the turning on your left hand,
The grocer at the corner's "Patch".

As you are walking up this lane
A brick wall stands upon your left,
The *Draba verna* grows upon it
Luxuriantly in every cleft.

A row of houses stands before you,
The road you want behind them goes,
A well you'll then see on your left,
And near it a fine hop-plant grows.

You pass some houses on your right,
Which seem to try the varied dodge,
One is taller than the others,
See, there's its name, "Benbraden Lodge".

Then passing by the "Spotted Cow",
A pond you'll notice on your right,
Proceed along this country lane,
And Shooter's Hill will come in sight.

Now pause, for here *two* roads you meet,
To th' right or else you will be wrong,
Th' road goes sweeping round two corners,
But follow it, the way's not long.

To th' left an oaken gate you see,
The entrance to a gravelled drive
Open this gate and walk up boldly,
And thus at Mountsfield you'll arrive.'[28]

Stainton was by no means alone in the kindly attention he gave to novices and the young. Edward Newman agreed to rearrange the Lepidoptera collections of the Entomological Club 'solely for the purpose of assisting beginners, who are almost daily applying to me for names. I purpose being at home at six o'clock every Thursday evening for this especial purpose'.[29] His soirées became a byword for pleasant and instructive 'evening classes'.

When they were not meeting in the field or corresponding by letter, entomologists recorded their captures and debated the issues of the day in one or other of the journals of the Victorian era. There were several of these. *The Entomological Magazine* was among the first, founded in 1833 and edited by Edward Newman, who also contributed no fewer than 15 of the first 63 papers. It lasted just five years.

Maintaining an entomological magazine was difficult, especially in the first half of the century before butterfly collecting became really popular, and when the numbers subscribing to them fluctuated greatly. *The Entomologist* came to grief after only two years. Edward Newman told readers that 'the spirit of the work' would now continue in the pages of *The Zoologist*, which, between 1843 and 1856, was the sole outlet available for the publication of papers, correspondence and records. No wonder entomology passed through a temporary 'phase of depression' during which Stainton doubted whether the number of active collectors numbered more than about 500 – evidently only a fraction of what they had once been. 'To what extent this may have risen from the Volunteer Movement, the deleterious effects of which have been so great', he wrote, 'it is impossible to say'. This Movement had been organized in 1859, during a period of alarm over the plans of Napoleon III, to strengthen the force available for home defence.

Even the cheap (at a penny a copy), chatty and once popular *Entomologist's Weekly Intelligencer* suffered from falling sales at this time. In 1861 it ceased publication and its place was taken by *The Weekly Entomologist*, published by the Bowdon and Altrincham Entomological Society at tuppence a copy. This venture, too, failed to prosper, and in 1863 the publishers despairingly concluded that there was not 'sufficient energy among entomologists' for its continuance.

Stainton, mindful of the 'phase of depression,' used the '*Intelligencer*' as a vehicle to attract new members for the Entomological Society of London. He suggested that, as timidity might have prevented some from putting themselves forward for election, he would sketch an imaginary scene in which a young entomologist attends his first meeting in the company of a kindly member prepared to point out the 'Who's Who' of the entomological world:

> 'Open the door quietly, for the meeting has already begun. You see everybody is so intent on the business of the meeting that hardly any one looks at us. Now, hang up your hat and coat on one of the pegs behind the door, and come and stand near the fire-place, while I point out the principal people present.
>
> 'That gentleman who sits between the table and the wall, with his chair more elevated than the rest, is the President; he is Mr. Saunders. You will observe when he speaks he has a very pleasant smile.

On his right you see a sallow-complexioned gentleman, with a great deal of hair about the lower part of his face; that is the senior Secretary, Mr. Douglas.'

'What? *The* Mr. Douglas, who wrote that jolly book, "The World of Insects"?'

'The same: now beside him you see a smaller, fair-haired gentleman; that is the other Secretary, Mr. Shepherd, of whose collection you have heard so much.'

'But, who is that lively-looking gentleman that sits behind him? now he is busy taking notes; but a few minutes ago he was talking very briskly.'

'That is Mr. Westwood.'

'What? Westwood of the "*Introduction to Modern Clas.–*"

'Yes. But now look on the other side of the President, you will see a gentleman very bald and with a careworn expression of countenance, who every now and then has some money handed to him, and he sends a slip of paper across the table in exchange; that is Mr. Stevens, the Treasurer; it is a very anxious post that of Treasurer to a learned body.'

'But who is that little gentleman with a chin as hairy as Mr. Douglas?'

'That is Mr. Janson, so learned in Staphylines.'

'Then who is that now talking to Mr. Westwood, with long dark hair and with no whiskers?'

'That is Mr. Stainton; you will see he is always ready to laugh at a joke, and he is a most hearty laugher.'

'Then who is that very pleasant looking gentleman, rather bald, who solaces himself now and then with a pinch of snuff?'

'That is Mr. Waterhouse, and he is not only pleasant looking, but really a very pleasant person.'

'What is the reason that, while apparently some scientific business is going on, every one at this end of the room keeps talking?'

'Well, I think we mustn't complain of that, as we are contributing our share: but the fact is the Secretary is only reading the minutes of the last Meeting, and it is looked on as a mere form.'

'Who is that gentleman sitting with his back to the window who might pass for the effigy at a tobacconist's shop, he is so continually taking a pinch?'

'Oh, that is Mr. Desvignes, so deep in the literature of the *Ichneumonidae*: they say he has so much snuff in his insect drawers that it drives away all the mites.'

'Is that Mr. Wollaston there?'

'Oh no! He very rarely comes; besides now he is out of town.'

'Then who is that?'

'That is Mr. Francis Walker, a perfect ambulatory encyclopaedia of entomological knowledge, you will find him very agreeable, and always ready to impart information. But we mustn't talk so loud now as the reading of the minutes in finished.'

'Now who was that? He has his back to us, but he said something so excessively droll, with the most solemn air, and set every body in a roar.'

> 'Oh! that's Mr. Newman, I thought you had known him.'
>
> 'Ah! yes: I remember now, he said in his "History of Insects" (Fig. 9), "It is important to avoid sitting on pill-boxes, as it must interfere with their structure;" and I thought I should have died of laughter, it was put so gravely.'[30]

History does not suggest that Stainton's venture was particularly successful. Although the Entomological Society elected eight new Fellows in 1856, and fifteen in 1857, these numbers were similar to those of previous years. However, at about this time fashion swung round once again, and there was a resurgence of interest in entomology, especially Lepidoptera. The year 1864 saw the appearance of a new journal, *The Entomologist's Monthly Magazine*, with a distinguished editorial panel that included H. T. Stainton, H. G. Knaggs, T. Blackburn, R. McLachlan, and E. C. Rye. The 'E.M.M.' proved a successful venture, and the journal continues to this day. By 1865, the new interest in entomology had prompted Edward Newman to revive *The Entomologist.* In his first editorial, Newman emphasized how approachable entomological magazines had become, compared with the stiff style of an earlier generation:

> **PRESERVATION OF INSECTS. 267**
>
> **succeeded in producing a most unexceptionable article.**
>
> **857. Finally, pill boxes, obtainable of any druggist, complete the outfit of the entomologist. There is now an excellent kind manufactured, of which the tops and bottoms never come out, owing to a little management in avoiding the usual pressure : it is important to get these. It is important also to avoid sitting on pill boxes, as it must interfere with their structure : to avoid this, the author carries them in a breast pocket.**

Fig. 9a. 'It is important to avoid sitting on pill boxes, as it must interfere with their structure', advised Edward Newman in his *History of Insects* (1841).

Fig. 9b. The pill-boxes, pierced with small holes, were useful for collecting insects. This drawing from Morris's *History of British Butterflies* shows how twelve were carried in a larger box which could be slipped into a pocket or knapsack.

> 'Like Rip Van Winkel, it awoke after a twenty years slumber, rubbed its eyes, and stepped forth amongst its living namesakes with all the formality of its pristine appearance. A few months have altered this and *The Entomologist* of 1865, although commenced with due solemnity, is as different from *The Entomologist* of '42 as good sound Saxon is from the canine Latin in which I formerly had the misfortune to rejoice. It became obvious that I had mistaken my calling. I had no skill in that very peculiar language which, like the Revd Edward Irving's, owes its popularity to its obscurity.
>
> 'No sooner was the changed character of *The Entomologist* apparent than entomologists came fluttering around me like moths to sugar.'[31]

FIELD CLUBS AND LOCAL SOCIETIES

Except for those who lived in London, and who could therefore attend the Coffee House gatherings, the early entomologists probably worked in isolation. Butterfly collecting became a truly social activity from the first half of the nineteenth century. One reason was the opening up of previously inaccessible wild places as a result of improved carriage ways, and, later on, the railway network. But equally it was a consequence of the growth of clubs and societies devoted to natural history, first in the capital cities, but soon in the provinces too. The field club was a feature of the Victorian age. The very first, the Berwickshire Naturalists' Club, was founded in 1831, a little earlier than

Queen Victoria's accession to the throne, by three Scottish members of an Edinburgh-based scientific society. It had at least two novel features. First, it was a county-based club and so harnessed the loyalty of local naturalists, who were every bit as keen to compile records for their county, or even their parish, as to discover a new species. It was in consequence of the activities of field clubs that certain counties, such as Cheshire and Norfolk, became extremely well recorded, whilst others remained *terra incognita* for all but a few. Secondly, this was a new kind of club, members of which did not merely meet formally in the evening, as did those of the scientific societies in London and Edinburgh, but spent a full day together outdoors, topped and tailed by a hearty breakfast and a dinner. Such gatherings were often quite intimate – the Berwickshire Club at first had only nine members and no permanent premises or library, though it did publish a journal. The main activity was the field excursion, for the purpose of collecting and observation. For that reason, subscriptions were low, and the membership was less firmly based on the professional middle classes than were those of the grander urban societies. Moreover wives were permitted to accompany their husbands.

Journals and field expeditions brought collectors together. These local societies were by no means lacking in distinguished members. The Norwich Entomological Society, founded in 1810, was the first provincial society devoted entirely to entomology. It owed a great deal to the inspiration of the distinguished lepidopterist, John Curtis, at whose home some of its meetings were held. The tiny Swaffham Prior Natural History Society, founded in 1834, attracted a company of leading scholars from Cambridge, including Darwin's mentor, J. S. Henslow, Professor Whewell, who has been credited with the first use of the word 'scientist', and the greatest field botanist of the day, C. C. Babington, who had also been a founder member the previous year of the Entomological Society. In their company were a host of leading entomologists, including the Revd William Kirby, John Curtis, J. F. Stephens, J. O. Westwood, and John Power, all probably attracted by the prospect of companionable outings to the Fens and the Broads. The Woolhope Naturalists' Field Club, founded in 1851, was another successful endeavour, in this case based in Herefordshire in the West Country – so successful, indeed, that it survives to this day. Among its first members were three Fellows of the Royal Society, including Professor Sedgwick from Cambridge, evidently an ardent collector of local field clubs.

Gradually, and mostly between the 1830s and 1870s, local clubs were established across England. There were particularly large and active ones at Manchester, Liverpool and Birmingham; there were other north-country clubs at Tyneside (founded 1846), and no fewer than six in Yorkshire, which eventually became federated as the Yorkshire Naturalists' Union – another survivor, whose journal, *The Naturalist*, has achieved an influence far transcending local boundaries. In the West Country there was the well-known Cotteswold Naturalists' Field Club (1846), and comparable clubs in the Malverns and at Bath. In East Anglia there was the popular Norfolk and Norwich Naturalists' Society (1869), and, at Oxford, the Ashmolean Natural History Society (1828). Sited near the rich hunting-grounds of Kent and Surrey was the Greenwich Natural History Club (1859), while north of the Thames, close to Epping Forest, was its counterpart, the Essex Field Club (1880). By 1873, there were at least 169 locally-based scientific societies, of which no fewer than 104 were explicitly field clubs.[32]

Victorian field clubs had a broader range of interests than similar societies today. Their members collected fossils and minerals, waded through rock pools with jam jars, packed hampers with flowers and ferns, and marvelled at majestic waterfalls and gorges. Butterflies were always among the most popular groups, especially if there was a rarity nearby like the Large Coppers and Swallowtails of Swaffham Prior. It was not all seriousness and lectures. A spirit of amusement and good fellowship often invaded the journals as, for example, in the following report of a trip to the Fens, written up by a precocious schoolboy:

> 'June 18th, 1836: Mr. J. A. Power, Mr. Broome, papa, Hugh, Tenny and me (R. F. Jermyn, Secretary, aet. 9) went to the Reach Chalk Pits and though we were there as soon as 9 o'clock a.m. almost all the petals of the *Glaucium* were fallen off. Then we went into a beer-shop and took two gallons of beer into Burwell Fen in a boat with Jim Retchy to punt us along and papa fell into a ditch up to his neck, and Tenny fell out of the boat into the water, and Hugh also fell into a ditch, but not in consequence of the beer, by no means. The worthy patron Mr. Power also got bogged which was great fun. We saw a great many *Machaons* and only caught two or three of the best.'[33]

Arrangements did not always run smoothly. The Richmond and North Riding Field Club once missed the return train at the end of their excursion to Flamborough Head. It says much for the magnanimity, not to say munificence, of the President, Edward Wood, that he chartered another train – entirely at his own expense. In 1877, the Lancashire and Cheshire Entomological Society held its first official meeting at the home of the President, S. J. Capper. This meeting was remarkable for the fact that, in his Presidential Address, Mr Capper admitted that he was unaware of the inaugural meeting until after it had begun, and was also 'mightily surprised to learn of the Society's formation and that [he] had been nominated as President'.

As well as discussing the captures of the day, many clubs paid due attention to the subject of food and drink. The Cotteswold Club frequently dined in style as, on one day in 1856, when 'Members and friends, to the number of forty, sat down to a substantial dinner, which, for its abundance and small cost, was certainly convincing proof that our dinner bill need not be extravagant when we have a liberal landlord to deal with; at the same time, we must bear in mind that our long rambles make us good trenchermen, while our modification in imbibition tends much to limit the usual profit'. The Essex Field Club, whose members studied earthquakes and parish church architecture as well as entomology, concluded one field trip with 'an excellent meat tea in the garden of the Enterprise Temperance Hotel'.

Sometimes these outings were so popular – the Manchester Club members often turned out in their hundreds – that provision of sustenance became a problem. However the Victorians were not always to be satisfied with packed lunches. On one occasion, the Cotteswold naturalists accepted an invitation to dine at the stately home of one of its members, the Earl of Ducie. The minutes noted carefully the 'well-spread table, furnished with fruits from his Lordship's garden, and wines from his cellar, [which] formed no inappropriate or unwelcome termination to the day's ramble'. On another occasion, members expecting a glass of sherry and biscuits were agreeably surprised to be offered champagne and a cold collation instead.

In 1873, the membership of the Norfolk and Norwich Naturalists' Society included three peers and three baronets as well as 15 clergymen out of a total of 135, but this was exceptional. Natural history transcended and dissolved the traditional social barriers of class and income. In 1848, Mrs Gaskell described how the handloom weavers of places as far apart as Oldham and Spitalfield showed a thirst for knowledge matched by considerable physical endurance – they generally walked from their workplaces to collecting-grounds in the countryside. These 'common handloom weavers, who throw the shuttle with unceasing sound, though Newton's *Principia* lies open on the loom, to be snatched at in work hours, but revelled over in meal times, or at night', were also many of them familiar with the Linnaean system of classification. They studied botany, geology, and entomology, and might be seen 'with a rude-looking net, ready to catch any winged insect'. Mrs Gaskell found such men to be practical, shrewd, and hard-working, who would 'pore over every new specimen with real scientific delight'.[34]

Working people shared fully in the Victorian sense of wonder at the natural world. From the 1860s a new kind of entomological society sprang up, mostly in and around London, and the main manufacturing centres in the Midlands and the North. Subscriptions were necessarily low, and meeting places were often taverns or temperance hotels. The East London Entomological Club, founded in 1862, attracted many of the Spitalfields weavers to its meetings at the 'Bell and Mackerel'. The Haggerstone Entomological Society (1858) met initially at the 'Carpenters' Arms', North London, but, as the membership increased, it moved to larger premises at the nearby 'Brownlow Arms' in Haggerstone. Those attending meetings of the West London Entomological Society (1868) gathered at the sign of the 'Mason's Arms', near Hammersmith. Members of the then North London Entomological Society, originally founded in 1869 as the I Zingari Entomological Society (a name derived from *Zingaro*: an Italian gypsy), patronized a number of public houses.

Fig. 10. A 'South London' field meeting at Seal Chart, Kent, in May 1905. It was conducted by Robert Adkin, the President, seen here standing behind the deckchair in the centre. Others present included Richard South (centre right) sitting beside H. J. Turner (in deckchair, with net), and (far left) T. A. Chapman. The photograph was taken by fellow member, Edward Step.

The South London Entomological and Natural History Society was of a somewhat higher social standing and did not meet in a public house (Fig. 10). Founded in 1872, rather later than the other London societies, it considered that 'the social position of our members is higher than the preceding societies, but the working-man element is still represented'. Although its membership was never more than a few hundred, the South London is the longest lasting of all the local entomological societies, surviving to this day as the British Entomological and Natural History Society.[35]

A vital aspect of these urban clubs was the annual exhibition or show, which were often lively affairs. One such show, in 1860, was described by Henry Stainton in his guise as 'Inquisitor', the roving reporter of *The Entomologist's Weekly Intelligencer.* It was held on the first floor of the 'Woolpack', with an admission fee, payable at the bar, of tuppence. According to the advertisement, 'upwards of 60,000 Insects, in all devices' were on display, and the 'beauties of the Butterfly and Moth will be shown under a powerful Microscope'.

> 'True to the promise held forth in the bill, we found the insects principally arranged in devices; thus, we found glazed frames containing *stars of insects – crowns of insects – festoons and all kinds of devices.* Now in all this there was much taste displayed, not only in figures, patterns, &c., but also in the contrast of colours. These working men learn and study such things in the woods and fields: they love flowers, they admire butterflies. A little girl, daughter, we suspect, of an exhibitor, in her deep delight, clapped her hands and exclaimed to her mother, "Oh! how beautiful".
>
> 'This short notice in the Intelligencer may serve to teach these working men that their humble efforts are not unregarded; that there are those who, although occupying a higher walk in the Science of Entomology, yet sincerely wish them "God Speed". Success to their undertaking, and may their annual show continue to be such for many years to come.'[36]

Although 'Inquisitor' may sound extremely patronizing to a modern ear, his admiration was nonetheless sincere and spontaneous.

The first 'Grand National Entomological Exhibition' was staged by the South London Entomological Society at the Royal Aquarium, Westminster, in 1878. Although it was attended by 70,000 people, the Society showed a lack of imagination in its arrangement of case after case of set butterflies and moths – for all except the keen collector it was a tedious, uninformative show, and many must have come away none the wiser. Later, the Society seems to have found a better grasp of the ingredients for a lively, interesting evening. In 1886 it staged a more successful annual exhibition at new premises at Layman's Auction Rooms near London Bridge. Among the exhibits on display were 'jumping beans from Mexico, peculiar fungi gathered in the fog of the preceding day, even a giant crab from Japan which measured twelve feet from claw to claw; and there were birds, plants and insects from all over the world. Musical accompaniment was provided by "Miss Minnie Spary, aged twelve years, who played Holly Bush Polka and eleven other pianoforte solos". About 1,200 people attended.'[37]

In succeeding years the Annual Exhibition retained this high standard and most were considered a great success. Inevitably, however, there were a number of incidents. It is recorded that the redoubtable J. W. Tutt arrived at the exhibition in 1877 with a large number of boys from Snowfields Board School, Bermondsey. Tutt was the headmaster 'and keen to convert his lads to the ways of natural history'. However, his charges proved unruly and there were a number of complaints. The following year, when Tutt requested twenty complimentary tickets, he was told that they were for adults only, not for boys. Tutt, who could be intemperate, was incensed, and there was a furious exchange of letters which led to the Secretary tendering his resignation. It is recalled that, following this, Tutt, showing magnanimity, climbed down from his lofty perch and apologized, and the Secretary continued in office.[38]

Fig. 11. The headquarters of the Royal Entomological Society of London at 41 Queen's Gate, South Kensington, photographed around 1925.

At a national level, the Entomological Society of London continued to represent the most serious side of entomology. The patronage of TRH The Duchess of Kent and Princess (later Queen) Victoria was obtained in 1835, although after the death of King William IV in 1837 it lapsed. Upon the accession of Victoria to the throne, an address was presented praying the Queen to continue to be its Patron, but this was refused owing to the fact that the Society did not then possess a charter. A Royal Charter was granted in 1885 but it was not until 1914 that Royal Patronage was obtained once more.[39] At the very start, the Society rented rooms at 17 Old Bond Street but in 1852, when these premises were declared unsafe, it moved to 12 Bedford Row. The Society was soon badly in need of more spacious premises for its growing collections and library and so, in the late 1850s, negotiated first for rooms at Somerset House, recently vacated by the Royal Society, and then at the house in Soho Square vacated by the Linnean Society on its move to Burlington House. Both efforts were unsuccessful and the Society was obliged to remain at Bedford Row, though in extended accommodation, until 1875. Negotiations to merge with the Linnean Society had begun in 1872 and were continued fruitlessly until 1874 when they were finally abandoned and, after a one-year extension of its Bedford Row lease, the Society eventually moved to 11 Chandos Street as a tenant of the Medical Society of London. Here it remained until the final move in 1923 to its present headquarters at 41 Queen's Gate (Fig. 11).

Even more exclusive than the Entomological Society was the Entomological Club, founded in 1826 by George Samouelle, author of the popular *Entomologist's Useful Compendium.* Membership was restricted to eight leading figures and was regarded as an honour. Edward Newman recalled his first meeting: 'I was not, at the period of which I am writing, a perfect novice in Entomology', he wrote, 'but in the course of conversation, not a single name was mentioned that I had ever read ... I longed for the utterance of one sentence about "*Emperors*" or "*Admirals*", then I could have chimed in. But no, every word was entirely scientific. I resolved before another month, to furnish myself with a little more knowledge.'[40] On one evening every month members met to dine at a particular member's house or some other venue – an arrangement that has lasted to this day. When G. H. Verrall hosted one of these dinners at the Holborn restaurant, in 1887, he proposed the organization of an annual dinner, open to all entomologists. These 'Verrall Suppers' proved immensely successful and continue to be so (Fig. 12). During the early years it was Verrall himself who paid the bill for those attending.

The formal meetings of the Entomological Society were not without their arguments and clashes of personality. Some of the eminent Victorian entomologists could be difficult, especially when, like the headmaster J. W. Tutt, they

Fig. 12. 'Releasing the Lepidoptera' – a 'great moment at the Annual Dinner of the Entomological Society'. *Punch* cartoon of 17th December 1930 by George Morrow, now owned and cherished by the Royal Entomological Society.

regarded themselves as the fount of all scientific authority. Tutt was like a faulty radio set – he transmitted readily enough, but the receiver worked only periodically. Stainton's voluminous correspondence contains many harangues and 'acid drops' from the sharp pen of Tutt. For example, he had doubts about the choice of Richard South as editor of *The Entomologist*.

> 'I suppose you know that Mr. Leech has bought *The Entomologist* from Mr. Newman & that Mr. South, Mr. Leech's curator, will be his Editor in future. If Mr. South calls an insect *nemoralis* then *zetterstedtii*, then calls it a new species *taeniadactylus*, and then changes his mind and thinks it is not & finally winds up with a suggestion that the species may yet be *nemoralis* as he at first supposed – I don't know what *The Entomologist* will be like in a few months. Mr. South has so upset the "plumes" that I rather dread him as an editor.'[41]

Tutt's outspokenness so annoyed South that he ceased to attend meetings when Tutt was present. Tutt's relationship with Stainton was more cordial, although one detects in the correspondence a certain wariness on both sides.

The following exchange concerned certain insects collected by Tutt which he wanted Stainton to examine:

> *March 2nd,* 1887. Tutt: 'I should like you to look through my insects with me, to settle up the matter, but I have no leisure at all, being engaged up to 10.30 pm, except Saturday and Sunday – when I am entirely free, both day and evening.'
>
> *March 3rd.* Stainton: 'My Dear Sir, I shall be very glad to help you in your difficulties and will let you know when I have a spare Saturday afternoon which I could devote to helping you.'
>
> *March 5th.* Tutt: 'My Dear Sir, Thanks very much for your kindness in promising to be troubled with me. I shall be very pleased to come at any time you may send for me.'
>
> *March 22nd.* Stainton: 'I find I shall be disengaged next Saturday the 26th when I shall be glad to see you at 3 pm. I am about a mile and a half from the Lewisham Station but the Catford omnibuses can let you down at the George about half a mile from this house.'[42]

One also begins to understand how Stainton could have spent his life writing letters.

A different kind of correspondence, which occurred regularly in the entomological journals of Victorian England, concerned the delicate subject of truth. As butterfly collecting gained in popularity and the wealthier collectors proved willing to pay large sums for rare specimens, various lucrative supply businesses sprang up. Some of these were honest and legitimate and were run by people who were themselves highly proficient naturalists. A few, however, seem to have been tempted into 'planting' exotic specimens and claiming they were British. Proof or disproof in such cases is difficult. It was a matter of 'let the buyer beware', but it also cast doubt on the reliability of records of certain species. Were, for example, the ten specimens of the very rare Mazarine Blue advertised for sale by W. Farren really captured by 'a Reverend Gentleman in Somersetshire' (Fig. 13), or did they come from overseas?

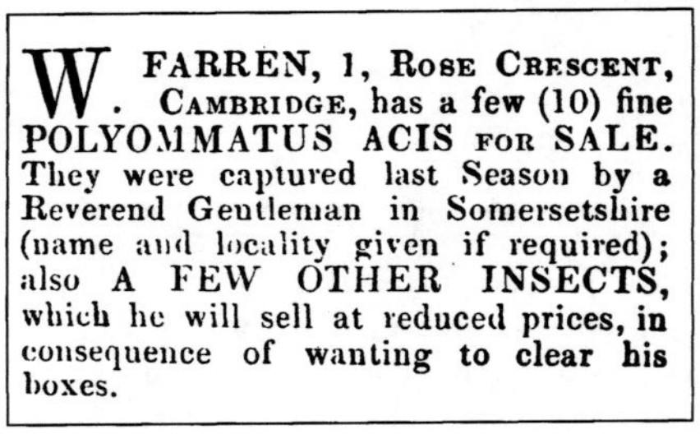

W. FARREN, 1, Rose Crescent, Cambridge, has a few (10) fine **POLYOMMATUS ACIS for SALE.** They were captured last Season by a Reverend Gentleman in Somersetshire (name and locality given if required); also **A FEW OTHER INSECTS,** which he will sell at reduced prices, in consequence of wanting to clear his boxes.

Fig. 13. Advertisement in *The Entomologist's Weekly Intelligencer* of 23rd February 1861, offering Mazarine Blues for sale. Their provenance is very dubious since the butterfly was considered extinct in the West Country by the 1840s.

As early as 1837, 'Inquisitor' had claimed that many 'rarities' in British collections were Continental specimens.[43] They had been imported by dealers, reset in the English style and sold to the naive or unwary as British specimens. His complaint was taken up by Henry Doubleday and Edwin Birchall. The latter went so far as to believe that no collector could be sure of the provenance of any specimens unless he had captured them himself. Birchall thought 'that the great majority of collections of Lepidoptera made in England over the previous thirty years had done little or no service to science or to their owners'.[44] His advice to British collectors was that they should study Continental butterflies as well as our native species.

The popularity of butterfly collecting outlasted Queen Victoria, but by the end of her reign most of the great names of the era were no more. Of the old guard – Edward Newman, Henry Doubleday, Henry Stainton, Professor J. Obadiah Westwood had all died – only J. W. Tutt, the formidably august 'Master of all things Lepidopterous', was left. The twentieth century, at the start of which the old queen died and which itself has now come to an end, is the subject of the last chapter in this book. It has witnessed many changes. However, history is seldom made up of watertight compartments, and many of the institutions and habits of the great Victorian butterfly hunters have proved long-lasting; the county-based natural history society; evening lectures and exhibitions; certain journals and grand suppers; even the equipment has not changed much. What has to some extent been lost is that sense of wonder and the urge to spread the gospel of natural history that the Victorians possessed. Or rather it has been transmuted into other forms, notably nature conservation. Today television has made voyeurs of us all, and the hands-on experience of the collector and naturalist has been replaced by a coloured image on a screen. Collecting is no longer fashionable. It is right that the rarest species should be collected sparingly, if at all, but have we not lost, in this politically correct age, some of the sunshine and convivial companionship of those far-off days?

In July, 1859, the front page of *The Entomologist's Weekly Intelligencer* carried a lively poem by J. W. Douglas. Written as the 'phase of depression' was approaching its end, it tells of a field trip to Reigate and the North Downs, in Surrey. It captures the spirit of the Victorian natural history outing as well as anything I have read.

The Reigate Gathering

The morning of the sixth July
A crowd of men came down
To Railway Station, London Bridge,
And booked for Reigate Town.

The booking clerk was quite amazed
That all asked for one place,
And wondered if they meant to go
To see a fight or race.

He wondered while they took a fare,
Though no affair of his,
And when one whispered "Butterflies",
He took him for a quiz.

The railway-carriages were filled
To quite an overflow,
Until at length the engine screams,
And then away they go.

There Smith and Stainton, Waterhouse,
And long-lost Walton too,
With Walker, Douglas, Stevens, all
Old hunters good and true.

There also Shepherd, Janson, Groves
(But not of Blarney Stone),
On whom I thought to make a pun –
'Tis better let alone.

And there were also Bowerbank,
And Pascoe, Bond and Weir,
With others whose unruly names
In rhyme will not appear.

Some showed the weapons of the chase,
The bottle, box and net,
And told how insect artifice
By artifice was met.

And how to kill the game when caught,
By leaves of laurel bruised;
Some cyanide or chloroform, –
One formic acid used.

And past New Cross and Forest Hill,
Then Croydon, Merstham run
These jolly souls who made the train
A vehicle of fun.

The jokes and mirth at railway pace
Still speeding bravely on,
Were suddenly brought to a stand,
By cries of "Reigate Town".

And all these men then left the train
To driver and to guard,
And thirty thirsty souls emerged
From Reigate Station yard.

There Wilson Saunders met the lot,
With welcome for each guest, –
For such these insect-hunters were,
Invited to his best.

The best of hearts – a "White Hart" too –
Was open there and then,
And at the latter knives and forks
Were laid for thirty men.

Soon nets were mounted, and the throng
Went out o'er hill and dale;
But all the dodges most in use,
Were found of no avail;

For insects all had learned the plan
For ravaging their race,
For over many miles of ground,
Scarce one would show his face.

Yet though they coyly hid themselves
In holes and chinks and grass,
Until the reconnoiterers
By their retreats should pass,

And so they mostly managed thus
To keep themselves from harm,
It happened that each visitor
Received reception warm.

For never was there greater heat
Evolved from solar ray
Than fell upon the luckless heads
Of our good friends that day.

At Wanham Park, at one o'clock,
Beneath the leafy trees,
The muster roll was called, and lo!
They also mustered cheese.

And as the panting multitude
Were literally baked,
Beer, wine and soda water ran,
Until their thirst was slaked.

Then for three mortal hours again,
They through the country ranged,
When white heat in the Reigate fields,
For "White Hart" was exchanged.

The party had been reinforced,
And fifty round the board,
Sat down, and with the good cheer there,
Their failing strength restored.

What though the captures had been few,
The intercourse of friends
Both out of doors and in, all felt,
Did more than make amends.

Then for the founder of the feast,
We cheer with three times three,
And if such gathering comes again
May we be there to see.[45]

NOTES TO CHAPTER ONE

1. Newman, L. H. (1953). *Living with butterflies*, pp.168–173.
2. Kirby, W. & Spence, W. (1815). *An introduction to entomology* **1**: 44.
3. Wallace, A. R. (1869). *The Malay archipelago* **2**: 51.
4. Harris, M. (1766). *The Aurelian, &c.*, p.26 [caption to The Grand Surprize, or Camberwell Beauty: pl.XII].
5. Darwin, F. (Ed.) (1887). *Life and letters of Charles Darwin* **1**: 539.
6. Lubbock, J. (1856). On the objects of a Collection of Insects. *Entomologist's Annu.*, **1865**: 115.
7. Linnaeus, C. (1753). *Species plantarum* (edn 1); and (1758) *Systema naturae* (edn 10).
8. Pope, A. (1746). Preface to Joseph Addison's *Dialogues upon the usefulness of ancient medals especially in relation to the Greek and Latin poets.*

9. Fitton, M. & Gilbert, P. (1994). *In* MacGregor, A. (Ed.), *Sir Hans Sloane, collector, scientist, antiquary*, p.112.
10. Baker, H. (1749). Preface to Benjamin Wilkes's *English moths and butterflies.*
11. Bristowe, W. S. (1967a). The Life and Work of the great British Naturalist, Joseph Dandridge 1664–1746. *Entomologist's Gaz.* **18**: 73–89.
12. Harris, *Aurelian*, Preface, p.1.
13. Allen, D. E. (1976). *The naturalist in Britain, a social history*, p.173.
14. Wilkinson, R. S. (1977b). The Great Cornhill Fire and the demise of the first Aurelian Society. *Entomologist's Rec.J. Var.* **89**: 250–251.
15. Fabricius, J. C. (1847). The Auto-biography of John Christian Fabricius (translated from the Danish by the Rev. F. W. Hope). *Trans.ent.Soc.London* **4**: i–xvi.
16. Haworth, A. H. (1803). *Lepidoptera Britannica*, pp.xii–xiv.
17. Haworth, A. H. (1812). A brief account of some rare insects announced at various times to the Society as new to Britain. *Trans.Ent.Soc.London* **1**: 232.
18. Neave, S. A. (1933). *The history of the Entomological Society of London, 1833–1933*, p.3.
19. Gage, A. T. & Stearn, W. T. (1988). *A bicentenary history of the Linnean Society of London*, pp.5–37, 50–58.
20. Barber, L. (1980). *The heyday of natural history*, p.35.
21. Emerson, R. W. (1841). *Essays: I. History.*
22. Allen, *Naturalist in Britain*, p.79.
23. Kirby & Spence, *Introduction to entomology* **4**: 525.
24. Newman, E. (1835). *The grammar of entomology*, p.259.
25. Knaggs, H. G. (1869). *The lepidopterist's guide*, p.76.
26. Stainton, H T. (1855). An address to young entomologists. *Entomologist's Annu.*, **1855**: 13.
27. Stainton, H. T. (1859). At Home. *Entomologist's Wkly Intell.* **6**: 73–74.
28. Anon. (1856). The Way to Mountsfield (A guide for Strangers). *Ibid.* **1**: 158–159.
29. [Newman, T. P.] (1876). *Memoir of the life and work of Edward Newman*, p.16.
30. [Stainton, H. T.] (1856). The Entomological Society [No. 1 and No. 2]. *Entomologist's Wkly Intell.* **1**: 65–66, 105–106.
31. Newman, E. (1865). Preface. *Entomologist* **2**: v–vi.
32. Allen, *Naturalist in Britain*, p.170.
33. MacKechnie-Jarvis, C. (1976). A history of the British Coleoptera. Presidential Address 1975. *Proc.Trans.Br.ent.nat.Hist.Soc.* **1976**: 101–102.
34. Gaskell, Elizabeth (1848). *Mary Barton* (World Classics edn), pp.40–41.
35. James, M. J. (1973). *The new Aurelians*, p.5 *et seq.*
36. Inquisitor [H. T. Stainton] (1860). An Entomological Show. *Entomologist's Wkly Intell.* **9**: 77.
37. James, *New Aurelians*, p.12.
38. *Ibid.*, p.13.
39. Neave, *History of Entomological Society*, pp.15–16.
40. Newman, E. (1836). Wanderings and Ponderings of an Insect-Hunter. *Ent.Mag.* **3**: 305–315.
41. Tutt, J. W. (*c.*1875). From collected correspondence of H. T. Stainton, in the Entomology Library, Natural History Museum, London.
42. Stainton, H. T. (1881). From his collected correspondence, in Entomology Library, Natural History Museum, London.
43. Inquisitor [E. Newman] (1837). Note on Butterflies questionably British. *Ent.Mag.* **4**:177–179.
44. Birchall, E. (1877). Collections of British Lepidoptera. *Entomologist's mon.Mag.* **13**: 279–280.
45. J. W. D[ouglas]. (1859). The Reigate Gathering. *Entomologist's Wkly Intell.* **6**: 121–122.

CHAPTER 2

'Weapons of the Chase'

PINS, NETS, AND COLLECTING BOXES

'There are few objects in Nature which raise the mind to a higher degree of admiration, than the Insect World.'
Abel Ingpen, 1827

'I HAVE FOR SOME YEARS TOGETHER BEEN A DILIGENT searcher out of Papilios, diurnal and nocturnal', John Ray told his friend Mr Derham, in 1703, 'and though I have found and described near upon 300 species, great and small, within the small compass of four or five miles; yet I came not to the end of them'.[1] Ray was among the first to realise how large and diverse was our insect fauna. We know relatively little about the collecting methods he used but he must have found an efficient means of acquiring specimens. It is recorded that Ray was supplied with specimens collected by other entomologists, but the '300 species, great and small' so close to his home he had probably collected himself with no little help from his wife and four daughters. How did he do it? He knew about assembling – or 'simbling' as some of his eighteenth-century contemporaries called it – the practice of using a freshly emerged female moth as a lure to attract free-ranging males. He seems to have owned a butterfly net, and certainly used collecting boxes to pin his captures. But, so far as is known, killing bottles had not yet been invented; nor had setting boards, 'sugaring', light traps or hurricane lamps. Ray did well to discover so many cryptic and tiny moths, especially when one considers he was about seventy-five years old at the time. However, many of his fellow spirits collected well into their eighties or even their nineties and continue to do so. Naturalists are remarkable for their longevity!

This chapter is about the 'weapons of the chase' used by the lepidopterists of the eighteenth and nineteenth centuries, from killing bottles and pincushions to setting boards and cabinets. It is also about the collecting trips themselves, and describes, as far as possible in the lepidopterists' own words, the delights and distractions of a day in the field (Figs 14, 15).

The organized field excursion, with the object of collecting specimens, seems to have originated in the early seventeenth century. Here the object was not insects but plants. Apothecaries needed regular supplies of wild herbs or 'simples' to supplement those grown in their gardens, and the existence, by 1620, of a Society of Apothecaries allowed them to meet at a prearranged time and foray into the countryside together. These 'herbarizing days' happened six times a year, and the party used to set off from St Paul's in the City of London at five o'clock in the morning. At that date the nearest countryside lay no more than a mile or two away, and choice places like Epping Forest and the North Downs were within reach of a mounted party. Then, as later, the collectors paid due attention to their creature comforts. D. E. Allen records that the day's proceedings were always rounded off with an impressive banquet, at which a haunch of venison featured regularly.[2]

Apothecaries had long collected wild plants – William Turner was doing so in the 1540s – and in recording the species and their localities in printed herbals they were establishing the foundations on which the study of British botany was later built. John Ray himself was one of the first to bridge the divide

Figs 14 & 15. 'The Chase' and 'Bagged': two small paintings by E. W. J. Hopley (1816–69), an English artist of portraits and domestic genre as well as of historical, allegorical and fairy subjects. Their present whereabouts is not known.

between practical herbalizing and scientific botany with his *Catalogus plantarum circa Cantabrigium nascentium* (1660), the first local flora ever published. Entomologists, with at first no society of their own, lagged behind the botanists. There may well have been organized collecting trips of which we have no knowledge, since there was no magazine to describe them and record their findings. But we do know that the Society of Aurelians organized field trips for members from 1740 onwards.

In his *Insectorum Theatrum*, Thomas Moffet describes a field trip in which the party was attacked by a swarm of wasps.[3] Such parties may well have been collecting insects, including butterflies. Moffet himself had first-hand knowledge of such phenomena as meconium, the often blood-red fluid excreted by emerging butterflies, and which had been a topic of speculation and discussion since the time of Aristotle. However, his experience with the wasps may hint at one powerful reason why entomology lagged behind botany in the seventeenth century. Wasps were considered to be of ill-omen, and not only because they could sting. Wasps and hornets were widely believed to be agents of the devil – Dante's Hell contained stinging hornets – and an angry swarm of them could be interpreted in ways quite unconnected with nature. Superstition and fear may have delayed the dawning light of scientific entomology.

The first documented details of collecting methods in Britain appear some years later than the earliest known specimens of butterflies and moths. By the time James Petiver was issuing his instructions for collectors, and corresponding

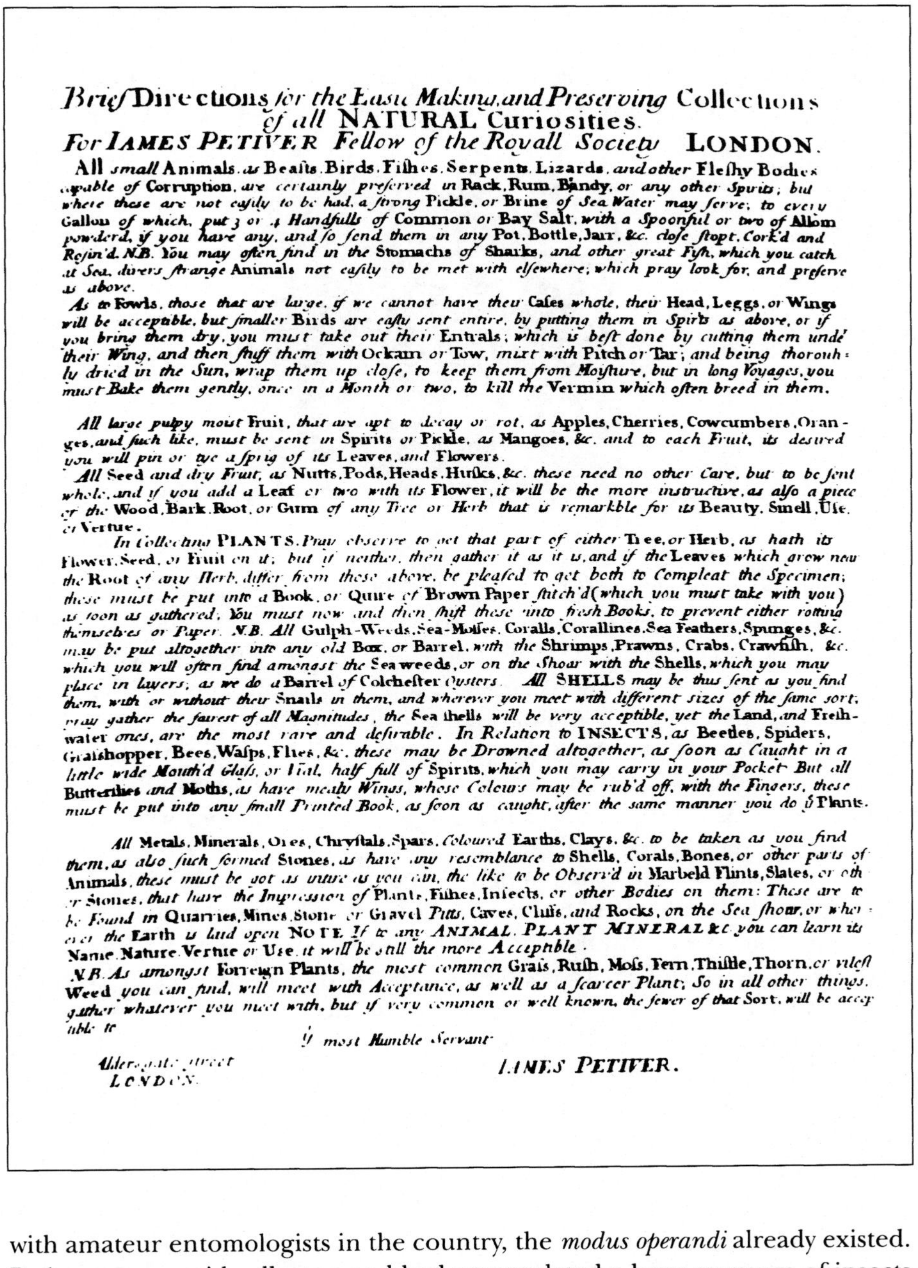

Brief Directions for the Easie Making, and Preserving Collections of all NATURAL Curiosities.
For JAMES PETIVER Fellow of the Royall Society LONDON.

All small Animals, as Beasts, Birds, Fishes, Serpents, Lizards, and other Fleshy Bodies capable of Corruption, are certainly preserved in Rack, Rum, Brandy, or any other Spirits; but where these are not easily to be had, a strong Pickle, or Brine of Sea Water may serve; to every Gallon of which, put 3 or 4 Handfulls of Common or Bay Salt, with a Spoonful or two of Allom powderd, if you have any, and so send them in any Pot, Bottle, Jarr, &c. close stopt, Cork'd and Rosin'd. N.B. You may often find in the Stomachs of Sharks, and other great Fish, which you catch at Sea, divers strange Animals not easily to be met with elsewhere; which pray look for, and preserve as above.

As to Fowls, those that are large, if we cannot have their Cases whole, their Head, Leggs, or Wings will be acceptable, but smaller Birds are easily sent entire, by putting them in Spirits as above, or if you bring them dry, you must take out their Entrals; which is best done by cutting them under their Wing, and then stuff them with Ockam or Tow, mixt with Pitch or Tar; and being thoroughly dried in the Sun, wrap them up close, to keep them from Moysture, but in long Voyages, you must Bake them gently, once in a Month or two, to kill the Vermin which often breed in them.

All large pulpy moist Fruit, that are apt to decay or rot, as Apples, Cherries, Cowcumbers, Oranges, and such like, must be sent in Spirits or Pickle, as Mangoes, &c. and to each Fruit, its desired you will pin or tye a Sprig of its Leaves, and Flowers.

All Seed and dry Fruit, as Nutts, Pods, Heads, Husks, &c. these need no other Care, but to be sent whole, and if you add a Leaf or two with its Flower, it will be the more instructive, as also a piece of the Wood, Bark, Root, or Gum of any Tree or Herb that is remarkble for its Beauty, Smell, Use, or Vertue.

In Collecting PLANTS, Pray observe to get that part of either Tree, or Herb, as hath its Flower, Seed, or Fruit on it; but if neither, then gather it as it is, and if the Leaves which grow near the Root of any Herb, differ from these above, be pleased to get both to Compleat the Specimen; these must be put into a Book, or Quire of Brown Paper stitch'd (which you must take with you) as soon as gathered; You must now and then shift these into fresh Books, to prevent either rotting themselves or Paper. N.B. All Gulph-Weeds, Sea-Mosses, Coralls, Corallines, Sea Feathers, Spunges, &c. may be put altogether into any old Box, or Barrel, with the Shrimps, Prawns, Crabs, Crawfish, &c. which you will often find amongst the Seaweeds, or on the Shoar with the Shells, which you may place in layers; as we do a Barrel of Colchester Oysters. All SHELLS may be thus sent as you find them, with or without their Snails in them, and wherever you meet with different sizes of the same sort, pray gather the fairest of all Magnitudes, the Sea shells will be very acceptible, yet the Land, and Fresh-water ones, are the most rare and desirable. In Relation to INSECTS, as Beetles, Spiders, Grasshopper, Bees, Wasps, Flies, &c. these may be Drowned altogether, as soon as Caught in a little wide Mouth'd Glass, or Vial, half full of Spirits, which you may carry in your Pocket But all Butterflies and Moths, as have mealy Wings, whose Colours may be rub'd off, with the Fingers, these must be put into any small Printed Book, as soon as caught, after the same manner you do ye Plants.

All Metals, Minerals, Ores, Chrystals, Spars, Coloured Earths, Clays, &c. to be taken as you find them, as also such formed Stones, as have any resemblance to Shells, Corals, Bones, or other parts of Animals, these must be got as intire as you can, the like to be Observ'd in Marbeld Flints, Slates, or other Stones, that have the Impression of Plants, Fishes, Insects, or other Bodies on them: These are to be Found in Quarries, Mines, Stone or Gravel Pitts, Caves, Cliffs, and Rocks, on the Sea shoar, or wherever the Earth is laid open NOTE If to any ANIMAL PLANT MINERAL &c you can learn its Name, Nature, Vertue or Use, it will be still the more Acceptible.

N.B. As amongst Forreign Plants, the most common Grass, Rush, Moss, Fern, Thistle, Thorn, or vilest Weed you can find, will meet with Acceptance, as well as a scarcer Plant; So in all other things, gather whatever you meet with, but if very common or well known, the fewer of that Sort, will be acceptible to

Yr most Humble Servant

Aldersgate street
LONDON.

JAMES PETIVER.

Fig. 16. James Petiver's 'Brief Directions' for collectors of 'Natural Curiosities', which he gave to ships' captains and others travelling abroad.

with amateur entomologists in the country, the *modus operandi* already existed. Petiver was an avid collector, and had accumulated a large museum of insects and other natural objects for the purpose of description. These included foreign as well as native species, and his printed instructions were intended to enable those embarking on sea voyages and expeditions abroad to collect for him. He stated, in his *Brief Directions for the Easie Making and Preserving Collections of all Natural Curiosities* (*c.*1700) – a single, engraved folio sheet (Fig. 16) – that '*In* *relation* to INSECTS, *as* Beetles, Spiders, Grasshopper, Bees, Wasps, Flies, *&c. these may be Drowned altogether, as soon as Caught in a little wide Mouth'd Glass, or Vial, half full of Spirits, which you may carry in your Pocket. But all* Butterflies *and* Moths, *as have mealy Wings, whose Colours may be rub'd off, with the Fingers, these must be put into any small Printed Book, as soon as caught, after the same manner you do y^r^* Plants'.[4]

Most of the specimens collected at that time have been lost or have perished. Fortunately Petiver's collection is one of the very few exceptions. He preserved many of his own specimens by placing them between thin sheets of mica, which were then bound with gummed paper or placed in shallow wooden

Plate 3
This is probably the oldest specimen of the Brown Hairstreak in existence. It once belonged to James Petiver and was taken near Croydon on the last day of August 1702. Petiver was the first to name and figure this elusive butterfly, although he blundered in assuming the contrasting sexes to be different species. This is a female which in 1703 he named 'The Golden brown double Streak'.

Plate 4 *(facing page)*
The Revd Adam Buddle's pressed Brimstone and Small Copper butterflies, incorporated into Volume XII of his herbarium, now in the Botany Department, Natural History Museum, where they have only recently been discovered. These specimens were probably collected around London between 1699 and 1715.

frames after the fashion of lantern slides (Plate 3). This was considered to be an advance on the method used by other early Aurelians such as the Revd Adam Buddle (*fl.*1660–1715) and Leonard Plukenet (see Chapter 3: 4), who, following Petiver's 'Directions', pressed their specimens between the leaves of books. Buddle was a well-known botanist. His name is commemorated by the genus *Buddleia* (represented in many gardens by the ever-popular 'butterfly-bush'). What is not so widely known is that he was also interested in insects. At the time of his death, Buddle's manuscripts and collections were in the possession of James Petiver, and Sir Hans Sloane, who inherited Buddle's herbarium, had 'some considerable difficulty in obtaining the material to add to his own collection'. When he eventually did so, he discovered that Buddle had gummed butterflies, moths and other insects on to some of his pages of pressed plants (Plate 4). Today, they can be seen in Volume 12 of his herbarium which is still preserved in the Botany Department of the Natural History Museum, London. This volume contains Buddle's collection of grasses, sedges and rushes, together with an inscription on the first page which tells us that '*In hoc volumine multa Insecta intersperguntur*' [In this volume many insects are interspersed]. Unfortunately the butterflies are in poor condition – often badly rubbed or lacking wings and bodies – but 31 different species can be readily identified, mostly collected in the countryside to the north and west of London. A close examination of Buddle's collection of butterflies has shown that it contains two surprises. Almost all the specimens have Latin descriptions, and five also have common names: 'The Admiral', 'The Peacock's Eye', 'The Greater Sylver Spotted Fritillary', 'The half Mourner' and 'The Painted Lady'. These all derive directly from Petiver's *Musei Petiveriani*, Centuria IV and V, of 1699, a work which, in the case of his Black-veined White, Buddle cited directly. The first discovery is that, although it had previously been

Papilio minor croceus or nigro praemaculatus M. pet. 317

Gramen glumis variis C.B. Theat. Icon 158

H Icon 1152 Fig. 6

Papilio sulphureus Mus: pet Fig 1

8 (herb.)

Gramen spicatum montanum asperum C.B. 253 idem est cum praecedente viz: spica crassiore purpureo-caerulea brevi ejusdem

a D. Newton

8 (herb.)

Gramen parvum montanum spicâ crassiore purpureo caeruleâ brevi C.B. Pr. 253

R. Supp. 603.

Col. 7. No. 2.

Gramen Dactylon latiore folio C.B. T. Icon 114

Ischaemon sylvestre latiore folio P. 1178

R. H. 1272.

Papilio sulphureus pallidus M. pet Fig. 2

Cap. 7. No 1

1270. Gramen repens cum panicula Graminis Mannae J.B. R. Hist 1271

R. H. 1271.

thought that Petiver's 'Papilio *Fritillarius* major' – his Greater Silver Spotted Fritillary for which he cited Moffet as an authority – was the High Brown Fritillary, Buddle's actual specimen from Richmond Park, under the same name, is clearly a Dark Green Fritillary, a butterfly not mentioned in print up to that time. In 1758, Linnaeus, who first described the species and called it *aglaja*, also cited Moffet. In 1699, specimens of both might have been known to Petiver but thought to be of the same species, though Petiver's record was based on a different specimen from Leicester. The other discovery is the presence of a fragment of a Queen of Spain Fritillary wing. Petiver first described this butterfly in 1702 from a foreign specimen – his Riga Fritillary – and its occurrence in Britain was not known until shortly after that date. Buddle's specimen, which has no inscription, must have been one of the earliest known and could possibly have been received from those who first collected it, some time before 1704, at Gamlingay (see Chapter 4: 30) or even earlier.

In 1716, Plukenet's collections of plants and insects were also purchased by Sloane and his bound volume of pressed insects has also fortunately survived, though in a very fragile state (Plate 5), and is today safely housed in the Entomology Department of the same museum.[5] There, too, though not discovered until a few years ago, are a very few specimens collected by John Ray.

It was from such unsophisticated beginnings that the science of entomology and the art of collecting emerged in tandem. In time, butterfly collecting came to be a popular pursuit among gentlemen and ladies of leisure. It was a healthy activity, both physically and intellectually. To gentlemen of the eighteenth century it may have been considered a sport, akin to fowling and hunting. To ladies it seems to have become a more aesthetic, intellectual, and even domestic pastime. Amongst the earliest-known entomologists were women of rank, wealth and fashion who reared butterflies, painted them and embroidered their images on cloth. For scientific details, on the other hand, we are mainly dependent on the illustrated books about insects by Eleazar Albin, Benjamin Wilkes, Moses Harris and others. It was not until the nineteenth century, however, that the organized field-trip came into its own, as revealed in great detail in the entomological journals of the time. Today, when paradoxically increased leisure does not seem to have brought more spare time, we can enjoy browsing through these discursive accounts of another age when everyday cares were for the moment left behind in the quest for butterflies. Here are accounts of arcadian simplicity, of gaity or derring-do, some expressed in terms of stern scientific aridity, others in high-flown sentimental fashion, scattering classical allusions like confetti. Victorians could not only collect and study butterflies; they could also moralize about them without inhibition and hold them up as mirrors to themselves. At the heart of an organized excursion, though, was sociability and pleasant living. Those intent on braving the late autumnal mists could take heart from Sir Humphry Davy, who once remarked, 'The chills of Novr mornings are very unfavourable to ardor in the pursuit of science, and I conceive we should all think better after experiencing the effects of Roast Beef and wine than in preparing tea, coffee and Buttered Buns'.[6] The Aurelians of old would have applauded these sentiments.

BEFORE SETTING OUT

Many nineteenth-century collectors, like modern golfers, had their own ideas about the correct attire for field trips. William Kirby could not recommend to collectors the bag-wig and velvet court-dress depicted by Réaumur, the French naturalist.[7] More suitable for them was 'a plain fustian jacket, with one side and other pockets, as worn by English sportsmen, provided that the outside pockets were large enough to carry all the usual equipment – and that an inside pocket was positioned so as to conceal the forceps-net'. F. O. Morris recommended 'a common fishing jacket' with two large and two small pockets outside,[8] while J. W. Tutt advised the lepidopterist to wear shooting gaiters.[9] F. W. Frohawk always wore a Norfolk jacket, with extra pockets.[10]

Moses Harris was among the first to remind these sportsmen of the desirability of a repair kit: 'It may not be improper to mention some other Conveniences, which I have often found very necessary, such as a pretty large Clasp Knife, and some Needles and Thread. The First he will find useful on many Occasions, and the Second is necessary in mending the Nets, in case he should happen to tare them, and to repair other Disasters, which are incident to People who frequent Places where such sharp Things as Thorns and Briars grow.'[11]

The eighteenth-century Aurelians had an advantage over later entomologists in their large and capacious coat pockets. Most later excursionists would need a knapsack to hold all their collecting equipment, which was steadily increasing in elaboration. By 1822, William Swainson was recommending that the collector's gear should be carried by a 'little boy', like a caddie in golf.[12] There was, in fact, a much greater variety of entomological equipment available to the Victorian entomologist than there is to the field naturalist today. Many of these items survive only in museums or as illustrations in contemporary catalogues – even the faithful clap-net, once owned by every Aurelian, is today a great rarity. Fortunately, a fine collection of this equipment and apparatus has been made by Michael Chalmers-Hunt, and donated by him to The Natural History Museum where it is on display in the Entomology Library.[13] It contains some pieces whose precise function remains obscure. However, the overwhelming impression on viewing this curious array of items is that the Aurelians and their nineteenth-century successors were very well prepared indeed. They certainly intended that nothing should go wrong on their day in the country.

KILLING BUTTERFLIES: FIRE, BRIMSTONE AND LAUREL LEAVES

Killing a butterfly, and yet preserving it in perfect condition, was a problem for the earliest collectors. In the days of Ray and Petiver specimens would seem to have been killed by direct pressure to the head or thorax. Petiver explained how insects should be killed by 'gently crushing their head & body between yr fingers which will prevent their fluttering'. This method was said to be instantaneous and, if done carefully, did not damage the specimen, although William Kirby warned that 'the larger ones will live long after it'.[14] Abel Ingpen therefore recommended that any specimen which resisted this form of execution could be instantly despatched by dipping a pin in *aqua fortis* (nitric acid) and piercing its thorax. Small butterflies could be killed at once by the fumes of burning sulphur or by holding them in the steam of boiling water. This latter method called for an ingenious technique: 'Fix a piece of cork to the bottom of a gallipot, stick the insect on the cork, and invert the gallipot in a basin of boiling water: the steam produces almost instant death and does not injure the plumage'.[15]

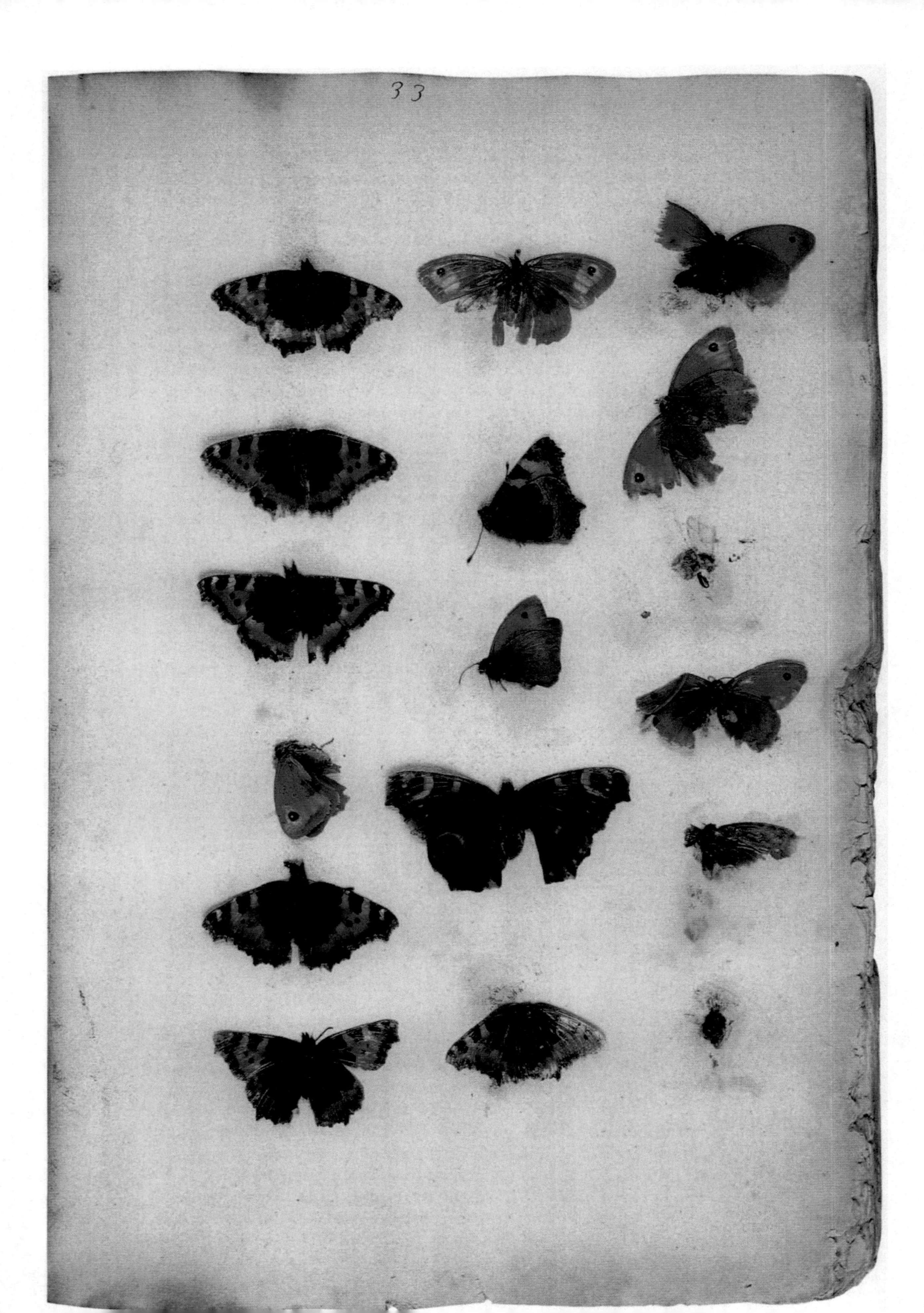

The Revd F. O. Morris found that burning sulphur – or brimstone, as it was called – was the best method for killing specimens. The collector should 'pile the pill boxes under a tumbler near the edge of a table, only do not let it be your best one; light two or three brimstone matches; draw part of the glass over the edge of the table; hold the brimstone matches underneath, so that the fumes of the brimstone can ascend into the glass, taking care not to touch the glass with the light, or it will be cracked; and as soon as you can see, or rather when you can see nothing in the glass for the smoke, replace it entirely on the table, to confine the vapour, and in a few seconds all the moths will be apparently dead, and by leaving them there for a little while they will become entirely so'.[16]

Unfortunately some insects were remarkably tenacious of life. Edward Donovan found that burning sulphur fumes failed to deprive a pair of musk beetles of their lives, even after half an hour. He also reported that a dragon-fly still showed signs of life twenty-four hours after being deprived of its head. For such difficult cases he thought 'the most expeditious method of killing ... is to run a red-hot wire up the body and thorax, for they will live a considerable time in agony if you attempt to kill them with *aqua fortis*'.[17] An equally barbarous (and drastic) method was used by Professor William Dandridge Peck (1763–1822), an American entomologist, who killed his specimens by holding them over the flame of a candle.[18]

Not surprisingly, entomologists were always seeking new and improved methods. In France, A. Ricord had used ether for killing insects as early as 1827, long before it was used for surgery.[19] But it was the advent of ether in 1842 and of chloroform in 1847 for use as anaesthetics, that led to the development of the killing bottle as we know it today. In 1852, Charles Barron, a surgeon at the Royal Naval Hospital, Haslar, was able to assess the effects of chloroform on his patients before deciding to use it as a killing agent for butterflies and moths. He described a wide-mouthed, glass-stoppered jar, containing a layer of perforated zinc to keep a chloroform-soaked sponge in place. The chloroform used to charge the jar was often carried in small metal bottles equipped with a screw top and a small nozzle that allowed the liquid to run out only in drops.[20]

The use of crushed laurel leaves to kill butterflies was introduced by J. F. Stephens, the poison released being prussic acid (hydrocyanic acid). Concerned over accusations of cruelty, Stephens hastened to dispel 'the apprehensions of those humane individuals whose fine sympathies are called into action by a practice which savours more of cruelty than humanity'. He went on to describe a 'not unpleasant mode of destroying vitality in the little objects of our research', that is, the use of crushed laurel leaves.[21] This proved a popular alternative to chloroform and ether, and was undoubtedly safer – generations of schoolboy collectors used nothing else. Guard Knaggs, however, warned that care was needed in the manufacture of an effective killing bottle. Young leaves of laurel were needed, and 'to prevent mildew it is very important that they should not be gathered in the dewy morning, or when the weather is at all wet'. They should first be wiped, then cut into strips, and bruised in a mortar or pounded with a rolling-pin, 'for unless this be done the two principles which go to form the poison will not act on one another'.[22]

Potassium cyanide was first recommended as a killing agent by G. Bowdler Buckton in 1854, and became the most important killing agent during the last half of the nineteenth century and well into the twentieth.[23] The cyanide bottle

Plate 5
Leonard Plukenet's pressed butterflies. These poorly preserved but still recognizable specimens of the Peacock, Small Tortoiseshell and Meadow Brown, adhering precariously to the leaves of a leather-bound volume, are perhaps the oldest surviving British butterflies, dating from 1692–95 and therefore older than Buddle's. The volume, kept in an airtight cabinet in the Entomology Department, Natural History Museum, is rarely displayed because of its fragility.

contained a few crystals of potassium cyanide covered by a thick layer of plaster of Paris on the bottom of the jar. It was important that the jar had an airtight stopper since the fumes were potentially lethal. In 1859, W. D. Crotch, an enthusiastic collector and regular contributor to *The Entomologist's Weekly Intelligencer*, described his own method of killing insects: a few drops of potassium cyanide were added to tartaric acid, which immediately effervesced, producing prussic acid fumes. 'A drop or two among the bruised laurel leaves is a great comfort on a field-day among the Coleoptera, some of which are so determined in their attempts to escape from the bottle.' Five weeks later he added a rider when he wrote: 'I must return to the charge, on the subject of cyanide of potassium as a means of killing insects; but first to refute the objection of its danger. It is already in the hands of every dabbler in photography, and will not therefore be new to many; it is also, if anything, less dangerous than oxalic acid, so generally employed'.[24] Surprisingly, cyanide was readily available at most chemists until the 1950s, and collectors could ask the local chemist to make them up a cyanide jar, which cost no more than a few shillings.

Guard Knaggs, a general practitioner who should have known better, put himself and others at risk with his own decidedly dangerous method of killing specimens. He advised collectors to take a strip of blotting paper and, holding it with pincers, dip it into 'a bottle containing a solution of cyanide of potassium, allowing it to absorb little or much poison, according to the size of the patient; then, opening the lid of the box a trifle, drop and shut in the prepared piece of paper'.[25] The unwary or inexperienced were in some danger of inhaling the cyanide fumes or contaminating their fingers. However, the dangers seem to have been well known, and there is no record of any entomologist poisoning himself accidentally.

One disadvantage of cyanide was that it could discolour insects. C. G. Barrett recorded that a male Brimstone, left undisturbed for two years in a cyanide bottle, developed blotches of a rich crimson colour. These affected the margins of the forewings and much of the hind wings, and oddly enough the only portion of the wings to remain free of the colour was that part of the forewings which in The Cleopatra (*Gonepteryx cleopatra*) 'is clouded with crimson'.[26]

Although the use of cyanide and volatile substances like ether or chloroform had become almost universal by the end of the nineteenth century, other killing agents such as ammonia, camphor, carbon tetrachloride and boiling water have all been used at one time or another. The disadvantage of ammonia is that it may discolour the specimen – as for example, the Peacock, which, if recently hatched, is likely to develop a greenish tint after exposure to ammonia fumes. Today, most lepidopterists use ethyl acetate, which allows specimens to be set straight away, whereas carbon tetrachloride, widely used in the 1960s, causes stiffening of the wings so that specimens may require relaxing before they can be set. It was because many smaller species developed *rigor mortis* quickly, especially on hot summer days, that the early Aurelians set their specimens in the field. However, some microlepidopterists still use crushed laurel leaves which have the great advantage of keeping the specimen relaxed.

The necessity for collectors to kill insects, especially attractive ones like butterflies, provoked a certain amount of adverse comment from people who would not have thought twice about swatting a wasp. Some were concerned that killing methods should be decent and humane; many were unable to see what was being achieved by killing and pinning insects. Some entomologists replied in print to the charges of cruelty. Lætitia Jermyn following her mentor, William Kirby, pointed to the torture of shot and wounded birds, of fish that swallow the hook and break the line, or the hunted hare that is torn to pieces by the hounds, with which the killing of insects hardly compared.[27] Abel Ingpen suggested that the person who would feel shocked at the idea of killing a beetle, 'would the next moment be eating a live oyster, or partake of eels, skinned, and very probably fried alive, without a symptom of remorse: so much for consistency!'[28]

Further comment on this contentious subject came from two churchmen. The Revd F. O. Morris contributed an 'Aphorismata Entomologica' as an appendix to his *History of British Butterflies*, entitled 'Death in the bottle'. He considered insects to have some 'if not a very high degree of feeling', and that collectors should 'make it an unfailing rule to kill them as instantaneously as possible'.[29] The Revd William Kirby, however, refuted the idea that insects might feel pain or distress. He had observed how injured insects showed no apparent awareness of their condition, and that sleeping moths, when transfixed with a pin against the trunk of a tree remained motionless, and apparently unaware of the calamity that had struck them.[30]

PINCUSHIONS AND HATS

One of the first problems for the Aurelian was to bring his captures home undamaged. Since it was seldom possible to keep them alive in good condition, the usual practice seems to have been to kill the butterfly by 'pinching' its thorax, and then impaling it on a pin. This meant that, in addition to a net, the collector needed somewhere for his pins – pins lying loose in a coat pocket was a sure recipe for punctured fingers. He also needed a receptacle for the pinned specimens. If James Petiver is typical of the early Aurelians, the pins would originally have been secured to a pincushion, which was worn like a medallion around the neck on a ribbon. Petiver's pincushion was 'fully stuck with pins of severall sizes'.[31] The reason for having a choice of different pins was obvious enough. As a later Aurelian, Moses Harris explained, you needed 'to be careful not to stick a small Fly [i.e. a butterfly] or Moth, with too large a Pin, which will certainly destroy it, by putting the Joints of the Wings out of place, for such Insects as are disjointed, will never set well, and fall to pieces in a short Time'[32] – advice as true and valuable today as it was then.

The pincushion was in fashion until the mid-nineteenth century, though other ways of carrying it were found. Edward Newman sewed his into the 'stuff of his coat'.[33] William Swainson decided to suspend his from his buttonhole, like a pocket watch.[34] As recently as 1911, Furneaux was still advising collectors to carry a supply of pins 'in a small pocket cushion, or arranged neatly on a strip of flannel which can be rolled up in the waistcoat pocket'.[35] Although no longer used by British entomologists, the pincushion continued to appear in catalogues issued by French naturalists' suppliers until the end of the nineteenth century. As late as the 1920s, the pincushion, or *pélote de chasse*, was used by older entomologists in France, although by then it had become an anachronism.[36] The problems with pincushions were twofold. Pins could fall out or be brushed off in the heat of the moment; and finding exactly the right-sized pin in field conditions could also be awkward. For that reason, dealers began to manufacture pocket boxes with compartments for the different pins. These boxes, often made of mahogany, can still be purchased today.

We do not know when insects were first pinned. None of the specimens in the Sloane collection are mounted on pins, although close examination has revealed that some have holes in them, suggesting that they were at one time pinned.[37] In the 1680s, William Courten (1642–1702), a noted collector of antiquarian and natural history specimens, instructed his cousin, Posthumus Salwey, that butterflies should be 'fastned with pinns to a box'.[38] At about the same time, James Petiver, in manuscript instructions for collectors travelling overseas by ship, advised them to preserve butterflies by 'thrusting a pin thr their Body and s[t]ick[ing] them in yr ha[t] until you get a board then pin them to ye wall of your cabin or ye inside lidd of any Deal Box so yet they may not [be] crushed'.[39]

There is an interesting story associated with this William Courten – known also as Charlton or Charleton – which is worth recounting here. One of the earliest pinned specimens, which came from his collection, was the subject of a celebrated hoax or practical joke that he may have perpetrated to fool Petiver, and which later fooled Linnaeus and many others. In 1702, Petiver wrote that it '*exactly resembles our* English Brimstone Butterfly ... *were it not for those black spots, and apparent blue Moons on the lower Wings*, adding that it '*is the only one I have yet seen*'. He also noted that it was given to him by his '*late worthy Friend* Mr William Charlton ... *a little before his Death*'. Linnaeus named it *Papilio ecclipsis* in 1763 and included it in the 12th edition of his *Systema Naturae* of 1767, but it was not so long before it was realised that it was a forgery, being no more than a skilfully decorated Brimstone (*Gonepteryx rhamni*), the species with which Petiver himself had compared it. In 1793, Fabricius, having examined it, referred to it as being '*arte tantum maculatus*' – the black spots being merely painted on – and dismissed it as a fake. It is believed that the two extant examples of this 'doctored' butterfly, both of which are now in the Linnean

Plate 6
'Charlton's Brimstone Butterfly'. This surviving example in the Linnean Society collection may be a 'replica' made in the late eighteenth century by William Jones, the man who revealed the original forgery which had deceived Petiver and led Linnaeus to give it the name *Papilio ecclipsis.*

Collection (Plate 6), are copies of the original specimen that may have been made in the 1790s after Fabricius exposed the deception. It is just conceivable that one of them is the original, though there is an account by J. E. Smith that Dr E. W. Gray, Keeper of National Curiosities at the British Museum from 1787 to 1806, 'indignantly stamped the specimen to pieces' following William Jones's disclosure of the fraud to him. However, it seems highly probable that William Jones (see Chapter 3: 17), who was a good lepidopterist and a brilliant artist, decided to replicate the historic forgery, and himself decorated two

Brimstone specimens. These are sometimes referred to as 'The Charlton Brimstones'.

The earliest pins, which resembled those used by drapers, were much thicker than their modern counterparts. They were made of steel with wire twisted around one end to form a head, and were very susceptible to corrosion and the formation of verdigris. To overcome this problem, the Victorian collectors used gilt or silvered pins, but these were expensive. Modern entomological pins are made of stainless steel and headless but those manufactured prior to the 1960s were made of brass and often covered with black enamel.

Having pinned his insect, the Aurelian needed somewhere to put it. Petiver, who wore the broad-brimmed hat of his day, suggested that it was as good a receptacle as any, though he also carried 'a long box for Insects with 2 or 3 Smaller for wt odd things may come my way'.[40] His contemporary, Ray, also pinned specimens in boxes.[41] But the convenience of the hat appealed at least to the unselfconscious. Linnaeus's student, André Sparrman, amazed the natives of South Africa by impaling insects along the brim of his hat – they thought he was a magician. In reporting this, William Kirby did not 'recommend such an exhibition in a civilized region', and went on, 'it has often struck me that the cavity of the modern hat, if lined with cork, might be made very useful as a receptacle for these animals in a long excursion'.[42] This practice was later adopted by fly-fishermen, pinning their lures to their hats (as well as to their lapels), as they still do.

By Kirby's day, the broad-brimmed hat had given way to the tall stovepipe hat of the Regency period. In some ways this was an even more convenient receptacle, since there was now much more space on the inside for the wearer to pin his captures out of sight. For those interested in breeding larvae, the hat was also a splendid receptacle for the transport of foodplants. Abel Ingpen lined his hat not only with cork but with white paper, like the drawer of a cabinet. In passing, he also tells us of the content of his coat pockets: 'Hard eggs and a flask of weak spirits and water will be found acceptable in the woods'.[43] As will be seen later, certain Victorian collectors were made of sterner stuff, and would most certainly have scorned *weak* spirits and water.

As hats became smaller – the later Victorian collectors habitually wearing bowlers or caps – the search for an alternative became a pressing matter. The Aurelians were already using cork-lined collecting boxes. Petiver's were made of deal, but Kirby and Spence described much grander boxes made of solid mahogany and opened by a spring catch.[44] In the well-known frontispiece to *The Aurelian*, Moses Harris is shown displaying an oval box in which his specimens (which include a Red Admiral and a dragonfly) are pinned. These boxes came in all shapes and sizes, the limiting factor being the size of the coat pocket or satchel. The pocket collecting-box is rarely used today, many collectors preferring to carry their captures home in pill boxes.

'FLYCATCHERS' AND BUTTERFLY NETS

Who used the first butterfly net? By the time of the first known collectors, in the last years of the seventeenth century, nets were already in use and are referred to as though they were familiar objects. Thomas Moffet described how he was once in an Essex wood in the company of his friend and fellow naturalist Thomas Penny (*c.*1532–88), and fortunately carrying switches of broom '*quibus insecta comprehendere soliti fuimus*' [with which we were accustomed to catch insects], perhaps by knocking them down – though not the best way, it would seem, to obtain perfect specimens! They were suddenly attacked by a swarm of fierce wasps and were able to fend them off with the broom as they ran for their lives.[45] It is also possible that the branches were being used to beat vegetation, the emerging butterflies and other insects then being captured with nets. Ray certainly used such a method when collecting – stick in one hand, net in the other – like many others since.[46] At the beginning of the eighteenth century, Petiver was undoubtedly using a net on his occasional collecting trips around London with Ray and others. One of his correspondents, a John Starrenburgh, had written asking him for a 'Small net to Catch butterflies and glass Vials wide mouthed to breed em from catterpillars'.[47]

More equivocal are scenes illustrated in *The Romance of Alexander*, a Flemish manuscript from about 1338, apparently showing people chasing and catching butterflies with a club or what seem to be rackets (Figs 17a, b). Possibly they were trying to swat them, or perhaps these stylized drawings represented racket-sized nets. The butterflies depicted are large and colourful, clearly not

Fig. 17a, b. These illustrations by Jehan de Grise, from the mid-fourteenth century Flemish manuscript *The Romance of Alexander*, were discovered by W. S. Bristowe in the Bodleian Library, Oxford. They are the earliest known depictions of people chasing butterflies. Their purpose in doing so is unknown, but fairly accurately drawn butterflies can be found in late mediaeval art, and must have been copied from actual specimens. Perhaps butterfly collecting has a longer history than we know.

cabbage whites or other pests. Why, at the outbreak of the Hundred Years War, these Flemish peasants were chasing butterflies is a mystery. Were there collectors even then, completely passed over by history? Or did butterflies have some domestic or commercial use – objects for artists of illuminated manuscripts to copy perhaps, or to incorporate into decorative jewellery? Following the Renaissance, more accurate paintings of butterflies appeared, as for example the Flemish artist Hieronymus Bosch's representations of a Meadow Brown among the crowd of figures in *The Garden of Worldly Delights* (*c.*1500) is remarkably accurate and must have been drawn from an actual specimen.[48] This is certainly a lot more convincing than the crudely executed woodcuts depicting butterflies in Moffet's work one hundred and fifty years later. Many seventeenth-century Dutch still-life artists, more or less contemporary with Moffet, included a number of recognizable species of Lepidoptera in their works.

The earliest recorded date in the history of the butterfly net is 1711, when Petiver visited Holland and returned with the first-known 'muscipula' or bag-net.[49] This was a ring-net: a hoop attached to a stick on which a bag of muslin netting was suspended. The design was not immediately popular, possibly because it was expensive to make. Richard Morris, an apothecary from Rugeley, Staffordshire, got his local blacksmith to make up a few dozen, but his price of four shillings a dozen was regarded as excessive. The man's pride was stung: the nets were well made and 'much lighter than ye London pattern, and as cheap as one can expect'. Whose then were these heavy London nets? Perhaps they were Petiver's, for he is known to have had some made to supply to his overseas collectors. As part of his instructions for their use, he advised they should be wielded by servants. Presumably they were considered too weighty for a gentleman-collector to carry![50] Petiver, incidentally, referred to them not as nets but as 'Flycatchers' ('fly', of course, meaning butterfly). In a letter to a plantation owner in Antigua, he asked for a list of butterflies '& other Insects yr Negroes have gott with my Flycatchers'.[51] Slaves as well as servants were then employed as collectors.

In the early days of entomology, the English collector's preference was for the clap-net or 'bat-fowler'. This was a strip of muslin held between two poles, which the user would 'clap' together to entrap the specimen. The hazel poles used by Benjamin Wilkes were five feet long.[52] The original use of this net was for catching birds or bats, and they were widely available. In a curious little seventeenth-century volume, *The Gentleman's Recreation*, William Cox described how 'Sparrows were to be taken at roost in the Eaves of Thatcht-houses by coming in the night with a Clap-net, and rubbing the Net against the hole where they are flying out. You clap the net together and forsake them: the darkest night with a Lanthorn and Candle is the chiefest time to take them'. Moses Harris recorded that these nets could be purchased at fishing-tackle shops where they were known as 'Butterfly Traps'.[53]

The clap-net would seem to have been a cumbersome device, and certainly looks it in the frontispiece to *The Aurelian* where Moses Harris demonstrates its use (Fig. 18). However, Wilkinson found a home-made 'bat-fowler' remarkably effective and easy to use after a little practice.[54] Moses Harris gave very precise instructions for using one, in phrases reminiscent of those in late-Victorian sporting books teaching schoolboys how to execute various cricketing strokes:

> 'On seeing the Insect come flying toward you, you must endeavour to meet it, or lay yourself in its Way, so that it may come rather to

the right Side of you, as if you intended it to pass; then having the Net in your Hands, incline it down to your right Side, turning yourself a little about to the Right, ready for the Stroke; not unlike the Attitude in which a Batman in the Game of Cricket stands, when he is ready to strike the Ball, only his Bat is lifted up, but your Nets must incline rather downwards: When the Fly is within your Reach, strike at it forcibly, receiving the Fly in the Middle of your Net, as it were between the two Sockets of the Benders, that being the Part of the Net which best receives the insect; and not only so, but should the Fly strike against the Belly or wider Part of the Net, the Course of Air caused by the Motion of the Nets, would carry the Fly with it out of the Net between your Hands, which I have often experienced. The Motion of your Hands in catching, must be from your right Hip to your left Shoulder, not at all retarding the Motion, 'till 'tis as it were spent, closing the Nets in the Motion.'[55]

Fig. 18. A detail from the frontispiece of *The Aurelian* showing the bat-fowling net in use. Obviously practice made perfect, and Moses Harris gave very precise instructions on its handling.

Moses Harris's net was smaller than that used by Benjamin Wilkes, and could also be dismantled and concealed beneath a top-coat. This might have been very useful for, in those days, the collector could be misunderstood, especially in places where superstition and rumours abounded. In 1710, the people of Cadiz, watching Jezreel Jones, one of James Petiver's collectors 'following butterflyes', suspected him of practising 'witchcraft, necromancy and [of being] a maddman'.[56]

By the end of the eighteenth century, lepidopterists had become so adept in the use of the 'bat-fowler' that there was simply no incentive to abandon it in favour of a continental 'muscipula'. During the eighteenth and early nineteenth centuries, the clap-net had become increasing popular and few of the principal collectors of the time used anything else. Lætitia Jermyn was an exception. In the second edition of her *Butterfly Collector's Vade Mecum* she informed readers that: 'For catching the butterfly in its flight a Bag-Net [ring-net] is essentially necessary'.[57] She did, however, use the clap-net as well. Edward Newman used nothing but the clap-net. He described it as 'the grand weapon of the entomologist', adding that it was 'the best for pursuing moths and butterflies on the wing'.[58] F. O. Morris too, writing in 1853, regarded it as the 'best kind of net'.[59] However, during the 1860s it was losing favour, and according to Henry Guard Knaggs at the end of the decade it seems 'to have quite gone out of fashion'.[60] By 1896, J. W. Tutt could refer to it as 'the clumsy old clap-net', which by then had 'given way almost entirely ... to the ring-net'[61] (Fig. 19). Indeed, Watkins & Doncaster, the suppliers of natural history equipment established in 1874, never listed the clap-net in their catalogues, selling only ring-, kite- and umbrella nets (Fig. 20).

The ring-net was initially unpopular. Collectors may have regarded it as a useless foreign invention. Kirby and Spence suggested a use for it, but not for catching butterflies so much as sweeping vegetation at the side of ditches. As entomology was a companionable exercise, 'this employment of brushing the

Books for the Country Advertiser.

WATKINS & DONCASTER,

Naturalists,

36, STRAND, LONDON, W.C.

(Five doors from Charing Cross.)

Every description of Apparatus and Cabinets of the best make for Entomologists, Ornithologists, Botanists, &c.

Wire or Cane Ring Net and Stick, 1s. 8d., 2s., 2s. 3d. Umbrella Net (self-acting), 7s. 6d. Pocket Folding Net (cane), 4s. 3d. and 4s. 6d. Corked Pocket Boxes, 6d., 9d., 1s., and 1s. 6d. Zinc Collecting Boxes, 9d., 1s., 1s. 6d., and 2s. Chip Boxes, nested, 4 doz., 8d. Entomological Pins, mixed, 1s. per oz. Pocket Lantern, 2s. 6d. and 5s. Sugaring Tin (with brush), 1s. 6d. and 2s. Naphthaline, 1½d. per oz., or 1s. 6d. per lb. Best Killing Bottles, 1s. 6d. Store Boxes, 2s. 6d., 4s., 5s., and 6s. Setting Boards, from 6d.; Complete Set, 10s. 6d. Setting Houses, 9s. 6d., 11s. 6d., and 14s. Steel Forceps of the best quality, 2s. per pair. Also Scalpels, Scissors, etc., for skinning birds, etc., etc. Larva Boxes, 9d., 1s., and 1s. 6d. Breeding Cages, 2s. 6d., 4s., 5s, and 7s. 6d.

A Large Stock of BRITISH and FOREIGN BUTTERFLIES, BEETLES, BIRDS' EGGS, etc.

Throughout the Autumn, Winter, and Spring, a large selection of CHRYSALISES OF BRITISH AND FOREIGN BUTTERFLIES AND MOTHS, including the gigantic Atlas and other Exotic Moths.

COLLECTIONS OF NATURAL HISTORY OBJECTS, carefully named and arranged.

NEW AND SECOND-HAND WORKS ON ENTOMOLOGY.

Label Lists of every description. THE COMPLETE LABEL LIST OF BRITISH LEPIDOPTERA (Latin and English names), 1s. 6d. post free.

One each of all the BRITISH BUTTERFLIES IN A CASE, 21s.

A magnificent assortment of PRESERVED CATERPILLARS always in Stock.

BIRDS and ANIMALS Stuffed and Mounted in the Best Style by skilled workmen.

A full Catalogue (pp. 56) will be sent Post Free to any Address.

Fig. 19. Some entomologists were reluctant to give up their clap-nets in favour of more modern designs. This group photograph, taken around 1900, is the last known picture of the 'bat-fowler' in use in the field, and no full-sized nets of this type are known to survive.

Fig. 20. Watkins & Doncaster, established in the Strand, London, in 1879, advertised a wide range of naturalists's equipment and specimens. This 1892 advertisement differs little either in content or prices from a very similar one published thirty-two years later, indeed some prices had gone down – a sobering thought in our inflationary times!

grass, etc. may be carried on if you are walking with any friend not interested in Entomology, without much interruption of conversation'.[62] They were clearly anticipating the sweeping net so widely used by coleopterists today.

The early collectors employed a number of other nets as well. A 'racket-net' was used for catching moths at rest on a tree or wall or a butterfly sitting on a leaf. It was made from wire, 'about the Size of a Raven's Quill, turned round to a Circle' about six inches in diameter. This ring, which was covered with gauze, was attached to a handle two feet in length. The collector carried a pair of them and clapped them over his specimens. Moses Harris advised that 'these Sort of Nets are what an AURELIAN should at all times carry about him'.[63]

The scissors-net or 'Scithers-Net' as Harris called it, was a variation of the Racket-net. It consisted of a pair of rackets fixed to two pieces of iron rivetted across each other and with their ends turned round in the form of rings for the 'Admittance of Thumb and Finger'. Harris wrote that 'a Pair of Toupee Irons or Curling Tongs, such as is used by a Hair-Dresser, are very well adapted for this purpose'.[64] This was sometimes called the 'forceps-net', after the manner in which it was used.

Other more complicated nets were devised. Around 1820, a Dr McLean of Colchester constructed one such out of two pieces of stout cane, connected by a joint at each end and with a rod that lay between them. A cord was passed over a pulley, which was attached to the rod, and fastened to the two canes. A long cane with a ferrule received the lower end of the rod, forming a handle, and a net of green gauze was attached to the two canes. When the collector pulled on the cord, the canes would bend in the form of a hoop and the net opened. As soon as the specimen was in the net, the cords were relaxed and the net closed. Kirby and Spence claimed that McLean had 'scarcely ever found this net to fail'.[65]

In *The Magazine of Natural History*, John Perry described a long-handled net 'for catching all kinds of flying Insects'. It was essentially a shallow-bottomed bag-net, similar to a shrimping net, to which was attached a moveable half-ring of wire, also covered with netting. The net was to be used 'with a slight jerk; and with a little dexterity, it will generally enclose the insect between the bags'. From his accompanying diagram (Fig. 21), this would seem to have been a small net, approximately six inches in diameter, carried on a three-foot-long 'light stick'.[66]

In 1853, under the heading 'Clap-trap', the Revd F. O. Morris described an enormous net which he called the Emperor Net. It was the forerunner of the high net. He likened the shrouds or stays of his fifty-foot net to those of the royal mast of a man-of-war (Fig. 22). He admitted that a fifty-foot net could be unwieldy, but added, 'I have built a castle in the air in the shape of a very long fishing-rod, made of light bamboo ... The net is to be made of very light open net-work, so as not to catch the air'.[67] It is not certain that he ever used it since he wrote that he hoped 'to find it as effective in practice as it is in theory on paper.' Perhaps it remained a 'castle in the air'! A more practical net in use throughout the nineteenth century for catching Purple Emperors was the 'high net'. Mounted on a pole fifteen feet in length, it could be closed with a draw-string once the specimen had been captured. There was much debate as to how 'high' it should be. Stainton used a ring-net attached to a pole thirty or forty feet long. If a ready-made pole was not available, Guard Knaggs advised cutting one from a twenty-foot mountain-ash. With one of these

A Description of an Instrument for catching all kinds of flying Insects. — *Fig.* 92. *a* represents the front view when open; *b* represents the side

92

a b c

view partly closed; *c* the front view when closed. The instrument is to be constructed in the following manner: — To a ring of stiff iron wire is to be attached a bag of spider-net to form the bottom, which is to be covered at pleasure by means of a movable half-ring of wire, filled in with spider-net, and hinged so as to close the top securely when flapped together: the whole is to be fixed firmly to a light stick about 3 ft. long. It is to be used with a slight jerk; and, with a little dexterity, it will generally enclose the insect between the bags. — *John Perry, jun. July* 15. 1831.

Fig. 21. Perry's novel contraption for collecting insects in flight.

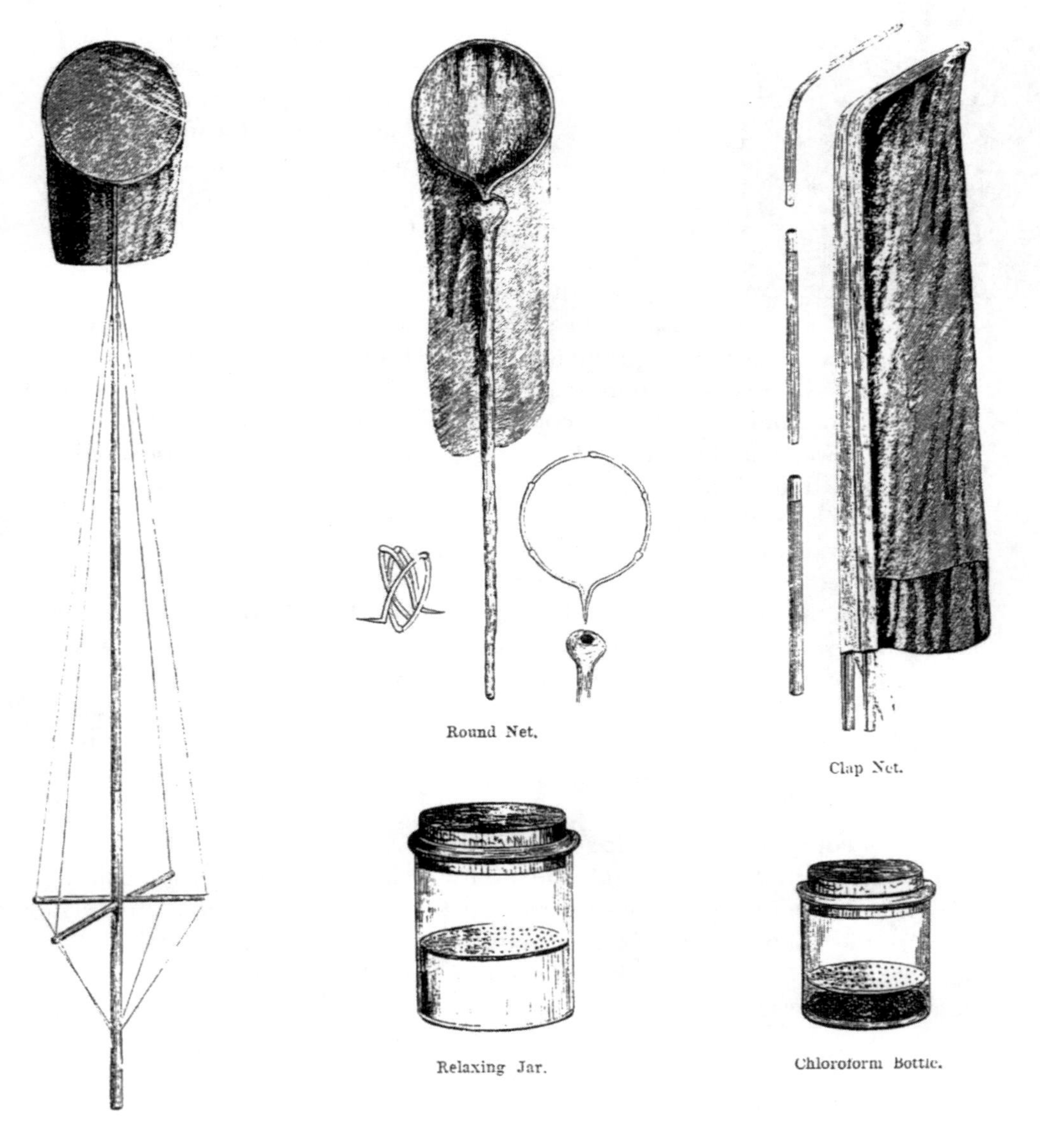

Fig. 22. Some of Morris's recommended nets and other essential equipment illustrated in his *History of British Butterflies*, 1853.

'Mr [C. G.] Barrett and a lady entomologist, managed to catch all [the Purple Emperors] they could get sight of last summer –about sixteen in number.'[68] Tutt preferred a shorter net – a hop-pole twelve to fourteen feet in length being quite sufficient. In the 1870s, he saw nine amateur collectors armed with such nets, standing three or four yards apart, catching each Purple Emperor as it appeared.[69] In 1881, two professional collectors with similar nets apparently captured between two and three hundred specimens. Perhaps these nets were a little too efficient. To Victorian entomologists they must have delivered wonders. It was still in occasional use as recently as the 1950s (Fig. 23).[70]

Fig. 23. Ian Heslop, snapped in 1955, with his 'High Net', based on the design of those used by Victorian collectors.

Guard Knaggs also described an umbrella-net. In size, it was 'rather less than the length of the collector's arm', and let into it was a circular piece 'in the manner of my grandmother's cap'.[71] When newly acquired, the umbrella-net should be 'christened' by taking it out on the first wet evening – this would remove the 'dressing'. This net did have certain obvious disadvantages, notably the mortification felt by a collector on hearing someone remark 'Look at that fool! Why doesn't he put his umbrella up in the pelting rain!' Coleman admitted that 'villagers may be astonished at the insane spectacle of a man scuttling along through the torrent and getting drenched through, while he carries a good-looking umbrella carefully under his arm for fear it should get wet; and if, on the other hand, the weather be fine, carrying such a protective would seem an equally eccentric whim'. He added, however, that 'only the *very* thin-skinned would be driven from the use of a good weapon by such a harmless contingency as I have here supposed'.[72]

D. E. Allen commented that self-conscious eccentricity and the umbrella-net on occasion went hand in hand. 'Why else', he asked, 'except as a complicated private joke, should Professor D'Arcy Wentworth Thompson, the author of the classic work *On Growth and Form*, have insisted on regularly using a battered old umbrella for chasing Lepidoptera – when butterfly-nets of every size and description could be had for almost nothing?'[73]

What was the colour of the early butterfly nets? The Victorian collectors favoured green or white. Edward Newman insisted on green muslin, as it was less conspicuous in country lanes where the the collector was likely to attract attention. In his experience, 'white never fails to attract a crowd, which causes some slight inconvenience to the entomologist, as well as loss of time, for he is invariably under the necessity of explaining to the bystanders exactly what he is doing'.[74]

Today's collector has a choice of several different nets. The kite-net – so called from its shape – is probably the most popular. The ring-net is little different from that used one hundred years ago, but a smaller version is available, in which the hoop is made of spring steel and can be folded into a small ring no more than four or five inches in diameter and then carried conveniently (and unobtrusively) in the pocket.

SETTING AND SETTING BOARDS

Although the early collectors preserved their specimens between slips of mica sealed with passe-partout, as already described, this process tended to damage specimens, as well as making them less available for examination. Many Aurelians recognized this and sought other ways to preserve their specimens, so that in the course of the eighteenth century they abandoned the practice and started to pin their specimens. At the same time, early collectors such as Moses Harris and J. R. Forster realized that in order to present pinned specimens to the best advantage the wings needed spreading, but that while their specimens were drying the wings needed to be held in place – often for days or even weeks. Once they had appreciated this, lepidopterists learned how to 'set' their captures.

One of the earliest descriptions of setting was given by Benjamin Wilkes in his *Directions for making a Collection.*[75] He used a plain cork-veneered board about ten inches by sixteen in size for at that time setting boards did not have a central groove to hold the body of the specimen. The wings were brought forward with the use of a setting needle and held in place by card or 'cork bracers' pinned into the board. The bracers were left in place on the wings of butterflies for a fortnight, or in the case of 'great Moths' for a month. Wilkes recommended setting smaller specimens in the field on account of the likelihood of early *rigor mortis*, adding that 'it is proper to carry always about you a little Box of Cork Bracers for this Purpose'. Moses Harris did likewise and in Plate XX of *The Aurelian* illustrates a Lime Hawk-moth set with card bracers. Other early collectors used thread to hold the wings in place. Rather surprisingly, Henry Stainton did not think that specimens need be left on the setting boards for a fortnight – two or three days was sufficient – 'till they are quite stiff and dry', but he was possibly referring to 'micros'.[76]

The main difference between the setting of these early collectors and that of modern lepidopterists is the position of the wings on the boards (Fig.24).

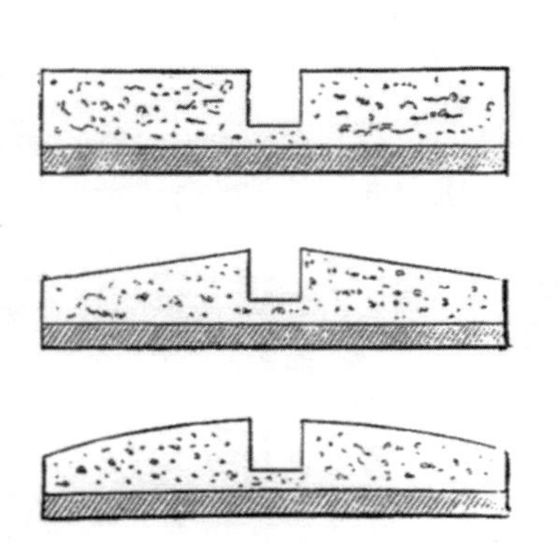

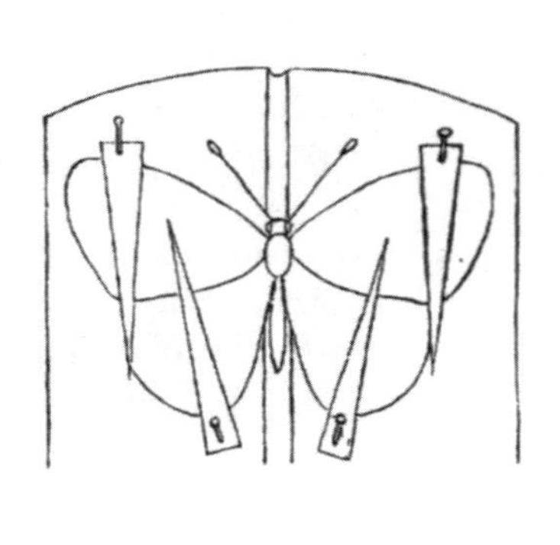

Fig. 24. Setting boards were made in various styles and the method of setting likewise differed. British collectors were scornful of continental practices but adopted them eventually, except for the length of the pins.

The Aurelians tended to set theirs with the forewings sloping backwards in a natural manner, while present-day collectors invariably set the dorsum (the trailing edge) at right-angles to the body to expose more of the pattern of the hindwings. The latter method was employed on the Continent long before the British collector abandoned the old style. The British collector was also slow to use the new grooved setting-boards, which arrived from Europe during the 1860s. These were curved rather than flat, and were often called 'saddles' from their shape. They were kept in specially designed travelling cases, or 'drying houses', which could contain up to a dozen saddles and were ventilated with windows of perforated zinc. The early saddles were small, circular, wooden blocks capable of holding no more than two or three specimens. The English first started using flat setting-boards during the 1930s although many older collectors, brought up to use saddle-boards, continued to use them until well after the Second World War.

Continental collectors have traditionally set their specimens on extremely long pins, and so the drawers of their cabinets are considerably deeper than their English counterparts. W. D. Crotch, ever a champion of lost causes, had no time at all for Continental flat-set specimens, impaled on their long pins. On opening such a box at an entomological sale, he exclaimed: 'No. 3 is a tremendous box – fourpence additional to pay, "Open Sesame", and "Oh! skewers and pancakes!" Here is *Miana expolita* on a spit fit for the eye of Polyphemus, and *Calocampa vetusta* with his wings on a dead level, as if set by a theodolite and artificial horizon. Those who have not seen a Continental cabinet, can better afford to go into rapture on the subject than those who have, and to their transports I leave them. We do not ask our Continental cooperators to change their mode and habits, nor do I think the "Long-Pin Club" at home will do more than cause an infinite amount of subsequent relaxing, – at first to the correspondents and afterwards to themselves.'[77]

The problem of *rigor mortis* and the methods for relaxing butterflies prior to setting were known early in the nineteenth century. W. D. Peck relaxed his specimens by holding them with forceps in the rising column of steam from a fireside kettle.[78] Abel Ingpen found that pierced insects were easily relaxed by sticking them on pieces of cork which were floated in a basin of water. 'If the bason be covered with a damp cloth, they will sooner be relaxed. If the insects are not pierced with a pin they should be lain in the lid of a tin box, and placed on the floating cork. Some insects require a very short time to relax, others require several days.'[79]

William Kirby advocated a variation of this method: 'Fill a basin more than half full of sand, and then saturate with water; pour off the superfluous water, and cover the sand with blotting paper: into this stick the insects you wish to relax, and covering the basin closely, leave them two or three days, according to their size; and the evaporation will render them sufficiently flexible for expansion or any other purpose'.[80]

Since then, relaxing tins and a variety of relaxing fluids – water-soluble products containing a mould inhibitor – have become available from entomological dealers.

COLLECTIONS AND CABINETS

Despite the 'annual cremations' of the rapidly deteriorating insect and other zoological specimens in the Sloane Collection, organized by Dr George Shaw (1751–1813), the Keeper of Natural History and Modern Curiosities at the British Museum, more of these old specimens have survived than was first thought.[81] Apart from Petiver's collection, and those of Leonard Plukenet and Adam Buddle, referred to above, there still exist some of the insects collected by Sir Hans Sloane (Fig. 25), Joseph Dandridge and even John Ray. None of those very early specimens are on pins, though some show pinholes. All are preserved either between thin wafers of mica, or in sealed glass-topped boxes – some with glass bottoms. It was recorded that in Sloane's collection 'instead of glass moonstone or muscovy glass is used, which is much more delicate and light but more costly'.[82] We also know that Henry Baker (1698–1774) employed 'flakes of isinglass upon cards' to protect his specimens but the fate of his collection, which was sold at auction in 1775, is not known.[83]

Fig. 25. Portrait of Sir Hans Sloane (1660–1753), engraved by John Faber after a painting by Sir Godfrey Kneller, 1716. We owe nearly all the earliest specimens of butterflies to his capacious pockets and zeal for collecting 'the Productions of Nature'.

Pehr Kalm, a student of Linnaeus, visited Sloane's Natural History Collections at Chelsea in May 1748 with several others and wrote a full and painstaking account of all he saw. Sloane himself, by then over eighty-eight years old, honoured his guests by spending two hours with them. Kalm described the contents in much detail and room by room. In a little narrow room among 'an endless number of other items were noticed ... a large number of various kinds of insects which filled many cabinets. They were all mounted in large drawers. Each species or individual was laid in a rectangular box the bottom of which was wooden. But among them, some had both the cover and the bottom of the box made of a crystal-clear glass, while some had only a transparent glass lid. At the joints where the glass ran up to the sides of the box it was sealed tight with paper in such a way that no air let alone any moth or other insect could get inside to damage the contents. Where the box had a glass bottom, the insects were firmly fixed to it.' There was also a larger room filled 'with a collection of fishes, birds, insects ...' among other things 'all kept in spirit'. Kalm did not examine these as he had no time to do so.[84]

Petiver, whose own collection of 'the Productions of Nature' was itself of very great importance, was asked by Sloane in 1711 to visit The Netherlands on his behalf in order to purchase an important collection of plants from southern Africa for his museum. Petiver, who so far as is known had never travelled abroad before, wrote to Sloane in some trepidation: 'In case I should dy before my return from Holland, I make you sole possessor of all my collections of naturall things whatsoever ...'.[85] Petiver's fears were fortunately groundless as he was to live for several more years. However, on his death his collection was left not to Sloane but to a sister from whom Sloane eventually bought it for a large sum of money. He found it in a very poor condition and observed that Petiver was careless in curating his specimens, as he 'put them into heaps, with sometimes small labels of Paper, where they were many of them injured by Dust, Insects, Rain &c'.[86] The remains of Petiver's collection can be seen today in two large, leather-bound volumes. Petiver had pressed his butterflies, in botanical style, between two sheets of mica. These were then sealed with gummed paper, and attached to the pages by a paper hinge. Each specimen now possesses a Sloane catalogue number, as well as a small amount of additional information concerning localities and dates, thought to be in Petiver's hand.

The early collectors soon encountered problems preserving their specimens. The cost of mounting individual specimens was high, and the methods used made examination of specimens far from easy. Many specimens were damaged by the very process intended to preserve them as Charles Koenig

(1774–1851), successor to Shaw as Keeper of Natural History and Modern Curiosities at the British Museum, explained to a Parliamentary Select Committee: 'Sir Hans Sloane had a method of keeping his insects which was very injurious to them. He squeezed them between two laminae of mica, which destroyed the specimens in most cases, even the wings of the butterfly.'[87] Consequently the technique was soon abandoned in favour of pinning, which was not only far cheaper but also much safer, provided the specimens were kept inside a box or cabinet. By the mid-eighteenth century most specimens were pinned in cork-lined, glass-topped drawers. From then on the use of setting boards and pins and the possession of a cabinet became more or less obligatory for the maintenance of a collection.

Lepidopterists have always enjoyed a choice of cabinets in which to store their collections. Although most cabinets contain ten, twelve, twenty, thirty, or as many as forty cork-lined drawers, lined also with a good white paper to show off the contents, there are some with fifty or even one hundred drawers, which was the size that Ingpen considered by 1827 that the then 'present state of British entomology' required.[88] The early cabinets were often ornate, individually made, and with drawers protected by sliding glass tops – the drop-on lid was yet to come. Drawer knobs might be made of wood, brass or ivory. Some cabinets had elaborate drop-handles of brass. Many of the early cabinets were made of deal or cedar, which sometimes spoilt collections by exuding resin. These were eventually replaced by harder woods, especially Brazilian mahogany, or a cheaper mixture of mahogany and deal, with plywood for the back. For the better cabinets, close-fitting drop-on glass lids gradually replaced sliding glass as they were far more successful in preventing the incursion of dust, mites or beetles into the drawers.

Edward Newman described his ideal cabinet. It would be made of the best mahogany. It would have two tiers of drawers, with fifteen or twenty in each tier, though 'if the number be fifteen only, there is abundance of room for a book-case to stand above them, which is not only convenient, but has an agreeable effect'.[89] The drawers should be enclosed in front by folding doors, the edges of which would be covered with velvet, to keep out the dust. Each drawer would be from fifteen to eighteen inches square, and two inches deep. Such a cabinet might have cost up to a guinea a drawer (£1.05) but the purchase would even then have been a good investment, as cabinets coming up for sale at auction realized their original cost and sometimes more. Today, a first-rate twenty-drawer cabinet by one of the well-known Victorian manufacturers would cost at least £1000.

According to R. L. E. Ford, the former proprietor of Watkins & Doncaster (Figs 26, 27), the finest nineteenth-century cabinets were manufactured by the Bradys, father and son, of Edmonton, London. Brady was a professional 'Shop-Fitter and Entomological Cabinet Maker'. It appears he started making mahogany cabinets, advertised at 10 shillings (50p) a drawer, in about 1870. Most Brady cabinets were of twenty drawers. It was not however possible to order a cabinet directly from Brady. It was necessary to meet him and remark that you had dined with a Mr So-and-So, and having seen his wonderful cabinets, you wished that you were lucky enough to possess one. All being well, Brady might then offer to make one for you.[90]

There were other cabinet makers producing work of varying standards. Thomas Gurney, like Brady, made mahogany cabinets of exceptional quality, which are still much sought after. One with 40 drawers, in good condition,

Fig. 26. The interior of Watkins & Doncaster's shop in the early 1950s. The late R. L. E. Ford, who took over the business in 1940, is photographed serving an unknown customer.

Fig. 27. 'At the sign of the Swallowtail' – the premises of Watkins & Doncaster at 36 The Strand, photographed in 1953, just before the firm's removal to Kent.

would fetch at least £2,500 today. Crockett's cabinets were considerably cheaper. He used plywood for the base of his drawers and deal for the main framework. They tended to have poorly-fitting drawers which were often not interchangeable. On the other hand, the Hill interchangeable 'storage units' were cabinets of ten drawers and drop-front doors and were purchased in quantity by museums, including the then British Museum (Natural History). They have always proved popular, and cabinets using the 'Hill' design continue to be manufactured today although they no longer match the quality of former times. During the past sixty years, the specialist cabinet-making industry has declined dramatically – it is no longer cost-effective to manufacture entomological cabinets for speculative sale, and today they are made to order.

For collectors who preferred store boxes, there was an enormous variety from which to choose. Nineteenth-century boxes were often designed to imitate leather-bound books and were intended to be stacked on shelves. Apart from entomological specimens, such boxes could also be used to hide small domestic treasures – letters, jewellery, or even perhaps a bottle of spirits! Modern store boxes are more mundane, being plain and made of deal and plywood. Cabinet drawers and store boxes have traditionally been lined with cork, although during the early years of this century, when cork was less easily obtained, a composition material was used. This material had the disadvantage that, after a number of years, it tended to crumble. Cork is less often used today, and is often replaced by Plastazote©, a man-made material, which, being white, does not require papering.

The earliest cabinets were not airtight and therefore, particularly as they usually had sliding glass tops to the drawers, were not pest-proof either. Eleazar Albin was one of the first to appreciate that the museum beetle (*Anthrenus museorum*) would ravage cabinet specimens. He advised collectors to rub the bottom of their store boxes with 'Oyl of Spike' [oil of lavender].[91] Moses Harris, who was also well aware of the problem, advised that the sides and base of cabinet drawers should first be warmed in front of a fire before a small amount of *Unguentum Serulium*, or Ointment of Russet, which contains mercury, was rubbed all over the drawer with a woollen rag.[92] Eleanor Glanville's collection suffered extensive damage. She wrote to James Petiver in December 1702, saying that because she had been 'so long neglecting to clean my butterflys being almost 2 years ye mites have done me much m[is]chefe. I [have] lost above a 100 Species of my finest ... wch I put up closest and safest for fear of Spiders or mice I believe for want of aire not being [(?)fresh] ye mites breed ye more, and ye Bettles was molded over with a whit crusty mould wch when I went to clean broke al to peeces ...'.[93] In the same year, however, Petiver was simply advising his friend Richard Richardson, a Yorkshire physician and naturalist, that if he put his specimens into 'frames with glasses over them ... they will keep for many years; and if at any time you find lice or worms in them, you may easily take out the glass and clean them'.[94]

William Kirby warned against earwigs. 'In sultry weather these animals will often then attack [the specimens] and spoil them'.[95] Edward Newman inveighed against mice, cockroaches, and wasps that could attack specimens drying on setting boards: 'Woe to the Butterfly that is attacked by either of these enemies! It is curious that [these] animals, that are not very likely to eat Butterflies and moths in a state of nature, should prefer them, when the entomologist has prepared them for his cabinet, to every other kind of food, however delicious. I call it a depraved taste; depraved, indeed, it certainly is, for how can these creatures reconcile this conduct with the laws of morality and honesty?' !![96]

Another who was troubled by mould and 'small animalcules', as she called them, was Lætitia Jermyn. She recommended that at the first sign of infestation, tobacco smoke should be blown through the small end of a pipe into a hole drilled through the back of the drawer or store-box. This operation 'not only corrects the putrid and stagnant air, but kills those formidable enemies which often destroy whole cabinets of insects – this process will preserve them for twelve months, when it will be necessary to repeat it ...'.[97]

In addition to these natural predators, collectors have sometimes had to face domestic problems in pursuit of their hobby. Children, spouses, and cats, have all destroyed valuable specimens unintentionally or out of curiosity. Birds, especially blue tits, have been known to enter through an open window and take specimens elegantly displayed on setting boards. All such hazards have to be overcome until the specimens reach the comparative safety of the cabinet drawer. Perhaps the worst indignity on record was experienced in

Fig. 28. An impressively whiskered Dr David Sharp, dressed for the field in late-nineteenth-century fashion, with a long-handled net for catching water beetles.

1840 by Dr David Sharp (Fig. 28), who had been taken on by a wealthy patron as medical 'guardian' to a mental patient. Apparently the doctor had been setting a number of beetles of a species so rare that it had been recorded only twice previously, and had relaxed his specimens in a tumbler of warm water. While mounting the third specimen, he was horrified to see his patient seize the glass of water and down it, beetles and all. Dr Sharp's agony of mind was such that he at first considered administering an emetic to the new possessor of his valued insects, but he pulled himself together and refrained.[98] Another of the disaster-prone was Thomas Edward of Banff in Scotland, who once lost 916 specimens from his insect collection to marauding mice. Undaunted, he at once set about collecting replacements. This poor man seems to have been plagued by small mammals. It is recorded that on another occasion, while sleeping rough after a collecting trip, he was awoken by weasels climbing over his face to get at his treasured trophies![99]

The Victorian practice of collecting a short series of each species changed from the latter part of the nineteenth century onwards. Emphasis was increasingly placed on aberrations and geographical forms, and some collectors who specialized in these amassed hundreds of specimens of a single species.

The placing of small labels under the set specimens, giving date, locality and other information of particular interest, did not become general practice until the middle years of the nineteenth century. Collections made before that time frequently lacked data. However, as early as 1819 George Samouelle had proposed that 'every entomologist should keep an exact journal of the insects he collects; with an account, as far as possible, of the place, food, times of appearance, etc., and place to each insect a number corresponding to that of his journal'.[100] That was fine in theory, but if the journal were lost or destroyed, the numbers attached to each specimen became meaningless.

One possible reason for the reluctance of British collectors to use data labels was the shortness of their pins. Stainton observed that individual labels were all very well when placed under specimens on long Continental pins, but when placed under specimens mounted on short English pins they were 'out of sight and we have to lift the insect up to see when and where it was caught'. He too recommended keeping numbered specimens and a descriptive journal and commented: 'What a mass of errors and blunders would have been avoided if everyone had done this', adding wryly that a collection in which each specimen 'could speak for itself, and say whence it came and when captured, would be a vast improvement on any of our existing collections'.[101]

Tutt was scathing about the lack of data in early collections. Referring to the magnificent collection of Henry Doubleday, at that time housed in the Bethnal Green Museum, he complained that 'there is nothing to tell us whether any particular insect came from the North Pole or the Sahara; nothing to help us draw any conclusions from one of the greatest masses of Lepidoptera ever collected together. What a different value the collection would have if some system of labelling had been adopted'. Tutt admitted that as a young collector he had himself not realized the value of data labels. 'We now repeat with shame, that in our early days we often received insects from well-known entomologists with such little labels and wondered what one wanted more than the insect, and systematically – we are ashamed to own – took them off before putting the insect in our cabinets. In time, however, we began to find that a series was not everything, and that a man, to be an entomologist, wanted something more than a fine collection, and that he might be a first-class entomologist, with a miserable collection – in the generally accepted sense of the word. We soon wished we had not destroyed these labels, and that all our correspondents would follow the example of these gentlemen.'[102]

Tutt was exasperated by the habit, especially of 'London lepidopterists', of concealing the localities from which they had taken their most treasured specimens. Without labels the specimens revealed nothing. 'Asking for information, we were treated to courteous explanations as the ground being private, etc., and leading in every case to the undoubted conclusion that we wanted the knowledge for poaching purposes. This was annoying, but, as we are probably more pachydermatous, we persevered, and eventually got the desired information; this difficulty vanishing in a short time, the facts being now most courteously given, when there is no doubt about the purpose for which they are required. But still it was a difficulty at first, and one which had to be overcome. We have frequently had pointed out to us that there is a real danger in allowing exact localities to be known.' This caveat would appear to be even more pertinent today.

COLLECTING TRIPS AND EXPEDITIONS – 'A DAY IN THE COUNTRY'

The joys of the field outing were evocatively related by the Revd J. G. Wood in his popular volume *Insects at Home*, quoting a friend's recollection of a particular summer day twenty years earlier. 'Saturday, July 10, 1871, will always be a red-letter day in my annals of natural history', his friend D. J. French wrote. He and his brother had gone 'Machaon-hunting' in the Cambridgeshire fens. They drove their pony and trap 'along one of those everlastingly flat Cambridge roads', eventually reaching an old whitewashed public house, with a conspicuous signboard on which was painted 'Five Miles from Anywhere'. A few minutes later and they were stepping off to the fens just as the sun began to peep through the mist. They cleared the ditch with the help of a jumping-pole, kept at the inn for the use of entomologists. 'My brother was over first, and before I could follow a hearty shout informed me that a Swallowtail was captured. In less than three minutes at 1.5 p.m. three hurrahs rang through the fens, as I beheld, for the first time in my life, a self-captured *Papilio Machaon* within my net. Oh, 'twas a pleasant sight to see! The next two hours were exciting ones indeed, for no sooner had I pinned an insect than another was seen.' Although nearly exhausted after tiring chases through the fen, the brothers continued to dash about 'disturbing from their nests and passing unheeded whinchats, stonechats, whitethroats, and other sedge and grass loving birds'. Soon another butterfly appeared and another. They were pursued with vigour. 'The fall into the sitting posture was the natural, necessitous and happily convenient move, for directly the capture was made our legs lost all power, and suggested a seat on the fen as the right position for pinning the prize; and so it was.'[103] Such behaviour would rightly be regarded with extreme disapproval today but at that time would appear to have been quite unexceptionable.

The early Aurelians collected widely, in spite of the difficulties and dangers of travel. Elaborate planning was often necessary, and complications sometimes arose on what might seem to us to be the simplest of journeys. In 1713, Joseph Dandridge and James Petiver, together with a group of friends, set out on an expedition, by river boat, to Gravesend. For this, a special permit, similar to a passport, was necessary. Issued by the Lord Mayor of London, it directed constables and watermen that the bearers were to be allowed to travel unhindered, 'it being represented to me that they have extraordinary occasions for their so doing'. The issue of a permit in the year that the Treaty of Utrecht brought peace to Europe is puzzling, but travel may have been restricted for fear of a Jacobite uprising.

As late as 1837, J. C. Dale required two guards for his journey to the north of England, on account of possible attack by robbers. The standard of hygiene in inns also left much to be desired, guests sharing their cramped sleeping quarters with vermin of all shapes and sizes. By the 1840s, however, the advent of an efficient railway system shortened journey times, and a trip from London to Edinburgh, which during the mid-eighteenth century would have necessitated ten days on the road being bumped around in a mail-coach, could be accomplished in a fraction of that time. Some, however, expressed regrets at the demise of the stage-coach which, after all, had provided passengers with splendid views of the countryside and even the chance of netting unexpected specimens while in transit. At least one paper was published which detailed insects taken en route by this means. Nevertheless, naturalists and natural history societies, field clubs and school societies all benefited from the new-found freedom of travel, and lepidopterists could now finish work on a Friday evening, assemble their 'weapons of the chase' and jump on the early morning train for the New Forest or the Downs of Kent.

D. E. Allen has observed that 'all through British Science a sudden thirst arose in the 1830s for massive amounts of factual material, the reflection perhaps of less rationally perceived requirements than of emotional convenience – but coinciding with the birth of a new professionalism'.[104] Before this time entomologists collected in comparative isolation, but the publication of new journals to cater for their interest and the proliferation of natural history societies and field clubs all over Britain provided far better opportunities for intercourse. Entomology suddenly became extremely popular.

Instructions for the novice collector proliferated. Some, such as those given by Abel Ingpen, state the obvious, but the obvious is sometimes imperfectly recognized. 'The collector', he wrote, 'when he makes an excursion should have three principle objects in view, for which he ought to be duly prepared. The first is to *find* insects. The next is to *catch* them, and the last, when taken, to bring them *safe* home. In exploring their haunts, he must also recollect that some will be *reposing*; others *feeding*; others *walking*; others *running*; others *flying*; others *lurking* in various places of concealment, and in different states of existence.'[105]

Henry Guard Knaggs's advice, on the other hand, erred on the side of eccentricity. To capture the Purple Emperor, the ultimate quarry of every Victorian lepidopterist, he suggested that: 'A brilliant idea is to shoot, with dust shot or with a charge of water (as they do Humming Birds in S. America), the Emperor as he sits in state!' Since the Emperor was very pugnacious, 'another dodge is to shy up shining pieces of tin, or stones with bits of white paper attached, when his dignity being wounded, he will sometimes chase the offending object to the ground'. Knowing that the Purple Emperor is attracted from its tree-top perch by carrion, Guard Knaggs once tried to tempt it with a long-dead cat or, as he put it, 'a decomposed specimen of the feline race'. However, he jocularly added, 'although we let the cat out of the bag, and H. I. M. evidently smelt our game, we could not induce him to come within reach – perhaps the game wasn't high enough'. For sedentary specimens on trees he recommended that 'Trunk beating by the mallet' would put them to flight, although he was forced to admit that the weight of the mallet militated against carrying such a weapon. 'Pumping by means of a powerful garden engine'; 'Pelting with stones'; 'Fumigating and vaporising'; 'Simply walking about', and 'gently disturbing the bushes' were other means of frightening insects into flight.[106]

Little is known about collecting trips before the nineteenth century since there were few journals in which the details could have been recorded. With the publication of *The Entomological Magazine* during the 1830s, and of subsequent journals such as *The Entomologist* (1840–42), which from 1843–63 was absorbed into *The Zoologist* before demerging in 1864 and continuing independently, *The Entomologist's Weekly Intelligencer* (1856–61), and *The Entomologist's Record* (founded in 1890 and still going strong), all kinds of collecting activities, records of captures, and details of excursions were described, often in meticulous detail. Victorian entomological journals differed greatly from those of today and space in them was never at a premium. Many words were written and published where only a few would now be considered sufficient or even desirable. Edward Doubleday and Edward Newman, for example, set out in 1832 on a three-week collecting excursion to the Welsh Marches and North Wales. They described the discomfort of their travels in *The Entomological Magazine.* From London they took roof seats on the Worcester Mail stagecoach for the fourteen-hour journey. Having breakfasted at Worcester, they continued north towards Leominster, where they stayed for two days. Unfortunately it rained incessantly. Indeed, the weather was so awful that 'we were glad to get dry shoes and stockings, and crowd to the fires which we found our friends everywhere enjoying'. They then boarded the Liverpool Mail. This took them to Shrewsbury and a brief stop, which was memorable for 'a hasty and bad cup of coffee', before they continued onward to Llangollen and 'a capital breakfast'. At Llanberis, near Snowdon, the weather turned nasty and Newman wrote: 'scarcely had we arrived when rain, hail, and snow, or a compound of the three, began to fall around us in torrents, and very speedily wetted us all to the skin'. The dauntless collectors proceeded 'with sundry falls, and divers bruises occasioned thereby, for about an hour and a half', eventually reaching a little stone hovel, erected by the workers of a copper mine as a shelter. Here, cold and drenched, they held a council of war – 'the usual consequence of a defeat'. On returning to Capel Curig they were overcome with fatigue. 'We were glad to get rid of our wet clothes and go to bed', he confessed. These were certainly not the ideal conditions for collecting! On leaving Wales some days later, Newman and Doubleday continued on their journey, returning via Shrewsbury and then Leominster. It was four miles from there, at Olden Barn, that they took five specimens of the Mazarine Blue (*Cyaniris semiargus*) in 'a rich meadow, on a hill-side' which belonged to Newman's father.[107]

Henry Stainton was anything but a private man. He advertised his whereabouts, his 'At Homes', and his holidays in the entomological journals. He even published a small book entitled *The Entomologist's Companion*,[108] in which he described quite intimate details of a collecting trip to Western Scotland and the Isle of Arran with his wife:

> 'The 17th, I revisited the boggy place near Sandbank; but I should premise, I was not living like a hermit all this time. Mrs Stainton and one of her sisters, then rather an invalid, were with me, and we were visiting a very pleasant lady, the wife of one of my cousins. On the 17th, as I was saying ... when lo! a flash of lightning and a clap of thunder and some unpleasantly large drops of rain, said pretty plainly it was time to retreat; accordingly we were setting off towards a cottage that was near the ferry, when Mrs Stainton stopped, in order to provide better for the safety of her bonnet.' Stainton went on to describe what happened to the bonnet as Mrs Stainton tried hard to protect it from the rain: 'Mrs Stainton's device was ingenious: not thinking the best place for her bonnet was on her head (to be sure, they are worn *behind* the head now), she took it off, and carefully pinned it beneath her petticoat; we then made for the cottage, where several, like ourselves, had been driven to seek shelter. Mrs Stainton sat very carefully on the edge of her chair, fearing to crush her bonnet; but on the storm abating a little, after we had been under cover for half an hour, she thought of putting on her bonnet, when to her dismay it was not in the secure haven where

> she had placed it! Of course there was but one solution of the mystery – it must have dropped soon after it had been pinned there; and on my running back to look for it, there sure enough it was on the grass, without any shelter, and nicely drenched. After waiting for another half an hour it was quite fine, and we recrossed the ferry and arrived at Strone, much to the delight of our hostess, who had been in a peck of troubles at our being out in the storm. She was sure we would either be killed by the lightning, or catch our deaths of cold from getting wet' ... 'We all laughed at her and then she laughed too, for she's the best tempered creature: but every now and then she said, with a solemn shake of her head, "Well, ye just don't know what an awfu' fright I had". The 18th, we went up the valley to Loch Eck, but here I found no Tineina, only a few *Crambus* [now *Catoptria*] *margaritellus*, so the day was rather a blank.'

There were numerous butterfly collectors among the clergy in the nineteenth century. In one of the earliest numbers of *The Magazine of Natural History*, J. C. Loudon shrewdly observed that 'a taste for Natural History in a clergyman has great advantages, both as respects himself and others. It is superior, in a social point of view, even to a taste for gardening ... the naturalist is abroad in the fields, investigating the habits and searching out the habitats of birds, insects, or plants, not only invigorating his health, but affording ample opportunity for frequent intercourse with his parishioners. In this way their reciprocal acquaintance is cultivated, and the clergyman at last becomes an adviser and friend, as well as a spiritual teacher.'[109]

That is one explanation why the passion for collecting became widespread and infectious in the nineteenth century. With such an upright and respectable model as the local parson, naturalists could reassure themselves that to obtain a series of fifty or even a hundred specimens of a single butterfly was indeed a worthy achievement. Certainly Abel Ingpen's opinion on the subject would be deplored by present-day conservationists although it was widely subscribed to in his day: 'Young collectors are also advised, when they are so fortunate as to meet with rare, or local insects in abundance, not to fail availing themselves of the opportunity to take more than they may require, in order not only to oblige their friends by the gift of specimens which they may not possess, but also to exchange their duplicates for other species which may be desiderata to themselves; by which means both parties are obliged and enriched by the labours of each other. To those who collect for their friends, it is suggested, that they should send them as many different sorts, and as many of a sort as possible.'[110]

J. W. Tutt, too, proposed that whenever the collector found a species, even a common one, in a new locality where he had never seen it before, he should take a few specimens. Publication of *The Origin of Species* had, he noted, opened up to students 'a new field of philosophical enquiry'.[111] Collectors could now appreciate that when a species was reared under different environmental conditions it would often show a number of variations.

Of all the rare butterflies that were the target of collectors, the Purple Emperor received the greatest attention. This magnificent butterfly appears to have evoked awe and reverence in many collectors of the time, partly because of its resplendent colours, but also because of its elusive and, as it were, haughty habits. The Revd F. O. Morris remembered one Emperor hunt in

Northamptonshire with a fellow clergyman, the Revd William Bree, curate of Polebrook, 'to whom I had no introduction but that which the freemasonry of Entomology supplies to its worthy brotherhood'. The Emperors were soaring round the tops of tall oak trees and Morris first beheld them 'at a most respectable distance; they at the "top of the tree" and I on the humble ground.

'The next day, in the same wood, at Ashton Wold, near Oundle, Northamptonshire, during my absence in successful search of the Large Blue, ... Mr Bree most cleverly captured one. ... That specimen, a male, ... now graces my cabinet, together with the first female that its captor had ever taken, both obligingly presented by him to me. ... I hope that Her Most Gracious Majesty [a reference to Queen Victoria] has no more profoundly loyal subject than myself, and I may therefore relate that, without any reference to what is now going on in France, or any allusion to Louis Napoleon, my toast that evening after dinner was, with as much sincerity as in the minds of the French, "*Vive L'Empereur*!".'[112]

Another 'Imperial' experience was described, in high-flown language, in 1857 by the Revd Walter Wilkinson, Rector of Hyde, near Fordingbridge, in the New Forest:

> 'My brother and I were returning from bathing one hot day, a fortnight ago, and had just arrived at the meadow in question, when an exclamation from my brother drew my attention to a splendid large butterfly sailing about among some alders growing in a damp corner. With my net in hand I made a rush at the spot; the creature made a lordly swoop in my direction, which had the singular effect of taking all the breath out of my body; a furious but futile dash of my net sent him careering off amongst the trees and we saw him no more. What was it? A day or two after I received the following tidings from my brother:– "On returning home from bathing yesterday I again encountered the Emperor, not far from the spot where we saw it before, in the ragwort field. He flew *bang* at me. I aimed a blow with my towel, and sent him sprawling to the ground, but before I could hit him again he was up and away. I pursued him in a most excited chase, but of no avail – he gained every step, and left me exhausted in the rear. Had you not better come over some morning and take this 'Royal City of Waters' lest it should be called after my name?" 'A few days later came the following letter: "The Royal City is taken! I went down and looked for it every day, saw it twice, knocked it down again with my towel, but all in vain. So I took the net, went and waited two hours without seeing him: at last he came. I saw him on the other side of the field – raced across and took him easily. Strange to say, as soon as I had caught him, I felt a pang of sorrow that I hadn't got two, so I waited an hour more, but alas in vain." '[113]

Such racy accounts are, however, fairly rare. The truth is that most field outings were not the stuff of romantic fiction, and exciting adventures were unusual. Although numerous field trips were reported in the journals, they tended to be about the same few places – the New Forest for Fritillaries, Chattenden Roughs and Bentley Wood for Purple Emperors, Royston Heath for Chalk-hill Blues. Some published reports with titles such as 'Doings at Darenth Woods' suggested unusual, even furtive, happenings, but most consisted of little more than long lists of the various species captured. During the main collecting season – July and early August – the direct rail link between the metropolis and the New Forest was very popular. An annual influx of several hundred collectors provided the basis for an important local industry in which local inhabitants competed to offer bed and breakfast (and packed lunches), as well as the hire of dog-carts and help with collecting (Fig. 29). W. H. Hudson, who strongly disapproved of killing things, described Lyndhurst – the centre

Fig. 29. Henry William Gulliver, a retired forester, shown here with his wife at the gate of their New Forest home in 1931. He was much in demand as an entomological guide between the World Wars.

of all this activity – as the place where 'London vomits out its annual crowd of collectors'.[114]

Lepidopterists who joined field trips soon discovered hospitality at a number of celebrated inns, where they could relax and discuss their morning's catches in congenial company. During the nineteenth century, the Bull Inn at Birch Wood, Kent (Fig. 30), was particularly favoured. More recently 'The

Fig. 30. An engraving of the Bull Inn at Birch Wood near Darenth, in Kent, a regular meeting place of the Entomological Club in the 1830s.

Chequered Skipper' at Ashton, Northamptonshire (built and named by Charles Rothschild), and 'The Chequers' at Royston, have proved popular.

Another meeting place for collectors was 'The Square and Compass' at Worth Matravers, Dorset. Here, a former inn-keeper recounted how Lord Rothschild sometimes appeared during the collecting season. Tramping the hillside that leads down to the sea, he was invariably accompanied by his chauffeur who carried a golf bag over his shoulder containing various nets.

H. C. Huggins gave an account of collecting in Kent, entitled 'The Old Days at Chattenden Roughs'. Visiting collectors made the most of Blake, the local 'Cadger'.

> 'Many Victorian and Edwardian households had a kind of cadger and odd-job-man, with no particularly defined duties, of which Blake was a sample. He was expected to do rough digging in the garden, carry coal, chop wood, carry bags to the station, exercise the dogs, feed the fowls, drown she-cats and generally make himself useful to mankind and an ornament to society. Blake, however, was a man of some little intelligence and in a rough way a naturalist. He could skin animals and birds and set them up in a somewhat terrifying fashion (the ghastly grin of my aunt's late cat Toby, was legendary in the family), and was a professional collector of butterflies. There were a number of these in the district at the time, of whom the chief were Doran, who caught [A. B.] Farn's famous black *galathea* [Marbled White] at Lodge Hill, then a part of Chattenden, and Packman, a celebrated *iris* [Purple Emperor] catcher.
>
> 'Packman always wore a top-hat when collecting and pinned his Iris in it. On one occasion when Farn met him at Chattenden he proudly showed him a very fine [var.] *iole* pinned in the crown. Farn at once offered him £5 for it, which he accepted on condition that he might take it home and set it. Of course Farn demurred to this and found his reason was that unless he showed them the butterfly Packman thought none of his friends would believe him, so Farn gave him a letter describing the insect and naming the price paid and carried it off safely.'[115]

J. W. Douglas, an avuncular and splendidly bearded lepidopterist and versifier, described one golden occasion with the Pale Clouded Yellow:

> 'Just through Croydon the embankment of the old tramway runs parallel and close to the turnpike road; it is a capital place for many insects. One of my earliest expeditions to this place was in 1842, with Mr. Lambert, who was one of my instructors in collecting; and I shall never forget his excitement when, as we were going leisurely along the Purley Oaks, on a broiling afternoon in August, a *Colias Hyale* [Pale Clouded Yellow] came dashing along, settled for an instance on a flower, and, as if sunshine was too precious to waste, rushed on again. "By Jove!" said Lambert, "it's ... ", He could say no more, nor did he stay to remember the name, but with hat off, coat flying and net extended, I saw him coursing like mad up a hill. After a while he returned flushed with victory and heat, and told me what a prize he had: in a short time we had captured several of the rare butterfly.'[116]

Not everyone was successful in catching this particular fast-flying butterfly. Edward Newman painted a very different picture of hunting it across the same part of Surrey: *Hyale* 'has led me the merriest dance among the blooming lucerne: it was where the Croydon rail now intersects those Surrey hills which constitute the first glimpse of country as we emerge from the fuliginous sea of London habitations: it was here, in market-gardens forbidden to the public, that I made her acquaintance. Here were employed a multitude of female Hibernians in the healthful pursuit of horticulture. On one occasion my quarry led me into their midst, when lo! they abandoned their occupation, and pursued me with the very same energy that I was wasting on the yellow-robed nymph'.[117]

These vivid accounts evoke a variety of emotions in the butterfly-lover of today: envy at the plentitude of butterflies to be found in the countryside then, at the same time a certain revulsion at the eagerness of the collectors to amass their bounty in such vast and quite unjustifiable numbers. However, it is only through such evocative reports that one can today appreciate what much of the countryside was like even as recently as sixty years ago. The basic hunting instinct is undoubtedly innate, but the 'weapon' of today is more likely the camera than the net. The blood may not race for many of us in the challenge of hot pursuit of a specimen for the cabinet, but it no doubt thrills not a little when the subject of the chase settles for a once-in-a-lifetime photographic opportunity.

NOTES TO CHAPTER TWO

1. Dale, C. W. (1890). *The history of our British butterflies*, pp.vi–vii.
2. Allen, D. E. (1976). *The naturalist in Britain, a social history*, p.6.
3. Moffet, T. (1634). *Insectorum sive minimorum animalium theatrum*, p.45.
4. Petiver, J. (*c.*1700). *Brief directions for the easie making and preserving collections of all natural curiosities*, one-sided folio.
5. Fitton, M. & Gilbert, P. (1994). Insect Collections. *In* MacGregor, A. (Ed.), *Sir Hans Sloane*, p.118.
6. Allen, *Naturalist in Britain*, p.58.
7. Kirby, W. & Spence, W. (1826). *An introduction to entomology*, **4**: 525–526.
8. Morris, F. O. (1853). *A history of the British butterflies*, p.4.
9. Tutt, J. W. (1896). *British butterflies*, p.92.
10. Chatfield, J. (1987). *F. W. Frohawk: his life and work*, pp.98, 157.
11. Harris, M. (1766). *The Aurelian, &c.*, Preface, p.7.
12. Swainson, W. (1822). *The naturalist's guide*, p.43.
13. Chalmers-Hunt, J. M. (1994). Entomological bygones or historical entomological equipment and associated memorabilia. *Archs Nat.Hist.* **21**(3): 357–378.
14. Kirby & Spence, *Introduction to entomology* **4**: 529.
15. Ingpen, A. (1827). *Instructions for collecting, rearing and preserving British insects*, pp.68–75.
16. Morris, *British butterflies* (Appendix): Aphorismata entomologica, p.7.
17. Donovan, E. (1805). *Instructions for collecting and preserving various subjects of natural history* (edn 2), p. 43.
18. Kirby & Spence, *Introduction to entomology* **4**: 530.
19. Ricord, A. (1827). Nouveau moyen pour faire mourir promptement les insectes. *Bull.Sci.nat.Géol.* **12**: 295.

20. Barron, C. (1852). Method of employing Chloroform for killing Insects. *Zoologist* **10**: 3435.
21. Stephens, J. F. (1835). Mode of killing insects. *Ent.Mag.* **1**: 436.
22. Knaggs, H. G. (1869). *The lepidopterist's guide*, pp.104–105.
23. Buckton, G. B. (1854). On the application of cyanide of potassium to killing insects for the cabinet. *Zoologist* **12**: 4503–4506.
24. Crotch, W. D. (1859b). Killing insects. *Entomologist's Wkly Intell.* **6**: 37–38, 92.
25. Knaggs, *Lepidopterist's guide*, p.103.
26. Barrett, C. G. (1884). Effect of cyanide upon colour. *Entomologist's mon.Mag.* **21**: 23.
27. Jermyn, L. (1827). *The butterfly collector's vade mecum* (edn 2), p.8.
28. Ingpen, *Instructions for collecting*, p.68.
29. Morris, *British butterflies*, p.17.
30. Kirby & Spence, *Introduction to entomology* **4**: 54–59.
31. Wilkinson, R. S. (1975c). The Rise and Fall of the Pincushion. *Entomologist's Rec.J.Var.* **87**: 142–146.
32. Harris, *Aurelian*, Preface, p.7.
33. Newman, E. (1841). *A familiar introduction to our history of insects* (new edn of *The grammar of entomology*), p.92.
34. Swainson, W. (1840). *Taxidermy*, p.16–17.
35. Furneaux, W. (1911). *Butterflies and moths* (2nd impression), p.79.
36. Wilkinson, 'Pincushion', pp.142–146.
37. Fitton & Gilbert, 'Insect collections', p.120.
38. Courten [Charlton], W. [1680s], Brit.Mus. *Sloane MS* 3962, *f.*186r.
39. Petiver, J. (*c.*1700). *Sloane MS* 3332, *f.*2r,v.
40. Petiver, J. (*c.*1700). letter to Samuel Brown, undated. *Sloane MS* 3332, *ff.*176r–177r.
41. Ray, J. (1710). *Historia insectorum*, p.524.
42. Kirby & Spence, *Introduction to entomology* **4**: 526.
43. Ingpen, *Instructions for collecting*, p.8
44. Kirby & Spence, *Introduction to entomology* **4**: 524.
45. Moffet, *Insectorum theatrum*, p.45.
46. Ray, *Historia insectorum*, p.144
47. Wilkinson, R. S. (1966b). English Entomological methods in the seventeenth and eighteenth centuries. Part 1: To 1720. *Entomologist's Rec.J.Var.* **78**: 143–151.
48. Martin, G. (1978). *Hieronymus Bosch*, Pl.22.
49. Wilkinson, R. S. (1978). The History of the Entomological Clap-Net in Britain. *Entomologist's Rec.J.Var.* **90**: 127–131.
50. Petiver, J. (1712). Letter to Thomas Grigg of 25th March. *Sloane MS* 3338, *f.*38r.
51. Petiver, J. (1713). Letter to Rachel Grigg of 20th October. *Sloane MS* 3339, *f.*81r.
52. Wilkes, B. (1742a). *Twelve new designs of English butterflies*, 13pp. of engraved plates including title page.
53. Harris, *Aurelian*, Preface, p.6.
54. Wilkinson, 'Clap net in Britain', p.131.
55. Harris, *Aurelian*, Preface, p.6.
56. Wilkinson, R. S. (1968). English Entomological Methods in the Seventeenth and Eighteenth Centuries. Pt. 3: Moses Harris *The Aurelian*. *Entomologist's Rec.J.Var.* **80**: 193–200.
57. Jermyn, *Vade mecum*, p.48.
58. Newman, *History of insects*, p.93.
59. Morris, *British butterflies*, p.4.
60. Knaggs, *Lepidopterist's guide*, p.75.
61. Tutt, J. W. (1902). *British moths*, p.357.
62. Kirby & Spence, *Introduction to entomology* **4**: 517.
63. Harris, *Aurelian*, Preface, p.7.
64. *Ibid.*
65. Kirby & Spence, *Introduction to entomology* **4**: 520.
66. Perry, J. (1831). A Description of an Instrument for catching all kinds of flying Insects. *Mag.nat.Hist.* **4**: 436–437.
67. Morris, *British butterflies*, p.25.
68. Knaggs, *Lepidopterist's guide*, p.75.
69. Tutt, *British butterflies*, p.386.
70. Heslop, I. R. P., Hyde, G. E. & Stockley, R. E. (1964). *Notes and views on the Purple Emperor*, p.227, pl. xxi..
71. Knaggs, *Lepidopterist's guide*, p.75.
72. Coleman, W. S. (1860). *British butterflies*, p.42.
73. Allen, *Naturalist in Britain*, p.153.
74. Newman, E. (1860). *A natural history of all the British butterflies*, p.22.
75. Wilkes, B. (1742b). 'Collecting Directions' for Moths and Butterflies. Supplement to *Twelve new designs*. Folio broadside.
76. Stainton, H. T. (1867). *British butterflies and moths*, p.46.
77. Crotch, W. D. (1859a). Pinning and setting Lepidoptera. *Entomologist's Wkly Intell.* **6**: 30–31.
78. Kirby & Spence, *Introduction to entomology* **4**: 530.
79. Ingpen, *Instructions for collecting*, p.75.
80. Kirby & Spence, *Introduction to entomology* **4**: 535.
81. Stearn, W. T. (1981). *The Natural History Museum at South Kensington*, p.17.
82. Quarrel, W. H. & Mare, M. (Eds) (1934). *London in 1710: from the travels of Zacharias Conrad von Uffenbach*, p.120–121.

83. Chalmers-Hunt, J. M. (1976). *Natural history auctions 1700–1972. A register of sales in the British Isles*, p.3.
84. MacGregor, A. (1994). The life, character and career of Sir Hans Sloane. Appendix 3: Sloane's Museum at Chelsea, as described by Per Kalm, 1748. *In* MacGregor, A. (Ed.) (1994). *Sir Hans Sloane*, pp.31–34. [Translated by Professor W. R. Mead from transcript of original manuscript, of which copy is in library of Linnean Society, London.]
85. Trimen, H. & Dyer, W. T. (1869). *Flora of Middlesex*, p.384.
86. MacGregor, 'Life of Sloane', p.23.
87. House of Commons (1835). *Report from the Select Committee on the conditions, management and affairs of the British Museum.*
88. Ingpen, *Instructions for collecting*, p.18.
89. Newman, E., *History of insects*, p.23.
90. Ford, R. L. E. (1979). Entomological cabinets. *Entomologist's Rec.J. Var.* **91**: 308–310.
91. Albin, E. (1720). *Sloane MS* 3338, *f.*2r.
92. Harris, *Aurelian*, Preface, p.4.
93. Wilkinson, R. S. (1966c). Elizabeth Glanville, an early entomologist. *Entomologist's Gaz.* **17**: 149–160.
94. Fitton & Gilbert, 'Insect collections', p.120.
95. Kirby & Spence, *Introduction to entomology* **4**: 535.
96. Newman, E., *History of insects*, p.23.
97. Jermyn, *Vade mecum*, p.18.
98. MacKechnie Jarvis, C. (1976). A history of the British Coleoptera. Presidential Address 1975. *Proc.Trans.Br.ent.nat.Hist.Soc.* **8**: 91–112.
99. Smiles, S. (1891). *Life of a Scotch naturalist*, pp.111–113; 146–147.
100. Samouelle, G. (1819). *The entomologist's useful compendium*, p.323.
101. Stainton, *Butterflies and moths*, p.47.
102. Tutt, *British butterflies*, pp.103–104.
103. Wood, J. G. (1892). *Insects at home*, pp.387–388.
104. Allen, *Naturalist in Britain.*
105. Ingpen, *Instructions for collecting*, p.1.
106. Knaggs, *Lepidopterist's guide*, pp.90, 100 *et seq.*
107. Doubleday, E. & Newman, E. (1833). An entomological excursion. *Ent.Mag.* **1**: 50–59.
108. Stainton, H. T. (1854). *The Entomologist's Companion* (edn 2), pp.104–111.
109. J. C. L[oudon]. (1835). Preface. *Mag.nat.Hist.* **8**: iii–iv.
110. Ingpen, *Instructions for collecting*, p.53.
111. Tutt, *British butterflies*, p.93.
112. Morris, *British butterflies*, pp.85–86.
113. Wilkinson, W. G. (1857). *Apatura Iris. Entomologist's Wkly Intell.* **2**: 148–149.
114. Hudson, W. H. (1903). *Hampshire days*, p.169.
115. Huggins, H. C. (1955). The old days at Chattenden Roughs. *Entomologist's Gaz.* **6**: 55–57.
116. Douglas, J. W. (1856). *The world of insects: a guide to its wonders*, p.178.
117. Newman, E. (1871). *The illustrated natural history of British butterflies*, p.vi.

CHAPTER 3

'A Company of Serious, Thoughtful Men'

THE BUTTERFLY COLLECTORS

'Every man is as Heaven made him, and
sometimes a great deal worse'
Cervantes, *Don Quixote*

THIS CHAPTER CONTAINS THE 'BRIEF LIVES' OF 101 MEN and women, arranged roughly in order of date of birth, who not only collected butterflies and moths but also contributed greatly to our knowledge of them. It spans more than three centuries and includes artists, scientists, writers, rich men and poor, people who anyone would have felt privileged to know, and others, perhaps, about whom we would be content to read. They had two things in common. First they were all gifted observers of nature – that is, through seeing they found understanding, and in most cases passed that understanding on through the literature. The other thing that they all have in common is that they are all dead. It was decided to omit living entomologists and concentrate on those whose work is necessarily completed, and so can be seen in the round. I have tried to present them as vividly as possible, while also including the main facts about their education, career and attainments. It is not possible, in a few pages, to give much more than a glimpse of the work of such luminaries as, say, Stainton or Tutt (though, for a few others, one could contain almost all that is known about them on a single side of paper). I would be more than content if the reader came away with a sense of what such people were like, and what they did to earn a place among the great lepidopterists. The choice is a personal one and is restricted to British naturalists.

I hope that, within this selection, there is enough human biodiversity to satisfy the variety hunter. Within this diversity a number of common strands do nonetheless emerge. Collecting and studying insects is a time-consuming hobby. To accumulate those huge collections of late Victorian and Edwardian times, not to mention the prodigious literary and scientific output and social duties of many entomologists, needed a large amount of leisure time. This was as true of the early entomologists as of the Victorian High Noon.

The beautifully illustrated books by Albin, Wilkes, Harris and others took years to complete, and one can only wonder at the devotion needed to rear Magpie moths on an industrial scale (Parson Greene) or Pug moths through more than twenty successive generations (Emma Hutchinson). Some provided time themselves by rising early, or forgoing the pleasures of marriage and children. Others, like some of the Victorian clergymen and squires had plenty of spare time, and a number of nineteenth-century entomologists had private means. Where our 'Aurelians' had to run a business, like Edward Newman or Henry Doubleday, that business was generally neglected. Some, like Chapman and Higgins, did most of their best work after retirement. Others cut the knot by working in museums or universities, or as professional collectors or, in the case of L. W. Newman, as a 'butterfly farmer'. In most cases, it is impossible to draw any hard and fast distinction between amateurs and professionals. In terms of their fieldcraft and passion for insects, nearly all of our selection of naturalists were amateurs, concerned with life-histories, distribution, variation and drawing accurately. Some of them would certainly have claimed that insect-hunting is in any case more of an art than a science; a few would perhaps have called it a sport. But many of them were scientists too, interested in the latest theories of evolution or advances in systematics, and sometimes contributing to them.

Personal eccentricity is an inescapable part of the history of entomology. It would be hard to imagine a more original set of whiskers and strange hair styles than displayed here, and we learn that their appearance in the field was equally singular, with top hats lined with cork or 'fungoid' jackets smelling pleasantly of tobacco or cheese. There are those unforgettable images of Lord Rothschild riding his zebra-drawn carriage to Buckingham Palace, of Poulton wiping his enormous moustache 'like some giant Dipteran grooming its antennae', and of Power, raising his top hat to a lady, only to have all his collecting bottles fall out with a clatter.

Inevitably, when we concentrate on individual biography, our list becomes biased towards the educated middle-classes, with a large proportion of clergymen, doctors, lawyers or civil servants. But, as many have commented, entomology was a great leveller, the hobby of rich and poor, worker or squire. Though their names remain mostly little more than names, we know, for example, that weavers from the Spitalfields factories were seriously interested in entomology and were prepared to walk a dozen or more miles through the night to arrive in time for a full day's collecting at Darenth Woods. One of them, Daniel Bryder, was so skilled at finding rarities that others paid him to collect for them. In late Victorian times, when the craze was at its height, there were working men's natural history societies in many of the industrial towns, from Cornwall to Aberdeen. The South London Entomological and Natural History Society was founded in 1872 on just such an ecumenical basis by William West, a brass foundry foreman, and Platt Barrett, a teacher of deaf-and-dumb children. Wealthier men, like H. T. Stainton, went out of their way to encourage entomology, especially among the young. A strong vein of evangelism runs through the history of entomology, with the notion that natural history not only adds to human knowledge but acts positively on the human spirit too.

Among our 'company of serious, thoughtful men' (and women) there are distinct groups. Many of those living in the eighteenth century, the original 'Aurelians', were primarily illustrators, whose coloured plates of living butterflies and moths are, and were intended to be, works of art. We know little about some of them apart from their works, although recent historical detectives have discovered a great deal about Joseph Dandridge and Eleanor Glanville, and even lifted a corner of the veil over the mysterious James Dutfield. Much of the early systematic work on butterflies and moths was based on privately or collectively-owned collections. In the early years of the nineteenth century, these 'Aurelians' became entomologists, spurred on by the works of Haworth, Curtis and Stephens, by the immortal *Introduction to Entomology* of Kirby and Spence, and by the formation of permanent institutions and their journals from the 1830s. It is surely no accident that this period produced so many advances in technique, like chloroform bottles, grooved setting boards and sugaring. Most of our smaller moths were discovered during the nineteenth century, and, unlike the butterflies, not a few of them were new to science and named and described by British authors, like Doubleday, Stainton and Barrett. Darwin's theory of evolution – hotly contested among British entomologists – produced a new focus on races, varieties and, eventually, genetics. The period from roughly 1870 to 1914 was the heyday of the field collector, when chasing butterflies with nets and sugaring for moths became almost popular enough to rank as a field sport – many of the most prominent collectors were equally dab hands with a gun or fly-rod. Today it is possible to deplore the killing of so many beautiful butterflies, but these preserved collections, in provincial museums and educational institutions throughout the country, have provided source material and inspiration to countless would-be entomologists ever since, whilst those in the Hope Department and the Natural History Museum are studied by scholars from all over the world. Few of our chosen entomologists were *just* collectors. F. W. Frohawk bred every British butterfly from egg to adult, for the first time, and drew every stage from life. E. B. Ford used the variable wing-patterns of butterflies and moths to study the processes of evolution. Captain Purefoy and Charles Rothschild made the first faltering steps towards conserving butterflies and their natural habitats. The wider history of natural history – rarely predetermined, and moving, by bumps and starts, in often unforeseeable directions – emerges through these 'brief lives of the great', and it reminds us that history is often made by individual people, in entomology no less than in politics.

1. THOMAS MOFFET
(variously spelled Moffat, Mouffet or Muffet) (1553–1604) (Fig. 31)

We are often told that schooldays should be the happiest days of our lives, and occasionally it may even be that they were. For Thomas Moffet, however, his student days at Cambridge were probably among the worst. He entered Trinity College in 1569, but after a change of heart moved to Caius College, where he read medicine under the watchful eye of two eminent physicians, John Lorkin and John Caius. It was while there that Moffet nearly lost his life from eating contaminated mussels. After graduating he chose to obtain his Masters degree from Trinity, for which offence he was expelled from Caius College by the Master, Thomas Legge. Ironically, Legge himself was later charged with, among other offences, having expelled Moffet without first having consulted the Fellows.

Moffet was the named author of the first book about insects published in Britain, the *Insectorum sive minimorum Animalium Theatrum* – The Theatre of Insects or of lesser living Creatures, which is referred to variously as *Insectorum Theatrum* or *Theatrum Insectorum* (Fig. 32). In it, descriptions in Latin of some eighteen more-or-less recognizable British butterflies appear for the first time: the Swallowtail, Clouded Yellow, Brimstone, Black-veined White, ?Large, Small and Green-veined Whites, Orange Tip, ?Common Blue, Red Admiral, Painted Lady, Small Tortoiseshell, ?Large Tortoiseshell, Peacock, Comma, Dark-green Fritillary, ?Speckled Wood and Wall, in addition to the continental Apollo and Scarce Swallowtail butterflies. Many of these were cited by John Ray in his *Historia Insectorum* of 1710, and by Linnaeus in his *Systema Naturae* of 1758.

Fig. 31. Enlargement of the engraved portrait of Thomas Moffet by William Rogers, intended for the title page of *Insectorum Theatrum* but never used.

Fig. 32. This splendid engraving by William Rogers (*fl.* 1589–1605), containing cameo portraits of Moffet, Gesner, Penny and Wotton, was prepared for the title-page of *Insectorum Theatrum* around 1600 but never published due to Moffet's death in 1604. Rogers also engraved title pages for Gerard's *Herbal* (1597) and Camden's *Britannia* (1600).

The publication history of the *Insectorum Theatrum* is somewhat complicated. It is a compilation of work by Conrad Gesner, Professor of Natural History at Zurich (described by a contemporary as 'a monster of erudition') and that of the English scholars Edward Wotton (1492–1555) (Fig. 33) and Thomas Penny (*c.*1532–88) (Fig. 34), as well as Moffet himself. Assisted by Moffet, Penny had begun to 'heap up material' for publication in England, based in

Fig. 33.
Edward Wotton (1492–1555), arguably the first man to study insects scientifically.

Fig. 34.
Thomas Penny (*c.*1532–1588), a great 'student of nature' whose researches 'heaped up material' on insects for Moffet to complete.

part on Gesner's greatest work, *Historiae Animalium*, which had been published in Zurich between 1551 and 1587, but died in 1588 before his task was completed. In his will, Penny left his printed books and papers to Moffet. Before the latter could collect the precious manuscript, however, it had been torn into tatters by Penny's ignorant niece. Moffet arrived in the nick of time to save it from being 'cast out of doors'. Painstakingly, he had the torn pages repaired, which cost 'a great sum of money', and completed the manuscript in March 1589. Moffet's own contribution to the content seems to have been mainly literary. 'I have forged the History to the best of my abilities' he wrote, 'adding ... the light of oratory which Penny lacked. I have amended the method, and the language, and have put out above a thousand tautologies, trivial matters and things unseasonably spoken.' Probably one of the 'trivial matters' he put out was Penny's details of localities and dates for the insects he collected. A pity: it might have added greatly to our knowledge of practical entomology in the age of the Tudors.

Though a lesser student of nature than Thomas Penny, Moffet's own knowledge of entomology was not inconsiderable. After graduating at Cambridge and Basle in Switzerland, he had established a medical practice in London (some idea of current medical practice may be learned from Thomas Penny's patent cure of crushed woodlice in wine; when that did not work, Moffet prescribed 'hot brimstone smoked through a funnel'). He was evidently an acute observer of the habits of insects and birds, and seems to have made a collection of insects, including specimens sent to him from overseas. Greatly interested in silkworm culture, which he studied while travelling in Spain and Italy, Moffet wrote a short tract in verse on the subject in 1599 entitled *The Silkwormes and their Flies.* He had a wide circle of acquaintances, including Sir Francis Drake, who apparently showed him a flying fish. One near-contemporary went so far as to call him the 'Prince of Entomologists'. Nevertheless, Moffet was a man of his times, and did not always try to distinguish between truth and fancy. His was still a world of monsters, devil's agents and birds that hibernate at the bottom of ponds.

Fig. 35. Sir Theodore de Mayerne (1573–1655), physician and scholar, who finally succeeded in publishing the *Insectorum Theatrum.* Engraved by William Elder after a portrait by Rubens.

Unfortunately, publishing a large illustrated work like *Insectorum Theatrum* in the sixteenth century was a difficult enterprise, since few printers were willing to take on themselves the very considerable financial risk. To make matters worse, another vast work on insects, the *De Animalibus Insectis* by the Italian Ulysse Aldrovandi, threatened at a stroke to make Moffet's work redundant. Moreover, although Moffet dutifully revised the manuscript and engaged a celebrated engraver to prepare the title page, with its fine portrait of Moffet, he had many distractions, not least his election, through the good offices of his patron, the Earl of Pembroke, as Member of Parliament for Wilton. He died in 1604, with the great work still not published.

The manuscript eventually came into the hands of the physician and scholar, Sir Theodore de Mayerne (1573–1655) (Fig. 35), who renewed Moffet's attempts to find a printer. At last, in 1634, the work was published in a small folio edition with Moffet named as the author, though its illustrations, 580 'rude but spirited' woodcuts, were much inferior to the original drawings. The work's popularity increased when an English translation was made and included as an appendix to Edward Topsell's *History of Four-footed Beasts and Serpents* in 1658. 'The Theatre' remained the 'standard work' on insects until the publication of Ray's *Historia Insectorum* in 1710, more than a century after the death of its author.

In Moffet's day, spiders were regarded as insects. Moffet clearly admired them, and relates some of their 'virtues', for example that swallowing spiders' webs helps to keep away the gout, or that spiders can foretell the weather. W. S. Bristowe made a convincing case for the 'Little Miss Muffet' of the nursery rhyme being Moffet's daughter, Patience.[1] The rhyme suggests the way in which his name was pronounced during his lifetime.

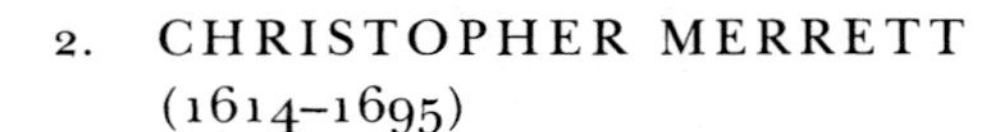

2. CHRISTOPHER MERRETT (1614–1695)

We need not linger over the details of Christopher Merrett's life, since he was primarily a physician and herbalist, not an entomologist. Nevertheless, his principal work, the *Pinax rerum naturalium Britannicarum,* published in 1666, was the first attempt to list British insects, together with plants, animals and fossils. It contains short Latin descriptions of twenty butterflies, mostly recognizable though none were given names and all but a few were borrowed from Moffet. They were: Silver-spotted and Dingy Skippers, Wood White, Brimstone, Black-veined, Large, Small and Green-veined Whites, Green Hairstreak, Common Blue, Small Tortoiseshell, Comma, Small Pearl-bordered Fritillary,

Speckled Wood, Wall, Marbled White, Gatekeeper, Meadow Brown, Ringlet and Small Heath. The list, which has some curious omissions, contains a description of a butterfly with wings '*externis purpurascentibus*' [becoming purplish at the edges], which could indicate the Purple-edged Copper. This, however, would be surprising. None of the other early entomologists referred to this species, although the fact that Merrett appears to have included it does raise the faint possibility that, despite doubt having been expressed as to whether it should be regarded as British, it might, at one time, have occurred here.

Almost as soon as it was published, by ill luck, the first impression of Merrett's *Pinax* perished in the flames during the Great Fire of London in 1666. It was reprinted the following year, but as a pirated copy, and Merrett probably never earned a penny from his celebrated work. Worse, the fire also destroyed the library and museum of the Royal College of Physicians of which Merrett had the charge. To his fury, he was relieved of his post and fobbed off with £50 compensation. Merrett took legal action against the College, and later published an angry tract, accusing it of 'all manner of frauds and abuses'. Merrett died, we are told, a bitter and disillusioned man. He was buried 'twelve feet deep' at St Andrew's church in Holborn.

3. JOHN RAY
(1627–1705) (Fig. 36)

'You ask what is the use of butterflies', wrote John Ray (though his original text was in Latin). 'I reply, to adorn the world and delight the eyes of men: to brighten the countryside like so many golden jewels. To contemplate their exquisite beauty and variety is to experience the truest pleasure. To gaze inquiringly at such elegance of colour and form designed by the ingenuity of nature and painted by her artist's pencil is to acknowledge and adore the imprint of the art of God'.[2]

Fig. 36. 'Our countryman, the excellent Mr Ray ...' wrote Gilbert White in a letter to Daines Barrington on 1st August 1771. Engraving by George Vertue after W. Faithorne.

John Ray turned to entomology relatively late in his career – too late, perhaps, for he never completed his most significant work on the insects, the *Historia Insectorum*. During his wide-ranging travels in the 1660s, the zoological collecting was done mainly by Ray's young patron and protegé, Francis Willughby (1635–72) (Fig. 37). Ray's remarkable career has been pieced together brilliantly by Canon Raven from his writings and surviving journals – the story of the blacksmith's son who 'brought forth order from chaos' by revealing the natural relationships of plants and animals.[3] His example established a long-lasting tradition of British natural history: the cataloguing of all forms of life through observation, identification, accurate description, and finally classification. Even Linnaeus was essentially perfecting Ray's 'Natural System', the discovery of a hierarchy of order in the teeming natural world.

Although insects were mainly Willughby's department, Ray certainly took notice of them. In his 'Cambridge Flora' (1660), for example, he noted that the Soapwort plant attracted a medium-sized 'papilio', with a long tongue, rapid flight and loud buzzing noise (probably the Hummingbird Hawkmoth). It seems that Ray did at least collect specimens of butterflies, a group in which Willughby took little interest. However his main commitments – treatises on ornithology, fishes and quadrupeds needed to be completed first. By the time he could give his full attention to insects, Ray was working under conditions of great difficulty. Isolated in rural Essex, he had no ready access to collections nor a library (and there were few reliable books in any case). He was ageing, in ill-health and living in some poverty. Even the dead Willughby's invaluable notebooks were withheld from him until almost the end of his life. In effect, Ray had to start from scratch, describing each specimen as it came to him. He decided early on to confine himself to British insects (with the exception of a few exotics in Hans Sloane's collection) which he estimated to number about 2,000, of which about one third occurred around his home. (The grossness of Ray's underestimate would not become apparent for many years; he also calculated there were about 20,000 insect species worldwide.) To catch insects in the vicinity of his home, 'Dewlands', at Black Notley, near Braintree in Essex, he enlisted the help of his wife and daughters, his gardener, and also his neighbour, the naturalist Samuel Dale. For specimens from further afield he depended on friends and correspondents like James Petiver and Joseph Dandridge from London, William Vernon from Cambridge, and the Bobarts, father and son, from Oxford. Ray himself collected butterflies and moths when health allowed him to do so, and also reared their larvae. He was perhaps the first to rear the elusive Purple Hairstreak butterfly, which Petiver named 'Mr Ray's Purple Streak', and also captured a streaked variety of a blue near London, which he gave to Petiver. Another species, 'Mr Ray's Alpine Butterfly,' or *Parnassius apollo* as it became known, was presumably collected during one of Ray's European tours. A few insects collected and mounted by Ray are believed to survive in the Sloane Collection in the Natural History Museum.

Fig. 37. Francis Willughby (1635–1672), Ray's younger colleague and benefactor. Engraving by W. H. Lizars after an unknown painting.

By the time he was ready to write an introduction to *Historia Insectorum*, Ray had described some 48 different kinds of butterflies (not all of them scientific species) and some 300 moths, mainly from around Black Notley – a remarkable achievement. But, he wrote, 'every year [was] offering new ones', and Ray despaired 'of ever coming to an end of them, much less of discovering the several changes they go through from the egg to the papilio [adult], and describing the erucae [larvae] and aureliae [pupae] of each'. Ray was by now often incapable of work. 'I am sadly afflicted with pain', he wrote to Sloane in 1702, 'w(hi)ch renders me listlesse and indisposed to any businesse and disables me with intention to prosecute any study.' Shortly before his death, he published a ten-page *Methodus Insectorum*, setting out some of the principles by which he was tackling the classification of insects. By this time he had recognized that 'The work which I have now entered upon is indeed too much for me. I rely chiefly on Mr Willughby's discoveries and the contributions of my friends'.

On Ray's death in 1705, the *Historia Insectorum* was an incomplete manuscript. It was published '*opus posthumum*' by the Royal Society five years later, after some editing, together with an appendix on beetles by Martin Lister (1638–1712). It represented a great advance in knowledge, both in terms of the insects themselves, and in the way they were classified – there was to be nothing comparable to it for nearly a century. With no known names for most of his insects, Ray provided brief but accurate Latin descriptions for each of them. The Red Admiral, for example, is recognizable as '*Papilio major nigricans alis maculis rubris et albus pulchre illustratis*' [a large black butterfly with wings spotted with red and handsomely marked with white]; the Brimstone as '*Papilio praecox sulphurea sive flavoviridis singulis alis singulis maculis ferrugineis notatus*' [an early yellow or greenish yellow butterfly with each wing marked with a single brown spot]. Of the butterflies described by Ray, twenty-nine of them for the first time in Britain, only the Mazarine Blue was new to science, though Ray's seems to have been the first published description of the Holly Blue.

Ray's memorial in Black Notley church apostrophizes him, aptly enough, as 'the bright wonder of the age to come'. The first generation of great British entomologists – Petiver, Dandridge, Vernon, Dale and others – owed much of their inspiration to him, as they acknowledged. Petiver paid him perhaps the most moving compliment. His *Gazophylacium* depicts 4,000 natural history objects, but only one human being: John Ray.

4. LEONARD PLUKENET (1642–1706)

Plukenet's place in this chapter is owed mainly to the survival of his insect collection, one of the oldest known. His specimens, collected towards the end of the seventeenth century, were simply pressed flat and then glued into an unlabelled leather-bound volume of some 140 leaves (see Chapter 2, Plate 5). Plukenet seems to have known no better method than to treat insects in the same way as botanical specimens. Examining them today is a matter of great delicacy, since opening the pages risks damaging the precious contents! However they number about 1,700 specimens, including a number of butterflies and moths, arranged roughly by 'Orders', with names such as 'Father Long Legs', 'Lady Cow' [Ladybirds], 'Fether'd Moths' and 'Moth Bees'. A notebook containing notes in Plukenet's hand also survives, and appears to relate to this collection.[4] If so, he caught his insects in 1696 and 1697, mostly in the London area.

Plukenet was a physician and botanist. In 1690, he was appointed in charge of the gardens of Hampton Court with the title of Queen's Botanist. Part of his herbarium of wild plants collected around London and Westminster also survives, having passed to the Natural History Museum from the collections of Sir Hans Sloane. Writing to Sloane in 1696, John Ray describes Plukenet as a rather disagreeable man, 'reserved, jealous of his reputation, and none of the best-natured, not to give him a worse character, being my friend ...'. On the other hand, Sloane must have held him in high regard to have acquired all he could of Plukenet's collections and papers after his death.

5. CHARLES DUBOIS (*c.*1656–1740)

Among the most precious objects in the Entomological Department of the Natural History Museum is a modest notebook which once belonged to Charles duBois, a silk merchant, amateur entomologist and Fellow of the Royal Society. Its contents, compiled in 1692 and 1695, include some of the earliest surviving descriptions of British butterflies and moths. From his notes and accompanying rough but accurate sketches, it is easy to recognize the species concerned, which include the Painted Lady (his one popular name), Small Tortoiseshell, Comma, Brimstone, Orange Tip, Small White and Small Copper, as well as various moths and other insects.

The Comma, for example, is described as follows:

> '6. Papilio alis laciniatis
>
> Taken 23th Septr. The head and body are Sad [i.e. sombre or dull] brown, the horns of the same colour except the tips of the knobs w[hi]ch are white. The proboscis and legs are brown, the latter pretty long, the wings are cut in and variously indented on the Edges, they Orange, marked with Sad spotts and clouds; Underneath they are finely marbled, with shades of sads and browns, the upper part Sadder, the lower part lighter, in the place where the Sad ends are, the under wings is set on each a very white mark like a c thus **C**' (Fig. 38).

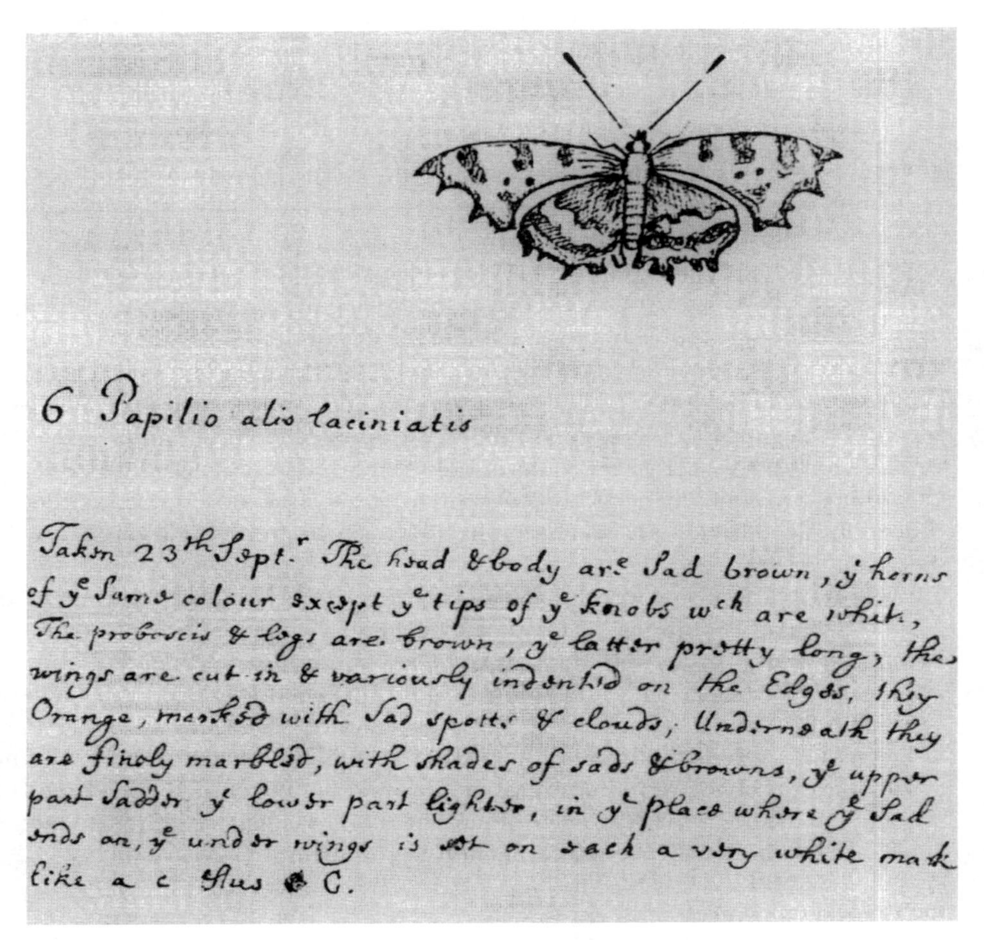

6 Papilio alis laciniatis

Taken 23th Septr. The head & body are Sad brown, ye horns of ye Same colour except ye tips of ye knobs wch are white, The proboscis & legs are brown, ye latter pretty long, the wings are cut in & variously indented on the Edges, they Orange, marked with Sad spotts & clouds; Underneath they are finely marbled, with shades of sads & browns, ye upper part Sadder ye lower part lighter, in ye place where ye Sad ends on, ye under wings is set on each a very white mark like a c thus C.

Fig. 38. The duBois notebook entry on the Comma butterfly, '*Papilio alis laciniatis*' [butterfly with jagged wings] is a forerunner of the traditionally fine-rendered specimen labels of butterfly collectors.

It seems likely that duBois based his drawing on the woodcut of the Comma in Moffet's *Insectorum Theatrum* (see Fig 155).

Despite the quaint seventeenth-century English, this is a well-observed, accurate description of a Comma butterfly – seemingly the first one ever made.

Of Huguenot descent, Charles duBois was a prosperous businessman in the silk and cloth trade, with premises at St Mary Aldermanbury in the City of London, and a country house set in fifteen acres of gardens at Mitcham in Surrey. He was a prominent botanist and naturalist, as his elevation to the Royal Society shows, and is known to have associated with men like Ray, Petiver, Plukenet and Sloane, and sometimes accompanied them on their excursions. Peter Collinson, a contemporary and close friend, referred to him as a 'very curious gentleman'. 'Curious', then a term of praise meaning observant and inquiring, was used to describe several other prominent early naturalists, like Dandridge and Albin. An early twentieth-century author pictured duBois as 'a kindly old gentleman in spectacles, riding down from London on a Friday evening, and reining in his nag to a walk as the lavender fields of Mitcham came into view; or later on, seated in his library over a cup of tea, telling his niece the news of the town ... or next morning, in a shabby coat and with a bandanna handkerchief tied round his head, pottering about his sunny garden and in and out of his greenhouses'.[5]

Fig. 39. Petiver's 'Hogs', from his *Papilionum Britanniae Icones*, 1717, showing the Small Skipper (figures 14,15) and Large Skipper (figures 16, 17). The oblique line of androconial scales in the males' forewings is clearly seen in figures 15 and 16. Engraved by Sutton Nicholls (*fl.* 1700–1740), also known for his views of London.

6. JAMES PETIVER (1663–1718)

If it is the act of a father to name his children, then James Petiver may be considered the father of British butterflies. Before Petiver, butterflies had no English names, so no one except the most learned could talk about them individually, and until people talked there could be no shared knowledge. Petiver supplied the means. The very first entry in his first known publication[6] – which predated Ray's *Historia Insectorum* by fifteen years – is of a yellow butterfly he called The Brimstone (perhaps the original Butter Fly?), after the solid yellow cake in which sulphur was then supplied. Another 48 names followed between 1695 and 1717, and although few can match the Brimstone for conciseness and appropriateness, many of them have their own charm. If we followed Petiver, we would have not skippers but 'hogs' (evidently pronounced 'ogs') (Fig. 39), not Walls but Enfield Eyes, not Swallowtails but Royal Williams, and not Marbled Whites but Half-Mourners. But, thanks to Petiver, we still have Admirals, Arguses, Hairstreaks and Tortoiseshells. He did not discover all these butterflies – many of them were already known – nor did he necessarily coin all the names himself ('Painted Lady', for example, was an already current folk-name). Nevertheless, by recording them for posterity, he deserves the credit and the appellation of 'father'. He was able to list and describe so many butterflies for one reason above all others. He owned a substantial collection, part of the 'Musei Petiveri', acquired directly or by gift, from Britain and from around the world.

James Petiver was born at Hillmorton, Warwickshire, the son of a haberdasher, and educated at Rugby Free School. He moved to London in the late 1770s, where he was apprenticed to an apothecary, and later set up in business at Aldersgate in the City, supplying medicines to St Bartholomew's Hospital and, later, Charterhouse. He is buried in St Botolph's Church close by.

We owe many of the first records of British insects to the broadsheets, pamphlets and other writings published by Petiver between 1695 and 1717, including his illustrated 'catalogue', the *Gazophylacium* (1702–06). Fortunately most of these ephemeral printings have survived, as in 1764 they were bound up into three volumes, as *J. Petiveri Opera*, by a London publisher and bookseller, John Millan, and reissued in 1767 with additional material. Petiver seems to have acted as the documentary spider at the centre of a web of correspondence with English naturalists, like Glanville, Sloane and Ray, and with travellers on ocean voyages, to whom he supplied instructions for preserving specimens (see Chapter 2). In the last full year of his life, Petiver amalgamated all his notes on British butterflies into a slim volume called *Papilionum Britanniae Icones*, containing 'the Figures, Names, Places, and Seasons of above Eighty British Butterflies' – more than there are today! Some of the Eighty are in fact males and females of the same species, then thought to be different, and there is also the engraving of that celebrated oddity, 'Albin's Hampstead Eye', now known to be an exotic species. Unlike most of his eighteenth-century successors, Petiver has little to say about the early stages, and he seems first and foremost to have been a cataloguer of natural objects, the prototype of the national museum curator. His personal collecting seems to have been confined to the Home Counties, and there is no indication that he ever travelled far from London. But to have known and described about two-thirds of our butterfly fauna at this early stage seems a remarkable achievement.

Although he was an educated gentleman, a Fellow of the Royal Society, and a colleague of the most celebrated scientists of the day, Petiver seems to have been constantly short of money, and evidently he neglected his medical practice to work in his museum. That this museum was, to the visitor at least, rather shambolic, is apparent from a diary entry by von Uffenbach, a German scholar, who called on Petiver in 1710 and came away disillusioned. Expecting to find a 'paragon of learning and refinement', he met instead a person 'wretched both in looks and actions ... speaking very poor and deficient Latin and scarce able to string a few words together ... As soon as he gets any object of the least value he immediately has printed a short and insipid description of it, dedicating it to any person with whom he has some slight acquaintance; and then he takes a present for it. Everything is kept in true English fashion in prodigious confusion in one wretched cabinet and in boxes ... He offers all foreigners who come to him a sample of his collection; but he takes care to ask a vast sum for it, so I declined with thanks ...'.[7] Corroboration of the chaotic state of Petiver's insects is provided by Sir Hans Sloane himself, who took 'as much Care as I can to bring his Collection and Papers out of the Confusion I found them in'. They had to be 'Methodized' by Sloane's curator, one Cromwell Mortimer.[8]

Uffenbach added that herbs from Petiver's garden of physic were as likely to enter his museum or be used in botanical exchanges than cultivated for their proper use! Petiver was clearly a somewhat mercenary individual, trying to scrimp a living from natural history objects. He borrowed freely and without acknowledgement from Adam Buddle's unpublished English flora and Dandridge's insect collections and unpublished lists. Petiver never married, and it must be said that he does not sound like a lady's man. As a scholar he was far inferior to Ray, and as an artist, though careful and accurate, he could not hold a candle to Albin, Dandridge or Wilkes. But without Petiver's talent for hoarding and documenting his objects, we would know very little about this formative period in British natural history.

After his death, Petiver's collections, books and manuscripts were purchased by Sir Hans Sloane for the then considerable sum of £4,000, and hence a portion of them now survive in the Natural History Museum. We can still touch the folio book in which Petiver pasted some of his specimens, and admire some of his butterflies, preserved between slivers of mica, almost as freshly coloured as on the distant day Petiver caught them.

7. JOSEPH DANDRIDGE (1664–1746)

Joseph Dandridge has been called the great forgotten figure of English entomology. Nearly all the early authorities mention him, often in terms of the highest praise. He seems to have been the template for the modest, unselfish type of English naturalist who spends much of his life helping other naturalists, sharing his knowledge and enthusiasms, and lighting a similar flame in hundreds of disciples. Dandridge collected and studied all kinds of natural object, and his home must have resembled a museum with its cabinets of insects, shells, fossils, minerals, birds' eggs and skins, paintings of spiders and other perishable objects, and herbaria of fungi, lichens, grasses and wild flowers. And yet, despite all this evidence of a passion for natural history, Dandridge never published a line. W. S. Bristowe, who pieced together much of what is known about him, comments that 'his happiness lay in the plants and animals he loved, in talking about them, and in helping enthusiasts from his wide store of knowledge'.[9]

Joseph Dandridge was born in Winslow, Buckinghamshire, the son of a barber-surgeon. He came to London in 1679, and served an apprenticeship as a 'drawer' to a merchant tailor. He became a pattern designer for silk weavers at Moorfields, which must have utilized his artistic sense and developed his ability to draw and paint natural objects faithfully. He joined the Merchant Taylors' Company in 1692, and near the end of his life, in 1740, was elected its Master. A contemporary recalled that Dandridge 'eminently distinguished himself and is deservedly remembered with great respect by the trade'. Some twenty of his designs are on display in the Victoria and Albert Museum. He died in 1746, at the then great age of 81, and was buried at Bunhill Fields, London. He was married, and left two daughters and a son.

Dandridge was one of the circle of collectors and naturalists flourishing in London at the turn of the seventeenth century.[10] He sent specimens of butterflies to Ray from collecting grounds near the City, such as Rotherhithe Marshes, and from excursions further afield to Box Hill, Dover and westwards to Henley and Maidenhead. He also reared moths and butterflies in boxes, in Albin's words, 'observing their several progressions from the egg through their worm state to the fly'.[11] Dandridge was probably the discoverer of at least two British butterflies, the Grizzled Skipper and the Marsh Fritillary, which Petiver called respectively 'Mr Dandridge's Marsh Fritillary' and 'Dandridge's midling Black Fritillary'. Like most naturalists of his day, he preserved his butterflies between slips of glass or mica. In his diary, Bishop Nicholas of Carlisle, who saw Dandridge's collection and described it as 'most complete', noted that each insect was preserved 'dried in double glasses etc'. Emanuel da Costa, who also saw it, remembered that his 'chief display was in insects', which were 'well kept and judiciously arranged' and 'shewed with great pleasure, and with instruction'.[12] Some 96 cabinet drawers of English insects from Dandridge's collection were purchased after his death by Sir Hans Sloane, and although most of them were lost in the 'Sloanian bonfires' carried out by Dr Shaw (1751–1813) of the British Museum, some specimens believed to be Dandridge's still survive, undetected for nearly a quarter of a century.[13]

Dandridge also drew and painted butterflies and other natural history objects. Originally there were twelve volumes by him of insects, caterpillars, plants, birds and fungi, all of which were supposed lost until Bristowe tracked down 119 of his 'delicately executed and often excellent' paintings of spiders, harvestmen and mites in the Natural History Museum – which, incidentally, prove that Albin borrowed Dandridge's work for his own published work, *The Natural History of Spiders.*

Dandridge was a member of the first Aurelian Society, which was in existence by 1738, and he may well have been its figurehead. The Society included a number of men from artistic trades, including Dandridge's colleague in silk patterning, James Leman. Benjamin Wilkes referred to him warmly as his 'principal mentor', whose 'noble collection' was the fruit of forty years' experience.[14] Dandridge certainly seems to have been a liberal and popular man and, by 1738, perhaps the senior figure in British natural history. Though no known portrait of him exists, another Aurelian, Emanuel Mendes da Costa, recalled him late in life as a jolly man, 'thick and of a middle size', full of 'anecdotes of the old collectors' and 'very merry and chatty'.

The then outlying village of Stoke Newington, where Dandridge had gone to live for health reasons in the 1720s, remembered him long after his death. One James Brown recalled that 'he pursued his [butterflying] sport with so much eagerness as to have given rise to stories which came down to my time'.[15] On one occasion he was spotted wildly lunging at the air for no apparent reason. Taking him for a lunatic, a farm labourer caught Dandridge by the arms and wrestled him to the ground. The labourer's suspicions seemed amply confirmed by Dandridge's wail of dismay: 'The Purple Emperor's gone! The Purple Emperor's gone!'

8. ELEANOR GLANVILLE (*c.*1654–1709)

In a letter to a friend, written in 1703, William Vernon mentioned a chance meeting in London with a remarkable lady from the West Country 'with the noblest collection of butterflies, all English, which has sham'd us'. That an important collection existed, unknown to Vernon and his contemporaries, was remarkable enough in that pioneer age, but that they had been collected by a woman, and a woman of property at that, was astonishing. The lady in question was Eleanor Glanville, until recently mistakenly thought to have been named Elizabeth Glanville, from a reference in one of Petiver's letters.

'Lady' Glanville's story has been pieced together with great skill by R. S. Wilkinson[16] and W. S. Bristowe,[17] from her surviving correspondence with Petiver and from other contemporary documents. She was born in Yorkshire in about 1654, daughter of Richard and Muriel Goodricke, still remembered by horticulturalists for their cultivation of the Ribston Pippin apple. From them she inherited considerable property, including Tickenham Court in Somerset, which later became her home. She was not, however, a titled Lady, as implied by Moses Harris in his reference to 'the ingenious Lady Glanvil' in *The Aurelian.* In those days it was often the custom to prefix a noun with a capital letter: all Harris meant was that she was a respectable woman of good family, in fact a lady.

Eleanor Glanville's interest in entomology seems to have begun in her maturity, after the breakdown of her second marriage to Richard Glanville, a violent man who once, 'presenting a pistoll loaded with bullets and cock't to her breast', threatened to shoot her dead. She corresponded with Petiver, whom she counted a dear friend, and also with Ray, Vernon, Dandridge, Buddle and other early collectors. Some of her letters accompanied cases of butterflies, among them specimens of what was to become the Glanville Fritillary, which she captured in Lincolnshire, of all places, in about 1702. Petiver figured it in his *Gazophylacium* as the Lincolnshire Fritillary, which in turn became Linnaeus's type specimen. Glanville also reared butterflies and moths, and may have been the first to refer to Geometrid larvae as 'loopers'. An incredulous bystander, one John Brewton, later described how 'she and her two female apprentice girls would carry a sheet out under the hedges and bushes and with a long pole beat the said hedges and catch't a parcel of wormes'. Glanville herself described the early stages of the High Brown Fritillary and Green-veined White in one of the first detailed references to rearing butterflies. On the other hand, she found it difficult to maintain her collection. In one of her quaint, wildly spelled letters, Glanville told Petiver that mites had done their worst that year and had eaten up a hundred of her best specimens. Her 'noble' collection probably perished long ago, but very recently three specimens she gave to Petiver, a butterfly and two moths, were refound in the Sloane collection in the Natural History Museum.[18]

Eleanor Glanville is best remembered for the dispute that arose over her will, when desperate relatives cited her interest in butterflies as an obvious sign of madness. Corroborative evidence was recalled by neighbours: she had been observed on the downs 'without all necessary cloathes', and on occasion she had dressed 'like a gypsey'. Quite possibly she had become a little unhinged by the behaviour of her parted husband, who, among other things, had organized a plot to kidnap one of her sons, his object being to persuade the lad to disclaim the property he stood to inherit and transfer it to him and his new mistress. Because of this husband's machinations, Eleanor Glanville had arranged to leave her estate in the care of trustees. By the terms of her will the principal legatee would have been her second cousin, Sir Henry Goodricke. When the will was published after her death, however, Eleanor's eldest son, Forest, entered a writ seeking to set it aside on the grounds that his mother had gone mad, and had wrongfully disregarded the rights of her children, believing, he alleged, that they had all been changed into fairies!

The case was brought before the judge at Wells Assizes in 1712. One hundred witnesses were examined. Apparently either Petiver or Sloane appeared as character witnesses for Eleanor Glanville (but, *pace* Moses Harris, not John Ray, who was dead), though no trace of their affidavits have been found. Local villagers, on the other hand, were full of stories of her entomological eccentricities, which were eagerly seized upon and highlighted by her relatives. The written verdict has not been traced, but the upshot was that the will was indeed upset in Forest's favour. It was agreed, at any rate in Wells, that no one 'not deprived of their senses should go in pursuit of butterflyes'.

Eleanor Glanville is remembered today as one of the most important of the entomological pioneers, and one whose name will live forever on the wings of a beautiful butterfly. Only forty years after her death the name Glanville

Fritillary was already current. Hers, as Bristowe remarked, is a 'sad story of a great entomologist, who gained happiness from natural history in the midst of sorrow'.

9. WILLIAM VERNON
(*fl.*1660–*c.*1735)

Among the entomologist J. C. Dale's voluminous papers was a letter from someone called Bunne, mentioning an extraordinary nine-mile pursuit of a butterfly by an elderly man called Mr Vernon. The athlete in question was probably William Vernon, one of the early collectors who supplied Ray and Petiver with specimens to describe. Though comparatively little is known of his activities in this country – for much of his significant work was done in America – Vernon is known to have been the first English naturalist to specialize in mosses, and he was also the probable discoverer of three rare butterflies: the Duke of Burgundy, Bath White and Queen of Spain Fritillary.

Vernon was a well-educated man, born in Hertfordshire, and educated at Cambridge, where he later lived as a Fellow of Peterhouse College. He is believed to have been elected to the Royal Society in 1702, though he never signed the obligation book. He was also a member of the Temple Coffee-House Botanic Club in London, then a regular meeting place of botanists and naturalists, which brought him into contact with men like Sloane, Petiver and Martin Lister. He also met Eleanor Glanville of whom he wrote: 'A lady came to town with the noblest collection of butterflies ... which has sham'd us'. He was a close friend of John Ray, whom he visited shortly before Ray died, finding him 'very old and infirm in body', but with his mind still 'very vivid'.

In 1698, Vernon voyaged to the American colonies on a lengthy expedition supported by the Royal Society and the Governor of Maryland, 'to make observations and Discriptions of all the Natural products of those parts'. Petiver related how Vernon, 'often over a Commemorative Glass wisht to arrive their [there] before the Moss-cropping Season'. Vernon was eager to discover new American butterflies, an easy enough task at that time. He remained in Maryland for about a year, on one occasion writing excitedly to Sir Hans Sloane of his discovery of 'several Curious parts of Naturall knowledge' which he was looking forward to relating when he arrived home. When he left America, he took with him nearly one thousand insects, many of them almost certainly new to science.

Vernon was evidently collecting butterflies by 1696, the most likely date of his capture of the Duke of Burgundy, or, as Petiver catalogued it, 'Mr Vernon's small Fritillary'. It seems that the Bath White, which Petiver named 'Mr Vernon's half-Mourner', and the 'Riga Fritillary' as the Queen of Spain Fritillary was then known, were captured at different times at the same place, White Wood, near Gamlingay, Cambridgeshire. Petiver's list suggests that Vernon captured his first Bath White in about 1699, after his return from America, and that he later caught several more. It is possible that one of them is the specimen in the Hope Department, Oxford, dated May 1702, regarded by E. B. Ford as one of the oldest butterflies extant (– it may be the oldest *pinned* and set butterfly).[19] Whether these rare immigrants had temporarily established themselves in Cambridgeshire, or whether someone, unknown to Vernon, was releasing them, is not known. But one still wonders what kind of butterfly could have sent him on that nine-mile chase when Vernon must have been at least sixty!

10. ELEAZAR ALBIN
(*fl.*1690–1742) (Fig. 40)

Albin was the first of the great entomological book illustrators of the eighteenth century, the author of the sumptuous *Natural History of English Insects*, first published in 1720, and also of the later *Natural History of Spiders and other Curious Insects* in 1736. Like most eighteenth-century entomologists, we know little about him other than what he chooses to tell us in his books. *A Natural History of Spiders* includes an engraving of Albin looking dignified and prosperous on the back of a fine white horse, perhaps to indicate his social aspirations. He had a large family to support by his painting, and one of his daughters, Elizabeth, later assisted him in hand-colouring the plates of his published works. He was a Londoner, at one time living next door to the 'Green Man neer Maggots Brew House' in Soho, at another near the Dog and Duck on Tottenham Court Road, at that time on the outskirts of London, with fields and marshes beyond.

Fig. 40. Eleazar Albin's equestrian portrait suggests the social aspirations of an artist whose works were patronized by royalty and leading members of the aristocracy. It formed the centre of the frontispiece engraved by J. Scotin for Albin's *Natural History of Spiders and other Curious Insects*, 1736.

Albin probably owed his interest in drawing and rearing insects to Joseph Dandridge, whom he refers to as 'a very ingenious man, and very curious in observing the works of nature'. Albin also acquired the patronage of Mary, Duchess of Beaufort, who helped him find wealthy subscribers for an illustrated book on insects. After issuing a written appeal for subscriptions on New Year's Day 1713, Albin set to work on the plates 'curiously engraven from the life' for *A Natural History of Insects*, and had completed fifty of them by the end of the following year. The delay in publication until 1720 might have been due to the death of the Duchess, and hence a falling off of financial support. Despite its expense – three guineas for a coloured version, thirty shillings uncoloured – the work was a success and went to five editions. Albin even proposed what we would now call a special offer, whereby 'If any Gentleman procure Subscriptions for Six Books they shall have the Seventh gratis, which reduces the price to Two Pounds Fifteen Shillings and Three Pence Colour'd'.

A Natural History of Insects was the first publication to show British butterflies and moths in lifelike poses and in full colour (see Plate 33). Albin's 100 fine copper plates include fifteen species of butterflies – by no means all that were then known – including the Black-veined White, Brimstone and 'The Admiral Butterfly'. The name of the Princess of Wales, to whom Albin 'humbly Dedicated' the work, adorns the plate showing 'the Cabbage White'! The majority of Albin's insects, however, are moths, most of which still lacked an everyday name. From his work it is evident that Albin was familiar with the written works of Ray, Merian, Goedart and Petiver, and that he personally reared many of his butterflies and moths. Wherever possible, he drew not only the adult insect with its foodplant, but also the early stages and even the parasites. His *Natural History* is a fine piece of work, scientifically and artistically, even if the colours seem a little flat compared with the work of later artists like Moses Harris or John Curtis.

Albin made a collection of butterflies and moths, and, like Eleanor Glanville, was plagued by mites, mould and other 'vermin'. His solution was to 'Rub the bottom of the boxes ... with Oyl of Spike' [lavender]. He also tells us something about his painting methods. In common with other artists of the day, he mixed his own colours, sometimes from very odd ingredients. His formula for preparing vermilion, for example, was to wash the dry pigment 'in 4 waters then grind it in boy's urine 3 times, [then] gum it & grind it in Brandy wine'. Most of his painting was done in watercolours. In the preface to *A Natural History of English Insects,* Albin remarks that 'teaching drawing and Paint in Water-Colours, being my Profession, first led me to the observing of Flowers and Insects, with whose various Forms and beautiful Colours I was very much delighted'. Sir Hans Sloane and Joseph Dandridge were impressed enough by the quality of his work to commission him to draw skins, caterpillars, stuffed birds and other natural history objects from their collections. W. S. Bristowe's researches[20] suggest that Albin learned a great deal about painting moths, butterflies and their larvae from Dandridge, and, indeed, seems to have copied Dandridge's spider drawings for his *Natural History of Spiders* without acknowledgement.

Whether Albin became a member of the Aurelian Society in his old age is not known. He died in January or February 1742. Today he is, rather unfairly, remembered in the name of the butterfly 'Albin's Hampstead Eye', a then unknown species he thought he remembered catching on Hampstead Heath some time before 1717 (see Chapter 4: 23). Its picture was copied from book to book, and for a long time was Britain's most mysterious butterfly. Eventually it was identified as the common Indo-Australasian species, *Junonia villida,* no doubt the result of either Albin or someone else muddling their labels!

11. BENJAMIN WILKES (*fl.* 1690–1749)

Until recently, *The English Moths and Butterflies* (1749), one of the most celebrated works of the eighteenth-century insect illustrators, was believed to have been wholly the work of Benjamin Wilkes. He was certainly the author of the excellent plates, showing some thirty butterflies and more than one hundred moths, including their early stages, in realistic colours and postures. The text, however, seems to have been completed for him by Henry Baker (1698–1774), a fellow Aurelian and Fellow of the Royal Society. 'The Natural History of Butterflies [as he referred to it] is in some sort my own child', confided Baker to a friend on 17th August 1749 in a letter discovered in the Natural History Museum by P. E. S. Whalley.[21] Wilkes had supplied him with all the necessary 'memorandums', but Baker provided the literary polish: Wilkes was 'indefatigable in his observations and faithful in minuting down every particular but for want of learning quite incapable of writing a book'. Nevertheless it is thanks mainly to Wilkes and his 'memorandums' that we know something about collecting methods and the general state of knowledge at the time of the first Aurelian Society.

Benjamin Wilkes was, like Albin, a professional painter, specializing, in his own words, in fashionable 'History Pieces and Portraits in Oyl'. We know almost nothing of his background. He was probably born before 1700 and died of a fever in June or July 1749. In the preface to *The English Moths and Butterflies,* Wilkes does, however, tell us about his induction into entomology.

He had often experienced great difficulty choosing the right colours 'to contrast *with* and set *off* each other to the best Advantage'. It so happened that a friend – who we do not know; it may have been Henry Baker – invited him one evening 'to bear him Company to a Society named the Aurelian ... And here he first saw such Specimens of Nature's admirable Skill in the Disposition, Arrangement and contrasting of Colours (particularly amongst the Moths and Butterflies) as struck him with Amazement, and convinced him, at the same Time, that studying them would turn greatly to his Advantage'. Wilkes joined the Society and, with the help of some of its members, soon acquired 'a tolerable Collection' of butterflies and moths. He tells us that for ten years past (i.e. from 1738), his 'leisure Hours have been chiefly employed in the collecting and making Drawings of the different *Caterpillars*, Aureliae or *Chrysalides*, Flies etc that he could any Ways obtain'.

It was perhaps with art and colours still uppermost in his mind that Wilkes started work on *Twelve New Designs of English Butterflies*, in which 'about three Hundred of the most beautiful Flies [i.e. butterflies] and Moths, of different Species, [are] dispos'd in a Picturesque Manner'. The work, which was dedicated to the members of the Aurelian Society, was completed in instalments and published at the end of 1742, 'Price plain One Guinea, colour'd Two Guineas and a half'. It was reasonably successful and was reprinted at least three times. In it, Wilkes gives some account of the foodplants, life-cycles, habits and localities of Lepidoptera. Hanging Wood, near Charlton in Kent was a favourite haunt, as was Coombe Wood near Kingston, Surrey, where Wilkes took a Purple Emperor.

Perhaps as a supplement to his *Twelve New Designs*, Wilkes also issued a broadsheet of '*Collecting Directions*', 'in order to oblige such Persons as may be desirous to make a collection of Moths and Butterflies, though unacquainted with the Manner how'. This invaluable publication, of which only a single copy has survived, forms a unique account of the collecting methods of the day, and shows how sophisticated they had become since the time of Petiver and Ray. By now, insects were being killed with a jab of a poisoned pen nib, mounted on pins and set with braces on cork-lined boards (which apparently lacked a central groove). Aurelians were using beating, pupa-digging and assembling to obtain their captures, and rearing larvae in willow pill boxes. Wilkes also provides detailed notes on the making and use of the double-handled clap-net, which at that time was even larger and baggier than the one described twenty years later by Harris (see Chapter 2).

The last and most important of Wilkes's publications was *The English Moths and Butterflies*, which he completed shortly before his sudden death. Baker calls it a 'laborious and elegant work'. Among the many species it illustrated for the first time are two of our finest moths, 'The Cleifden Nonpariel' [Clifden Nonpareil] and 'The Glory of Kent' [Kentish Glory]. By now, most moths illustrated had English names, often those we still use today, like the Peach Blossom, the Emperor and the Codling Moth. Among the butterflies Wilkes also introduces the Camberwell Beauty, under the rather prosaic name of 'Willow-Butterfly' (see Plate 34). Wilkes bred many of these insects himself, including such rarities as the Heath Fritillary, whose larvae he found at Tottenham Wood, feeding, apparently, on heather. Wilkes, like Moses Harris and John Curtis later, sometimes used artistic licence, depicting attractive blooms that had nothing to do with the insect. He explained, apologetically, that his patrons would soon tire of looking at similar bunches of grass! The work was deservedly successful, and new editions were published in 1773 and 1824, long after Wilkes's death, attesting to its lasting value.

12. JAMES DUTFIELD
(*fl.*1716?–1757?)

On 16th June 1748, the *Daily Advertiser* announced the publication of 'A New and Complete Natural History of English Moths and Butterflies ... drawn and coloured exactly from the Life ... '. The first fascicle of the work appeared four days later, and five others followed into 1749, each one consisting of two plates and two leaves of text. Most of the plates, which are of high quality, are of hawkmoths, tiger-moths and other large moths, and the artist, James Dutfield, includes only three butterflies, the Red Admiral (as 'The Admiral or Alderman Butterfly'), the Orange-Tip (as 'The Wood Lady, or Prince of Orange Butterfly') and the Glanville Fritillary ('The Glanvil Fretillary') (see Plate 38). Like Albin and Wilkes, Dutfield included 'all progressive States and Changes', 'together with the Plants, Flowers and Fruits, in their Seasons, on which they feed, and are usually found'. The work also includes an advertisement for a map and engraving of the aftermath of the great Cornhill fire of 1748 which destroyed the premises of the Aurelian Society. Yet his work never progressed beyond the sixth fascicle, and was abandoned after less than a year. Only one set of these is known today. Very probably, as R. S. Wilkinson suggests,[22] the work suffered from competition with Benjamin Wilkes's *English Moths and Butterflies* which also began to appear in 1748 and was already complete by July 1749. At that time, interest in Lepidoptera – or perhaps the lepidopterists' pockets – may have been insufficient to sustain two expensive and potentially similar works. Eighty years later, on the other hand, sufficient subscribers could be found for simultaneous large-scale works by James Francis Stephens and John Curtis.

Of the author of this ill-fated – and, in retrospect, ironically titled – project, little is known. James Dutfield collected in and around London, and his writing indicates that he bred some of his specimens, and obtained Emperor Moths and others by 'symbolling' (i.e. assembling). The late Robert Mays[23] discovered that he had made a clandestine marriage with Sarah Higgs in 1745, who bore him three children, Apelles, James and Dinah. He may have been the James Dudfield baptized at Kemerton, Gloucestershire in 1716, and either he or his son of the same name was laid to rest in Ealing churchyard in 1757. But for such an obviously talented artist, author and probable member of the Aurelian Society, he seems to have left remarkably little trace behind him.

13. GEORGE EDWARDS
(1694–1773) (Fig. 41)

Most of what is known about the early life of the naturalist George Edwards has been summarized by A. A. Lisney.[24] He was born at Stratford in Essex and educated at a school in Leytonstone before being apprenticed to a tradesman at Fenchurch Street, London. There he had access to a library of books recounting the many exciting discoveries in the natural world taking place at that time by men like Ray and Hooke. His literary browsings produced in the young tradesman a thirst for adventure, and instead of entering business he decided to travel abroad, and, by some means or other, visited many foreign countries between 1716 and 1733. In the latter year he settled in London and, through the influence of Sir Hans Sloane, was chosen as Librarian to the Royal College of Physicians, a post he retained until retirement in 1769.

Edwards's chief contributions to entomology are contained in his *Gleanings of Natural History*, published in three quarto parts between 1758–64. The work, which was written in English and French in opposite columns, 'exhibited Figures of Quadrupeds, Birds, Insects, Plants etc most of which have not till now been either figured or described'. The copperplate engravings and colouring were Edwards's own work and show him to have been an accomplished artist. The cost, in half-binding, was two guineas, and the original drawings, as well as some of their models, were put on display at the Royal College.

Fig. 41. George Edwards, author of the *Gleanings of Natural History*, published between 1758 and 1764. Engraved by J. S. Müller (an artist who published work on plants and insects and who, from 1759, adopted the name John Miller), this portrait is after a painting by Dandridge.

The Gleanings seems to have been the result of Edwards's happy hours in his library and those of the Royal Society and the Society of Antiquaries, where he was a Fellow. They do not include any new species of British butterfly, but Edwards does provide some original information on how butterflies were collected in the eighteenth century. In a remarkable essay, first published in 1770, he describes a novel method for 'taking the figures' of butterflies using gummed paper 'which may be cut out and stuck into other pictures by way of embellishment'. The method, which may well have been in wide use at the time, may be unfamiliar, and is worth describing. It worked as follows:

> 'Take Butterflies, or field Moths, either those catched abroad, or such as are taken in Caterpillars, and nursed in the house till they be Flies, clip off their wings very close to their bodies, and lay them on clean paper, in the form of a Butterfly when flying ...' The artist then spreads gum arabic, mixed with a little ox gall, on a thin sheet of paper 'big enough to take both sides of your Fly'. Once the gum starts to become 'clammy', the paper is laid over the detached wings: 'it will take them up; then double your paper so as to have all the wings between the paper; then lay it on a table, pressing it close with your fingers; and you may rub it gently with some smooth hard thing; then open the paper, and take out the wings, which will come forth transparent. The down of the upper and under side of the wings, sticking to the gummed paper, form a just likeness of both sides of the wings in their natural shapes and colours'.[25]

Afterwards you painted in the body using watercolours. No doubt, as in setting, practice made perfect. Edwards notes that the trick lay in just the right degree of gumming – too wet, and 'all will be blotted and confused'; too dry, and 'your paper will stick so fast together, that it will be torn in separation'.

George Edwards was primarily an ornithologist. In the engraved frontispiece of his greatest work, *Natural History of Uncommon Birds*, issued in four parts between 1743 and 1751, he is described as '*Ornithologia nova*'. Unlike most eighteenth-century naturalists, we have more than one fine engraved portrait of him – both in middle and old age, the earlier one based on a portrait by Dandridge (presumably our Joseph Dandridge). A contemporary described him as of medium stature, inclined to plumpness and of a cheerful, kindly nature 'associated with a charming diffidence'. He died of cancer on 23rd July 1773 and is buried in West Ham churchyard.

14. DRU DRURY
(1725–1803) (Fig. 42)

Dru Drury owned the most magnificent insect collection of the age, consisting of some 11,000 species 'collected from all countries with which Britain has any intercourse', including 2,148 species of Lepidoptera alone (at a time when the world's insect fauna was estimated at no more than 20,000 species). His material was arranged in mostly short series, for at this time there was little interest in variation. Drury is said to have spent some £4,000 on it, supposedly grading his payments to the size of the specimen. Sixpence was his rate for new insects larger than a honey bee. He also bought entire collections, including the butterflies and moths in the collection of the Duchess of Portland (1715–85), one of the finest of the age. Drury's cabinets were one of the reasons why Fabricius, the founder of modern insect classification, visited England more than once. It is said he inspected the collection as a connoisseur of such sights, with 'as much glee as a lover of wine does the sight of his wine cellar when well stocked with full casks and bottles'.[26] The two became friends, and in 1775 Fabricius named a micro-moth in Drury's honour.

Fig. 42. Dru Drury, silversmith and collector, from the engraving by W. H. Lizars for the *Memoir of Dru Drury* by C. H. Smith, published in Jardine's Naturalist's Library in 1842.

Drury's family owned a prosperous silversmith's business based at Wood Street in London, of which Drury took over the running in 1748. Late in life, he published his reflections on the trade under the sonorous title, *Thoughts on the Precious Metals, particularly Gold, with directions for Obtaining them, and Selecting other Natural Riches from the rough Diamond down to the pebble stone* (1801). Under Drury's directions the business prospered, and he opened a second shop at 32 The Strand, by coincidence just two doors away from the house where Watkins & Doncaster, the well-known natural history suppliers, opened for business in 1879. Late in life, Drury was appointed goldsmith to Queen Charlotte, wife of George III, allowing him to display the royal arms. He had need to work hard; his wife, Esther Pedley, bore him seventeen children!

Drury was a prominent member of the Aurelian Society, who corresponded with Linnaeus and, later, William Kirby, who both named insects after him. He produced two works of lasting significance. The first volume of Drury's *Illustrations of Natural History*, on exotic insects, was published in 1770. It contains upwards of 240 figures including 50 coloured plates by Moses Harris. When volumes two and three appeared a few years later, an impressed William Kirby described it as an '*Opus entomologicus splendissimus*!' Included for the first time was an illustration of a butterfly which Drury named *Nymphalis cardui virginiensis*, the American Painted Lady, from a specimen received by James Petiver from North America. Half a century was to pass, however, before the first British specimen was captured.

Drury was wealthy enough to pay people to collect for him and avoid the penny-pinching methods of Petiver or Albin. In 1800, he published *Directions for Collecting Insects in Foreign Countries*, intended for travellers who might supply him with specimens from overseas. He was not the first to issue instructions of this kind – both James Petiver and Benjamin Wilkes had done likewise. Drury is said to have contacted members of the crew and passengers of over 70 ships, providing them with equipment where necessary, and advising them on how to make a collection. To kill butterflies, Drury suggested that you should 'give the body a good squeeze ... and then stick it in one of the boxes'. 'Large pins should not be stuck into small insects', he implored. To preserve them for a long voyage, he recommended filling the storeboxes with ground pepper or tobacco dust (having lost at least one consignment of Indian butterflies from loosened balls of camphor). Occasionally he received more than he bargained for. Henry Smeathman sent him four very large collections of West African plants and insects in 1771. On inspecting them, Drury exclaimed: 'My house could not possibly contain one half the things when taken out of their packages'.

Drury kept an entomological diary, extracts from which dating from 1764–66 were transcribed by his admirer William Jones (q.v.) into his own notebooks. Judging from Drury's journals, much of his collecting was done in Middlesex, with occasional forays further afield into Surrey, Sussex, Kent and Epping Forest in Essex. Enfield Chase (which he often refers to simply as 'the Chace') was a favourite destination, and on the way he would pass by the still rural woods, fields and ponds of Hornsey, Tottenham and Southgate. In his references to the 'Black veind white Butterfly plentiful and fine' on 'the Chace', or 'Swallowtails very plentiful in that Country' [around Warnham in Sussex], we are given nostalgic glimpses of the wealth of butterfly life in the first half of the eighteenth century.

Drury died in 1803 at the then advanced age of 78, and was buried at St Martins-in-the-Fields, not far from his shop on The Strand. When his collection came to auction in 1805, said to have been the first such sale entirely devoted to entomology, it realized the surprisingly modest sum of £614, with £300 more for his cabinets and books. Many of his specimens were purchased by Drury's friends William Kirby and Edward Donovan, who had earlier prepared sale catalogues for some of Drury's exotic insects.

15. MOSES HARRIS
(1730–*c.*1788) (Fig. 43)

Moses Harris was the author of the most celebrated of all the early works about butterflies and moths, *The Aurelian*. Yet, in spite of the book's great success, we know very little about its author. In it, Harris described himself as a 'Painter, who has made this Part of Natural History his Study and has bred most of the Flies and Insects for these twenty years'. We know from other sources that he also did engravings and miniatures. From an original drawing signed 'Mo[s] Harris Pinx 1783. Aged this day 53. Ap[l] 15', we learn that he must have been born in 1730, but nothing is known about his family background or his personal circumstances. Nor are we sure where he lived, although Dru Drury, in a letter dated 1770, implied that it lay some distance from London. Harris was, by his own admission, not an educated man, but he had artistic talent, great enthusiasm and a capacity for learning. He had an entomological mentor in his uncle, Moses Harris Snr, who introduced him to the Aurelian Society at the Swan Tavern in the 1740s, when Moses Jnr was only twelve. The boy had earnestly wished to become a member, but he was considered too young to qualify for admission 'till Age should ripen and furnish [him] with sufficient Sagacity'.

In his preface to *The Aurelian*, Harris described the tragic end of the Society a few years later, when the Swan Tavern was burned to the ground in a great fire, destroying the Aurelians' collection, library and regalia, and nearly consuming the Aurelians as well (see Chapter 1). In about 1762, a second Aurelian Society was established, this time with Moses Harris, by now an entomologist of almost twenty years' experience, as its Secretary. His great work, *The Aurelian*, dedicated to the new society's 'worthy members', was already in production by then, its first fascicles having been dated to 1758.[27] His intention had been to publish one plate per month, at 2s 6d each. The price of the complete coloured volume was five guineas, which of course limited its circulation to the well-to-do.

Fig. 43. This engraved self-portrait of the 49-year-old Moses Harris surrounded by the tools of his trades – artistic and entomological – appears in the first edition of his *An Exposition of English Insects* ([1780]), 'done at the desire and expense of Mr I. Millan'. It portrays the same expressive eyes and distinctive eyebrows as the figure in the celebrated frontispiece to *The Aurelian* (Fig. 2).

The Aurelian was finally completed in 1766, though between 1772 and 1775 Harris incorporated a supplement with an anatomical diagram, three additional plates, a 'Table of Terms' and an Index, which he sold separately for 10s. 6d. [$52\frac{1}{2}$p]. The original edition contains 41 beautifully engraved plates, showing some 39 species of butterfly, 93 moths, 4 beetles and a single dragonfly. Each plate is dedicated by their 'most Obliged and Obedient Servant', Harris, to some, usually titled, person, who was probably a subscriber (see Plate 31; also Plates 25, 28). Harris drew his specimens from life, and, like Wilkes, must have reared many of them for he is scrupulous in showing the early stages of his subjects, sometimes at different stages of development. His arrangements are chosen for their aesthetic qualities, with moths, butterflies and other insects all thrown together, often with vases of flowers and other unrelated objects. But their eye-catching effect and fidelity to life is undeniable, and his beautifully coloured plates have given pleasure to entomologists and non-entomologists alike for nearly two and a half centuries. *The Aurelian*'s famous frontispiece (see Fig. 2) depicts a well-dressed man, whose features match the known portrait of Harris, with his collecting gear, a clap-net, pincushion, and small oval boxes containing a Red Admiral and Painted Lady, among other insects. This seated figure points to a second, more distant figure who wields his net by the banks of a stream.

The Aurelian did not add any new species of butterfly, but some well-observed details of their habits emerge, like the characteristic flight of the Large Skipper, with 'a kind of skipping motion .... by reason of their closing their wings so often in their passage', or that of the Wall butterfly that 'delights to fly along very low in dry Ditches ... [and], when it comes to the End of the Bank, will return back again, frequently settling against the Bank, or perhaps against the Side of a Wall; and is, for this Reason, called THE WALL FLIE'.

In 1775, Harris published an excellent little handbook, *The English Lepidoptera, or, the Aurelian's Pocket Companion*, unillustrated apart from a coloured anatomical plate and so more affordable. It provides the best information so far on the foodplants, habitats and times of appearance of the 400 or so species of butterflies and moths known at the time. Moreover it provides not only the English names (which, by now, include many still in use today) but also the new Linnaean names, taken from the 1767 (twelfth) edition of the *Systema Naturae*, which Harris was evidently acquainted with. Harris says he had carried a handwritten prototype in his pocket for many years, and that the

Pocket Companion was 'a compendium and repository of every new discovery of the author's researches for almost these thirty years'.

A few years later Harris contributed a now very rare illustrated 'Essay' on the hitherto neglected wing venation of butterflies and moths, and their use in identification. In the accompanying copperplate engravings, Harris numbers the tendons and membranes of a butterfly's wings for reference in the text. It marks a new departure from the aesthetic pleasures of observing and painting butterflies to a serious examination of their structure and form. Among the other artistic commissions undertaken by Harris were drawings of insects made for William Curtis, the founder of *The Botanical Magazine*, and plates of butterflies and moths for William Martyn's two-volume *New Dictionary of Natural History* of 1785. He also completed fifty coloured plates for the first volume of Dru Drury's *Illustrations of Natural History* published in 1770.

Moses Harris did much to encourage entomology at a time when the original dynamism of the age of Ray and the first Aurelian Society was waning. He was probably the prime mover in founding the second Aurelian Society; through his acute and observant eye, he contributed significantly to the study of butterflies and moths, and in the unsurpassed plates of *The Aurelian* he left a timeless classic to future generations. Though one senses a pleasant, engaging personality behind his writings, borne out by his self-portrait, it is symptomatic of how frustratingly little we know about him that even the year of his death is unknown, although it is thought to have been between 1787 and 1789 when Harris was about 58. Dandridge and Drury apart, few of the eighteenth-century Aurelians reached the great age of so many of their nineteenth-century successors.

16. JOHANN (JOHN) REINHOLD FORSTER (1729–1798) (Fig. 44)

Shortly after dawn on 13th July 1772, H.M.S. Resolution, under the command of Captain James Cook, slipped out of Plymouth Sound to commence her three-year voyage of exploration to Antarctica and the Pacific. This was to be Cook's second voyage, and among the 118 men on board was John, or Johann Forster, civilian-naturalist and the Royal Society nominee, charged with collecting and observing natural objects and phenomena. The voyage was hugely successful. It seems, however, that Forster proved an uncomfortable companion at sea. 'It is a puzzle', writes Richard Hough, Cook's biographer, 'how this tiresome man lasted the voyage.'[28] He seems to have made enemies all round 'by his pedantry, self-righteousness, vanity, habitual acrimony and downright rudeness'. He even, on occasion, challenged Captain Cook himself, questioning his judgement and criticizing his failure to divulge his plans in advance. On several occasions Forster came close to being thrown overboard. In fairness to Forster, the unaccustomed trials for a non-mariner were severe and few naturalists who had endured such long and arduous voyages ever volunteered for a second time. Preserving specimens in such conditions was a nightmare, and Forster had lost many hard-won items, not to mention notebooks and drawings, through salt-water damage in high seas. He remembered the voyage as 'a series of hardships such as had never been experienced by mortal man'. Recent biographical research has shown that Forster's scientific reputation, however difficult, not to say impossible, his conduct may have been on the voyage, was shamefully traduced in the reports that were put about afterwards. Although undoubtedly a proud and obstinate man, he was, in truth, a dedicated and honourable scientist whose achievements and discoveries, under cramped, crowded conditions of utmost difficulty on a long sea voyage over a period of three years, can now be seen in their true light. Who was he, and why did Cook take him?

Johann Forster (while resident in England he anglicized his baptismal name to John) was born in Dirschau in what was then Polish Prussia. His great-great-grandfather, a Yorkshireman of Scottish descent, had emigrated to Danzig around 1642 at the start of the English Civil War. The Forsters were royalists and on the losing side. In the course of several generations the Forsters became successful merchants in Dirschau – three of them also holding the office of mayor – and the young Forster was brought up in an educated family dedicated to learning. Despite a somewhat neglected early formal education due to his father's illness, he was sent at the age of fifteen and a half to the Joachimsthal-Gymnasium (high school) in Berlin where he excelled in Latin, Greek and Hebrew, and built on his self-taught knowledge of natural history. At the age of eighteen he enrolled at Halle University to study theology (but for his father's objections, he would have read medicine), but also found time to attend natural history lectures. In 1751, on graduating, he entered the Reformed Lutheran Church and was ordained two years later as pastor of a country parish at Nassenhuben, south-east of Danzig, where he gained a considerable reputation as a fine and scholarly preacher. The Seven Years War (1756–63) rumbled around Danzig, which was for a time occupied by Russia. This gave Forster the opportunity to make contact with the authorities in St Petersburg. In 1765 he was commissioned, with the signed authority of the Empress, Catherine the Great, to investigate the conditions of the German Colonies established on the River Volga, thus providing him with the chance to leave his badly-paid living, which he had held for twelve years, in order to study the natural history of that region of Russia. He took with him his ten-year old eldest son George, of whom he proudly wrote to a friend, 'His inclination for this study is indescribable, and he knows all the plants growing in our area systematically after Linnaeus.'

Fig. 44. This fine engraved portrait of Johann Forster by I. F. Bause (1781), after an oil painting by Anton Graff, suggests an intelligent and fairly amiable character, at odds with his reputation as an argumentative pedant.

Forster, who could be intemperate and outspoken because of his fiercely independent spirit, self confidence and considerable intellectual powers, was to fall out with authority throughout his life. It has been said of him that he 'never learned that self-righteous martydom brings but a hard moral victory'. Disappointed by the lack of recognition of his successful mission to the Volga, he had a dispute with his Russian superiors over remuneration and decided to seek an appointment elsewhere. Armed with letters of introduction from influential scientific colleagues in St Petersburg, he set off for England, reaching London in October 1766 where he was very quickly accepted into scientific circles. By 1778 he had been offered a position as Tutor at the noted Dissenters' Academy in Warrington, where he succeeded the famous 'father of modern chemistry', Joseph Priestley. While there he taught languages (of which he was to become proficient in at least fifteen) and natural history. As in all academic establishments there were tensions, and his success in his lectures on French and German languages and on Natural History, combined with the extramural teaching he undertook to help pay off his large debts incurred among other things by regular book purchases and general financial mismanagement, brought him into conflict with the Principal. Following a bitter dispute, his appointment was terminated in June 1769.

While at Warrington he had, however, greatly advanced his knowledge of English natural history, as the result of which, on moving back to London, he published, in 1770, *A Catalogue of British Insects* (Fig. 45) and, in 1771, *Novae*

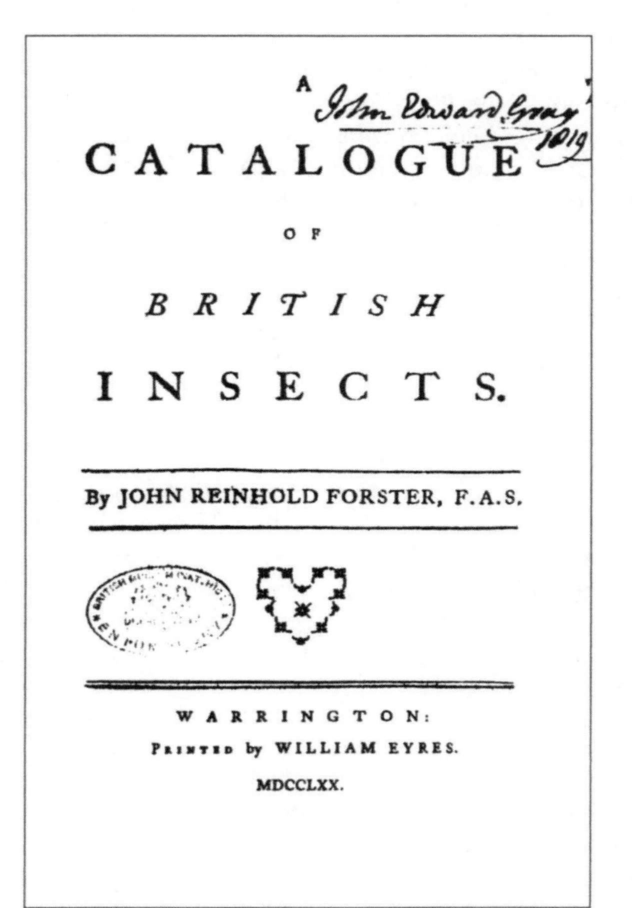

A

CATALOGUE

OF

BRITISH

INSECTS.

By JOHN REINHOLD FORSTER, F.A.S.

WARRINGTON:

Printed by WILLIAM EYRES.

MDCCLXX.

(2)

THE author of this catalogue intends to publiſh a *Fauna* of *Britiſh Inſects*; and as he thinks not to ſet out upon it, till he can offer to the public a work, as little imperfect as poſſible, and to give no other deſcriptions than from ocular inſpection: he preſents his moſt reſpectful compliments to all ladies and gentlemen who collect inſects, and begs them to favour him, if poſſible, with ſpecimens of ſuch inſects, as they can ſpare, and which he is not poſſeſſed of: for this purpoſe he has made this catalogue, and put no mark to the inſects in his poſſeſſion; thoſe which he has ſo plentifully as to be enabled to give ſome of them to other collectors, are marked with a (*d*); thoſe which he has not, are marked either *Berk.* ſignifying *Dr. Berkenhout's Outlines of the Natural Hiſtory of Great Britain*; or *B.* ſignifying a manuſcript catalogue of *Britiſh Inſects* communicated to the author; or *B. B.* which ſignifies *Berkenhout*, together with the manuſcript catalogue. *N S.* is put to ſuch inſects as have not yet been deſcribed by Dr. *Linnæus*, and are *new ſpecies* with new ſpecific names. If therefore the author ſhould be ſo happy as to be favoured with ſpecimens, he will acknowledge the favour publickly, and give in return to any collector who deſires to exchange, ſuch inſects as he can ſpare. This catalogue contains 1000 inſects; the *Swedes* have near 1700, it would therefore be an honour to this country, to ſcrutinize carefully into the various branches of *Natural Hiſtory*, and to give the public as perfect and extenſive catalogues of *Britiſh Animals* as poſſible. The four firſt claſſes, containing *Quadrupeds*, *Birds*, *Reptiles* and *Fiſh*, have been ſo well and perfectly deſcribed by *Thomas Pennant* Eſq; that it would be a diſparagement, if the inſects ſhould be deſcribed defectively. But as Mr. *Pennant* has given up this branch to the author, he will do his utmoſt to give his work all the perfection which lies in his power. Therefore he takes this opportunity to ſhew the ladies and gentlemen collectors, how far he is gone by three years aſſiduous collecting in the neighbourhood of *Warrington*, and to enable at the ſame time the ſpirited and patriotic promotors of knowledge, and eſpecially of Natural Hiſtory in *Great Britain*, to contribute towards perfecting the Natural Hiſtory of their own country.

DRAWINGS and ſcientific deſcriptions of ſuch inſects as are not yet deſcribed by *Linnæus* will be ſtill more acceptable, and engage the publiſher's thanks and acknowledgments.

WARRINGTON,
Auguſt the 1ſt.
1770.

Fig. 45. The cover and introductory page of Forster's *Catalogue of British Insects*, published within five years of his arrival in Britain from the Continent.

Species Insectorum. Brief as they are, these booklets are of historic interest not only because they attempt to include all the species known at this time but also because they are the first English publications to apply the new scientific names of Linnaeus and Fabricius to our insects. Despite their importance, they were not designed to last and are now very scarce. Forster's knowledge of the latest scientific advances on the Continent – he was a close friend of Fabricius, who named several insects in his honour – must have impressed contemporaries, especially when they learned that he was intending to catalogue the insect fauna of the entire world. The influential Joseph Banks had him elected to the Royal Society in 1772, and, after Banks was obliged to give up his place on Cook's forthcoming voyage, he proposed Forster as his replacement. Sailing around the world with Cook was the greatest challenge the world could offer a natural scientist. The ambitious Forster jumped at the chance but, although Cook allowed his talented son, George, to accompany him and share in the work, he was evidently already arguing about the pay and unsatisfactory conditions before the ship even set sail.

On his return, in 1775, Forster insisted that with his employment as naturalist he had been given the 'exclusive rights' to publish the natural history findings of the voyage as well as 'permanent employment for the remainder of [his] life'. This was disputed, and after a great many increasingly angry exchanges, the result was not one but two unofficial accounts: *Voyage round the World* (1777), written by George under his father's supervision, and *Observations made during a Voyage round the World, on Physical Geography, Natural History and Ethnic Philosophy* ... (1778) by Forster himself, which he dedicated to the President, Council and Fellows of the Royal Society. This work of some 650 pages was perhaps his finest achievement and he received widespread praise for its learning and discoveries, especially outside England where opinions were uninfluenced by personal animosities and official disfavour. Forster decided to return to Halle, where he was shortly made Professor of Natural History and from where he published *Zoologia Indica* (1781) with German and Latin parallel texts, as well as other works. He remained in Halle, contentious to the last, until his death at the age of 69. He was widely respected for his incontestable ability and erudition but loved by few.[29]

17. WILLIAM JONES (1750–1818) (Fig. 46)

Fig. 46. A contemporary silhouette portrait by T. Rider of William Jones of Chelsea – 'this great and insufficiently appreciated naturalist', E. B. Poulton wrote.

Compared with his contemporaries, Harris, Kirby and Lewin, William Jones of Chelsea is not widely known, since he published no sumptuous illustrated volumes, no celebrated textbooks, and only one scientific paper.[30] However, his collections, unlike theirs, have survived. Moreover Jones, a prosperous wine merchant, was a distinguished painter of butterflies and moths, of which he figured in watercolour some 1,500 in what have become known as the six-volume 'Jones Icones' (1783–85). In addition, he produced some 132 paintings of birds' eggs that are also in existence. His butterfly paintings were consulted by Fabricius, who described over 200 new species of Lepidoptera with the support of their evidence. They are now what are called 'Iconotypes', as the original specimens have been lost. Among them was a small butterfly, called by Jones 'the Brown Whitespot', but which is now known as the Northern Brown Argus (*Aricia artaxerxes*) (see Chapter 4: 20). He may possibly have obtained it from the Edinburgh naturalist, William Skrimshire, who also supplied A. H. Haworth with the species, since it is not thought that Jones ever visited Scotland.

A brief account of Jones, written eleven years after his death, remembered him as 'amiable' and 'mild' in his manner, and as a 'fluent and instructive' scholar. He was 'genuinely skilled in the Hebrew and Greek languages', and was a master of both oil and watercolour painting, using Chinese blocks that have preserved his colours better than those of many contemporary artists. Late in life he became a follower of the doctrines of the Swedish mystic Emanuel Swedenborg, who held that creation exists only through divine intervention.

Jones knew many of the leading naturalists of the day. He regarded Dru

Drury as his artistic and zoological mentor, and went so far as to transcribe pages from Drury's entomological notebooks into his own, no doubt with a view to visiting the sites that Drury had documented. Haworth, who was his neighbour in Chelsea, referred to him repeatedly as his 'tutor' and 'my very able and much respected friend'. He was highly enough regarded by J. E. Smith to be consulted over the formation of a new society to promote natural history – which became the Linnean Society, of which Jones was elected one of the first Fellows in 1791, Smith himself becoming its first President. From a close study of the wing-venation of butterflies – concomitant with his artistic labours – Jones had recognized that Linnaeus had placed several species in the wrong 'families'. His reclassification was accepted by Smith, who in 1797 wrote that Jones's knowledge of butterflies was 'perhaps unequalled'.[31]

Jones's collection of some 800 species of British butterflies and moths, housed in a 44-drawer cabinet, still survives. Judging from two of his manuscript notebooks, which also survive and which contain extracts copied from Drury's journals for 1764–66, it is likely that he visited and no doubt collected in Drury's haunts in south-east England accessible from London, and as far afield as Sussex.

The collection, notebooks and paintings were inherited by Jones's cousin, John Drewitt of Peppering, near Arundel. They remained with the family until 1931 when Drewitt's grandson, Dr Frederick Drewitt donated them to the Hope Entomological Collections in the University Museum, Oxford, where they can still be consulted. In a separate donation Drewitt also gave the Museum several insects preserved in amber and three Large Coppers (two males and one female) which had belonged to William Jones and which may be the oldest existing specimens of this extinct butterfly.[32]

18. JOHN BERKENHOUT (*c.*1730–1791) (Fig. 47)

Berkenhout, the English-born son of a Dutch merchant, was among the first authors to use the Linnaean system of classification. His was a versatile intelligence, whose writings ranged from medicine and the natural sciences to the classics. One of his shorter works is an essay on the Bite of a Mad Dog. He was apparently a good mathematician, and a talented musician and painter. His only entomological work is contained in the three-volume *Outlines of the Natural History of Great Britain and Ireland* (1769–72), in which Berkenhout attempted to provide 'a systematic Arrangement and concise Description of all the Animals, Vegetables and Fossiles which have hitherto been discovered in these Kingdoms'. In the section on Lepidoptera, he included not only the Linnaean and English names of some 39 butterflies but also details of their larvae and foodplants. Probably much of this information was secondhand, for Berkenhout often borrowed Petiver's by now outmoded English names, and also repeated several mistakes by Linnaeus.

Fig. 47. A characterful portrait of John Berkenhout, then aged nearly 60, engraved for the *European Magazine*, 1788.

Berkenhout's apparent intellectual restlessness is mirrored in his much-travelled life. Before he graduated in medicine at the Universities of Edinburgh and Leyden, he had toured much of Europe and seen active service in the Prussian and English armies. In 1775 he established a medical practice at Isleworth and, later, Bury St Edmunds. Three years later, however, he changed direction yet again and sailed to the rebellious American colonies on some mysterious government mission, where he was arrested as a spy. Granted a pension by the government, he devoted his declining years to writing, and died at Besselsleigh, near Oxford, in 1791. An engraved portrait in the *European Magazine* of 1788 shows an interesting face, with a high domed forehead, cautious, baggy eyes and a swooping Roman nose. Berkenhout was not an Aurelian but a wide-ranging intellectual of the age of reason, perhaps attracted to Linnaean classification by its logic and imposition of order on randomness.

19. WILLIAM LEWIN
(? – 1795)

Beyond his authorship of one of the most significant of the early works on British butterflies, and another work on British birds, almost nothing is known of the life of William Lewin. He was elected a Fellow of the Linnean Society in 1791. In 1792 he lived at Darenth in Kent (whose woods were a popular destination for London Aurelians), and two years later moved to Hoxton in the East End. He is assumed to have died in 1795, the year in which his book was published, for his name does not appear in the list of Fellows of the Linnean Society for 1796, and in their Transactions for the following year he is referred to as 'the late William Lewin'.

Lewin conceived the ambition to publish a comprehensive, illustrated work on 'The Insects of Great Britain'. In the event, his untimely death prevented the completion of all but the first volume on 'Papilios', which contains significantly more species of butterflies, together with their early stages, than works by previous authors: in present-day terms, some 62 species plus three recognized forms (see Plates 17 and 26). Lewin was also the first author to follow the strict order of families, and to use Linnaean names consistently and correctly applied (though, in some cases, misspelt). He may have acquired this information from the Danish entomologist, Fabricius, who had visited England on several occasions, most recently in 1787. Among the new species introduced by Lewin were the Large Heath (as 'the Manchester Argus'), Small Blue, Large Blue and Large Copper. He also believed the Mazarine Blue (as 'Papilio cimon', the 'Dark Blue') to be new, from several specimens taken on a hillside near Bath, though in fact it had already been described by Ray. Yet another supposedly new butterfly was his 'Scarce Spotted Skipper' or 'Papilio fritillum', in fact only the rare ab. *taras* of the humble Grizzled Skipper. Lewin coloured his plates personally, and although the quality varies from copy to copy, his work has rightly been acclaimed as some of the best produced in the age of the Aurelians. As Emmet suggests,[33] Lewin was probably first and foremost an artist who became attracted to entomology, for he faithfully copies several minor errors from other authors.

Three species illustrated by Lewin call for particular comment (see Chapter 4). His Large Coppers, accurately depicted though given the name of *Papilio hippothoe*, a Continental species, had been taken by 'a gentleman in Huntingdonshire, on a moorish [i.e. fenny] piece of land'. However Lewin and several other eighteenth-century authors also considered the Scarce Copper to be British. J. R. Forster had claimed that it was common enough near Warrington to enable him to give specimens away to other collectors, but he must have confused it with the Small Copper. Lewin, too, claimed he met with it one hot day in August, when 'two of these butterflies settled on a bank in the marshes, the sun at that time shining very hot on them', adding that 'they were exceedingly shy, and would not suffer me to approach them'. Unfortunately he does not say where. As for the Large Blue, it was then thought to be associated with 'high chalky lands', having been recorded at Dover, the Marlborough Downs, 'the hills near Bath' and 'near Clifden [Cliveden] in Buckinghamshire'. In its case there may have been some confusion with the Mazarine Blue, which also occurred near Bath, and whose modern name Donovan gives to the Large Blue. Lewin's lists of localities indicate that either Aurelians were by now more widely distributed than earlier in the century, or that they were travelling farther. Only a few years after Lewin's death, naturalists like Haworth, Donovan, Curtis and Weaver would extend the frontiers of entomology over the whole of Britain.

20. JOHN ABBOT (1751–1840) (Fig. 48)

John Abbot is remembered in the New World more than in Britain, but a handsome new biography should help to re-establish his reputation here as one of the great eighteenth-century insect illustrators.[34] His claim to fame rests primarily with the coloured drawings he made for *The Natural History of the Rarer Lepidopterous Insects of Georgia*, published in London in 1797, and the first significant work on American butterflies. His paintings of insects and other natural objects are extremely fine and compare favourably with the best of the work of Maria Sibylla Merian, Benjamin Wilkes or even Moses Harris. He also sent numerous cases of butterflies collected in Virginia and Georgia to Britain, though the first two shipments were lost when, by singular bad luck, both vessels foundered.

Fig. 48. John Abbot, from a drawing believed to be a self-portrait.

Abbot was a lawyer's son, born in London, and acquainted in his youth with many of the leading Aurelians, including Dru Drury, Emanuel Mendes da Costa and Henry Smeathman, from whom Abbot purchased a Purple Emperor for the then very considerable sum of one guinea (£1.05p). Drury's magnificent collection was the first he ever examined, and such was the young Abbot's 'pleasure and astonishment' at the sight that he immediately splashed out another six guineas on a 26-drawer mahogany cabinet.

What might have been a famous career as an English Aurelian, a second Moses Harris, was cut short when, aged only 22, Abbot sailed to the New World in 1773, never to return. He had been engaged as a collector by the naturalist Thomas Martyn, with commissions from two or three others, and with a letter of introduction from the Royal Society. His embarkation was delayed by repairs, and Abbot caught the ship after it had set sail from London unannounced only by means of a high-speed pursuit by post-chaise to Deal, Kent, its last port of call in England. What was first intended only as a collecting expedition and adventure became an emigration, as Abbot got married and settled down on a 200-acre estate in Georgia, collecting and painting every imaginable natural history object from shells and minerals to beetles and birds. He left Britain at the time of the second Aurelian Society, yet he was still hale, if stout and rather rheumatic, on the accession of Queen Victoria, still sending boxes and drawings of insects home to a London he could barely remember, and, if he had returned, would have barely recognized.

21. CHARLES ABBOT (1761–1817)

Unlike the London-based Aurelians, who could meet regularly and exchange ideas and information, the Revd Charles Abbot – no relation to John Abbot – lived in the country and so was obliged to develop his interests in isolation, or through correspondence. He received his education at Winchester and New College, Oxford, following which he combined a career as usher or under-master at Bedford Grammar School with church duties as vicar of the neighbouring parishes of Oakley Reynes and Goldington. He was one of the earliest Fellows of the Linnean Society and, like most eighteenth-century naturalists, an all-rounder, being author of a very creditable *Flora Bedfordiensis* (1798), only the third published county flora, and a series of insect records, some of them remarkable. He will be long remembered as the discoverer of the Chequered Skipper in Britain, at Clapham Park Wood, near Bedford, during his 'first season as an Aurelian' in 1797 (see Chapter 4: 1). On the same occasion he swiped at, but missed, a Swallowtail. Some time later he discovered a colony of the Large Blue near Bromham.

Unfortunately Abbot's entomological reputation has suffered from the much less likely butterflies he claims to have collected around Bedford. Several of these became enshrined in Haworth's *Lepidoptera Britannica*, who wrote in the introduction that 'my friend, the Rev Dr Abbot of Bedford' had taken the Scarce Swallowtail at Clapham Park Wood in May 1803, and Bath White and Queen of Spain Fritillary in White Wood, Gamlingay in Cambridgeshire a month later. Haworth remarked that there was not a single extant British specimen of these rare insects. Any faith in these captures, however, fades when we learn that Abbot's collection also included two exotic skippers, *Hesperia bucephalus* and *Pyrgus oileus*, also taken at Clapham Park Wood. At a time when the knowledge of our butterflies was still very imperfect, these discoveries did not at first seem implausible.

Charles Abbot's collection was saved by J. C. Dale, who, stopping by at the Swan Inn in Bedford one day in 1817, saw the collection and bought it on the spot, together with Abbot's notebooks. No doubt with their help, Dale later made special visits to Clapham Park Wood, Gamlingay and Bromham, but unsurprisingly failed to find any of Abbot's rarities. P. B. M. Allan, in *Talking of Moths*, suggested that the 'great' Dr Abbot had been duped, perhaps by the dealer Plastead, who seems to have specialized in planting exotic butterflies and moths on honest, unsuspecting lepidopterists.

Relatively little is known about Charles Abbot the man. However an unknown hand in one of Dale's journals makes the intriguing remark that Abbot was passed over for the headmaster post, when it became vacant, and that this 'had such an effect on him [that] he died of a broken heart, it was supposed'. The same authority adds that at some stage Abbot 'had cracked his skull and had a piece of silver plate let in and fancied odd things'.[35]

22. WILLIAM KIRBY (1759–1850) (Fig. 49)

Several distinguished persons have, at one time or another, received the appellation of 'father of entomology' – Petiver, Haworth and J. F. Stephens are leading candidates, but the honour is offered most regularly to the shade of the Revd William Kirby, rector of Barham and joint author of *Introduction to*

Entomology. Kirby was not only a great naturalist; he seems to have been regarded by contemporaries as someone who embodied the very spirit of entomology. When the Entomological Society of London was founded in 1833, one of its first actions was to elect an emotional, 74-year-old Kirby its Life President, an honour subsequently awarded to only two other men. They also decreed that Kirby's portrait should hang behind the President's chair from that time onwards.

Fig. 49. The Revd William Kirby, Rector of Barham, – the 'father of British entomology' – from an engraving by W. T. Fry, after the portrait by H. Howard, R.A.

A Life of William Kirby, written soon after his death,[36] includes a vivid portrait of the great man in his old age. Kirby was conservative in his habits and wore the costume fashions of a couple of generations earlier, with his shovel hat, buckled garters and antiquated coat, with its collar cut straight in front and huge flaps over the pockets. He had shoulder-length fair hair, which he sometimes tied back in a kind of pony tail, and small deep-set blue eyes which, 'when animated were bright and sparkling, and gave a peculiar lustre and brilliance to his countenance'. His working practices were orderly and generally conformed to the following plan on which he thrived until the ripe old age of 91:

> 'The time before breakfast was devoted to reading portions of the scriptures in Greek or Hebrew. After breakfast, [he read] one of the [Christian] Fathers until noon, with a classical author on alternate days, and this was followed by exercise until an early dinner. The afternoon was devoted to natural history, and the evening to miscellaneous reading, correspondence, etc. Wednesday and Friday were devoted to systematic visitation in his parish. These rules were observed with great accuracy for a very long period of his life: latterly his custom was to read the New Testament in Greek after breakfast (which he always did aloud) and it was rarely that this was neglected ... He would often rise early to ascertain, if he could, the proceedings of the insect world.'

William Kirby lived all his life in Suffolk. He was born at Witnesham Hall, son of John Kirby, a solicitor, and educated at Ipswich Grammar School and Gonville and Caius College, Cambridge, where he graduated in 1781 and was ordained the following year. He was presented with the living of Barham, close to his family home, and remained there as Rector until the end of his days. He had become interested in botany and entomology in childhood, the latter being kindled, so he tells us, after observing the habits of a ladybird on his study window. Early on he decided to keep a journal, which he maintained in often lively and humorous detail through much of his life, and which is quoted extensively by Freeman. Here, for example, is part of his characteristic account of a collecting expedition in summer 1797 with a distinguished friend and co-founder of the Linnean Society, Thomas Marsham (*c.*1748–1819):

> 'During the whole of the day *Tabanus rusticus* [a troublesome horse-fly] attacked our poor steed with the outmost fury, so as to occasion the blood to flow all over her ... At two we reached The George in the village of Fen Stanton. The landlord, an honest looking person, is concerned that we did not arrive before their dinner was over, but

> hopes we can make shift with cold ham and chickens: to these he adds stuffed leg of pork, roastbeef, green peas and cucumber. A Suffolk damsel, to my no small glee, waits upon us. We start for Cambridge at 5.'[37]

The account continues with details of overnight stops at public houses, manorial dwellings and sleepy villages, including Ufford, where Kirby spent a difficult night in a bed which 'in its dimensions did not fall far short of the Great Bed of Ware': 'I was forced to ascend my bed by means of steps, and then plunge into an ocean of down, not without some apprehension of suffocation'.

Kirby became an Associate of the Linnean Society, three years after its founding, in 1791 (he was elected a Fellow in 1815 and three years later a Fellow of the Royal Society). Wishing to keep entomology within the Society, he was reluctant at first to give his support for an independent entomological society. In 1822, he persuaded Haworth, Stephens and others to form instead a zoological club within the parent Society. Unfortunately, under the rules, only Fellows and Associates were eligible, which meant excluding many entomologists. It says much for the esteem in which Kirby was held that the opinion of this Suffolk clergyman was valued so highly. There were two good reasons. The first was Kirby's celebrated *Monographia Apum Angliae*, first published in 1802, a work that makes him, if not the 'father of entomologists', then certainly the 'father of hymenopterists'. The second, of course, was *Introduction to Entomology*, jointly written with his friend, William Spence, and published in four volumes between 1815 and 1826 (Volume 1: 1815, Volume 2: 1817, Volumes 3 & 4: 1826). The work had a profound effect on the future of entomology. The subject of most previous works had been dead insects on pins and their relationship to one another. Kirby and Spence, however, dealt also with living insects, and such topics as honey-dew, hibernation, delayed emergence and social behaviour, using modern deductive methods based on observation. In so doing, they demolished once and for all the superstition and myth that still surrounded the insect world. As Allan once commented, 'it is impossible to read their four volumes without admiration. There is original thought and observation in every chapter, and although, in the light of our present day knowledge, we can assert that they sometimes misinterpreted their observations, they advanced the science of entomology more than any previous writer had done'.[38] Probably no other work contributed more to the surge of interest in insects that led to the first lasting entomological society in 1833.

The work is also, of course, an invaluable source on collecting methods in the first quarter of the nineteenth century. They tell the reader how to make various nets, including a bag-net of the kind 'the French collectors use', though the preference of most collectors was still for the time-honoured 'fly-net' or bat-fowler. Kirby's nets were used not only for flying insects but as trays to catch larvae when beating overhanging branches. Kirby and Spence, it seems, wore tall hats when collecting, and suggested, with a seemingly straight face, that the readers lined theirs with cork as a useful receptacle during a long excursion! Among the more elaborate items they describe is a device shaped rather like a paraffin stove for killing large numbers of insects by steam from a boiling saucepan (for chemical killing agents were still some years away). Anticipating criticism from faint hearts, Kirby felt bound to defend entomology from charges of cruelty. He insisted that insects felt no pain and, producing various grisly examples to prove it, added that those who practised field sports were in no position to cast stones, since 'the tortures of wounded birds, of fish that swallow the hook and break the line, or of the hunted hare, [were] beyond comparison, greater than those of insects destroyed in the usual mode'.

In 1835, Kirby presented the whole of his insect collections, including all his many type specimens, to the Entomological Society – he may have wished to atone for his initially reluctant support. One of his types, the parasitic insect *Stylops kirbii* from the Order Strepsiptera, which Kirby erected in 1811, is the Society's emblem. Westwood drew the specimen for the Society's seal, and it is his drawing which adorns the Royal Entomological Society's club tie to this day, a symbol of the lasting respect and affection it had for a modest clergyman who laid the foundations for modern entomology.

23. ADRIAN HARDY HAWORTH (1767–1833) (Fig. 50)

With unusual felicity, the great naturalists of the mid-nineteenth century – Pickard, Westwood, Babington & Co. – who compiled the *Accentuated List of the British Lepidoptera*, summed up A. H. Haworth's sterling qualities as a systematist in two lines: '1st, that he described from Nature; and 2nd, that he described well'. Those simple but rare qualities went a long way to ensure that Haworth's great four-volume work, *Lepidoptera Britannica* (1803–28), was 'the most complete, most learned, and most useful [monograph] ever published on the Entomology of Britain, and one which will long remain an invaluable treasure to the Lepidopterist'.[39] Haworth was the first to deal scientifically with the whole of the British Lepidoptera. His *Lepidoptera Britannica* remained the most authoritative work on British butterflies and moths for nearly half a century, until Stainton produced his *Manual* in 1857. It is still a key historical source on distribution and other topics.

Fig. 50. A. H. Haworth is depicted in this charcoal drawing of a bust, draped in a toga with a Roman haircut in the best classical tradition, no doubt symbolizing his scholarship. No other likeness is known to exist.

Like most early entomologists, Haworth was not trained as a scientist. Born at Hull, the younger son of squire Benjamin Haworth of Haworth Hall, the young Adrian was educated at the local grammar school and was then articled to a solicitor. However he had little interest in the law, and since from the age of 21, and particularly after his inheritance of the family manor a few years later, he was given independent means, he was able to devote most of his time to his first love, natural history. In 1792, Haworth moved to Chelsea, then still an attractive village set among nurseries and market gardens, where he met and came under the influence of William Jones (q.v.). From there he could attend meetings of the Linnean Society (he became a Fellow in 1798) and use the library and herbarium of his friend Sir Joseph Banks, as well as visit the royal garden at Kew and the Apothecaries' Physick Garden in Chelsea itself. Haworth was greatly interested in botany, specializing in bulb plants and Mesembryanthemums, on which he was the authority. His greenhouses are said to have been full of exotic plants.

At that time, no entomological society existed, and the newly-founded Linnean Society was no real substitute. In 1801, therefore, Haworth established an Aurelian Society – the third bearing that name – but it was dissolved after only five years after members proved unwilling to abide by Haworth's high-minded ruling that they contribute their best specimens to the Society collection. Haworth was apparently a leading member of the new Entomological Society which soon arose from its ashes. In its first few years, the Society produced three parts of its *Transactions*, but later seems to have run out of energy. As Gage and Stearn comment, it 'seemed more or less awake up to 1812, and more somnulent from 1812 to 1822',[40] when it finally fell asleep, or, more accurately, fractured into two more short-lived Entomological Societies, no doubt as the result of some forgotten feud. Despite Haworth's best efforts, the time for a thriving entomological society had not yet arrived. Ironically the Entomological Society of London (later the Royal Entomological Society) was founded just a few months before his death.

Haworth must have been hard at work on the systematics of British Lepidoptera at around the same time as the establishment of the third Aurelian Society, for his first significant work, the now very rare *Prodromus Lepidopterorum Britannicorum* was published in 1802, followed a year later by the first volume of *Lepidoptera Britannica.* Unlike the work of the Aurelians of the previous century, Haworth's work was intended not so much for well-heeled patrons as for Linnaean scholars; it was unillustrated and full of terse Latin descriptions. It was the first comprehensive work on the British Lepidoptera based on Linnaean and Fabrician principles of classification. In the preface to the first volume, Haworth tells us how he came to write it:

> 'I began to collect, arrange and describe the natural productions of this fertile and happy island; but more especially its *Birds, Insects* and *Vegetables.* For these purposes I have diligently examined many parts of England personally, and usually on foot and alone; but sometimes accompanied by pedestrian friends [i.e. fellow walkers!] of congenial sentiments and taste. Industriously we have sought, and never once in vain, a great variety of woods and lawns, hills and vales, marshes and fens ... travelling in various journeys not fewer than a thousand miles. In spite of heat or cold, wet and drought, and various concomitant impediments.'

Lepidoptera Britannica introduced many new British species, including some that were new to science. Among the butterflies is the pale var. *helice* of the Clouded Yellow, then considered to be a distinct species, and the true name, *dispar,* of the Large Copper (in consequence, the now extinct British subspecies bears the type name for the species). Among the many moths named by Haworth are such relatively common ones as the Black Rustic (*Aporophyla nigra*) and Small Wainscot (*Photedes pygmina*), hitherto overlooked through their similarity to other species – and because of the disgraceful neglect hitherto shown for small brown moths. We can also note, in passing, that Haworth mentions for the first time some of the *English* names of the smaller moths, including the 'Pugs' (*Eupithecia*), which, as he explained, have unequal wings, the upper being long, the lower short, much like the lips of the Pug dog! Haworth's Minor (*Celaena haworthii*), on the other hand, was named by J. F. Stephens in honour of the man 'whose "Lepidoptera Britannica" and splendid Cabinet, so liberally opened to his friends, entitle him to the thanks of every one engaged in the study of this beautiful Order'.

Haworth's collecting zeal and characteristic generosity is mentioned by several of his surviving friends. For example, J. O. Westwood noted in his diary that in June 1823 Haworth had promised to give him some beetles he wanted. Evidently what Haworth had in mind was an exchange, for a month later he 'called to see my Ins[ec]ts when he picked out some which he had not got & promised me some in exchange for them'. Later that month, he kept his promise and handed over the beetles.[41] Haworth's Chelsea house was a regular meeting place for entomologists and botanists (in those days the two often went together), for tea, discussions and, of course, inspections of the finest collection of insects then available. Our portrait of Haworth, a drawing of a bust, shows an amiable looking man, noticeably well-cropped and clean-shaven compared to most entomologists of the nineteenth century.

Haworth fell victim to an epidemic of cholera which spread westwards from Russia in 1833, claiming some 22,000 lives. He was in his garden one evening, watering his favourite plants and in apparent good health, when he was suddenly taken ill. Less than 24 hours later he was dead.

Haworth's collection of some 40,000 specimens, full of type material from Britain and around the world, was possibly the most important ever to be sold at auction. More than half were Lepidoptera, and it included most of the species described in *Lepidoptera Britannica*, each one characterized by Haworth's trapezoidal-shaped labels. Every naturalist of note was at the sale, which lasted eleven days but realized only some £552.[42] Some of the exotic insects evidently failed to attract a bidder. Haworth's magnificent library was sold at Sothebys in a separate sale. Fortunately many of Haworth's types were purchased by his friends Stephens and Westwood (who compiled the sale catalogues), and have since passed respectively into the national collection and the Hope Collections in Oxford.

24. EDWARD DONOVAN (1768–1837)

Though little is known about his family background or early life, Edward Donovan was evidently a wealthy man, able to devote much of his life to writing, illustrating and collecting. Among his many works was one of the literary milestones of entomology, the *Natural History of British Insects*, while his collections of natural and antiquarian objects and library were second to none. And yet so much learning and labour brought him little financial reward. In a bitter '*Memorial Respecting My Works on Natural History*', published in 1833 near the end of his life, Donovan accused his publishers and booksellers of robbing him of his literary dues – which he calculated at more than £60,000 – leaving him 'well-nigh ruined'. He died, it is said, virtually destitute and in obscurity.

The Natural History of British Insects, originally intended as a ten-volume production but in the event requiring no fewer than sixteen octavo volumes to complete, was produced in instalments between 1792 and 1813. Judging from the large number of copies that survive, including reprints of volume one, it was a popular work, attesting to the increasing interest in entomology at that time. In the prospectus attached to the first volume, Donovan set out his aims. Because of the inaccuracies that often arose from copying figures from one book to the next, Donovan would scrupulously base his illustrations on actual specimens (on the title page he states living specimens). When complete it would form the most comprehensive work on insects yet attempted, comparable to the contemporary work on the British flora by Smith and Sowerby. It would describe their early stages and transformations, their food and 'economy', their distribution, and also include 'the History of such Minute Insects as require investigation by the Microscope'. The work would be illustrated throughout by coloured figures by the author.

This vast undertaking effectively forms a transition between the primarily artistic (albeit well-observed and accurate) productions of the eighteenth century and the more austere scientific works which were to follow. Donovan was a talented and faithful illustrator, and his figures are noteworthy for their vivid – if sometimes a little too vivid – colours. His *Natural History* does not treat insects systematically, however. His beetles, flies, moths and bugs follow one another in no particular order. In the course of it Donovan described a number of new species which often still bear the name he bestowed on them. The work includes all sixty-two British butterflies then known, and is the first to include the Chequered Skipper, which he refers to only as 'Papilio paniscus', 'deemed a rare insect by entomologists' (see Plate 12).

In 1794, Donovan published a much shorter volume, *Instructions for Collecting*, a handbook on techniques covering taxidermy, pressed flowers, preserving shells and corals, and other natural objects, together with 'a Treatise on the Management of Insects in their several States', that is, rearing. This work is particularly interesting for its insights into what constituted a complete collection of Lepidoptera at that time. Donovan saw no point in having long series of specimens: 'It is usual to put two specimens of each species of the Butterfly kind into the cabinet, one to display the upper, and the other the underside; for the underside is much more beautiful in most species, and differs entirely in appearance from the upperside.' His instructions for moths were slightly different. 'Sphinxes and Moths are generally disposed in pairs to shew the male and female, and as their under sides are seldom very beautiful, only their upper sides are shewn.'

Between 1800 and 1805, Donovan spent several successive seasons travelling in England and Wales in pursuit of wildlife and antiquities, and in the process was perhaps the first person to survey the insect fauna of South Wales. He published an account of these journeys in 1805, complete with coloured engravings from his own sketches. These must have been highly productive years. By the later volumes of the *Natural History of British Insects*, he was listing himself as 'E. Donovan, F.L.S., W.S., Author of the Natural Histories of British Birds [1792–97], Shells [1799], Fishes [1806], Quadrupeds [1823], &c'. Incidentally, the initials 'W. S.' may indicate that he was a Fellow of the Wernerian Society of Edinburgh, founded in 1808 by Robert Jameson, Professor of Natural History in the University, for the purpose of promoting the Science of Natural History. Donovan's survey of British natural history rivals, in output at least, that of Ray a century before, but he also added 'epitomes' of the insects of India, the islands of the Indian and South Pacific Oceans, New Holland [Western Australia], and China (learning Chinese in the process).

Donovan owned what must have been one of the most extensive natural history collections of the time, housed in his private museum, the well-known London Museum and Institute of Natural History. By 1817, he was presumably feeling the pinch, for the museum and its contents were advertised, and sold at auction the following year.[43] Some of his type specimens survive, and are now in the Natural History Museum or the Hope Collections in Oxford. Donovan's greatest talent lay in his skills as an artist. As a systematist, the advances of Latreille, Haworth and other contemporaries left him stranded, and Donovan's copy of Haworth's *Lepidoptera Britannica*, still preserved in the Hope Department, is 'full of once bitter, but now amusing remarks against his adversary'.[44] Even so, Donovan's achievement as a 'gentleman naturalist', working largely on his own, is impressive. Surprisingly, there is no known portrait of him.

25. JOSEPH GRIMALDI
(1779–1837) (Fig. 51)

Grimaldi is remembered as the archetypal clown of English pantomime. He first appeared on the stage at the age of two, and performed every season but one thereafter until 1828. A memoir of Grimaldi was published the year after his death in 1837, edited by none other than Charles Dickens, under his youthful pseudonym of 'Boz', the sketcher of 'every-day life and every-day people'. His most surprising revelation is that Grimaldi collected and painted butterflies. According to Boz, his cabinet contained no fewer than 4,000 specimens, collected, in Grimaldi's own words, 'at the expense of a great deal of time, a great deal of money and a great deal of vast and actual labour'. Among them was a Camberwell Beauty, which Grimaldi considered to be 'very ugly'. He was, however, entranced by a beautiful butterfly he called the Dartford Blue, after the locality where he had first found it, in June 1794. This, as F. O. Morris makes clear,[45] was the Adonis Blue, also known at that time as the Clifden Blue. Grimaldi must have told Dickens about his Dartford Blue excursions, and the latter's account of one of them is so interesting that it deserves to be quoted at length.

Fig. 51. An engraving of the clown, Joseph Grimaldi, from a portrait by George Cruickshank, published in Charles Dickens's *Memoirs of Joseph Grimaldi.*

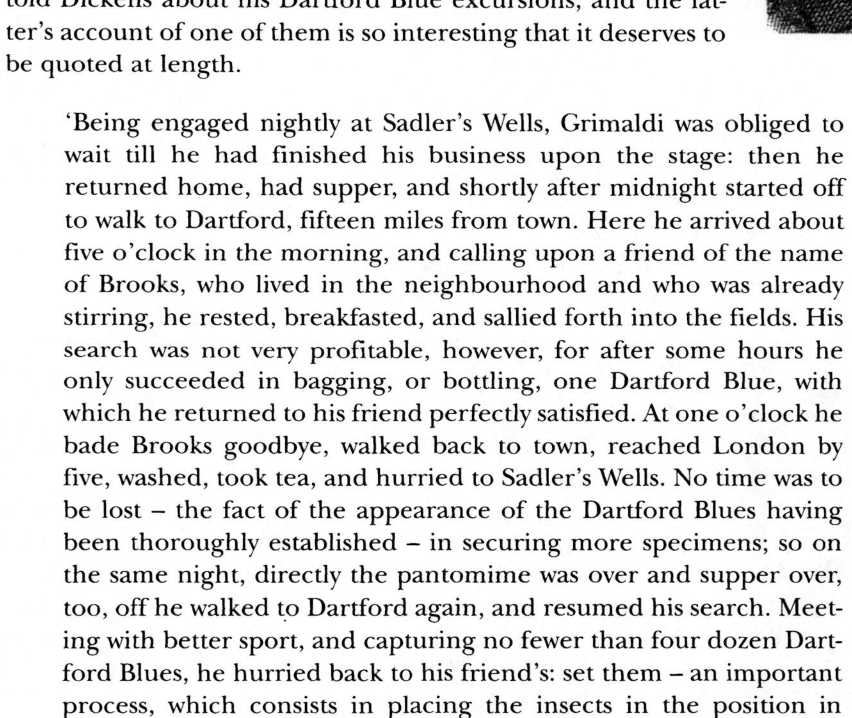

> 'Being engaged nightly at Sadler's Wells, Grimaldi was obliged to wait till he had finished his business upon the stage: then he returned home, had supper, and shortly after midnight started off to walk to Dartford, fifteen miles from town. Here he arrived about five o'clock in the morning, and calling upon a friend of the name of Brooks, who lived in the neighbourhood and who was already stirring, he rested, breakfasted, and sallied forth into the fields. His search was not very profitable, however, for after some hours he only succeeded in bagging, or bottling, one Dartford Blue, with which he returned to his friend perfectly satisfied. At one o'clock he bade Brooks goodbye, walked back to town, reached London by five, washed, took tea, and hurried to Sadler's Wells. No time was to be lost – the fact of the appearance of the Dartford Blues having been thoroughly established – in securing more specimens; so on the same night, directly the pantomime was over and supper over, too, off he walked to Dartford again, and resumed his search. Meeting with better sport, and capturing no fewer than four dozen Dartford Blues, he hurried back to his friend's: set them – an important process, which consists in placing the insects in the position in which their natural beauty can be best displayed – started off with the Blues in his pocket for London once more, reached home by four o'clock in the afternoon, washed, took a hasty meal, and then went to the theatre for the evening's performance.
>
> 'As not half the necessary number of Blues had been taken,

Grimaldi had decided upon another visit to Dartford that same night, and was consequently much pleased to find that, from some unforeseen circumstance, the pantomime was to be played first. By this means he was enabled to leave London at nine o'clock, to reach Dartford at one, to find a bed and supper ready, to meet a kind reception from his friend, and finally to turn into bed, a little tired of two days' exertions. The next day was Sunday, so that he could indulge himself without being obliged to return to town, and in the morning he caught more flies than he wanted; so the rest of the day was devoted to quiet sociality. He went to bed at ten o'clock, rose early next morning, walked comfortably to town, and at noon was perfect in his part at the rehearsal.'[46]

Grimaldi's cabinet must have contained rather a lot of Dartford Blues! Unfortunately it did not survive him. One night, around 1800, thieves broke into his house while Grimaldi was attending a rehearsal at Sadler's Wells. They soon found the cabinet, with its drawers of jewelled butterflies, inside a closet, and, perhaps because it was too heavy to remove, they contented themselves with smashing it to pieces, with, as Dickens puts it, 'the most heartless cruelty, and ... absence of all taste for scientific pursuits'. Only a single small box survived the vandalism, which encompassed not only the collection but even Grimaldi's drawings, paintings and models. This calamity seems to have broken Grimaldi's heart. He collected together his nets and cases, and the only box which was not destroyed, and gave them all away. It would, as Dickens comments, have taken years to replace the collection, and 'it would have cost at least £200 to have replaced them by purchase' – suggesting, incidentally, that entomology was a popular enough hobby by that time to be serviced by commercial dealers. And so Grimaldi gave up his beloved Blues and took up pigeon-rearing instead. A sad story, but one that illuminates how the quest for butterflies already attracted people from unlikely walks of life, and also the extraordinary lengths some were prepared to go to to secure their specimens. What a pity Dickens did not make use of it in one of his novels!

Fig. 52. A portrait of William Spence, co-author of *Introduction to Entomology*, engraved by W. T. Fry from a painting by J. J. Masquerier.

26. WILLIAM SPENCE (1783–1859) (Fig. 52)

William Spence is best known for his collaboration with William Kirby in the celebrated *Introduction to Entomology* (1815–26). He had begun corresponding with Kirby in about 1805, and the latter soon became very impressed with his youthful admirer who, 'so far from falling into the errors usual with beginners, determined his species with the judgement and precision of the most experienced naturalist'. The two men became the closest of friends. It was Spence who first broached the idea of collaborating on a textbook on entomology and who, after Kirby's agreement, set out a work-plan. The preparatory work took seven years until the first part of the four-volume *Introduction* was published in 1815, and it took them a further eleven years to complete it. Three years after that, the by now middle-aged Spence was taken ill from exhaustion and advised to take a complete break from entomology – that is, to lock up his books and collection, and take a long rest-cure.

Little is known of Spence's background and early life. He was born in Hull, like his distinguished contemporary, A. D. Haworth. At the age of ten he was in the care of a clergyman interested in botany who, Spence recalled later, led him 'from mere boyish imitation to collect and dry plants and to copy out the names of the Linnaean classes and orders'. This was the sole extent of his then botanical acquirements, he wrote, 'which were wholly interrupted by going to another school; and for the next seven years or eight years I never looked at a plant'. It was not, apparently, until his twenty-second year that he became seriously interested in entomology.

After the publication of the *Introduction to Entomology* Spence became a widely fêted entomologist. In 1834, he was made a Fellow of the Royal Society. He was also elected an 'Honorary English Member' of the Entomological Society, a unique honour, and was President of the Society in 1847–48, as well as serving it on its councils and committees. When in London, he was a regular attendant at the Society's meetings, even in old age and afflicted by deafness. His library was given to the Society by his son in 1884.

27. RICHARD WEAVER
(*fl.* 1790–1860)

The name of Richard Weaver is most often associated nowadays with a butterfly, *Boloria dia*, the so-called Weaver's Fritillary, which is not, and probably never has been, native to Britain. Weaver's much greater claims to fame – his explorations of Scotland and Wales, his discovery of a number of British insects and his remarkable private museum in Birmingham – are, by comparison, almost forgotten.

Weaver's eponymous Fritillary first turned up in Loudon's *Magazine of Natural History* in 1832, in a 'Notice of some singular Varieties of Papilionidae in Mr Weaver's Museum, Birmingham' contributed by the Revd W. T. Bree of Allesley Rectory. Both the upper- and undersides of the butterfly are very accurately figured, with the following note: 'a very interesting insect, allied to *Melitaea selene* [Small Pearl-bordered Fritillary]; of which perhaps, it may be only a variety. Mr Weaver possesses two specimens, both of which were taken in Sutton Park: one about ten years ago; the other, not more than five or six' (see Chapter 4: 29; Fig. 140; and Plate 15, figure 4). After a description, Bree invited 'the attention of entomologists to this insect, which deserves minute investigation. As Mr Weaver has taken two examples of it in the same place, in different and distant seasons, it may possibly prove a distinct species'.[47]

It did indeed. The specimens were taken to London and identified as *Melitaea* [*Boloria*] *dia*, a Continental butterfly known from the Alps and other places since the time of Linnaeus. The question was whether Weaver's specimens were genuinely British. Challenged about it publicly in 1842, Weaver denied any possibility of having mixed foreign specimens with his British ones, as had been suggested: 'No one can condemn that practice more strongly than myself,' he wrote to *The Zoologist*. 'Yet I really think that there is little justice or good feeling in accusing, or even suspecting, an entomologist of such unpardonable negligence, without something more than surmise of the fact. It has been well observed by an eminent entomologist with whom I have the honour of occasionally corresponding, "that if we disbelieve the existence of all which we do not ourselves see alive, our list (of British insects) will be small indeed". Now, I can very confidently state that, until long after the time when the two specimens of M. dia fell into my hands, I had never bought, nor exchanged, any insects; and that I possessed not one single foreign specimen: so that I could not possibly have confounded those of my capture with others. ... I must have taken them within ten miles of Birmingham: that being the utmost extent to which I had then travelled in search of insects'.[48]

'*Must* have taken them' [my italics] made it clear that Weaver could not remember the actual details of where and when he had taken them, but earlier in the same letter he had written 'I have the pleasure of positively stating that I had the good fortune to capture two specimens of that new insect in one of my early entomological excursions, more than twenty years ago'. In the 1820s, data labels were rarely attached to cabinet specimens so that the provenance would inevitably depend on the collector's memory or an entry in his notebooks. The result of Weaver's imperfect memory, according to Allan,[49] was disbelief, and 'he allowed a storm of discredit, abuse even, to break upon his head'. This was not helped by the release of this butterfly by fraudulent dealers later on, in the hope of passing it off as a newly discovered (and hence valuable) British butterfly. Whether Weaver's specimens were indeed captured in Birmingham in the 1820s, suggesting that the species was temporarily resident, or whether they were added to a case of native insects by someone else, will never be known, but we can surely acquit Weaver of deliberate falsehood.

Weaver's more solid contributions to British entomology were great. From the late 1820s until well into the 1850s, he explored localities like Aviemore and the Black Wood of Rannoch on behalf of entomological clients, discovering hitherto unknown British insects such as the Cousin German and Northern Arches moths (*Paradiasia sobrina* and *Apamea zeta assimilis*) and the Northern Emerald dragonfly (*Somatochlora arctica*). In 1827, he was also the first to capture female specimens of the Small Mountain Ringlet (*Erebia epiphron*) in its English localities of Cumberland and Westmorland (Plate 7), and in 1844 he discovered it in Scotland, in the hills near Rannoch.[50] Some of the specimens he captured on these excursions still survive in the Dale Collection at Oxford.

Plate 7 *(facing page)* Richard Weaver, the professional collector, discovered many species of Lepidoptera new to the British Isles. Among butterflies, he was the first to capture a female of the newly-found Small Mountain Ringlet in the Lake District in August 1827, and later discovered this species in Scotland in 1844. J. C. Stephens was the first to depict it, in December 1827, on this plate from Volume 1 of his *Illustrations of British Entomology*, under the name *Hipparchia cassiope*. Here both sexes are illustrated, the female as figure 2, probably from specimens obtained by Weaver. Engraved by C. Wagstaff from drawings by C. M. Curtis.

Even more remarkable was Weaver's own Museum of Natural History, housed in the Institution Rooms at Temple Row, Birmingham, which a contemporary described as comparing favourably 'in extent and splendour with any provincial collection in the world'. In Loudon's *Magazine*, a friend of Weaver's, A. J. Wallace,[51] describes how the former had begun studying and collecting insects in about 1818 'chiefly from an innate love of science, and an admiration of the splendour and fitness of natural forms'. Without private means or the help of learned institutions, 'but by the aid of persevering industry alone', Weaver travelled far and wide in pursuit of insects, and eventually succeeded in amassing a collection of some 5,000 species, scientifically arranged in cases and open to public view. Word got round of this unexpected gem in what was then a provincial town, and the Museum 'received great numbers of visitors from distant parts of the United Kingdom; and by noblemen and foreigners'. With the promise of help and royal patronage if he expanded the Museum, Weaver began to branch out into stuffed birds, fossils and minerals, and shells, housed in a lofty gallery, fifty feet long and lit by five lights on each side. The butterflies, home produced and foreign, were displayed in glass cases down the middle. It was, according to one contemporary, 'beautifully arranged' and 'one of the best and most extensive of any to which the public have ready access ... I would strongly recommend readers to pay a personal

London, Published by I. F. Stephens, 1 Dec. 1827.

C. Wagstaff Sc.

visit ... which, if I mistake not, will afford them a considerable treat'.

Such was the achievement of a comparatively poor but perseverant man, who, by a cruel stroke of fate, was destined to be forever associated with a dubious British butterfly. In his day, Weaver was rightly esteemed as one of the most industrious field entomologists who did much to unveil remote corners of Britain and display the treasures of a lifetime to the public. But whatever, one wonders, happened to his Museum?

28. LÆTITIA JERMYN (1788–1848)

In Jane Austen's England, fathers considered that their daughters should practise only the feminine virtues – 'Euclid, politics and dead languages were not for them'. It was agreed, however, that 'the female mind, so far as its capabilities go, should [nevertheless] be exercised' Hence, Lætitia Jermyn, the daughter of George Jermyn, printer and bookseller at the Butter Market in Ipswich, was allowed to exercise hers by botanizing and chasing butterflies, an interest encouraged by her influential neighbour at Barham, the Revd William Kirby.

Lætitia Jermyn, who liked to describe herself as the 'Fair Aurelian', is remembered chiefly for her charming book, *The Butterfly Collector's Vade Mecum*, dedicated to Kirby and published in 1824 by her stepfather, John Raw,

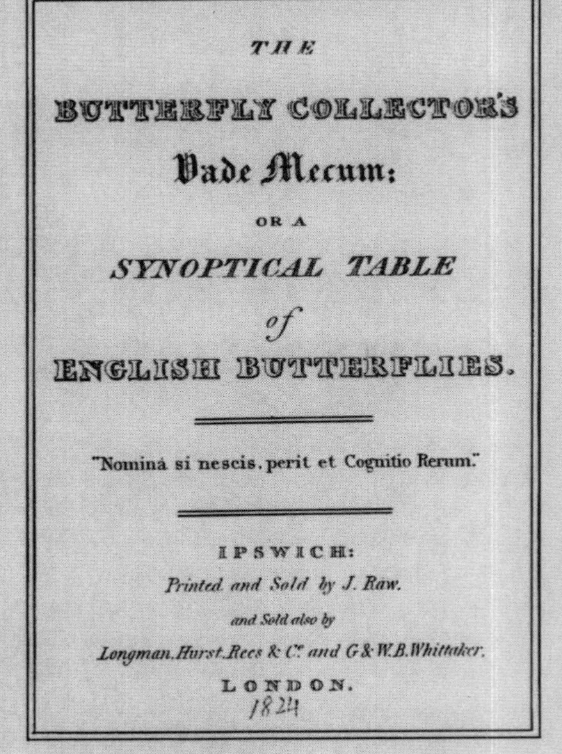

THE

BUTTERFLY COLLECTOR'S

Vade Mecum:

OR A

SYNOPTICAL TABLE

of

ENGLISH BUTTERFLIES.

"Nominá si nescis. perit et Cognitio Rerum."

IPSWICH:

Printed and Sold by J. Raw.

and Sold also by

Longman. Hurst. Rees & Co. and G & W. B. Whittaker.

LONDON.

1824

Fig. 53. There is no known portrait of Lætitia Jermyn, 'the Fair Aurelian' as she called herself, nor did her name appear on the title page of her famous work. However, at the foot of this charming frontispiece of a Swallowtail butterfly with its caterpillar and chrysalis appear the words '*Lætitia Jermyn, delin.*'.

who had taken over her father's business on his death in 1799. Although the author's name is not printed on the title page, the initials 'L. J.' and the place and date of publication are printed after the preface, and the frontispiece, depicting a Swallowtail butterfly with its larva and chrysalis, is clearly shown as the work of the author herself (Fig. 53). In it she spiritedly defends butterfly collecting 'against the scorn of those who attack the study of natural history as a trifling and worthless pursuit'.[52] Her 'synoptical table' provided a great deal of useful information on larval foodplants, times of appearance and localities, especially in Suffolk and Essex, which in those days were among the most heavily 'entomologized' parts of Britain.

Lætitia Jermyn married a local clergyman, James Ford of Navestock, a friend of long standing, at the then relatively late age of forty. He has been described as an ardent bibliophile, studious, punctilious, meditative, but also rather pompous and intolerant. One Sunday, Lætitia had been so engrossed in her collecting that she was late for the morning service. Her husband paused in the delivery of his sermon to lean over the pulpit and rebuke his wife with the words: 'I wonder where you will be Madam when the last trumpet sounds?'.

Both Lætitia and her husband died within a few months of one another, in 1848, and were buried in Navestock Church. A mural tablet in the nave commemorates Lætitia Jermyn under the Shakespearian-sounding name of 'Mistress Ford'. In her one all-too-brief book, she comes across as a lively, engaging personality, and an individualist who brightened the scene at a time when ladies were expected to conform to the social stereotype.

29. GEORGE SAMOUELLE (*c.*1790–1846)

George Samouelle has escaped the notice of biographers, and consequently very little is known about him save that he was originally a bookseller in the firm of Messrs Longmans & Co., and that he amassed a large collection of Lepidoptera. Today he is remembered mainly for his concise and often lively *Entomologist's Useful Compendium*, published in 1819 and full of sound advice. Among the accounts of the Linnaean system, collecting and preserving, and tips on fieldwork, it contains some sharply phrased observations that sound like personal experience. For example, Samouelle warns the collector that 'hedges in dusty roads are seldom productive', and that those considering a day's collecting at Norwood should beware of inconsiderate gamekeepers who will 'frequently interrupt and warn the unoffending entomologist to quit the wood immediately'. He castigates the 'vile practice' of repairing broken specimens with parts from other insects, and urges every entomologist to keep an exact journal recording his finds, together with 'the place, food, times of appearance etc'. The *Compendium* acknowledges the help of the distinguished zoologist, W. E. Leach (1790–1836), Superintendent of the Zoological Collection at the British Museum. It was a useful and successful contribution to entomology, though destined to be eclipsed by the mighty reputation of Kirby and Spence's *Introduction to Entomology*.

In 1826, Samouelle followed up his success with *General Directions for collecting and Preserving Exotic Insects and Crustacea*, which also sold reasonably well, and in 1832 published the first of a proposed periodical for collectors, *The Entomological Cabinet*. Only two volumes were completed, however, apparently through insufficient support. An attempt to revive it four years later was no more successful. Samouelle's more lasting innovation was the Entomological

Club in 1826, which he instituted with the co-operation of Edward Newman and two others, and which survives, as an exclusive dining club, to this day. The Club was always kept small, cosy and exclusive, the original idea being that each member would entertain the others on one evening per month. Newman's *Entomological Magazine*, launched in 1832, was effectively the Club's own journal.

In 1821, Samouelle had been appointed Assistant in the Department of Natural History in the British Museum, through the good offices of Dr Leach, where he was responsible for the care and arrangement of the entomological collections. Twenty years later, however, he was sacked for a variety of misdemeanors: 'he had taken to drink, neglected his duties, addressed his superiors with insulting language, and spited his fellow worker Adam White ... by deliberately removing the registration numbers affixed to the specimens, thereby creating utter confusion!'.[53] It seems that museum work, which had also worn down the health of Leach, eventually became too much for him. Five years later, he was dead.

30. JOHN CURTIS (1791–1862) (Fig. 54)

John Curtis was one of the finest insect illustrators of the nineteenth century – probably only Noel Humphries was his equal, and that in talent rather than industry. His sixteen volumes of *British Entomology*, with their 770 exquisite colour plates, are by any estimation one of the milestones of entomological publishing (see Plates 13 and 16). Among the many new species of moths and butterflies it introduced were the Black Hairstreak, in 1829, and the Lulworth Skipper, in 1833. These were great days for the travelling entomologist. In 1825, for example, Curtis and J. C. Dale had made a tour of Scotland, then hardly known entomologically, and discovered no fewer than thirty unknown species. In June 1827, they visited the Lake District together and, at Ambleside in Westmorland, were able to confirm Thomas Stothard's discovery nineteen years earlier of the Small Mountain Ringlet (*Erebia epiphron*). Stothard had provided Dale with information about the locality but they could find only males, it being too early for the females to emerge. The following month, Richard Weaver (q.v.) obtained both males and females in Westmorland and Cumberland (see Plate 7). Later in life, Curtis extended these tours to the Continent, especially Italy, combining his entomology with botany and cultural sightseeing.

Fig. 54. John Curtis, one of the greatest entomological artists, who owed much to his lifelong friend and patron, J. C. Dale.

Curtis was the son of an engraver, and inherited his father's artistic abilities. At school in Norwich he was given painting lessons by John Crome, and also discovered the joys of hunting Swallowtails on the Norfolk Broads. It is said that his eyes were first opened when his woolly bear caterpillar was transformed into a beautiful tiger-moth. Then and through his life, his interest in entomology ran hand-in-hand with a love of botany and meticulously detailed artwork; the resultant eyestrain, it was said, caused him to go blind in his last years. His artistic trademark was the depiction of insects on or by their corresponding foodplant (though sometimes straying from that ideal, as when he showed the newly discovered Blomer's Rivulet moth (*Discoloxia blomeri*) next to a Slipper Orchid!).

Having completed his schooling by the age of sixteen, Curtis unenthusiastically joined a firm of Norwich solicitors. Fortunately he was rescued from drudgery by a wealthy naturalist, Simon Wilkin, who offered him a home and much more pleasing employment as curator of his private museum. Through Wilkin, Curtis met William Kirby, and as a result some of Curtis's engravings were used to illustrate Kirby and Spence's *Introduction to Entomology*, his first published work.

In 1819, Curtis moved to London to be closer to the centre of the scientific world, and began to earn a living as an illustrator, first of shells and flowers, and later of insects. His work on *British Entomology* (issued in parts from 1824 to 1839) soon absorbed most of his time, especially after deciding to change its emphasis from genera to species, prompted by J. F. Stephens's parallel work on *Illustrations of British Entomology*. In 1829, Curtis published a *Guide to an Arrangement of British Insects*, essentially a label list and catalogue of all known British and Irish insects, again undercutting Stephens's own (and better) *Systematic Catalogue* of that year. After the completion of the last number of *British Entomology*, Curtis turned to economic entomology as an editor of *The Gardener's Chronicle* – an important post in which he was succeeded by J. O. Westwood. He also wrote some sixteen reports on injurious insects for the Royal Agricultural Society's journal, and numerous articles in *The Entomological Magazine* and the Annals of the Zoological Society and Linnean Society. On his retirement in 1857, Curtis was granted a pension from the Civil List of £100, a sum increased by £50 after the government had heard he had gone blind.

Curtis was a reserved man. An old friend who knew him well, a Mr Halliday, found him 'very loveable' and warm-hearted, though 'too much so perhaps for his own tranquillity'. His feud with Stephens, in which Curtis has to take most of the blame, suggests a certain professional jealousy. He was a punctilious and methodical man, rarely missing his publishers' deadlines. His library and collection were, like his pictures, a 'model of the greatest neatness and order'. The latter was sold after his death, and shipped halfway round the world to Melbourne. On one occasion, as was the routine, an engraving based on a Curtis drawing was shown to the great man for his approval. Curtis examined it carefully for some time, in silence. Finally he turned to the engraver and remarked: 'Sir, you have only put twelve hairs upon this fly's tail instead of thirteen!'[54]

31. JAMES CHARLES DALE (1792–1872) (Fig. 55)

J. C. Dale, the wealthy squire of Glanville's Wootton (Plate 8), which lies between Sherborne and Dorchester in Dorset, was one of the most assiduous entomological note-takers, letter-writers and diarists of all time. Among the treasures of the Hope Collections in Oxford are Dale's insect cabinets and boxes, still kept separately from the general collections, together with some 50 notebooks containing his journal, kept almost daily throughout his adult life, and over 5,000 letters, mostly to entomologists, as well as valuable manuscripts, catalogues and journals from early entomologists like Charles Abbot and Captain Charles

Plate 8
The Manor House, Glanville's Wootton, home of the Dale family and a popular meeting place for entomologists visiting Dorset and the New Forest during the mid-nineteenth century, photographed in 1976.

Blomer. It is among the richest historical legacies left by any British entomologist.

Dale was born at Iwerne Minster, Dorset, the son of James Dale of Glanville's Wootton. He was an educated man, who attended Wimborne Grammar School before graduating at Sidney Sussex College, Cambridge, (whence he collected in the Fens, discovering, among other things, the now extinct Reed Tussock moth, *Laelia coenosa*). Most of Dale's subsequent life, however, was devoted to managing the broad acres of the family estate in Dorset, and observing his social duties as a magistrate – at one of his sessions, some wag, knowing Dale's interest in matters entomological, released a mass of butterflies into the court.

Fig. 55. J. C. Dale, the tireless squire of Glanville's Wootton, 'thought nothing of riding to Lulworth and back, a round journey of more than forty miles.'

This left him plenty of time for fieldwork. Dale was a horseback naturalist (as far as we know, most of the early entomologists walked), often journeying forty or more miles a day to visit interesting spots along the coast, or on the downs or in the Blackmore Vale. In the 1830s, Dorset was virtually unknown entomologically – most of Dale's first records are new to the county – and he added many species new to Britain, as well as some that were new to science. His most famous discovery was the Lulworth Skipper, which he took for the first time on August 15th, 1832 at Durdle Dor, near Lulworth Cove, during one of his forty-mile rides from Glanville's Wootton.

The entries in his journal sometimes allow us to build up quite a detailed picture of the fluctuating abundance of certain species. The Mazarine Blue, for example, occurred near Glanville's Wootton (as, before 1816, did the Swallowtail). First found by the young Dale in 1808, it was very scarce in 1816, common in 1819, and again in 1825 and 1835. In the following very wet summer, however, only one was seen, then 'a few' in 1837, none in 1838, 'scarce' in 1839 and 1840, and only a single pair, the last recorded alive in the county, was seen in 1841.[55] Dale also received a few from near Ringwood in Hampshire. We learn that the butterfly was rarely seen much before June 10th, and that it continued on the wing until early August, by which time most were worn.

The student of everyday nineteenth-century life is given a feast of details, including the exact price Dale paid for everything: a bottle of brandy (a frequent purchase) 6s. 4d.; railway ticket for the family from London to Glanville's Wootton £3. 12s. 0d. (plus cab 5s. and turnpike 2d.); a week's lodging with sitting room and four bedrooms £1. 2s. 0d.; a punnet of strawberries 7d. More to the point, two small fly-nets cost him two shillings, a satchel, 3s. 6d., while a packet of pins proved surprisingly expensive at three shillings. Dale kept this up, together with lists of the most interesting insects taken, from the age of seventeen until, quite literally, the day he died. The last entry reads: 'Took *G. leucophaea* and female *Torticollis*', after which his son added: 'His last writing. Feb 6th – J. C. Dale died, thus at last his diary comes to an end after a brilliant career of 64 years. His end was peace. Aged 80 years and 2 months. C. W. Dale.'[56]

His letters, too, show us aspects of the great names of nineteenth-century entomology which normally escape attention. We learn, for example, that Stainton's 'at home' entertainments included champagne. Dale's son recalled a ridiculous 'I saw it first' squabble over a specimen with the great William Kirby. Dale spotted a Lobster moth, then regarded as rare, and was boxing it when Kirby claimed he saw it first. 'Take it', said Dale. Shortly afterwards, the latter captured something even rarer, the first British specimen of the large hornet-mimicking cranefly, *Ctenophora ornata.* 'Now Dale, I change with you' offered Kirby. 'No' replied Dale. 'The Lobster moth I may take a future day, but the fly, never.' This is, of course, also a reminder that Dale was as much a Dipterist as a Lepidopterist; indeed he collected all orders of insects.

Dale's funeral at Glanville's Wootton was well-attended, and his coffin, made from local oak and bearing a cross of snowdrops and poinsettias, was borne to the family plot by eight estate workers. The manor, and Dale's entomological mantle, passed to his son, Charles William Dale, who wrote the *Lepidoptera of Dorsetshire* (1886), based on his father's collection, articles and journals. He was not, however, the meticulous worker his father was, and was 'notoriously careless in identification and transcription of records' (and with 'abominable handwriting' to boot).[57] It was through the unmeticulous son, however, that Dale's collection and manuscripts were preserved, and ultimately left to the Hope Department in 1906, where they reside to this day, among the most important historical collections we have.

32. JAMES FRANCIS STEPHENS (1792–1852) (Fig. 56)

At the age of sixteen, James Francis Stephens decided to compile a list of all the known British animals. Four years later he had produced a 'catalogue' of 3,673 indigenous species of insects – all that were known at that period – including 1,367 Lepidoptera. However, such was the rate of entomological discoveries at this time, that over the next ten years the list more than doubled. By the time his famous *Systematic Catalogue of British Insects* was finally published in 1829, it contained no fewer than 10,116 species, two-thirds of which Stephens had collected himself. It was a remarkable achievement, especially as Stephens did it with little or no outside help. There were at that date, wrote Stephens many years later, 'not half-a-dozen collections named in England, and those only partially so'. The available literature in English was hardly better, and Stephens 'had to ferret out the names as best [he] could, frequently from works in the Teutonic and other European languages'.[58]

Fig. 56. An early daguerreotype photograph of J. F. Stephens, author of *A Systematic Catalogue of British Insects*, the only known portrait.

Stephens was born at Shoreham, Sussex, the son of a naval officer, and educated at The Bluecoat School, Hertford. Aged 15, he entered the Admiralty Office at Somerset House on the recommendation of his uncle, and worked for some years as a clerk. In 1818, however, he was given leave of absence for a much more congenial occupation, to help arrange the insect collections at the British Museum. This work gave him the impetus to produce the first major illustrated work on British insects since Donovan. The first volume of *Illustrations of British Entomology*, published in parts, was completed in 1828, and a further eleven volumes followed over the next eighteen years, four of them devoted to the Lepidoptera alone. Stephens commissioned several artists to draw the plates, among them the young J. O. Westwood, and species new to Britain were illustrated (see

Plate 8). Completing it proved too much, even for Stephens, and the work excludes the Diptera, Hemiptera and much of the Hymenoptera. Many of his descriptions are faulty, but even so it was by far the most complete work of descriptive entomology yet undertaken, and especially good on the habits of species. Unlike most other authors up to that time, Stephens eschewed English names, using only Linnaean scientific names. Initially the *Illustrations* brought Stephens at least as much trouble as acclaim. In 1832 he brought a lawsuit against James Rennie who, it was alleged, had plagiarized Stephens's work for his own *Conspectus of the Butterflies and Moths found in Britain.* Stephens lost the case and had to bear the legal costs, 'an affair,' wrote Edward Newman afterwards, which 'reflects anything but credit on the laws of this country'. He was saved from bankruptcy by a fund organized by Newman and other sympathizers. A year later, when Stephens's rival, John Curtis, rubbed salt in his wounds with a critical remark in his *British Entomology*, Newman again sprang to his defence with a tremendous rebuke, worthy of Cicero, concluding that 'We fear Mr Curtis will find he had better, far better have committed the whole copy of the tainted number to the flames than have ventured to risk it on the excited wave of public opinion'.[59] In truth this celebrated case divided the entomological world, although the majority sympathized with Stephens's plight.

His ill-luck was compounded on Stephens's return to the Admiralty. Apparently he encountered animosity from some of his seniors who had taken a dim view of his prolonged absence and the publicity resulting from his lawsuit. Stephens eventually resigned. He returned to the British Museum and worked there as an unpaid assistant until his death, in his sixty-first year.

Stephens was active as a Fellow of the Linnean Society and a co-founder with Kirby, Haworth and others of its Zoological Club, intended to replace the various defunct entomological societies. Like its predecessors, it too gave way to another with the formation of the Entomological Society of which he was also a founder member and later President.

Although Stephens adopted many new techniques, he steadfastedly refused to use a microscope. 'What you cannot see with a pocket lens', he used to say, 'would not be helped by a microscope'. He was, however, the first to publicize the effectiveness as a killing agent of chopped and bruised laurel leaves, which emit prussic acid gas. His rich collection was purchased by the Trustees of the British Museum and his fine library was acquired by Stainton, who maintained Stephens's practice of keeping open house for entomologists on Wednesday evenings, and making the library available to them.[60] Stainton's own library, including Stephens's books, was donated after his death to the Entomological Society.

33. JAMES RENNIE (1787–1867)

James Rennie is remembered mainly for the celebrated lawsuit which J. F. Stephens brought against him, alleging that he had pirated Stephens's *Illustrations* for his own *Conspectus of the Butterflies and Moths* published in 1832. Though Rennie won the case, the subscription launched by fellow entomologists to help Stephens pay the legal bill showed where their sympathies lay.

Rennie was a Scot, born at Sorn in Ayrshire and educated at Glasgow University, where he gained an M.A. in 1815. The *Conspectus*, his best-known published work, is a curious little volume, unillustrated apart from one engraving.

Rennie's nomenclature was idiosyncratic to say the least, using 'scientific names' ignored by most authors, and replacing many familiar English names with dreadful ones of his own, such as The Navew for the Green-veined White and The Turnip for the Small White. Fortunately they did not catch on.

Rennie was author of several popular works on insects, such as *Insect Architecture* and *Insect Transformations*, and was for some years Professor of Natural History at King's College, London, until the chair was abolished in 1834 owing to a dearth of students. Later he emigrated to Australia, to take up a new chair at Sydney, New South Wales. He died at the advanced age of eighty-one, long gone from English shores but not entirely forgotten. As his obituarist in *The Entomologist's Monthly Magazine* remarked, 'there are yet living those to whom the [Stephens vs. Rennie] litigation was a cause of much excitement and regret'.[61]

34. ABEL INGPEN (1796–1854)

What we know of Abel Ingpen's professional life could be summed up in a sentence: he was appointed a solicitor's clerk in Chelsea at the age of seventeen, and remained one until the day he died – of cholera – forty years later. More interesting is what he did with his free time. Ingpen was a keen gardener and contributed articles to the *Horticultural Magazine*; he was also a keen fisherman and, above all, he was of course a keen entomologist as well as being a founder member of the Entomological Society in 1833. In 1827 he published the work that established his reputation, *Instructions for Collecting, Rearing and Preserving British Insects*, subtitled, in the leisurely manner then fashionable, 'also for Collecting and Preserving British Crustacea & Shells; together with a description of Entomological Apparatus: to which is added, a List of New and Rare Species of Insects, etc with their Localities and times of Appearance'. The work was 'Intended for Collectors, and Residents in the Country'. Ingpen dedicated 'this trifle', as he called it, to James Stephens. The *Instructions* contains a great deal of useful information and is an important source for what we know about collecting methods in the first half of the nineteenth century. Butterflies were still being killed by pinching the thorax – 'Needles must never be used for piercing insects, as they always rust', counselled Ingpen. He used a curiously designed pocket net with a short handle for plucking resting moths from tree trunks or palings. For butterflies and other flying insects, Ingpen recommended 'a bag-net made of green gauze from two to three feet long' fitted to a circular hoop borrowed from a water net. When fixed to a pole twenty or thirty feet long such a net could bag even a Purple Emperor 'whose residence is generally on or near the tops of oaks'.

Ingpen's wooden storeboxes were made 'on the principle of backgammon boards, of about twelve by eighteen inches long, and four inches deep, and lined with cork at the top and bottom, and made to fasten close'. To capture tiny insects, Ingpen recommended using quills – stoppered at one end with cork and wax, and at the other end with a cork – for glass tubes were yet to come. The quills should 'be shaken every time a fresh insect is put in, to preserve those already captured'. Another of his ideas was a kind of miner's lamp fastened by straps to one's hat (the tall hat of the day): 'The light will be found to attract insects; and, it is obvious, render it easy for them to be secured' – though perhaps at some danger to one's hat catching fire.

Like others, the honest Ingpen was misled by fraudulent dealers into mentioning species which were not, or only doubtfully, British, among them the Scarce Swallowtail (*Iphiclides podalirius*), allegedly taken in the New Forest, and the 'Scarce Heath' (*Coenonympha hero*) from Ashdown Forest, where, no doubt not by coincidence, the notorious dealer Plastead lived (see Chapter 4).

35. FREDERICK WILLIAM HOPE (1797–1862) (Fig. 57)

The Reverend Frederick William Hope, whose deed of gift established the Hope Department and Entomological Collections at the University Museum, Oxford, was a slightly built man of delicate health but with plenty of money and a pretty wife (Ellen Meredith, who had, incidentally, rejected an offer of marriage from Benjamin Disraeli). He was born in the family house at No. 37 Upper Seymour Street, London, the second son of John Thomas Hope of Netley Hall in Shropshire. Educated privately, he graduated at Christ Church College, Oxford, and was 'presented with the family living' – a curacy at Frodesley, Shropshire. But, although the life might have suited Hope temperamentally, ill health forced him to abandon it. Instead, he concentrated his energies on his private museum at Upper Seymour Street. This was opened to the public on certain days, and became a meeting place for naturalists. Hope was a generous host, and the evening meetings there are said to have been convivial affairs, and 'a delight to all who attended' them. The young Charles Darwin was a regular caller, and he and Hope went on a collecting trip together through North Wales in 1829. Although there was only a dozen years' difference in their ages, Darwin referred to Hope as 'my father in entomology' and to himself as his 'disciple'.

Fig. 57. A drawing of the Revd F. W. Hope, the future benefactor of Oxford University, as a young man.

Hope's entomological collections came both from Britain and around the world. His own collecting was necessarily restricted, but he was among the few to have caught a Scarce Swallowtail in Britain,[62] close to the family seat at Netley in 1822 – evidently this butterfly had established itself there for a number of years, though whether as a result of an introduction or by natural means may never be known (for the full story, see Chapter 4: 7). In the 1830s, Hope travelled extensively in Europe, collecting insects, fishes, shells and other natural objects. In later life, however, he was obliged to restrict himself to sales catalogues, marking what he wanted and sending a representative to purchase the items.

The wealthy and influential Hope was a major entomological institution-builder. He helped to found the Zoological Society in 1826 and the Entomological Society of London in 1833, serving the latter as its first Treasurer, and as its Vice-President and President three times each between the years 1833 and 1847. For reasons that are unknown, but probably connected with his poor health and the consequent need to spend much of the year living on the Mediterranean, he and his wife and another member of his family all severed their links with the Society in that year. He was honoured by learned societies in Britain and Europe, and elected Fellow of the Royal Society in 1834. His greatest contribution, however, was to his *alma mater*, the University of Oxford. Hope had apparently discussed with his Oxford mentor, Dr John Kidd, the possibility of leaving his collections to the University. The arrangements took years to finalize, but a Deed of Gift was eventually accepted in 1850; the story is well told by Audrey Smith.[63]

Hope, however, did not leave it at that. By 1856, he was taking steps to found a new chair in the University, to be called the Hope Chair of Zoology. The department would be based at the new University Museum whence the Hope collections and library would also be transferred. Hope, who had even agreed to pay for the furnishings of the allocated rooms, was again irritated by the usual delays and misunderstandings, at one point writing in despair, 'Adieu to all my dreams ... Pray do not write any more I cannot bear it. I shall never reach Oxford again. It is all up'. By 1860, however, it was all resolved. Hope would invest an endowment of £10,000 for a Hope Professor of Zoology to supervise the collections and 'give Public Lectures & Private Instruction on Zoology'. Hope's nominee, J. O. Westwood, who had curated the collection for some years, gave his inaugural lecture in April 1861.

The next year, Hope died, aged 65, leaving a further endowment to Oxford in his will. His portrait, presented by his widow, now hangs in the Department's Reading Room, and his name lives on in this world-famous entomological institution. Hope, the man, appears more faintly. He had wide-ranging intellectual interests, his favourite entomological topics including such things as the economy of silk, human parasites and insects mentioned in the Bible. His surviving letters show him to be a most careful and precise observer, and necessarily patient in practical matters. Nor did he lack a sense of humour. One of his notes in *The Entomologist's Weekly Intelligencer* ends '*spes nunquam fallit*' (Hope never fails).

36. EDWARD NEWMAN (1801–1876) (Fig. 58)

Edward Newman was among the most multi-talented of the 'Aurelians' in this chapter. Entomologically speaking, his main interest was in Lepidoptera and injurious insects, but he was first and foremost an all-round naturalist and writer, as well as an authority on botany – especially ferns, birds and their nests, and mammals. Many of his wryly humorous pieces, closer to literature than science, must have helped to create a sense of social cohesion among entomologists (albeit a sometimes fractious one). He could write or parody verse and draw recognizable caricatures with a few strokes of his pen. Like his contemporary and one-time neighbour, H. T. Stainton, Newman was one of the great advancers of natural science, through his popular books on ferns and butterflies, his editorship of journals and his active service in some of the leading London societies. He had taken 'the lamp of knowledge from the hands of his forerunners' wrote his friend Dr Bowerbank, 'and passed it on brighter and better trimmed to those of his successors'.

Fig. 58. Edward Newman, of whom it was written: 'Few naturalists have done so much ... to diffuse a taste for the Natural History of his country'.

Newman was born in Hampstead, the son of Quaker parents. They were keen naturalists, and the boy was brought up on books like Bewick's *Birds* and Gilbert White's *Natural History of Selborne*. At boarding school, at Painswick in Gloucestershire, he was further encouraged to study all aspects of the natural world by a Quaker master. From the first though, Lepidoptera were among Newman's foremost passions: 'From my earliest years I had been a hunter of butterflies', he recalled. 'I may say even from my nurse's arms.' Newman had to leave school at the age of sixteen to join his father in his small business (he was a manufacturer of morocco leather before moving to Guildford to become a wool stapler). Newman worked with deep unenthusiasm, often absenting himself to tramp the Surrey lanes and fields with a gun or net. Many of these wanderings were later written up anonymously for the *Magazine of Natural History* or *The Entomological Magazine* as 'Letters of Rusticus'.

In 1826, Newman moved to Deptford, then still a rural part of Kent, to take over a rope-making business, though he left most of the work to a foreman. There he met many of the leading entomologists – friendship was another of Newman's talents – who included a fellow Quaker, Edward Doubleday, who became 'his nearest friend and second self'. Newman was a founder member of the Entomological Club, which often met at the Bull Inn at Birchwood, where Newman read his first scientific paper. In 1832, he was unanimously elected as editor of the Club's journal, *The Entomological Magazine*, and the following year he became a Fellow of the Linnean Society and one of the founder members of the Entomological Society of London. It was in *The Entomological Magazine* that some of Newman's most characteristically penned pieces appeared, notably 'Entomological Sapphics', professedly translations from Persian, Arabic or Greek (but which were, of course, his own work), 'Wanderings and Ponderings of an Insect Hunter' and 'Colloquia Entomologica' in which he purported to present the sayings and doings of some of his friends 'in a light and pleasant vein'. Edward Doubleday appeared as 'Erro' (a wanderer), Francis Walker as 'Ambulator', A. H. Davis as 'Venator' (a hunter) and Newman himself as 'Entomophilus', the lover of insects. Through these writings we can perhaps glimpse the camaraderie and shared sense of wonder that must have characterized the evenings spent at the Bull Inn, or at Newman's hospitable home.

Newman gave up his rope business in 1837, apparently in despair at his prospects. 'I am wholly without any definite prospect as regards business', he wrote, 'having entirely given up my own, which was a very small affair ... I am very indifferent as to any business engagement, as it is always so great a tie, and cannot be abandoned for any length of time without something like a dereliction of duty. Moreover', he added, mysteriously, 'I think that the opportunity for enjoying life will with me shortly expire, and I am desirous, while blest with strength and health, of visiting the country, and breathing the air of mountain-wilds unchecked by the necessity of returning on a certain day ... In the summer, if my life and health should continue, I propose wandering about the country'.[64] This melancholy attitude to life was short-lived, but it was to recur.

In fact this time marked an upward change in Newman's fortunes. He got married, and in 1840 published the modest first edition of what was to become a natural history classic, *A History of British Ferns*. He also discovered, at last, an occupation he could enjoy and combine with his natural history pursuits: he became a partner in a firm of London printers, Luxford & Co. (who changed their sign, a 'bouncing "B" ', so that it resembled a winged insect). Newman had found his metier as a printer and publisher of books on natural history and science. In his later years he also turned increasingly to writing and journalism, as natural history editor of *The Field*, editor of *The Zoologist*, and, after 1864, of *The Entomologist*. His two most significant works on the British Lepidoptera, the *Illustrated Natural History of British Moths* (1869) followed by its companion *Illustrated Natural History of British Butterflies* (1871), were written late in his career. The latter in particular has been much praised, and advances in printing using steel engravings meant that, unlike most of their predecessors, working people could afford to buy them. In all his writings, his son tells us in a fine memoir published in 1876, Newman 'became engrossed with his subject, as was always the case with everything he undertook'.[65]

Newman's last years were darkened by continuous ill-health and saddened by the death of most of his closest friends. That of 'Ambulator' Francis Walker in 1874 was 'a blow from which I can never hope to recover', he wrote to Jenner Weir. 'He was my right hand in all the *difficult* branches of Entomology; and more than that, he was a constant friend. I feel alone in the world now that he is gone – alone as regards literary labour.' Then, a year later, 'our poor friend', Henry Doubleday died. Newman himself followed in 1876, peacefully and 'without care or anxiety'. He is buried in Nunhead Cemetery.

The shiftless impression given by Newman's early years masks his immense energy. In summer, at least, he was said to rise regularly 'by five, four, or even three o'clock' to find 'uninterrupted quiet' to devote to his collections and writing. He loved mountainous country, tramping the hills of western Ireland and North Wales, and referring with some scorn to those 'brother entomologists of Cockney-land whose researches are confined to Copenhagen brick-fields, or the wilds of Battersea cabbage-fields'. Like several Victorian 'Aurelians', he was a keen cricketer, a member of two clubs, in which he excelled as a batsman, and regularly attended the big matches at Lords or the Oval. He had, very clearly, a well-developed dry sense of humour, and reputedly sometimes had the dignified members of the Entomological Society in stitches. He was also, to a degree, unselfish. As a Quaker, he took a great interest in parish affairs, and for many years spent a day a week curating and building up the Entomological Club collections. That this Dickensian energy and heartiness was combined with a streak of melancholy that occasionally prostrated him is also clear from his letters. All in all, there are few long-dead naturalists whose personality survives in print quite as well as Edward Newman's. He must have been excellent company.

37. JAMES DUNCAN (1804–1861)

There are rather too few truly Scottish 'Aurelians' in this chapter (the only other one being Francis Buchanan White) – fewer than Scotland's long and fine tradition of natural history study deserves. Several others were Scots by birth but lived elsewhere or, like E. C. Pelham-Clinton, were English but lived in Scotland. The reason is that not many Scottish naturalists were primarily lepidopterists, though one should not forget Professor John Walker (1731–1804), the eighteenth-century Edinburgh naturalist, whose unpublished notebooks on Scottish butterflies survive in the University Library (see Chapter 4: 34). Before George Thomson's excellent book, *The Butterflies of Scotland*, published in 1980, only two Scots had written books about British butterflies, Alexander Morrison Stewart[66] (1861–1948) and the Revd James Duncan.

James Duncan is undeservedly neglected. P. B. M. Allan, one of the few writers to show much familiarity with his work, referred to him as 'that excellent old Scottish entomologist'.[67] His interests were wide, and also included geology, botany and agriculture. Duncan wrote several early works on Scottish insects, especially butterflies, moths and beetles, and also contributed items of interest to the entomological journals, including P. J. Selby's experiments at sugaring. They all suggest he was a careful and discerning naturalist. Educated in Edinburgh, Duncan followed his father into the Scottish church, a not altogether happy choice of career. One obituarist, Sir William Elliott, suggested that he was 'a genius who mistook his calling' and that 'had he allowed free course to the original bent of his mind [i.e. for writing and natural history], he would probably have earned a higher reputation, and certainly would have achieved a greater amount of happiness.[68]

Duncan's first work, originally published by the Wernerian Natural History Society of Edinburgh (named in honour of the German geologist, A. G. Werner), was a list of insects found near Edinburgh, where records were available from as long ago as the previous century. In 1835, he wrote a useful volume on *The Natural History of British Butterflies* for Jardine's Naturalist's Library, which contains much interesting information on the distribution of Scottish butterflies at that date. Through Duncan we know, for example, that the Grizzled Skipper, Comma and Gatekeeper butterflies then occurred further north than they do today, and that the Large Tortoiseshell occurred here and there as far north as Dunkeld. Duncan was the first to recognize the Dark Green Fritillary in 'the middle and northern districts' of Scotland, and to show that the Large Heath was widespread there, following its belated discovery in 1833 by Curtis and Dale. He even saw an Apollo flying near the west coast during the summer of 1834. He followed up this work in 1843 with a companion volume on *British Moths* (among which he includes the skippers!) and an *Introduction to Entomology*.

He seems to have retired early from church duties, and worked for Edinburgh publishing firms, among other things preparing an index for *Encyclopaedia Britannica*.

38. JOHN OBADIAH WESTWOOD (1805–1893) (Fig. 59)

In his entertaining reminiscences of his good friend, the late Professor Westwood, Octavius Pickard-Cambridge related the following story, illustrative of the former's supposed lack of any sense of humour. The Professor was about to deliver a lecture, and was 'making a final disposition of his beautiful drawings and specimens'. But, by the allotted hour, no one had turned up. 'Half an hour passed; still no arrivals. But the Professor was hopeful (was he not *Hope* Professor? but such a horrible joke could not occur to *him*): "They will come presently; they are often rather late". A gentle knock is heard at the door at last. "Come in"; but no one coming in, the Professor goes to the door. "Is this Professor Westwood's lecture-room?" asks a little timid voice. "Yes, ma'am; we are all waiting." And the Professor returns, followed by a little, rather elderly, frightened-looking lady, who is duly placed in a front seat; whereupon, without moving a muscle of his countenance, the Professor begins, and goes through an excellent and interesting lecture, with this little old lady as his whole audience; for it was only by being employed in assisting him with his drawings and specimens that I could restrain myself from exploding at the absurdity of the whole thing. First and last the Professor was as serious as if the whole University were before him ... I was informed, later ... that the Professor's

Fig. 59. J. O. Westwood, first Hope Professor of Zoology at the University of Oxford, and nomenclator of '*Pulex Imperator*'. His net, spectacles and lecturing shoes are lovingly preserved in the Departmental library.

lectures were not infrequently attended (or rather *not* attended) as on [this] occasion.'[69]

J. Obadiah Westwood was born in Sheffield, the son of John Westwood, a medallist and die-maker, and educated at a Quaker school in that city, and later at the grammar school in Lichfield. At the age of sixteen, he was sent to London to be articled to a firm of solicitors, with the intention of forging a career in law. It was at about this time, however, that the young Westwood became passionately interested in entomology. In his words, he 'devoured' every text he could lay his hands on, haunted the London collecting grounds at Hackney Marsh, Battersea Fields and Coombe Hill, Kingston, and drank tea, named and exchanged specimens, and talked entomology with fellow enthusiasts. In 1827, at the early age of twenty-two, Westwood was elected a Fellow of the Linnean Society, and, four years later, abandoned his neglected legal practice to engage himself in entomological and antiquarian pursuits. To his abiding good fortune, his wife Eliza, whom he married in 1839, fully shared his many interests.

Westwood was a talented artist, and acquired a great reputation for his ability to draw rapidly but accurately, in a style that became known by some as 'Westwoodian'. By the age of eighteen he was already drawing specimens from the cabinets of leading lepidopterists of the day, such as A. H. Haworth. A few years later, J. F. Stephens engaged him to draw some of the plates for his *Illustrations of British Entomology*. In all, he produced thirteen plates of Lepidoptera, including one of butterflies, and many for the volumes of Mandibulata in the same work. He was also a prodigious and versatile writer, author of hundreds of papers and books on subjects as diverse as sessile crustacea, oriental insects, ancient inscriptions and manuscripts, and most orders of insects, as well as being 'entomological referee' for *The Gardener's Chronicle* for nearly half a century. His two-volume *Introduction to the Modern Classification of Insects* (1839–40) was an important and original work that gained him the Royal Society's gold medal in 1855 (though Westwood always refused election to the Society, possibly because he considered its views too advanced, and, after Darwin's *Origin of Species*, even seditious). His best known work today is the two-volume *British Butterflies and their Transformations* (1841), with its exquisite colour plates by H. N. Humphreys (see Plates 22, 27), followed a few years later by a companion volume on moths. Westwood's own delicate drawings and wood engravings illustrate his *Entomologist's Text Book* (1838), which provided an authoritative introduction for the general reader (in Germany it was known as 'the Entomologist's Bible').

Westwood was one of the founder members of the Entomological Society of London in 1833. He served as its Secretary between 1834 and 1847, and was elected President on no fewer than three occasions, in 1852, 1872 and 1876, followed, in 1883, by the conferment of Honorary Life President, an honour at that time received by only one other person, the Revd William Kirby. Westwood's long association with the Revd F. W. Hope also secured him the latter's nomination first as curator of the Hope collection, later as professor when the Hope Department of Zoology was established at Oxford in 1860. The University had procrastinated; Westwood was, after all, formally unqualified. But he was made an Honorary M.A. and Fellow of Magdalen College by decree, and finally offered the appointment in 1861.

The first Hope Professor was remembered by friends in a series of unusually vivid character sketches after his death. Under a brusque and argumentative manner, he 'concealed a hearty sympathy for all real workers'.[70] The walls of his rooms were furnished with 'quaint adages'.[71] Though educated at a Quakers' school, Westwood was a regular churchman with no sympathy for religious sects. When asked what sect he belonged to, he replied, 'Sir, I am an Insectarian'. He was apparently unable to pronounce the latter 'h'. A former pupil recalled one of his sayings thus: 'If I could 'op as 'igh as a flea – in comparison with our 'eights I could 'op to the top of 'Eadington 'ill in an 'op and an 'arf' – it sounds as though he retained his South Yorkshire accent, too.[72]

His parsimony was proverbial. His letters were generally enclosed in already used envelopes, and his drawings and writings were often made on the back of bills, circulars and the like. 'Waste not, want not' was Westwood's motto. By the same token, he was an accomplished mender of broken insects, as Pickard-Cambridge recalled: 'I have seen him with his little pot of dirty gum, a bit of an old match, two or three needles and pins, and a paper of the veriest *fragments* of an insect, and in a brief space of time the insect would appear built up in a most marvellous way, almost defying the power of any ordinary pocket-lens to discover that it had ever been otherwise ... I ventured to hint at the chances of some insects obtaining bits to which by nature they might not have been entitled, and so tending to confuse future entomologists. He repudiated the idea with scorn ...'.

Pickard-Cambridge thought Westwood's secret lay in an 'abundance of self-confidence'. When attending to his bee hives, he never wore protective clothing, relying instead on the fumigatory properties of his cigar. He once exhibited what he pronounced to be a gigantic flea, found in a bed at Gateshead, and named it *Pulex imperator*. The 'flea' turned out to be only a young cockroach, squashed flat by the bed's incumbent. Such a mistake might have crushed a lesser man, thought Robert McLachlan, but Westwood did not seem unduly embarrassed at this public demonstration of his fallibility 'though the name he imposed on the supposed flea clung to him for years'.

Westwood died aged 87, active almost to the last, though he 'had been in feeble health for some time, and we believe ... died from decay of nature'. His last visits to the University Museum were made in a Bath chair. He is buried in St Sepulchre's Cemetery at Jericho, Oxford, and a commemorative plaque is in St Andrew's Church, Sandford-on-Thames.[73] His cabinet, instruments, spectacles, and even his shoes are preserved in the library at the Hope Department, together with some of his amusing doodles – which suggest that J. Obadiah Westwood was by no means as devoid of humour as some of his friends supposed.

39. HENRY DOUBLEDAY (1808–1875) (Fig. 60)

40. EDWARD DOUBLEDAY (1811–1849) (Fig. 61)

The Doubleday brothers, the sons of an Epping grocer, are credited with an invention that transformed the nature of moth-hunting – 'sugaring'. Until the 1840s, moth-hunting had, more often than not, really meant hunting for larvae and rearing them on. Sugaring – painting a mixture of treacle, brown sugar and alcohol on tree trunks and other convenient places to tempt the appetites of adult moths – was seen, in Stainton's words, as 'a revolution in our cabinets'. Many species formerly considered rare, especially among the noctuids, were found in surprisingly large numbers by this method.

The origin of sugaring is contained in the famous 'Note' which the young Edward Doubleday sent to *The Entomological Magazine* in 1832.[74] The brothers had noticed that moths were attracted to empty sugar barrels in the yard

Fig. 60. Henry Doubleday, the Quaker of Epping, who, according to Edward Newman, knew 'more about Butterflies and Moths than all the other Entomologists in the Kingdom'.

Fig 61. Edward Doubleday, possibly the first to use 'sugar' to attract moths. He had the makings of a great entomologist, but died young.

behind their warehouse, and in his note Edward recommended rolling such barrels into 'mothy' places – scarcely a practical proposition. The note apparently caught the eye of a P. J. Selby of Twizel House, Bedford, who experimented further with honey and empty beehives. By 1841, Henry Doubleday had dispensed with barrels and beehives by simply 'brushing a mixture of sugar upon the bark of trees where moths are likely to abound'[75] and it was not long before entomologists discovered that fermenting fruit or a dash of old Jamaica increased its attractiveness still further. Much later, in 1881, when neither Doubleday was alive to contradict him, their assistant, James English, claimed the credit for himself, although his account, which was based on memory, is demonstrably inaccurate in some respects.[76] At any rate, the Doubledays were using the technique with success from 1842 onwards, as the records in Henry's journal attest, and it was from their information that H. N. Humphreys introduced it to a wider readership in his *British Moths and their Transformations.*

The Doubledays were Quakers, a sect which produced many fine naturalists in the nineteenth century. The family lived in a large brick house, formerly the Black Dog inn on Epping High Street (not the Black Boy, as formerly thought),[77] with Epping Forest, then a famous hunting ground for birds and insects, close by. The boys attended the local Quaker school and received the usual classical education of the day. They spent much of their childhood in the Forest, searching for larvae and hunting rare specimens with gun and net. They soon owned a fine collection of bird skins, insects, and pressed plants, as a school friend, Henry Deane remembered, years later.

Despite their closeness and shared interests – and the fact that both remained bachelors to the end of their days – the brothers were contrasting personalities. Henry, the elder, was chronically shy, and in later years lived as a virtual recluse. It was he who ran the family grocery and hardware business after the death of their father in 1847 – less than assiduously, since natural history generally had priority. He disliked travelling, and went abroad only once, to visit the great French entomologist Guenée, in Paris. Edward, on the other hand, was more adventurous and out-going. He was a member of the Entomological Club, those most sociable of entomologists, and a close friend of the convivial Edward Newman with whom he collected in Wales and the New Forest, and on the south coast. He made a two-year insect-collecting expedition to the United States in 1835, on his return from which he was appointed assistant in the British Museum, responsible for its collections of Lepidoptera. By 1848, he had completed the Museum's first entomological catalogue, *A List of Lepidopterous Insects in the British Museum.* That same year, he was made Secretary of the Entomological Society, a demanding job requiring tact and perseverance. For a while he also looked after the Entomological Club's collection, though he proved a most unsuitable curator, as Edward Newman recalled later in a letter to Jenner Weir: Doubleday's 'generous disposition' had reduced the collection to a skeleton – 'he gave and lent to everyone whatever they asked of him'.

It was Henry, however, who became famous. Edward Doubleday had been a frequent contributor to *The Entomological Magazine* and other journals, and was working on an important publication, *On the Genera of Diurnal Lepidoptera*, with beautiful coloured illustrations by W. C. Hewitson, when he contracted a spinal tumour and died in December 1849, aged only thirty-eight. The book was completed after his death by J. O. Westwood. Henry Doubleday found himself suddenly alone, with his brother and both parents dead. He had himself just completed the work that established his reputation, the *Synonymic List of the British Lepidoptera* (1847–50). In it, Henry Doubleday, with his friend Guenée's help, established a lasting system of generic and specific names for Lepidoptera (excluding the Tineina) that swept away many previous errors. Ten years later Doubleday produced an extended edition that covered all the microlepidoptera. With supplements in 1866 and 1873, it brought the number of recognized British species to some 2,100 – treble the number known at the start of the century. Doubleday himself named a great many new species, taken personally or by acquaintances, such as Pigmy Footman (*Eilema pygmaeola*), Ashworth's Rustic (*Xestia ashworthii*) and the Marsh Oblique-barred (*Hypenodes humidalis*). He is himself remembered in the name of a little tortricid, *Olethreutes doubledayana.* As far as Edward Newman was concerned, Henry Doubleday knew 'more of British Butterflies and Moths than all the other Entomologists in the Kingdom'. Moreover, he had 'acquired all his knowledge solely to gratify his ardent love of the science, and for the purposes of instructing others'. The modest Doubleday was, it seems, entirely without personal ambition, so long as he had a roof over his head and space to continue his natural history work.

Henry Doubleday's collection, perhaps the finest in the land at that time, still survives intact, and in the exact order he left it, at the Natural History Museum, London, where it is kept separately from the general collections. Doubleday's setting of both macros and micros is of a standard that prefigures the great collections of late Victorian and Edwardian times. He has been criticized for not labelling his insects, but few collectors did at that time, when interest lay in typical forms of species – the study of races, genetic strains and ecotypes lay in the post-Darwinian future. Doubleday was, however, concerned about foreign insects being passed off as British. He exposed the true nature of supposedly British specimens of the Queen of Spain Fritillary and Purple-edged and Scarce Coppers being sold by a dealer, Seaman, by relaxing them, and showing that their wings reverted to the characteristic 'straight-winged' setting of Continental collectors.[78] 'I believe nine tenths of the rare British *lepidoptera* in the recent collections are continental', he wrote to *The Zoologist* in 1856. 'Some people appear to have more money than wits, and will give five pounds at King Street for what they could buy for fourpence.'[79]

Doubleday's last years were clouded by illness and debt. The possibility of having to close down the business, sell the house, and part with his collections and books haunted him. 'Everything has gone against me the last four years', he wrote to a friend, 'and I see no prospect of brighter times. I must part with everything and I am quite broken-hearted.' In December 1870, suffering from anxiety and stress, he entered 'the Retreat', a Quaker mental hospital, for a few months' convalescence. His friends, led by Edward Newman, organized a subscription to save him from destitution, to which the entomologists of Victorian England, to their great credit, responded with generosity. Henry Doubleday died at home on 29th June 1875, aged sixty-six. 'With him has died a larger amount of natural history knowledge than has ever been possessed by any one of our countrymen', wrote Newman. 'What a pity that it should be lost for ever. He was so reluctant to write ...' For Doubleday's memorial card, Newman composed a fine Latin panegyric: 'He cared little for gaining wealth or title, preferring to deserve them rather than to acquire them. Worthy of a more distinguished fate, he grew old beneath his own roof, contented with his lot.' His funeral, at the Friends' Burial Ground in Epping, where his plain headstone can still be seen, was attended by entomologists from all over the country. A well-researched account of his life was written by Robert Mays, a fellow Quaker.

41. JOHN ARTHUR POWER (1810–1886) (Fig. 62)

John Power was famous for his uncanny ability to divine where a rare and little-known insect would be found, and to guess intuitively what its habits would be. On one occasion he was shown a specimen of the rare, beautifully coloured ground-beetle *Drypta dentata* and 'twitted that he had never taken it'.[80] 'Well, I will go and get it', was the reply, and get it he did, eventually tracking the beetle down in its only known haunt near Portsmouth. It took him five journeys to do so, totalling perhaps a thousand miles, but the result was the fine series of the beetle mentioned in the Proceedings of the Entomological Society for 1857. No wonder Edward Newman called him 'the Indefatigable Power'. Given a problem, he would not stop until he had solved it.

For most of his non-entomological career Power was on the teaching staff of Westminster Hospital. He was born into a family of medical men, and himself studied medicine at Addenbrooke's in Cambridge, as well as the Hebrew and

Fig. 62. Dr J. A. Power, who collected wearing a tall hat in which he kept his tins and bottles, only to have them fall out with a clatter when he raised his hat to a lady!

Arabic languages in preparation for a hoped-for scientific expedition to the Near East. At about the time of his graduation, in 1832, he and J. F. Stephens made a special expedition to Holme Fen on the edge of the still-undrained Whittlesey Mere, and his account of it is one of the few we have of collecting the extinct Large Copper. They must have seen plenty, because Power says he collected twenty-seven, and Stephens thirty. Power is best remembered however as a hemipterist and coleopterist. It is said of him that he always collected in formal dress, wearing a tail coat and top hat, in which he kept his bottles and boxes wrapped in a bag. His daughter remembered the bag falling out once, as he raised his hat to a lady. He was always 'the cheery little Doctor', full of chat and anecdote, and inevitably, at some stage, producing some interesting bug or beetle from his pockets. Fortunately Power's wife shared his enthusiasm, making her own collection of butterflies by the novel method of gumming their wings on to paper.

Power suffered a paralytic stroke at the age of seventy-two, but recovered sufficiently to be able 'to walk four or five miles, dress myself, feed myself and all that', as he told a friend. However, within six months he suffered another setback: 'I trod on my dressing-gown in going downstairs, and fell head foremost nearly from top to bottom. I thought I was smashed, but found no great damage done beyond a considerable amount of shake and bruise, which for the present puts me quite *hors de combat*, and with muddled brain; but I have no doubt that I shall recover in a few days'.

'It would be easy to raise a laugh at this wiry active little man', wrote his friend, J. W. Dunning, for 'regardless of appearances, and oblivious of all but the immediate object of his quest, – perched at the top of a rotten willow-tree; crouching for hours together in a ditch; standing up to the knees in a river, scooping the water upon the bank with his hand; poking about in ants'-nests, with string round his wrists and ankles; trotting off with a blacking-bottle and a bit of meat to set a cunning beetle-trap; or coming home at night with his hat full of *débris* from the bottom of a haystack, and scarcely waiting to sup before the contents were emptied on the dining-room table for careful examination.' It was this genuine and hearty quality that commanded the affection of his fellows: 'His enthusiasm silenced the scoffer; his earnestness compelled respect'.[81]

42. FRANCIS ORPEN MORRIS (1810–1893) (Fig. 63)

The Revd F. O. Morris was the author of two well-known books on Lepidoptera, *A History of British Butterflies* (1853) and *A History of British Moths* (1871), as well as of illustrated volumes on birds, birds' eggs and other subjects of no interest to us here. Like many entomologists, he was a vigorous opponent of Darwin and his followers. He was also an enthusiastic moral crusader campaigning against vivisection and cruelty to animals in pamphlets like 'The Cowardly Cruelty of the Experimenter on Living Animals' (1890) and 'A Defence of our Dumb Companions' (1892). Among the last of his many letters to *The Times* was one recommending the simple pleasures of putting out food for garden birds, then regarded as an eccentricity.

Morris was born at Cobh in County Cork, son of Rear-Admiral Henry Morris, and educated at Bromsgrove School and Worcester College, Oxford. He was ordained in 1834, and served the church initially in Northamptonshire before moving to Yorkshire as vicar of Nafferton, and later rector of

Nunburnholme, where he died, aged eighty-two. During this time he acted as editor of *The Naturalist*, Yorkshire's excellent natural history journal, founded by Morris's brother Beverley in 1851, and sometimes known as 'Morris's *Naturalist*' during this period. A natural history polymath and litterateur, Morris collected, studied and wrote about moths and butterflies from his youth. He frequented many of the classic collecting sites of the Midlands and North, including the colony of Large Blues at Barnwell Wold, Northamptonshire. A new wainscot moth he captured at Charmouth, Dorset, was named *Photedes morrisii* in his honour by C. W. Dale. His fine collection was broken into lots and sold after his death in a house sale by the auctioneers Richardson & Trotter.

Fig. 63. The Revd F. O. Morris, the indefatigably productive parson-naturalist whose books were hugely popular in their day.

Morris's *History of British Butterflies*, an impressive octavo volume with seventy-one coloured plates, was successful enough to go through numerous editions over some thirty years, helped by the advent of relatively cheap machine colour-printing. His style, larded with now obscure classical and poetic allusions and wordy, orotund phrases, shows the gulf in literary taste between his age and our own. For example, instead of simply saying that butterfly collecting is a great leveller, Morris preferred the following form of words: 'The Butterfly-collector's pride of race is centred in one which is alien to his own. "There is my friend the weaver," says the excellent poet, Crabbe, speaking of an entomological one; and the honest artizan or mechanic will be "hail fellow well met" with the "Proud Duke of Somerset" himself, if both should meet together on common ground, in the kindred pursuit of a rare species'. Yet at the time Morris was dubbed 'the Gilbert White of the North' and his works on moths and butterflies were among the most popular of the day.

Despite their many *longeurs*, Morris's books were full of information and good advice, written in an accessible way, like his then novel suggestion of procuring larvae by shaking young trees. He was also among the first to advocate the use of the chloroform bottle, though to the end of his days he stuck by his, by then old-fashioned, clap-net, and 'as to forceps, *et id genus omne*, I exclude them altogether from my vocabulary of entomological apparatus'. Morris was credulous about some of the mysterious foreign butterflies appearing in the countryside, probably as the result of release by fraudulent dealers. In his *History* he gives Albin's Hampstead Eye another airing, as well as admitting three other very dubious species, the Venus Fritillary (*Argynnis aphrodite*) from America, the European Silver-bordered Ringlet, today known as the Scarce Heath (*Coenonympha hero*) and Weaver's Fritillary (*Boloria dia*) (see Chapter 4: 29).

43. FREDERICK BOND (1811–1889) (Fig. 64)

Frederick Bond used to say that a naturalist ought to have three lives: 'Seventy years for collecting, seventy years for studying his collection, and seventy to impart his knowledge to others'. Bond was quiet and self-effacing by temperament, but, 'Take a few specimens to him', wrote J. W. Dunning, ask him a few questions, and 'nothing more was required; he would pour out all he knew, take you to his cabinet, ... tell you where and when he captured this or bred that, interspersing the whole with quaint anecdote and homely story, – his eyes twinkling, and his rugged features beaming all over with merriment'.[82]

Bond was one of the old school of collectors, who learned more from the field than from books; one who published little beyond a few dozen short notes, and whose stock of entomological lore largely went with him to the grave, though he was willing enough to share it with anyone. He was the son of an army officer, born at Exmouth and educated at Brighton where he was intended for a medical career. After a few queasy experiences of dissecting rooms, however, he settled instead to enjoy the life of a country sportsman and naturalist of private means. A lifelong bachelor, he was looked after for much of his life by a widowed half-sister and her family. He fished; he shot and stuffed birds and made a collection of their eggs; and he cultivated ferns and other plants in his garden and greenhouse. His splendid collection of moths and butterflies was regarded as among the finest in private hands, 'rich in curious varieties' and, being often bred by himself, were generally in perfect condition. Bond showed no interest whatever in foreign insects, and, indeed, never seems to have gone abroad, though he travelled extensively in Britain in search of insects. Bond's Wainscot, now relegated to a subspecies, was discovered by him at Folkestone and named in his honour by Bond's old friend, Henry Guard Knaggs.

Fig. 64. Frederick Bond, whose disinclination to marry or follow a career left him 'free to indulge his propensity for sport and natural history'.

Bond was a Fellow of the Entomological Society of London for forty-eight years until his death, serving as Vice-President in 1851, and for a time was assistant editor of *The Entomologist*, as well as a Fellow of the Zoological Society (1854). He was remembered by friends as 'a thorough Naturalist and genial friend'.[83] His life was 'peaceful and uneventful', save for his pursuit of butterflies, birds and game with rod, gun and net.

44. JOHN WILLIAM DOUGLAS (1814–1905) (Fig. 65)

At the age of fifteen, J. W. Douglas was seriously injured by a schoolboy prank. On bonfire night one of his schoolfellows dropped a match into his pocket, which was full of crackers. In the resulting explosions, Douglas's leg was burned so severely that he was confined to his bed for the next two years. To while away the time he took up botanical drawing, and became so absorbed in the subject that, once on his feet again, he applied for and obtained employment at the Royal Botanic Gardens, Kew. While there, he began to collect insects.

His first papers were devoted mainly to butterflies and moths. He became sufficiently expert in the microlepidoptera to join the great H. T. Stainton as a junior author of *The Natural History of the Tineina*, and to publish a revision of the Gelechiidae. He was a gifted writer, and his book *The World of Insects* (1856) proved a popular success. His greatest work, however, was *The British Hemiptera*, written with John Scott, and published by The Ray Society in 1865.

Douglas was born at Putney, the son of an expatriate Scot from Edinburgh. For most of his professional career, until his retirement at seventy, he was employed in the Customs House, rising eventually to a senior position. At one point he was sent on an inspection tour of European vineyards by the then

Prime Minister, Gladstone, who, on his return, thanked him personally and awarded him a special Treasury grant of £100. Douglas was among a group of prominent entomologists, which included Stainton, Newman, Tutt, West, McLachlan, Beaumont and Jenner Weir, who at one time or another lived at Lewisham. With neighbours like these it is not difficult to imagine the many fruitful and convivial entomological evenings they must have enjoyed. 'JWD', as he was known, was a generous, kind-hearted man, fond of doggerel verse, much of it in parody of the style of Longfellow's *Hiawatha.* One of his closest friends, Edward Newman, once turned the tables on him in 'The Song of Bugfliwatha':

> 'Should you ask me whence this story
> With its music, with its magic,
> With its wonderful perfection,
> With its deep and wholesome teaching,
> With its learning, with its science,
> With its wild conglomeration,
> I should answer, I should tell you –
> Jolly Douglas told me of it,
> He who writ the "World of Insects",
> He who lives at pious Kingswood –
> Kingswood by the Blackheath Station –
> Station of the North Kent Railway;
> Douglas with the lots of children –
> Wondrous Alice, silk-haired Laura,
> Laughing Polly, fattest Harry,
> And a new and perfect baby;
> Jolly Douglas told me of it ...'

Fig. 65. J. W. Douglas, known as 'Jolly Douglas', was the author of *The World of Insects* and, in a lighter vein, a prolific entomological versifier.

(There was a great deal more, involving the silk-haired Laura and various bugs.)

Like many entomologists, Douglas tended to be forgetful. On one occasion he had left the type specimen of a capsid bug lying about in a box, which one of the children, perhaps 'fattest Harry', promptly purloined for a pen-box. By the time Douglas had recovered it, the valuable specimen had been 'smashed to atoms', and the great hemipterist had to crawl cap in hand to its owner to explain what had happened.[84]

Douglas was one of the last survivors of the great generation of British entomologists that began with Curtis and Stephens. At the time of his death he was the oldest Fellow of the Entomological Society of London, having been first elected sixty years earlier in 1845. He left his 'splendid special library' on the Hemiptera to the Society, which he had served in many capacities over the years, including President from 1860–61.

45. WILLIAM BUCKLER (1814–1884) (Fig. 66)

For the last quarter-century of his life, William Buckler devoted his energies to drawing and describing the caterpillars of British moths and butterflies, most of which he reared through from eggs or larvae sent to him by correspondents. Yet, apart from a series of short unillustrated pieces in *The Entomologist's Monthly Magazine,* he never lived to see any of this work in print. His great endeavour was finally published after his death by the Ray Society between

1886 and 1901 as the nine-volume *Larvae of the British Butterflies and Moths.* This work was rightly hailed as the finest set of such illustrations ever made. It still is.

William Buckler was born at Newport in the Isle of Wight, and, showing an early aptitude for drawing, became a student at the Royal Academy. He went on to enjoy a modest career as a portrait artist, specializing in watercolours, based first at Portman Square, London, and later at Emsworth in Hampshire. Like many, he initially turned to entomology 'as an amusement for his leisure hours', but was brought firmly into the mainstream when, in 1857, H. T. Stainton found himself needing the services of a first-rate artist who also knew something about insects. He added, characteristically, that 'those individuals who are afflicted by the bump of unpunctuality had better not reply to this appeal'.[85] Buckler, whose portrait business had by then been badly undermined by photographers, offered his services, and was accepted. Over the next three years he drew and coloured from life some 120 figures of larvae of micros for Stainton's *Natural History of the Tineina.* Though his work was exceptionally fine, he was unable to complete the commission, due partly to a condition he suffered from called 'Scrivener's thumb', a kind of cramp, when he could barely write let alone draw. During these attacks he took up carpentry, and, among other things, made his own 24-drawer entomological cabinet.

In 1858, Buckler met the Revd John Hellins (1829–87) (Fig. 67), at that time a master at Exeter Grammar School, and the two decided to make common cause and to tackle the neglected larvae of native butterflies and moths in order, as Hellins put it, 'to wipe off what seemed a blot on the fair fame of

Fig. 66. *(left)* William Buckler, who made the drawing of larvae of British Lepidoptera his specialization. He loathed photographers, who had taken away his livelihood as a portrait painter, and, in the words of his close collaborator John Hellins, 'got his revenge by being photographed in all sorts of still attitudes and sullen expressions'.

Fig. 67. *(right)* The Revd John Hellins, diarist and larva-finder *extraordinaire.* With Buckler, he set out 'to wipe ... off a blot on the fair fame of British Entomologists' by describing and figuring larvae of all the British butterflies and moths.

British Entomologists, since the larval descriptions in Stainton's newly published *Manual of British Butterflies and Moths* (1857) were 'all taken from foreign authors'.[86] The two made a good match: while Buckler was a fine artist, his poor eyesight restricted his ability to find larvae; this task therefore fell to Hellins, one of the best field entomologists of the day whose entomological diaries ran continuously from 1857 until the week of his death. The two corresponded for years while the portrait gallery built up, eventually to cover more than 850 species in some 5,000 figures showing the various states of growth for each species. Buckler was a perfectionist, always working from live specimens, and burning the original drawings if he found some error in them – as he did for many in the large genus, *Agrotis*, despite the weeks of painstaking work they had taken. Many rare species were provided by Continental entomologists, and to correspond with them Buckler, at the age of sixty-eight, taught himself German from a grammar book. Hellins recalled the 'broods of growing larvae' in countless jars and flower pots at Buckler's home 'on the window seats of his bedroom and sitting room, or else in German test tubes of very thin glass'. Buckler, 'lens in hand, would sit for hours, alternately observing and recording [their] habits'. His constant use of the magnifying lens eventually showed in his features, where one eye became more prominent than the other. Buckler sketched his larval portraits on the grey and white paper interleaving Stainton's '*Manual*', from which he would then make his final figures on white drawing paper. Often he would sit up into the small hours, poring over tomes lent to him by friends, and making notes. And so the good work went on, year after year, until Buckler's sudden death, from bronchitis, in 1884.

William Buckler evidently possessed a sardonic sense of humour. Hellins recalled how, 'still sore from the injury which photography had done him', he got his revenge by being 'photographed in all sorts of stiff attitudes and sullen expressions'.

46. CHARLES STUART GREGSON (1817–1899) (Fig. 68)

Fig. 68. C. S. Gregson, the famously irascible doyen of 'the old school of Lancashire collectors'.

C. S. Gregson was the last survivor of 'the old school of Lancashire collectors', an alternative 'centre of entomological excellence' to the mainly London-based entomological establishment. In their heyday in the 1850s, there was keen rivalry between these northern lepidopterists – Gregson, Hodgkinson, Greening and others – and their London equivalents who gathered around the feet of the great H. T. Stainton. In Gregson's numerous papers and correspondence in the journals, an uncompromising personality emerges: strongly parochial, forthright, energetic, testy and argumentative. He added much to our knowledge of distribution, especially among the 'micros', and discovered several new species, including the Northern Footman (*Eilema sericea*), the British subspecies of The Grey (*Hadena caesia mananii*), and the doubtful vagrant noctuid known as Gregson's Dart (*Agrotis spinifera*).

Gregson was born in Lancaster, and for many years worked as a ship painter in Liverpool. He supported many of the local natural history societies, and, as secretary of one of them, printed its reports himself, 'the orthographical peculiarities of which caused him to be good-humouredly bantered by some of his southern brethren, a procedure which he resented'.[87] His tight-lippedness about good sites was legendary, as were the lengths he was prepared to go to secure a particular rarity. On one famous occasion he purchased several birch trees believed to harbour the rare Welsh Clearwing (*Synanthedon scoliaeformis*), had them felled and then sent to Liverpool by rail. On another occasion, while out rabbiting, he saw a Clouded Yellow and, not having a net, took aim and shot it. In place of a watch-dog to guard his house, he kept for thirty-five years a tame African eagle.

Gregson's huge collection, rich in varieties, was sold to another famous collector, Sydney Webb, at a time when Gregson feared he was going blind. On recovering his sight, however, he decided to start again from scratch, and had formed a second collection of some 5,500 specimens before he died. He took a strong subsidiary interest in beetles, and had enough energy left over from his entomological activities to marry three times and rear a large family. His obituarist in *The Entomologist's Monthly Magazine* noted that in Gregson there were all the requirements for an outstanding field entomologist, but one 'perverted by educational deficiencies in the first instance, combined with an excess of egotism, this latter largely being a corollary from the former'.

47. EMMA SARAH HUTCHINSON (1820–1906) (Fig. 69)

Emma Hutchinson is remembered by the pale, summer form of the Comma, named *hutchinsoni* in her honour. In her day, however, she was renowned for her skill in rearing butterflies and moths from the egg, including many rarities. Through her work, the life-cycles of some Lepidoptera were worked out for the first time, including the double-brooded nature of the Comma butterfly, with its contrasting generations.

Emma Hutchinson (born Emma Gill) spent most of her life in rural Herefordshire, where her husband Thomas, whom she married in 1847, was vicar of Grantsfield near Kimbolton. Unlike so many entomologists, she showed no special interest in insects until well into adult life. What ignited the spark was the capture of a humble Swallow-tailed Moth (*Ourapteryx sambucaria*) by her five-year-old son, also Thomas. Henceforth, in the words of her obituarist, E. R. Bankes, she 'devoted herself with entiring zeal and energy to the study of the *Lepidoptera*'.[88] One of her relatively few publications was a piece entitled 'Entomology and Botany as pursuits for Ladies', which suggested that instead of simply collecting butterflies they might study their 'habits'.[89] Emma Hutchinson's own skills in rearing larvae were almost legendary – she was 'wondrously adept with her fingers'. She bred one small moth, the Pinion-spotted Pug (*Eupithecia insigniata*), continuously for thirty-one years from 1874 to her death, without loss of fertility or diminution in size, introducing 'fresh blood' on only one occasion, thirteen years after the start of her experiment.[90] This suggests quite exceptional care and devotion.

Fig. 69. Emma Hutchinson, indefatigable breeder, immortalized in the scientific name of the summer form of her favourite butterfly, the Comma, *Polygonia c-album* f. *hutchinsoni*.

Although she travelled little, and never, as far as we know, visited London, Mrs Hutchinson corresponded with many of the well-known entomologists of her day, including Newman, Henry Doubleday, Buckler and Stainton. In 1881, a letter of hers was published in *The Entomologist*, commenting on the belief that the Comma butterfly was declining and perhaps in danger of extinction.[91] Emma Hutchinson, who probably knew more about Commas than anyone alive, disagreed. 'Herefordshire is its great stronghold in this kingdom,' she wrote. 'I am an old entomologist, and have lived in this county and noted the habits of *Vanessa c-album* for fifty years; and

I can safely say that I never remember this species so common in any autumn as the present one, except in the year 1875, when every blackberry bush was covered with specimens of this lovely and distinct species until late in the autumn.' She suggested that the reason for its decline in its former stronghold of Kent was that the Comma's foodplant, cultivated hop-bines, were being burned after the harvest, resulting in the destruction of countless larvae and pupae. In Herefordshire, Mrs Hutchinson had 'for many years bribed those over whom I have no control in this parish to collect for me every larva and pupa they can find, and by this means I have preserved many thousands of this lovely butterfly'. She ended by saying she had sent hundreds of larvae and pupae for liberation in Surrey and elsewhere, in an attempt to reintroduce the species. Although her endeavours had been without success, 'hundreds have gone to gladden other naturalists in their collections'.

Emma Hutchinson suffered more than her share of sorrows and bereavements. Of her large family, all keen naturalists, four children and her husband predeceased her. Thomas, her eldest son, and two other surviving children all became distinguished entomologists. John made a remarkable collection of butterflies from Natal in South Africa and lived until 1945. All are buried in Kimbolton churchyard, along with their mother. Emma Hutchinson's historically important collection of over 20,000 Herefordshire Lepidoptera was presented in 1937 by her daughter to the Natural History Museum in London. Her notebooks and diaries are in the library of the Woolhope Naturalists' Field Club in Hereford.

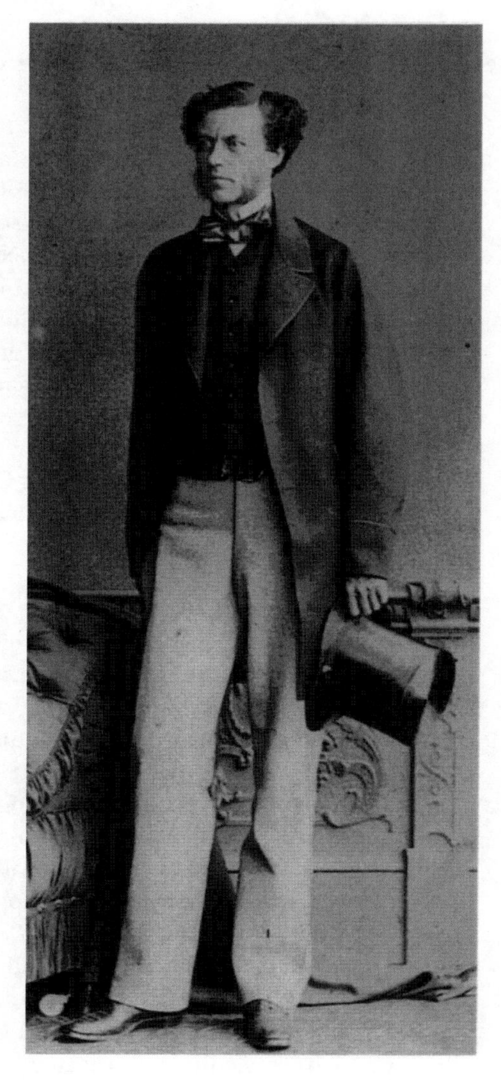

Fig. 70. T. V. Wollaston, an unusually tall man of six feet seven inches, his life was one of 'constant warfare between physical incapacity and will'.

48. THOMAS VERNON WOLLASTON (1822–1878) (Fig. 70)

Chronic ill-health forced Thomas Wollaston to spend each winter in a warm equable climate, and it was on islands in the Mediterranean and Atlantic that he made his entomological reputation. The most important of his written works is the monumental *Insecta Maderensia* (1854), a quarto volume of 677 pages, illustrated with fine coloured drawings by J. O. Westwood. He also studied the insects of the Canaries, the Cape Verde Islands and St Helena; the series of 'catalogues' and papers he wrote about them between 1854 and 1867 were unparalleled in their meticulous detail. He discovered 255 new species in one family alone, the Cossonidae (Coleoptera), in which he specialized. Some of the places he explored, encumbered with nets and collecting gear, require considerable agility even today, and must have required no little determination and courage from an invalid who was forced to spend much of his life in bed. 'The constant warfare between physical incapacity and will', he wrote, attended him even in the sunshine of Madeira, 'where half my work was actually written in bed, and when suffering more or less from bleeding of the lungs.'[92]

Thomas Wollaston came from a distinguished family which included William Wollaston (1659–1724), author of the *Religion of Nature*, and William Hyde Wollaston (1766–1828), discoverer of the elements rhodium and palladium. He was the tenth son and fifteenth child of the Revd Henry Wollaston, rector of Scotter in Lincolnshire, and was educated at the Grammar School, Bury St Edmunds, and Jesus College, Cambridge, where he graduated in 1845. It was there that he developed an interest in insects. A shared interest in island faunas and beetles gave Wollaston and Charles Darwin plenty to talk about – Wollaston was present at the crucial meeting at Down in 1856 when Darwin discussed his theories with T. H. Huxley, J. D. Hooker and himself. Wollaston remained a firm believer in divine creation, although, in his own work, *On the Variation of Species*,[93] published in that year, he dimly reflected Darwin's theory of evolution through natural selection. He deduced that species of insects had arrived on the Atlantic islands from land bridges which had since been submerged by the sea. In these circumstances, the species had come to differ from their closest relations on the mainland. The distinctive butterflies and moths of Madeira were, he thought, a product of the island's unusually high humidity. However, mainland species could not so change – they would always remain as God had created them. His book was dedicated to Darwin, and their friendship seems to have survived despite his scathing review of *The Origin of Species*, published three years later.

Wollaston excelled in taxonomic work. He was a diligent collector and cataloguer of insects and his ability to find, distinguish and describe closely-related species puts him in the highest rank of nineteenth-century entomologists. He was elected a Member of the Entomological Society of London in 1843 and a Fellow of the Linnean Society in 1847, remaining a contributor to their journals until his death in 1878. His many type specimens are now in the Natural History Museum, London, and other parts of his collections are preserved in the Hope Department at Oxford. His last years were spent in Devon, which inspired a volume of lyric poetry, *Lyra Devoniensis* (1868). He is buried at Teignmouth.

49. HENRY TIBBATS STAINTON (1822–1892) (Fig. 71)

H. T. Stainton, editor of *The Entomologist's Annual* and *The Entomologist's Weekly Intelligencer*, and much else besides, had a strongly developed social conscience. 'Possessed of an ample fortune, he was conscious that "property has its duties as well as its rights" '.[94] Stainton saw his duty primarily in terms of the advancement of entomology, in which respect, possibly no single person has done more. He held open house on Wednesday evenings, when anyone could come in and inspect his collection and superb library. He founded and edited some of the best entomological magazines of the Victorian period, and wrote a two-volume *Manual of British Butterflies and Moths* (1857–59) that P. B. M. Allan, writing in 1937, still regarded as indispensable.[95] When Stainton died, his friends suffered 'from a sense of void'.

Stainton was born, and lived all his life, in Lewisham. Of a wealthy family, he was educated privately and then at King's College, London. For part of his life he was engaged in what was vaguely termed 'commercial occupations', to which some of his friends attributed Stainton's lifelong instinct for orderliness, and 'habits of accuracy of observation, method and punctuality'. He was perpetually busy. Stainton made a habit of rising early, at 5 o'clock it is said, and much of his best work was done between then and breakfast time. He wrote innumerable letters, many of which survive, since he copied each one in his fastidious handwriting before filing them along with the replies he received (one of which, written by J. W. Tutt on Christmas Day, asked him to copy out a number of scientific descriptions from some tome or other – which doubtless Stainton attended to in the small hours of Boxing Day). Stainton also filed his bills (from one of which we learn that a new brougham had cost him £198. 10s. 0d. in 1872), memoranda and everything else appertaining to a crowded and busy life. Any future biographer will not lack material.

His famous 'at homes' have already been mentioned in another chapter. There was nothing unusual about home-based entertainments in the Victorian era, but Stainton went further than most in announcing them in specialist journals, and welcoming anyone, young and old, rich or poor, so long as they were interested in entomology. He particularly welcomed requests from 'young collectors' (over 14), and, in a rather charming 'Address to young Entomologists', claimed to be 'more pleased at receiving his enquiry than *he* would be at obtaining my answer', adding that 'I have no sedateness about me, and am as full of fun as anyone'.[96] Visitors could bring specimens for identifying or exchange, browse in Stainton's fine library, and meet like-minded people. Lewisham was at that time the social centre of the entomological world – Stainton's near neighbours included J. W. Douglas and Edward Newman – and Stainton's home, Mountsfield, was the great gathering ground of the day, just as Hope's Museum on Upper Seymour Street had been a generation earlier.

Fig. 71. The great H. T. Stainton, whose published output was prodigious and whose fastidious, 'bird-like' features masked an iron sense of duty and an unfailing kindliness towards the less fortunate.

Stainton's record of service to the scientific societies of the day is probably unique. Though naturally diffident and unobtrusive, and often in indifferent health, he was Secretary of the Linnean Society (he was elected a Fellow in 1859), the Ray Society, the British Association for the Advancement of Science and, of course, the Entomological Society, which he joined in 1848 and served in one capacity or another, for over forty years. He was elected a Fellow of the Royal Society in 1867 and served on its Council for two years. As for entomological journals, he founded and edited *The Entomologist's Annual* (1855–74), whose twenty volumes contain much of his work on micros, as well as the excellent and still very readable ten volumes of *The Entomologist's Weekly Intelligencer* (1856–61). He helped to breathe new life into *The Entomologist* in 1864, and was one of the founders of *The Entomologist's Monthly Magazine*, which he edited from 1864 until the year he died. On top of all that, he helped found the Zoological Record Association, and acted as its Secretary until its functions were taken over by the Zoological Society in 1886.

His separate works include the legendary *Manual* on the British Lepidoptera referred to earlier, which included descriptions and localities of all the species of Lepidoptera 'interspersed with readable matter'. His *Natural History of the Tineina*, published in 13 volumes between 1855 and 1873, is distinguished for its beautiful hand-coloured plates of larvae, many of them drawn by Buckler. (For Stainton, the Tineina included all the micros, except the tortricids, pyralids and pterophorids or plumes.) The text is uniquely in four languages, English, French, German and Latin – Stainton was fluent in all of them – arranged in parallel columns. In the 1860s and 1870s, he extended his accounts of the Tineina to southern Europe, Asia Minor and North America, describing many new species in the process. His work also included the section on Tineina in the series of volumes known as *Insecta Britannica* (1854), and a systematic catalogue of the tineids and plume moths held in the Natural History Museum, London. In his later years Stainton's collecting was increasingly restricted to micros, both in Britain and Europe, and his collection – mostly in Continental straight-set rather than the traditional English droop – was one of his major sources of material. Using it, Stainton completely revised the genera and species of micros found in Britain, and the large number of new ones described by him show how acutely he observed, bred and studied them. It is interesting to speculate whether the introduction of electric lighting in museums helped to advance the study of micro-moths; previously work on their collections was more or less restricted to the daylight hours.

Stainton's greatest prop in his ceaseless labours was his wife, Isabel, who shared his entomological travels and collecting. She also nursed him during his long final illness. The correspondence still poured in, but, as Stainton replied to a typical request, 'I was taken suddenly ill last January & have never since been able to recover my ordinary health ... A month ago I was so reduced by sickness, I had to have Dr Wilks to see me. I doubt whether I could accomplish the inspection of the Linnean Insects – at any rate I should be very glad to be excused ... I must not be surprised if I lack the elasticity of more youthful days'. He died, at home, on 2nd December 1892, aged seventy-one, leaving, as his friends mourned, only 'a remembrance of his geniality and striking individuality', or of 'his kindly bird-like face, his charm of manner, and above all, his unfailing kindness'.[97] Alas, Stainton's once famous house, Mountsfield, in Lewisham, has gone, though, as some small recompense, there is a Stainton Road close by which runs past Mountsfield Park.

50. JOHN JENNER WEIR (1822–1894) (Fig. 72)

J. Jenner Weir was a man of lively intellect and many interests. Outside a busy professional life as a senior customs official he found time to serve in most of the leading entomological and natural history societies of the day, keep up with the most advanced scientific ideas, and even judge regularly at the cat, dog and bird shows held at Crystal Palace. He was among that minority of Victorian entomologists, a keen experimenter. Indeed, he claimed that the reason he published so little original work was that 'his time was too much occupied in testing the observations of others'. Jenner Weir was justifiably proud of the help he had given to two of the greatest British naturalists, Charles Darwin and Alfred Russel Wallace. He had corresponded at length with Darwin, whose views he shared, and advised him on the entomological content of his works. And for Wallace, Weir undertook a series of experiments to test whether the edibility or otherwise of butterflies and their larvae was reflected in their colour, and whether this was recognized by insectivorous birds. It must have helped that Jenner Weir was greatly interested in birds, and kept a small aviary in his 'pretty' garden in Camberwell and, later, Lewisham.

Weir was born in Lewes, East Sussex from a family of Scottish descent. Though his family soon moved to Camberwell, Weir put Lewes on the entomological map when, in his first published paper, he announced the discovery there of the first known colony of the Scarce Forester moth (*Adscita globulariae*).[98] He was educated at a private school in Camberwell before joining the customs service in 1839, where he spent his entire career, rising to the position of Accountant and Controller-General in London by the time of his retirement in 1885.

Jenner Weir was a popular man, numbering among his friends most of the leading London entomologists, such as Douglas, Stainton, and, later, Tutt and Robert McLachlan. They remembered his modesty, but also his vigorous conversation 'exhibited in a strikingly emphatic manner ... but never tedious'.[99] Sometimes Weir would seem to abandon entomology altogether for a while, to pursue his other interests in botany and ornithology. Initially he took a keen interest in micros, until an accident, in which he lost the top of his left thumb, left him unable to set small insects. From then on he became more interested in evolution, and particularly in insect mimicry and protective resemblances, often exhibiting cases of exotic butterflies like the Danaidae as examples. In the 1870s, such topics were still controversial and hotly debated. Tutt recalled one occasion when an anxious Jenner Weir solicited his support for a paper he was reading at the South London (now the British) Entomological and Natural History Society: 'I do hope you will be present tonight. I have some notes to read which will interest you, and I want your support. It is difficult for a man at my age to understand that comparatively young men publicly delight in expressing their disbelief in evolution, and almost in the same breath inform you that they have never read the main works thereon, whilst at the same time pretending to do scientific work.' Tutt himself commented that 'Old views die hard, and in talking the matter over afterwards we agreed that it was good so much had been accomplished in such a short time'.[100]

Fig. 72. J. Jenner Weir, the customs official who corresponded with Darwin and Wallace.

Jenner Weir served on the Council of the Entomological Society of London for a number of years, and was Vice-President on two occasions as well as Treasurer for four years. He was also a sometime Council member of both the Ray Society and the Linnean Society. Shortly before his death, he was elected President of the 'South London', among whose members Weir's record in advancing science and 'the progressive welfare of all' was considered quite outstanding.

In his last years, Weir was afflicted by *angina pectoris*, and his heart finally gave up in 1894, in his seventy-second year. From the unusual length and admiring tone of his obituaries there emerges an attractive personality, sociable, conscientious and intelligent, more interested in understanding nature than in amassing large collections of dead insects for their own sake. His life, wrote a contemporary, had been 'well-filled'.

51. JOSEPH GREENE (1824–1906) (Fig. 73)

'Parson' Joseph Greene is best remembered today for his celebrated essay 'On pupa digging', which was originally published in *The Zoologist* in 1857,[101] and later included in Greene's popular book, *The Insect Hunter's Companion.* Greene had not invented pupa digging, for the technique had already been described by Benjamin Wilkes as long ago as 1748, but he undoubtedly helped to popularize it. In his anxiety to extol its merits, however, he may, on occasion, have been a little too enthusiastic in his claims, thus allowing P. B. M.

Allan to have some fun at his expense: 'Parson Greene's luck was of ... the "cornucopia" kind. It was the sort of luck which befell old Lewin when he went out to look for *antiopa* and found a field swarming with them, all in perfect condition. And it never deserted him. If you asked him for an *anceps* pupa, he would take up his little trowel and return in half an hour with a couple of dozen. Parson Greene was unique. "I once," says he, "had a thousand pupae of *incerta*" [The Clouded Drab].' Allan calculated that to obtain so many by pupa digging in the right season, Greene would have had to work five hours a day (except Sundays when he must have been otherwise employed) for five months on end. 'What a craftsman! And all for the sake of *incerta*. What an enthusiast!'[102]

Fig. 73. Parson Joseph Greene, the popularizer of 'pupa-digging', of whom P. B. M. Allan wrote, his 'luck was of what I may call the cornucopia kind ... and it never deserted him'.

Greene's luck with butterflies was just as remarkable, though it seems to have grown in the telling. In 1849, he took two Mazarine Blues at Guiting Power, in Gloucestershire. Fifty-three years later he had changed the date to 1850, by which time the two specimens had somehow increased to eight![103]

Greene graduated at Dublin University, and while there published one of the first comprehensive lists of Irish Lepidoptera.[104] He served as a parish priest, first in Derbyshire, then at Guiting Power in the Cotswolds and finally at Halton in Buckinghamshire, where his nearest clerical neighbour was the butterfly-collecting Henry Harpur Crewe. His local lists of Lepidoptera, such as the 'List of Lepidoptera occurring in the county of Suffolk',[105] added significantly to our knowledge of distribution. Greene was a frequent contributor to entomological magazines and supplied his friend Edward Newman with local lists for the latter's *Illustrated Natural History of British Butterflies.* When ill-health at last limited his active insect hunting, Greene contented himself with rearing enormous numbers of the Magpie Moth (*Abraxas grossulariata*) to obtain a fine series of 'vars'. He was one of the last survivors of the 'generation of the 1850s'. A shy, retiring man, he was remembered for his 'high sense of probity, his liberality, and willingness to give any information or assistance in his power to his brother Entomologists'.[106]

52. GEORGE CARTER BIGNELL (1826–1910) (Fig. 74)

George Bignell is perhaps best remembered for the collapsible 'Bignell beating tray' for larvae, which he designed and which is still on sale. In his day, however, he was considered one of the pre-eminent entomologists of south-west England, 'attacking all Orders of insects', as one obituarist put it, but specializing in the parasites of lepidopterous larvae. He discovered no fewer than fifty-one of them for the first time in Britain, of which nineteen were new to science. Two hymenopterans, *Apanteles* (now *Cotesia*) *bignellii* and *Mesoleius bignellii* (since synonymized with *Perispudea sulphurata*), were named in his honour.

Bignell was born in Exeter and educated there at St John's College. As soon as he could he joined the Royal Marines and saw action on board H.M.S. Superb off the coasts of Spain and Portugal. On leaving military service, he was, from 1864 until his retirement, the Registrar of Births and Deaths and Poor Law Officer, based at Stonehouse in Devon. Fortunately he seems to have had plenty of time left for field work, and specialized in breeding and drawing larvae. Bignell was also fortunate enough to find an entomological wife, who herself discovered a new species of insect on Corsica.

Fig. 74. G. C. Bignell who, as his obituarist J. H. Keys observed, possessed 'a rich fund of humour, and delighted to jocularly entangle a questioner'.

An important contribution to the study of Lepidoptera was his 'Lists of Parasites' which appeared in each of the nine volumes of Buckler's *Larvae of the British Butterflies and Moths* (1886–1901), published by the Ray Society. He also contributed regular pieces to *The Entomologist*, and was an editor of the *British Naturalist*. Much of his collecting equipment was home-made and, as his obituarist J. H. Keys noted, 'for the most part, either original in scheme, or distinct in some clever departure from the ordinary'. Keys remembered Bignell as 'a man possessed of strong will power, combined with ingenuity and enthusiasm, and, above all, he had the ability to plod'. He meant by this that Bignell was not daunted by failure, but rather stimulated by it to try again but in a slightly different way. He had 'a rich fund of humour, and delighted to jocularly entangle a questioner'.[107]

He was honoured in his native Devon with the presidency of the Plymouth Institution, the leading learned society of the South West at that time. His unique collection of insect parasites and hyperparasites was left to the Natural History Museum, London, while the rest of his collection was acquired after his death by Plymouth Borough Museum.

53. HENRY GUARD KNAGGS (1832–1908) (Fig. 75)

Henry Guard Knaggs was one of the best-known Victorian entomologists – a big, bluff, jovial man with a hearty, seemingly dashed-off style which survives well in his writings. During the 1860s, fellow entomologists would convene at Guard Knaggs's hospitable house in Camden Town after Entomological Society meetings, where he would entertain the younger members with riddles, 'with ever a wink and a nod at us old stagers'.

The son of a medical man, he was educated at University College School, and afterwards trained as a doctor at University College Hospital. He succeeded to his father's busy north London practice, where he was soon 'universally esteemed as an able and skilful general practitioner'.[108] Knaggs devoted much of what leisure time he had to the cultivation of a large kitchen garden. It was thought that continuous overwork was the reason for the gout and ill-health that assailed him from middle-age.

Guard Knaggs had been an active Fellow of the Entomological Society from 1858 to 1863, then resigned but became a co-founder of *The Entomologist's Monthly Magazine*, which he edited from its inception in 1864 until 1874 when he was forced to resign owing to pressure of work. His legacy from this, his most active period, includes *A List of the Macro-Lepidoptera of Folkestone*,[109] a town of which he was very fond and where he owned a seaside house, and a celebrated book, *The Lepidopterist's Guide* (1869), which began as a series of 'Notes on Collecting, Management, etc' in the 'E.M.M.' In its pages, wrote his obituarist, 'the genial Doctor is seen at his best'. To find pupae, for example, Knaggs offered the following sound, if strenuous advice: 'Out-houses, stacks of all descriptions, should be poked about and investigated, thatch inspected, old beehives seen to, ledges, copings, and other overhanging structures peered under, posts, palings – especially such as are rotten, defective or lichen-covered; old walls surveyed for the artfully concealed pupa of *Bryophila*, and all places, conceivable as likely, ransacked for pupae'.[110]

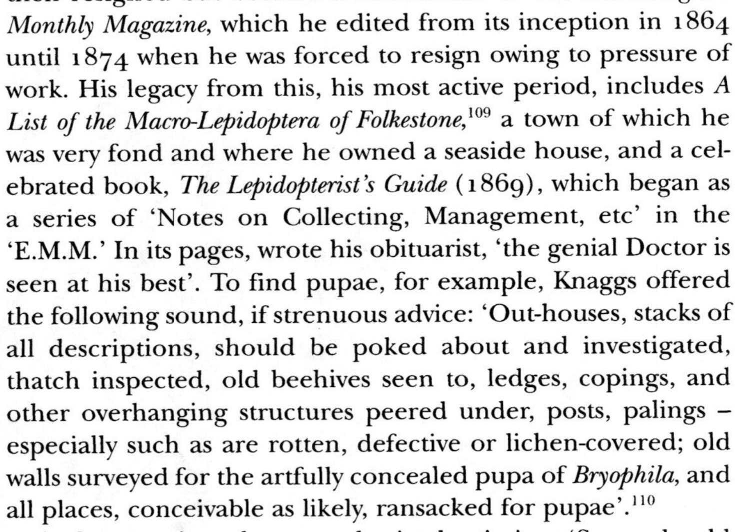

Fig. 75. Dr Henry Guard Knaggs, moth-hunter and raconteur. 'The grand secret of successful collecting, whether by day, dusk, dark or dawn, lies in one little word – WHY?'

As for sugaring, the secret lay in the timing: 'Sugar should be got on before dusk, but not too soon before, or its virtues and sweetness will be expended on the desert air; when the first cockchafer or "lousy watchman" booms past us, we should be reminded that it is at once time to begin to lay on our sugar'. Yet another useful tip was to make friends with the neighbourhood lamp-lighter, for in those days streets were gas-lit. Though at first he was likely to 'bring you enough Poplar hawks and Tiger moths to make a breakfast off ... if he is an intelligent man, this will soon wear off, and after a time he will know a "good un" from a "duffer" as well as you do'. Not surprisingly, perhaps, *The Lepidopterist's Guide* remained in print for forty years.

Knaggs was responsible for many notable discoveries, especially in Kent. They included a wainscot new to science which he named after his friend, Frederick Bond (*Chortodes bondii*, now regarded as a subspecies of *C. morrisii*), and the first known breeding colony of the Scarce Chocolate-tip (*Clostera anachoreta*). He eventually retired to his beloved Folkestone, leaving his practice to his son Valentine, and continued to collect at the Warren as far as his gouty legs would let him. His collection was a very fine one, though restricted to the macrolepidoptera. He died in Folkestone after a long illness but was buried in Highgate Cemetery.

A Knaggs entomological philosophy emerges from his writings: the field worker is a truer scientist than the museum worker because, for all that he is despised as 'a poor fly-catcher', he is busy acquiring and diffusing 'sound knowledge' through observation, instead of merely 'piling up a synonymy for the bewilderment of future generations'. And the best kind of observer, thought Knaggs, is the one who continually asks himself: 'Why?'

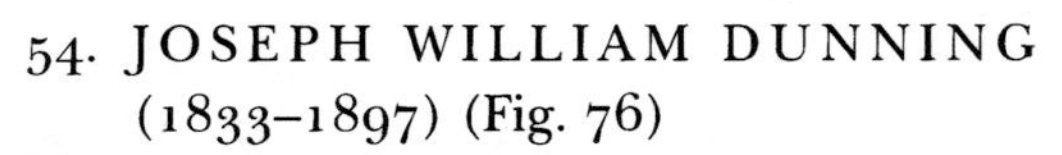

54. JOSEPH WILLIAM DUNNING (1833–1897) (Fig. 76)

J. W. Dunning did a great service to the Entomological Society of London by rescuing it during a low period when it had become 'crippled for want of funds, and disturbed by recent internal dissensions'.[111] Dunning took over the Secretaryship in 1862, and by editing the Society's publications 'fearlessly and impartially', helped to restore their high reputation. Dunning also helped the Society financially and, while serving as its President, personally defrayed the considerable costs of the Society's Royal Charter of Incorporation in 1885, when it became the Royal Entomological Society.

Dunning, the son of a Leeds solicitor, was educated at a private school near Huddersfield and at Trinity College, Cambridge, where he graduated in 1856, becoming a Fellow of his college. He took his M.A. in 1859. In 1861 he was called to the Bar at Lincoln's Inn, and became an expert on conveyancing and an authority on wills. His love of entomology had begun during his school holidays at Brandon, Suffolk, where in 1845 at the age of eleven he was lucky enough to rediscover the little Spotted Sulphur moth (*Emmelia trabealis*) and so, in Stainton's words, 'awoke and found himself famous'.[112] He joined the Entomological Society of London when aged only sixteen and, in 1858, before leaving Cambridge, did the lion's share of the work for the '*Accentuated List of British Lepidoptera*', a translation of all the Greek and Latin generic and specific names, for which his classical training proved invaluable. The work also included biographical notes of the principal describers of butterflies and moths. Though he was not a prolific contributor of entomological notes and articles, Dunning's critical intelligence and literary polish show in his papers, like that on *Acentropus* (now *Acentria*), a genus of aquatic pyralid moths, and in the obituaries of a number of his friends. At the age of fifty-seven, Dunning suffered the first of a series of strokes that eventually killed him. The scrupulous bent of his mind can be judged from a line from his paper on *Acentropus*, cited in his obituary in *The Entomologist's Monthly Magazine*, and used as the motto on the title page of the issue containing it: 'Believing that, by simply asking an abstract question, I am less likely to provoke investigation and discussion than by expressing an opinion that can be contradicted and disproved, I will conclude by expressing an opinion to which I am not wedded, and from which I shall be glad to be converted, but still an opinion founded on such evidence as I have been able to obtain.'[113]

Fig. 76. J. W. Dunning, barrister, editor and saviour of the Entomological Society.

55. OCTAVIUS PICKARD-CAMBRIDGE (1828–1917) (Fig. 77)

When, one morning in the Paris of 1863, his future wife Rose Wallace first saw the thirty-five-year-old Revd Octavius Pickard-Cambridge 'with his big stride and coat-tails fluttering in the wind', she clearly liked what she saw. But from the clergyman, Miss Wallace drew only this comment in his diary: 'A lady and her two daughters [are] in the hotel – English evidently, from having an urn at breakfast'.[114] Within days, however, she succeeded in attracting his attention in other ways. A few years later in 1866, after a decorous courtship, the couple married and embarked on a lengthy honeymoon tour which took in most of the cathedrals of England.

Octavius Pickard-Cambridge was a leading British arachnologist of the nineteenth century, an authority not just on native spiders and harvestmen but also foreign species, especially those from the Near East, Central America and from various major expeditions. He was not himself a wanderer, however, and, apart from a long tour across Europe to Palestine and Egypt, lived nearly all his life in Dorset, where for most of his career he was, like his father before him, rector of Bloxworth. He was born in Wareham and educated privately (by the poet William Barnes), and at the University of Durham where he studied theology, graduating in 1858. His first appointment was to the curacy at Scarisbrick in Lancashire, where unfortunately the landowner was a Roman Catholic who refused to allow Pickard-Cambridge to live on the estate, which forced him to lodge several miles away from the parish. In 1860 he left to rejoin his father at Bloxworth, succeeding him in 1868.

Fig. 77. The Revd Octavius Pickard-Cambridge, rector and squire of Bloxworth, an authority on British spiders but also blessed with the knack of finding new British species of Lepidoptera.

Thanks to Pickard-Cambridge, the name Bloxworth has become attached to both a moth and a butterfly. What was for some years the only British specimen of the Bloxworth Snout moth (*Hypena obsitalis*) was the one found at rest on a doorpost in the rectory garden in 1884. The Bloxworth Blue, now usually known as the Short-tailed Blue (*Everes argiades*), he discovered the following year – one of the great years for immigrant butterflies and moths – while 'butterflying' with two of his sons on Bloxworth Heath on 18th August. They took a female and, two days later, a male on the same site. A third specimen was taken the day after that near Bournemouth, some ten miles away, by a schoolboy, Philip Tudor. They identified the species readily enough from W. F. Kirby's *European Butterflies and Moths*, but although Pickard-Cambridge examined over 500 more blue butterflies at Bloxworth Heath, he found no more 'Bloxworth Blues'. The Short-tailed Blue is one of our rarest migrant butterflies; since then, only a dozen or so have been recorded from Britain, though the existence of two earlier specimens, caught at Frome, Somerset, in 1874, was uncovered as a result of Pickard-Cambridge's published account (see Chapter 4: 19).

Pickard-Cambridge was obviously a man of broad interests and enthusiasms. He was a sporting man and at University acted as steward at the steeplechases. He was also very keen on cricket. In later years his son recalled how each year his father would make a pitch, 'roll it for hours and bowl cunning underhand balls'. Eventually, when a more permanent cricket field was established, he 'entered with zest into all the arrangements for matches, would put up the tents himself, and delight in welcoming the visitors'. On one occasion an unexpected guest arrived at the vicarage in the person of the Earl of Moray, an old friend and son of the late Revd Edmund Stuart, rector of the neighbouring parish of Winterbourne Houghton. Accepting an invitation to lunch, the Earl found himself sitting down with the Pickard-Cambridges to a dish of squirrel pie, the result of a shooting trip the previous day! His interests extended to bee-keeping and he possessed, among other things, a large collection of stuffed birds. He also loved music: he played the violin and was president of the local choral society. In the last full year of his life Pickard-Cambridge fulfilled a long-held ambition when he was finally carried aloft in a flying machine!

His collection of butterflies and moths – micros as well as macros – was beautifully set and said to consist almost entirely of specimens captured or bred by himself. In later life he concentrated on the micros, and managed to discover in Bloxworth the extremely rare plume-moth *Trichoptilus* (now

Buckleria) paludum, whose larvae subsist on that normally insectivorous plant, round-leaved sundew (*Drosera rotundifolia*). He regularly collected with his near neighbour Frederick Bond, was a close associate of Stainton and Westwood, and knew J. C. Dale in the last years of Dale's life. Despite his clerical calling, Pickard-Cambridge was quickly convinced of the truth of Darwin's theories, agreeing from his own observations that species were still evolving, and supplying notes and information to both Darwin and Wallace.

Pickard-Cambridge was one of the most accomplished amateur naturalists that ever lived. Several of those who knew him attest to his phenomenally accurate memory, which he retained into old age. He was made a Fellow of the Royal Society in 1887 for his work on spiders, and for a short time in his younger days, from 1855 to 1861, he had been a Fellow of the Entomological Society of London. Like many nineteenth-century naturalists, he was an all-rounder as well as a specialist. To be able to combine all that with his many cultural and sporting interests, his church duties and his devotion to his large family, it is scarcely surprising that the one word nearly everyone used to describe him was 'indefatigable'.

56. CHARLES GOLDING BARRETT (1836–1904) (Fig. 78)

C. G. Barrett began writing his great work, *The Lepidoptera of the British Islands*, in 1892, but did not live to complete it. Unhappily, the missing portion – most of the microlepidoptera – was the one that needed his guiding hand most. Even so, the work, which remained the standard on moths until far into the twentieth century, lived up to its author's high standards as expressed in his preface: 'My aim is ... not only to furnish original and accurate descriptions of the perfect insects, and the most reliable descriptions obtainable of their larvae and pupae, but also such particulars of their habits and ways, drawn from personal experience and the most reliable records, as shall present them to the reader as creatures which enjoy their lives, and fill their allotted positions, before they take a more permanent place in the museum or the cabinet'.

Fig. 78. C. G. Barrett, author of the eleven-volume standard work on British butterflies and moths, although he did not live to see its completion.

He was described in an obituary as being a 'lucky' collector, but his luck depended on the practical application of a profound knowledge acquired over many years as a field naturalist with exceptional powers of observation. Among the rarities he discovered were Barrett's Marbled Coronet (*Hadena luteago barettii*) at Howth, Dublin, the Lesser Belle (*Colobochyla salicalis*) at Haslemere, the Marsh Moth (*Athetis pallustris*) close to the City of Norwich, and the Scarce Pug (*Eupithecia extensaria occidua*) at King's Lynn. He also described six British tortricid species new to science. Most of the Lesser Belles then in collections came from Barrett's rearing cages. His obituarist, F. D. Wheeler recalled how Barrett had 'encouraged him – then an enthusiastic boy – to pick whatever he liked from the store boxes that seemed inexhaustible, and introduced him freely to his choicest collecting grounds'.[115]

Born at Colyton in Devon, Barrett entered the civil service in 1856, eventually rising to senior positions in Excise and later in the Inland Revenue, from which he retired as Collector of London South in 1899. His tours of duty stationed him in several different parts of the country, many of them, either by accident or design, noted for rare butterflies and moths. From the age of twenty he was a regular and frequent contributor to entomological journals. He was invited to join the editorial board of *The Entomologist's Monthly Magazine*, on which he served from 1880 until his death in 1904, contributing some 330 separate pieces. In 1884, he was elected one of the first Fellows of the Entomological Society of London (the term 'Member' had always been used prior to the granting of the Society's charter in that year), and in 1896 became secretary of the committee appointed by the Council 'to consider the protection of British insects in danger of extinction' which in 1897 proposed the formation of an Association to discourage over-collection of Lepidoptera. From this body all subsequent Protection Committees have derived.[116] He was elected Vice-President of the Society in 1901. He was also President, in 1892, of the South London Entomological Society. Barrett's 'genial, energetic and hearty manner', was much appreciated, especially in contrast to 'the hectoring approach of some of our other great lepidopterists' (a shaft probably aimed at J. W. Tutt). One story about Barrett, which amply illustrates his tireless enthusiasm, concerns two Clouded Yellows he spotted from a railway carriage. Jumping out at the next station, he ran back down the line – and bagged them both!

The *Lepidoptera of the British Islands*, begun when he was already fifty-six, was an undertaking which only someone as indefatigable as Barrett could hope to bring to a conclusion. However, he had reached only the ninth volume on his death twelve years later, though he left sufficient notes for the tenth and eleventh volumes on 'Pyralidina' and 'Tortricina' to be compiled by Richard South (q.v.) soon afterwards. The eleventh and final volume was published in 1907. Barrett was among the first to illustrate the larvae of every species, as well as a significant number of aberrations. And, as he had intended, the work was about living insects, their habits, distribution and biology, as much as descriptions of dead ones on pins.

Barrett's collection was auctioned by Stevens of King Street, London in 1906 and 1907. Fortunately many of the most important specimens were purchased by Lord Rothschild, and are now in his collections in the Natural History Museum, London.

57. WILLIAM WEST
(1836–1920) (Fig. 79)

William West was one of the founders, in 1872, of the South London (later the British) Entomological & Natural History Society. Then known as the 'South London' and today as the 'Brit. Ent. Soc.', the Society seems to have been formed in an ecumenical spirit, welcoming all social classes, and was less overtly middle-class than the senior Entomological Society. Its guiding light was Edward Newman, whose regular social group attending his 'at homes' formed the impetus for the new Society. Until his death, half a century later, West was perhaps the 'South London's' most loyal member, who 'continued from first to last to take the same enthusiastic interest with which he helped to found it'.[117] He rarely missed a meeting in his forty-eight years of membership. West himself contributed the Society's first small reference collection of 100 species, and, as for their curator, 'who could be a better keeper', wrote H. J. Turner, 'than the plodding, steady, field-worker W. West'. Who indeed? He was appointed the 'South London's' Hon. Curator in about 1878, and remained so until the day he died.

As the foreman of a brass foundry, part of the large engineering firm, John Penn & Sons, William West embodied the 'South London's' spirit of uniting aristocrats and artisans. He was principally a field man, initially specializing in Lepidoptera (first macros, then micros) and later in beetles, bugs and grasshoppers. Born in Rotherhithe, he lived all his life in Greenwich, and the still open country nearby – Greenwich Park, Blackheath, Kidbrook and the Thames marshes – formed his main hunting grounds, with occasional forays further afield to Wicken or the New Forest. He was generous in the disposal of perfectly-set duplicates, and over the years he donated hundreds of rare specimens to the Society. His important collection of nearly 3,500 Homoptera was left to the Natural History Museum, London. West contributed little to the journals – he was one of those whose vast store of field knowledge is forever lost – but he did provide lists of records and localities to others, especially from around Greenwich, Woolwich and Lewisham. West enjoyed good health all his life – his portrait photograph, taken when he was eighty-four,

Fig. 79. William West, founder member and lifelong curator of 'the South London'.

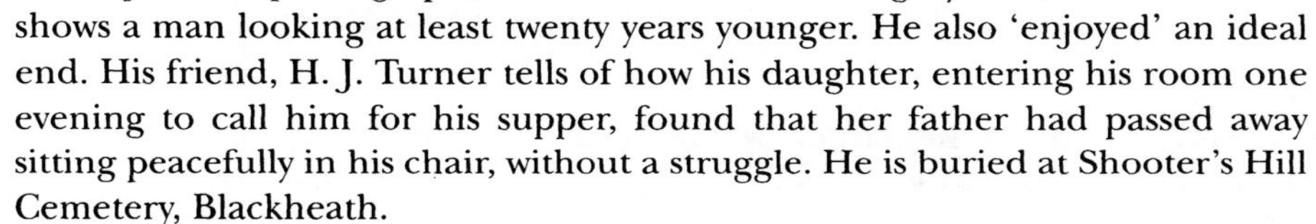
shows a man looking at least twenty years younger. He also 'enjoyed' an ideal end. His friend, H. J. Turner tells of how his daughter, entering his room one evening to call him for his supper, found that her father had passed away sitting peacefully in his chair, without a struggle. He is buried at Shooter's Hill Cemetery, Blackheath.

58. JAMES PLATT BARRETT (1838–1917) (Fig. 80)

J. Platt Barrett (unrelated to C. G. Barrett) was another of the founders of the 'South London' in 1872, along with William West, G. C. Champion and five others; indeed, as the earliest meetings were held in Barrett's house in Peckham, he can probably be regarded as its chief instigator, though little record of the Society's early days survives. Platt Barrett was an avid collector and breeder of butterflies and moths, especially after his retirement in 1908. Few, it is said, knew the classic London and Kentish hunting grounds, like Chattenden Roughs, Dulwich Woods, Box Hill and Folkestone Warren as well as he, and his collection contained rarities like the Rest Harrow Moth (*Aplasta ononaria*) which few others managed to find.

Platt Barrett, it seems, had hearing problems, though no contemporary seems to have remarked on the fact. He entered an institution for the deaf and dumb in his native Yorkshire at the age of thirteen, and spent his whole professional life – some fifty years – teaching in those same institutions in Yorkshire, London and Kent. Whatever the degree of his own disability, he seems to have been able to communicate with his fellow entomologists readily enough. Fascinated by Lepidoptera since boyhood, Platt Barrett late in life extended his travels to Sicily, which he visited eight times between 1894 and 1914. On one visit, in 1908, he narrowly escaped death when a severe earthquake struck the town of Messina where he was staying. Tragically, his daughter-in-law and a grandchild, who were sleeping

Fig. 80. J. Platt Barrett, deaf-and-dumb teacher and co-founder with William West of 'the South London'.

in the room next to his, perished in the collapsing building.

Platt Barrett was the first Secretary of the 'South London' from 1872 to 1876, and the following year was made its President. He also became a Fellow of the Entomological Society in 1911, and regularly attended its meetings. He left his fine collection to The Horniman Museum at Forest Hill, London.[118]

59. WILLIAM HENRY HARWOOD (1840–1917)

W. H. Harwood was one of the first people to use sleeves – tubes of muslin tied at either end – to rear insects outdoors in relatively natural conditions (the method itself was not new – it had been described by Ingpen as long ago as 1827 but had apparently never been widely used). He perfected the method in his modest back garden on North Station Road in Colchester, and used it to rear many British species of moths and butterflies, some of whose early stages were then unknown. Harwood even went to the lengths of planting Spanish Catchfly (*Silene otites*) in the sandy crevices of the walls of Colchester Castle, to have a handy supply of food for the rare Viper's Bugloss moth (*Hadena irregularis*), which, despite its name, feeds only on this plant. Rearing ensured that Harwood's specimens were always in pristine condition, with a good number of rare 'vars'. In a letter to a friend, Wilfred Appleby recalled how it all looked:

> 'There was a big bush in the garden where WHH used to sleeve some of his caterpillars. His collection was in the drawers of specially-made mahogany cabinets in a first-floor room, which he used as his study. I was deeply impressed by the immaculate tidiness and perfect orderliness of everything ... many [specimens] had been bred or reared from eggs or wild-gathered caterpillars. All were perfectly set and labelled. My excitement grew when he slid open each drawer to reveal rows of butterflies and moths I had never seen before. I remember there were purple emperors which WHH had caught locally, and he told me he had lured them from the tops of oaks by putting out dead rabbits. The male emperors would come down to feed on the decaying juices. I also saw a Camberwell beauty and several large tortoiseshells, the latter species was still common around Colchester before the First World War. I also spied an orange-tip male with lemon yellow tips to its forewings instead of the usual orange. WHH said he had caught it at Friday Wood on a May morning.
>
> 'Then came the great hawk moths – the sinister death's-head and the wonderful pink and olive elephant hawks. There were also some bee hawks, again the first I had ever seen.'[119]

Harwood was born in Colchester and educated at the local grammar school. He was first apprenticed to a firm of chemists, but, taking his doctor's advice to find some form of outdoor work, Harwood managed to establish himself as a professional supplier of natural history specimens. In his day, with fine woods, heaths and salt-marshes all within easy reach, some forty-six species of butterflies occurred around Colchester, including such present-day rarities as Heath Fritillary, Large Tortoiseshell and Purple Emperor. Harwood supplied collectors with specimens captured or reared locally, but he also made a practice of rearing and releasing insects, especially rarities. Later on, he was aided by his two sons, Bernard and Philip, the latter eventually bequeathing his own fine collection of British insects to the Natural History Museum, London. Harwood was a highly regarded entomologist who corresponded with many of the leading entomologists of his day, including Newman, Henry Harpur Crewe and South. Buckler used his unique knowledge extensively for his work on larvae, and Lord Rothschild bought varieties from him. Harwood also advised his brother, the Hon. Charles Rothschild – the 'father of nature conservation' in Britain, on local sites of wildlife interest, and it was through him that Ray Island on the Essex coast was secured as one of the first nature reserves.[120] Harwood published no large-scale works, but the notable 42-page section on Lepidoptera in the *Victoria History of the County of Essex* (1903) is his work.

60. ALBERT BRYDGES FARN (1841–1921) (Fig. 81)

A. B. Farn formed what was, by general agreement at the time, the finest private collection of British butterflies and moths in the country. It was especially rich in butterfly aberrations, including such extreme ones as the unique all-black Marbled White captured at Chattenden, Kent in 1871, and the extraordinary silvery-white 'ghost of a Comma', which he caught in South Wales. Farn was a frequent attender of auction rooms, and in 1906 purchased the entire Sabine collection, with its lengthy series of Small Copper varieties and its three dozen British Large Coppers. Many of Farn's finest aberrations were illustrated by F. W. Frohawk for his published works.[121]

The owner of this fabulous collection was a noted all-round naturalist and country sportsman, a man of vigour, courage and rather boisterous good humour. 'He was an incessant and incorrigible jester', wrote one obituarist, quick to see the ludicrous side of things 'but never wearisome'. Once, to gain entry to a particular wood where Purple Emperors flew, Farn caught up with the owner and his party, and cried out, "I say Sir, have you seen the ladies *Iris* and *Iole*, cousins of the Emperor, you know? They must be lost in the wood!" The wood's owner, assuming Farn to be on intimate terms with royalty, gave strict orders that he was to be admitted to the property at all times, and himself set off in quest of the missing ladies.[122] Farn's charm, generosity and love of practical jokes were matched, however, by a long memory for a slight or injury, real or imaginary and a complete incapacity to forgive or forget it. On occasion he could lose his temper explosively and without warning, as he did when a certain dealer tried to palm off some manufactured 'vars' as genuine wild captures – an experience which the incautious dealer probably did not forget in a hurry.

Fig. 81. A caricature by G. B. Kershaw of A. B. Farn, sportsman, 'King of the vars' and an 'incorrigible jester'.

Farn published only occasional notes and critical pieces (though neither Stainton nor Westwood escaped his sharp eye for errors). He wrote a lengthy article on Japanese silk culture and revised and extended his friend Joseph Greene's *The Insect Hunter's Companion* in 1880. He much preferred, though, to spend his leisure time in the field. He made a large collection of stuffed birds (he did the stuffing) and of birds' eggs; he later made coloured drawings of the latter, with Frohawk's help. He was a legendary shot, and his bag of thirty snipe with thirty consecutive shots on Lord Walsingham's estate established a record which has probably never been equalled. While sitting outside his room-sized moth trap in the evening, he would 'pot' at bats with a .22 rifle, and would expect to down about three out of five shots. He once challenged an acquaintance to drive tandems [two wheeled horse-drawn carriages] from his house at Swansfield to Rochester and then play billiards non-stop for twenty-four hours. Farn won easily; his opponent collapsed well before the time-limit was up.

Farn studied medicine at University, and went on to become an inspector and administrator of vaccines at Whitehall, first at the Public Health Department and then on the Local Government Board. He managed to find houses close to some good localities, first at Dartford, then Greenhithe, and, on retirement in 1906, near Hereford 'chiefly with the hopes of turning up, in some of its old haunts, the now supposed extinct Mazarine Blue'. In 1906, he moved further west to Ganarew in the Wye Valley, where he caught many a fine salmon, and, it is said, deliberately extirpated a colony of the introduced Map butterfly (*Araschnia levana*) in the nearby Forest of Dean because he disapproved of the release of foreign butterflies. Very little is known about this episode, however, and Farn's own notes and diaries revealed only the lines: '*A. levana* introduced surreptitiously about 1912. Fortunately only survived two years' (see Chapter 4: 28).

Farn suffered his share of sorrow, losing his wife and beloved son comparatively early. In the last twenty years of his life he was looked after by the daughter of an old friend, described as a companion. Francis Jenkinson remembered Farn in the last few weeks of his long life, now weak and ill but with mind and memory as alert as ever. 'He spoke of old times; and I looked through many of the drawers in his collection, while he from his chair directed my attention to this or that specimen, the position of which he seemed to know by heart.' He left no directions for the disposal of his butterflies, and the peerless collection was sold by auction over three days the year after his death, though some of the most outstanding specimens are now in the Natural History Museum in London. 'It is a pity', he wrote in 1909, 'unless they could be sold *en bloc*. However, many happy times have gone into them'.[123]

Fig. 82.
Dr T. A. Chapman, F.R.S., best known for his scientific work on lycaenid butterflies, who unlocked the secret of the Large Blue.

61. THOMAS ALGERNON CHAPMAN (1842–1921) (Fig. 82)

T. A. Chapman was born in Glasgow on 2nd July 1842, the son of 'a parent genuinely devoted to natural history', as his obituarist put it.[124] He qualified in medicine at both Edinburgh and Glasgow Universities and worked as a resident physician and surgeon at the Glasgow Royal Infirmary until moving south, taking up an appointment at the Abergavenny Asylum, before becoming medical superintendent of the County and City Asylum in Hereford where he remained until his retirement in 1897. A bachelor, he then moved to live with two unmarried sisters in Reigate.

From this time on, until the outbreak of the First World War in 1914, Chapman spent a large part of every year on the Continent where he made a particular study of the early life history of a number of species of Lepidoptera. His principal interest was in the Lycaenidae, in which area lies his main claim to fame, and he put forward an interesting theory explaining the extinction of the Mazarine Blue (*Cyaniris semiargus*).[125] He also studied microlepidoptera, especially the Tortricidae and Psychidae, as well as publishing articles on Diptera, Coleoptera and Hymenoptera – a number of his early papers were written jointly with his father. His approach to his subject was always scientific. His Continental travels were not random but aimed at the haunts of particular species he wished to study – in the Alps, the Pyrenees, Spain, Sicily or Norway. After breeding them, making detailed examinations of their genitalia under the microscope and writing up his findings, he did not keep his set specimens but gave them all away to friends, leaving no collection of his own. He described the life cycles of various European butterflies, two of which, *Callophrys avis* and *Plebicula thersites*, were respectively named after him as Chapman's Green Hairstreak and Chapman's Blue.[126] But it was his discovery of the nature of the relationship of the fourth instar of the Large Blue larva with species of *Myrmica* ant that solved the riddle that had been puzzling British lepidopterists until that time (see Chapter 4: 22).

Algernon Chapman was elected a Fellow of the Entomological Society of London in 1891 and, after his move to Reigate, was a regular attendant and active participant at meetings where, according to his obituarist, 'His cheery manner and attractive personality soon brought him many friends.' Although he served as Vice-President of the Society on several occasions, he could not be persuaded to accept the office of President 'much to the regret of his friends'. He was highly regarded by his fellow entomologists for the quality and thoroughness of his writings. To P. B. M. Allan, for example, he was the acme of entomologists: 'so distinguished, so erudite, [and] so perspicacious a lepidopterist'.[127] He was elected a Fellow of the Royal Society in 1918 in recognition of his entomological work and was also a Fellow of the Zoological Society. He died in 1921 and is buried in Reigate Cemetery.

62. THOMAS DE GREY, 6TH BARON WALSINGHAM (1843–1919) (Fig. 83)

Lord Walsingham's earliest contribution to entomology was the following manuscript note on an envelope, made when he had just turned eight: 'I have found out that the catipillars hind feete are different to its froant ones'.[128] He went on to become one of the leading authorities on microlepidoptera, whose collection from England, Europe and the New World was one of the largest and most important ever made. It included larvae mounted in natural positions on their foodplants. The ether-bellows used to inflate their skins, seem to have been used for the first time by Walsingham.

Fig. 83. Thomas de Grey, Lord Walsingham, sometime Member of Parliament, country sportsman and proud owner of over 250,000 micro-moths, on which group he was a great authority.

He was born on Stanhope Street in Mayfair, the family's London house, and educated at Eton and Trinity College, Cambridge. He initially chose a career in the army, but, becoming impatient with delays in receiving his commission, he went into politics instead, serving as the Conservative M.P. for West Norfolk from 1865 until succeeding to the title and estates of his father, the fifth Lord Walsingham, in 1870. It was widely thought that he had missed his real calling – the law. For the rest of his life Walsingham ran his family estate at Merton, Norfolk, served as a trustee of the British Museum and undertook numerous other public duties. He also enjoyed himself fishing, and shooting innumerable pheasants, partridges and grouse. In a single day on his Blubberhouses Moor, North Yorkshire (where A. B. Farn was an occasional guest) he shot a record 1,070 grouse – which might not seem so admirable now, but was certainly admired then. Lord Walsingham was also very keen on cricket, since the palmy days when he captained his eleven at Eton and played for his University.

We are, of course, interested mainly in his entomological pursuits. Walsingham had collected butterflies and moths almost since the day he discovered that 'catipillars' had two kinds of feet, but in his maturity specialized in the microlepidoptera, and eventually switched his allegiance from collecting *per se* to rearing. Perhaps his greatest achievements were in rearing so many – perhaps thousands – of the species for the first time, and in describing new ones. He also wrote an important work on the Tortricidae of North America. Though not a systematist to rank with Stainton or Meyrick, he amassed one of the world's finest collections of microlepidoptera, full of type material, which, when augmented by his purchase of the Zeller, Hofmann and Christoph collections, contained over 260,000 specimens. With typical generosity, he donated the whole magnificent collection together with his library of 2,600 books to the British Museum (Natural History), now The Natural History Museum, London, in 1910.

Fig. 84. Francis Buchanan White, the great Scottish naturalist and institution builder. Note the partially hidden sweep-net.

Lord Walsingham was a member of numerous British and foreign entomological and agricultural societies. He was elected a Fellow of the Royal Society in 1887, and was a loyal member of the Entomological Society, which he joined in 1866, and served twice as its President. He died suddenly from a virus infection and was buried at Merton Park, where he had so often hunted moths, fish and game. Though he married three times, he left no heir, and his title passed to his half-brother.

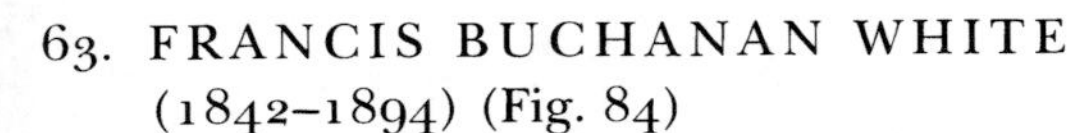

63. FRANCIS BUCHANAN WHITE (1842–1894) (Fig. 84)

Despite his relatively early death at fifty-two, F. Buchanan White was one of the greatest naturalists Scotland has ever produced. To an extent rarely matched south of the border, he combined his all-round knowledge of nature with 'the eye of an artist and the feeling of the poet', as his obituarist in *The Entomologist* expressed it.[129] He saw nature in the round, as well as in its separate components.

White was the eldest son of a leading Perthshire doctor, and it was intended that he should follow his father's career in medicine, for which he did in fact qualify at the age of twenty-two. White's doctoral thesis at Edinburgh University, however, was 'On the Relations, Analogies, and Similitudes of Insects and Plants'. Since he had sufficient private means to do so, White decided to devote his life to the study of natural history rather than medicine. He was founder, in 1867, and the most energetic supporter of the Perthshire Society of Natural Sciences, whose admirable *Transactions and Proceedings* developed under White's influence into *The Scottish Naturalist*, one of the finest long-running journals in the field. White edited both journals until 1882. In 1884, he founded the Union of Naturalists' Societies in Scotland, and served as its first President. He was also, from 1868, a Fellow of the Entomological Society of London and, from 1873, of the Linnean Society to which he contributed scientific papers.

Apart from his abilities as an institution builder, White was pre-eminent as a field naturalist, a great observer and an authority on subjects as various as British willows (on which he wrote a 112-page revisionary paper), mosses and lichens, land and freshwater molluscs, birds, mammals and geology in addition to entomology. Among the insects, his favourite orders were the Lepidoptera, Coleoptera and Hemiptera. His papers in *The Entomologist's Monthly Magazine* include three on insect migration, and *The Entomologist* contains another on the fungal parasites of insects. He had collected butterflies in Europe in 1866 after completing his university studies, and many well-known Scottish localities, especially in Atholl and around Braemar, were discovered by him. He discovered several new species of bugs and moths, the latter including the Scotch Burnet (*Zygaena exulans*) and two micros, *Kessleria saxifragae* and *Eana argentana*. He wrote two major works on the Scottish Lepidoptera. His *Insecta Scotica – Lepidoptera*, published in *The Scottish Naturalist* between 1872 and 1879, was the first full account of the distribution of Scotland's butterflies and larger moths. *Fauna Perthensis*, published in 1871 in the Perthshire Society journal, has been acclaimed as one of the most thorough studies of Lepidoptera of any single county.[130]

White was a good-humoured, powerfully-built man, a great hill-walker and genial companion. Though he held strong opinions, he was always modest enough about his attainments and was endowed with a sense of humour. In a memorial address in 1894, the then President of the Perthshire Society fondly remembered how 'in the midst of serious work, some trivial circumstance would strike him in a ludicrous aspect, and his merry laugh could not but infect those round about him'.

64. WILLIAM FORSELL KIRBY (1844–1912) (Fig. 85)

W. F. Kirby, no relation of the great Revd William Kirby but father of the prolific entomological author W. E. Kirby, was the privately-educated son of a Leicester banker, and a precocious entomologist. His father died when he was ten years old and, in 1856, the family moved to Wandsworth from where, at the age of twelve, he contributed a note to the twelfth number of *The Entomologist's Weekly Intelligencer* on the Scarce Vapourer (*Orgyria recens*). In 1857, the family moved again, to Burgess Hill, and thence to Brighton the following year when, at the age of fifteen, his list of local dragonflies was also published in *The Entomologist's Weekly Intelligencer*, a journal to which he continued to be a regular contributor until it ceased publication in 1861. In 1859 he himself published a *List of British Rhopalocera*, on sale for threepence, of which only a single copy survives. A concise *Manual of European Butterflies* followed in 1862 when Kirby had by now reached the relatively mature age of eighteen, perhaps the first significant work on European butterflies by an English author. The work which established W. F. Kirby's reputation, however, was his *Synonymic Catalogue of Diurnal Lepidoptera* (1871, with a

Fig. 85. W. F. Kirby, museum man, linguist and popular author. 'He always started on his travels armed with a butterfly net and collecting box.'

supplement in 1877), which covers some 10,000 genera and species published since Linnaeus's time, and brought Kirby into contact with many of the best-known entomologists of the day.

Kirby's professional life was spent in museums, initially for twelve years as curator in the Museum of the Royal Dublin Society (later the National Museum). From 1879 to his retirement in 1909, he worked as Assistant in the newly opened British Museum (Natural History) in South Kensington. 'His kingdom', wrote N. D. Riley, included the Hymenoptera (132,000 specimens, of which 34,000 were unidentified!), in addition to the 'Orthoptera, Neuroptera, and indeed all the Orders except Lepidoptera, Coleoptera and Diptera'.[131] Kirby also held office as Hon. Secretary of the Entomological Society of London from 1881 to 1885. W. T. Stearn described him as 'a quiet, retiring man of prodigious industry'.[132] A contemporary recalled his 'genial kindliness, tact and quiet amiability' and readiness to help others. He had a talent for languages, including Finnish, reputedly one of the hardest to learn, from which he produced an English translation of the great Finnish epic, the *Kalevala*.[133] He travelled, often with net in hand, in Europe and North America. He was more widely known, however, as the author of popular works including *European Butterflies and Moths*, published in monthly parts between 1878 and 1882, and the *Elementary Text-book of Entomology* of 1885. An early disciple of Darwin, he also dabbled in chemistry and theology, compressing his own views on the origin of the cosmos into a small book titled *Evolution and Natural Theology*. His Museum work included a succession of catalogues on various groups of insects, all, writes Stearn, 'esteemed for their reliability'. He is buried in Chiswick cemetery.

65. WILLIAM HOLLAND (1845–1930)

In Reading Museum is a fine (though mostly unlabelled) collection that once belonged to William Holland, who has been called 'the most famous lepidopterist [Berkshire] has produced'.[134] Holland was the man who, as P. B. M. Allan recalled, had 'once picked up a larva of *Stauropus fagi*, the Lobster Moth, from a path in a beechwood. Said he, "it very closely resembled a curled-up leaf like those beside it on the path" '. As Allan remarked, 'A man who can do that sort of thing can do anything, anything'![135]

Holland was born and lived in humble circumstances. In the words of H. M. Wallis, who knew him well, 'while supporting himself as a working shoemaker ... [he] succeeded in learning almost all that could be learnt of the butterflies and moths of the South of England and more especially of Berkshire, Oxfordshire and Hampshire. Under extreme difficulties of narrow means and a house shared by others, he worked at his life-long study and accumulated the really remarkable collection, which ... has been one of the most constantly used and most treasured exhibits of the [Reading] Town Museum'.[136] In 1893, Holland was appointed by Professor E. B. Poulton as Assistant to help organize and arrange the insect collections of the Hope Department, serving there until his retirement in 1913.

Holland's dedication was legendary: it was normal for him to walk the twelve miles from his home to Pamber Forest, collect there all day, eating his sandwiches as he scanned the trunks, and then walk all the way back. More than one private estate grew suspicious of Holland's night visits with lantern and treacling tin. One put up a notice, 'Caution to Butterfly Catchers', now also in the Reading Museum, threatening collectors – Holland in particular – with prosecution. Eventually, however, Holland seems to have won the confidence of most local landowners, and also of his rival collectors, and few have done so much to reveal the entomological riches of the Thames valley. Among many notable captures were one of the earliest known specimens of the Golden Plusia moth (*Polychrysia moneta*) at a gas-lamp in Reading, a male Gypsy Moth (*Lymantria dispar*), which flew up while Holland was chasing after Large Tortoiseshells, and the first British specimen of the beetle *Gynandrophthalma affinis* at Wychwood in 1899. He submitted numerous pieces and 'Practical Hints' to entomological journals, including a lengthy note on entomological pins for the *Entomologist's Record*, which received some hard-won praise from J. W. Tutt. Holland was still visiting Pamber Forest into his eighties. He was one of the most distinguished working-man naturalists, much in demand in his later years for his unique knowledge and ability to track down rare insects.

66. HENRY JOHN ELWES (1846–1922) (Fig. 86)

The Natural History Museum contains many collections of butterflies and moths, but few are bulkier than that of H. J. Elwes, who over the years donated some 11,370 specimens of Palaearctic butterflies, all collected personally, to the Museum. He also specialized in Oriental skipper butterflies (Hesperiidae), on which he was probably the world authority. Like many collectors of the late Victorian and Edwardian period, Elwes was an outdoors man, keen on field sports, and well known as an arboriculturalist and gardener. N. D. Riley described him as 'a giant of a man and a very dominating character'.[137] W. T. Stearn recalls his domineering 'booming voice which carried well across his Gloucestershire estate, but was very disconcerting elsewhere' – for example at the Royal Horticultural Society flower show, where someone was overheard wondering: 'Is that a man or a foghorn?'[138] Unfortunately for the nerves of the Museum workers, Elwes insisted on rearranging the collections himself.

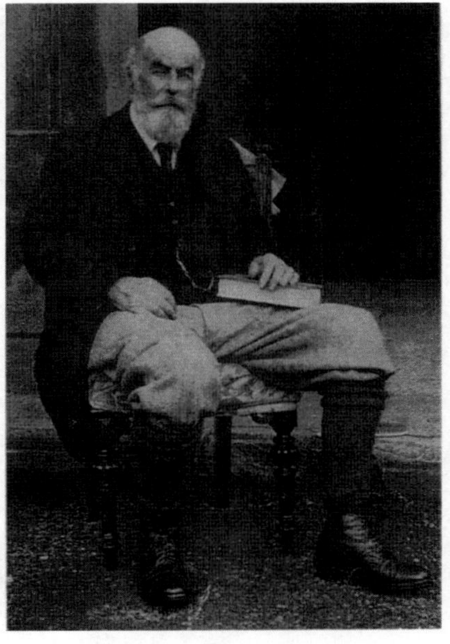

Fig. 86. H. J. Elwes, whose giant frame and fog-horn voice proved disturbing to entomological scholars.

67. HENRY HARPUR CREWE (1828–1883) (Fig. 87)

68. SIR VAUNCEY HARPUR CREWE (1846–1924) (Fig. 88)

The Harpur Crewes of Calke Abbey[139] were an ancient family with more than their fair share of eccentrics but they all had a profound love of the countryside around them and its natural history. The ancestral estates in Derbyshire provided an ideal setting within which this affection was nurtured during childhood, there being every opportunity to study the birds, insects and plants as well as to engage in the more conventional pursuits of the landed gentry, such as shooting.

Henry Harpur Crewe, nephew of the 8th baronet, Sir George Crewe of Calke, was the eldest son of the Revd Henry Robert Crewe, rector of the nearby parish of Breadsall, and it was there that the earliest of young Henry's natural history observations were made. He followed his father into the church, eventually becoming rector of Drayton-Beauchamp, near Tring, where he was the incumbent from 1860 up to the time of his death in 1883. Apart from his parish duties, he lived the comfortable life of the well-to-do clergyman, much time being apparently available for intellectual pursuits – in his case, ornithological,

Fig. 87. The Revd Henry Harpur Crewe, rector of Drayton-Beauchamp, an archetypal English parson-naturalist.

Fig. 88. Sir Vauncey Harpur Crewe of Calke Abbey, an eccentric recluse whose collection rivalled Lord Rothschild's.

entomological, botanical and horticultural. He also found time for many a good day's game shooting.

At the age of nineteen, Henry Harpur Crewe was already contributing observations on natural history to *The Zoologist* and he was a regular correspondent in the course of his life to journals such as *The Entomologist's Weekly Intelligencer*, *The Entomologist's Annual*, and *The Entomologist's Monthly Magazine*. He was interested in rearing the larvae and developed a particular expertise with Pug moths (*Eupithecia*) which so fascinated him that in 1858 he announced his determination 'if possible to get the larvae of the whole family illustrated'.[140] He advertised for larvae and began writing what was eventually a series of eighty-four accounts of the early stages of these difficult little moths that were classics of their kind and treated all of the fifty-one species and subspecies known at the time. His last paper was published in 1881.[141] He described the Jasione Pug and the Ash Pug, giving them the scientific names *E. jasioneata* and *E. fraxinata*. Unfortunately for him *jasioneata*, though an accepted form, was downgraded to a subspecies, and *fraxinata* has been considered by most lepidopterists to be a synonym of another species.

Henry Harpur Crewe was also a good botanist and gathered records which appear in W. R. Linton's *Flora of Derbyshire* (1903) and G. C. Druce's *Flora of Buckinghamshire* (1926). He was a keen horticulturalist too, and the results of his cultivation of crocuses, in search of which he more than once travelled to unfrequented parts of Europe, were illustrated in *Curtis's Botanical Magazine* (1875, t. 6168). He was also an expert on wallflowers, one named after him still appearing in gardening catalogues.

The Revd Henry Harpur Crewe suffered from poor health for most of his life and died after a protracted illness at the early age of fifty-four. After his death his herbarium was deposited at the British Museum (Natural History), and his correspondence in the library of the Royal Botanic Gardens, Kew.[142] It is not known where his collection of Lepidoptera went – perhaps to Calke Abbey?

By contrast, Henry's first cousin once removed, Vauncey Harpur Crewe, who succeeded his father as 10th baronet, represented the less conventional side of the family. 'There goes Sir Vauncey!' the country people would say, when they saw a crazily driven carriage, rushing along like the north wind. Named after a distant mediaeval ancestor, he displayed some of his family's more extreme traits of eccentricity, before time took its toll and he became a recluse.

Sir Vauncey, too, had a passion for the countryside and its wildlife but in his case this developed into an obsession. His early years were those of a delicate child who loved to watch birds and butterflies in the park and surrounding woods. He was educated privately and although his tutor no doubt taught him the three Rs, Latin and Greek, he also learned how to stuff birds, many of which he had shot himself. At the age of twelve he wrote the first of a two-volume *Natural History of Calke and Warslow*, in which he listed more than a hundred species of birds he had seen in the park or on the lake at Calke. He treated the Calke Abbey estate as his private game reserve, and year by year the house came to look more and more like a museum. During his lifetime Sir Vauncey amassed a huge collection of butterflies and moths said, in its later years, to be second only in size to the collection of Lord Rothschild at Tring. In 1890 Sir Vauncey was elected a Fellow of the Entomological Society of London, remaining so until his death. He was also a long-time member of the British Ornithologists' Union.

In 1876, he married the Hon. Isabel Adderley, daughter of Lord Norton, the couple being driven from the church to Calke by way of the surrounding villages and fêted at each stop with flowers, bunting and triumphal arches. Unfortunately that day – remembered locally long afterwards – was perhaps the high point of the marriage. It appears that Lady Harpur Crewe had been unaware of her husband's obsessions. One of her complaints, which recurred again and again, was his management of the fires, which he insisted must burn morning, noon and night in each room in order to maintain stable temperature conditions for his various collections. If one of these fires became too hot, or alternatively was allowed to go out, he would send a written message by a footman to the culprit ordering their instant dismissal. Fortunately, it is said, Sir Vauncey hardly knew one servant from another, so these dismissals were generally ignored, and servants managed to survive two or three sackings without their employer being any the wiser. It was not long before Sir Vauncey and his wife started leading separate lives – though she nonetheless bore him four daughters.

Apart from serving for a short period as High Sheriff of Derbyshire, Sir Vauncey played little part in public life. In later years he became more and more reclusive. His Aunt Isabel noted 'how completely he is losing, or rather has lost, all position in the County. It vexes me terribly, I can't understand him. He does not seem to know how to behave like a gentleman'.[143]

Sir Vauncey's constant companion was Agathos Pegg, the head gamekeeper, who would accompany him on collecting trips. On one particularly sunny day, Sir Vauncey decided to collect butterflies near Repton Park, a small country house on the Calke estate, occupied at that time by his cousin, John Harpur

Crewe. The cousins were not on good terms, and the sudden appearance of Sir Vauncey and Agathos Pegg with their butterfly nets led to an altercation. Enraged, Sir Vauncey returned to Calke and ordered his agent to have Repton Park razed to the ground. The task was accomplished within a week. On another occasion, Sir Vauncey caught his daughter Airmyne smoking. She was immediately banished from Calke and not allowed back during her father's lifetime. Sir Vauncey was the last baronet of Calke. The greater part of his collections of Lepidoptera was auctioned in 412 lots at Stevens' Auction Rooms after his death in 1924, but the contents of Calke Abbey remained virtually untouched for almost sixty years (Plate 9), until the house was taken over by

Plate 9
Part of Sir Vauncey's huge collection of Lepidoptera escaped the saleroom and was uncovered by the National Trust when it took over Calke Abbey in 1985.

the National Trust in the 1980s and completely restored. It is now open to the public, and Sir Vauncey's extraordinary collections of stuffed birds and other natural history exhibits can be seen there more or less as he left them.

69. RICHARD SOUTH (1846–1932) (Fig. 89)

N. D. Riley recalled how, as a small boy, he would enlist the help of his next-door neighbour, who happened to be the great Richard South. Riley particularly remembered 'the quiet, unfailing patience, with which all his innumerable questions were answered, often many times a day, and the care with which his earliest captures, frequently almost unrecognizable on account of their rough handling, were identified'. He often spent time discussing entomological matters in 'Mr. South's "den"', and one of his 'greatest treats at this time was to be invited by Mr. South for a day's collecting. In this way he learned to appreciate his courtly and altogether delightful character, to catch some of his enthusiasm for his "bugs" and to honour him as a staunch personal friend'.[144] Not many entomologists can have been blessed with such a neighbour!

South's name will always be associated with the two books he wrote for Frederick Warne's Wayside and Woodland library: *The Butterflies of the British Isles* (1906) and the two-volume *Moths of the British Isles* (1907–08), which covered all the species of macrolepidoptera. Although South himself acknowledged his debt to C. G. Barrett's great works (q.v.), South's books added much new information, especially on distribution, and were, moreover, revised regularly (after South's death by H. M. Edelsten). They were the standard popular works for much of the twentieth century, the butterfly volume remaining in print until the 1960s, while those on moths were last reprinted in 1980, nearly 50 years after the death of the author! South's famous books were among the first to include colour photographs of all the British butterflies and larger moths (only in later editions were the pictures of actual specimens replaced by coloured drawings). Norman Riley was not alone in considering them 'by far the best thing of their kind ever produced' and exactly the kind of book that had been needed for so long. They were innovative in so many respects: they could fit into the jacket pocket; and gone was the customary Victorian prolixity and classical preciousness, to be replaced by brisk notes on habits, distribution and life cycles. And they were mass-produced, and so cheap. No books have been kinder to the collector and field excursionist. South had pioneered a new kind of book, a pocket guide illustrated by photographs, which set the style for the new century.

Fig. 89. Richard South, part-time museum entomologist of independent means, editor of *The Entomologist,* and author of the standard works on British butterflies and moths which, in Norman Riley's estimation, were 'the best of their kind ever produced'.

With his dignified manner, handlebar moustache and formal attire, South hardly looked the part of a revolutionary. He was born in Marylebone of a fairly prosperous family who counted among its forebears a celebrated divine, Dr Robert South, chaplain to James, Duke of York (later King James II). Educated at a private school in Reading, Richard South early married his cousin Sarah, and secondly, after her death, Evelyn Urquhart, daughter of a former mayor of Paddington. South lived in London throughout his life. With private means he was able to devote much of his time to entomology, serving as editor of *The Entomologist,* as a special assistant at the British Museum (Natural History), and as an active member of the leading societies, including a thirty-year stint as Secretary to the exclusive Entomological Club.

South did not contribute an entomological paper until he was nearly thirty, but from then onwards he wrote many short notes and longer articles, mainly for *The Entomologist* which he owned for many years, with occasional excursions into *The Entomologist's Monthly Magazine.* Although much of his early work was on British butterflies, he later specialized in the micros, especially the pyrales, plumes and tortricids and developed a 'marvellous aptitude for naming in the field'.[145] From 1887 he published many notes on the butterfly genus *Lycaena*, which led him into conflict with the ever-formidable J. W. Tutt, whose concept of what constitutes a species differed from his own. Similar arguments followed South's publication of his *Synonymic List of British Lepidoptera* (1884) in *The Entomologist*; few entomologists take kindly to changes from the old familiar names. South made a dignified reply to his critics, though with some barely disguised barbs presumably directed at Tutt, in his address as the newly-elected President of the 'South London' in 1885:

> 'We should remember that it is not by mere verbosity and liberal use of the first person that we are most likely to convince others of our being right, and their being wrong ... Difference of opinion leads to discussion but the legitimate outcome of argument should be elucidation of the truth. At the same time, however, the ordinary courtesies of debate oblige us to treat opinions of others with respect ... To essay the overthrow of an opponent by a flood of sarcastic criticism and personal invective is altogether unseemly, and quite out of place either in writing or speaking on subjects pertaining to science.'[146]

South collaborated with J. H. Leech on oriental butterflies and, after the latter's death in 1900, became the recognized authority. He spent much of his time in the first years of the twentieth century arranging, at the request of the Trustees, the world collections of Lycaenidae and Hesperiidae at the BM(NH), as well as the British collections of butterflies and moths (in which he was succeeded by his protegé, Norman Riley). South's reign as one of the most active and respected lepidopterists in the country spanned about thirty-five years, from 1880 to 1914. The Great War affected him, both personally and in his efforts to maintain *The Entomologist* in the new climate of rising costs and falling subscriptions. He parted with his fine collection – his second – in 1924, having sold his first in 1898; parts of it are now owned by the City of Birmingham, while South's important collection of Tortricidae is now in the Natural History Museum, London. His death, wrote Riley, seemed like 'another milestone – the end of another chapter'.

70. ROBERT ADKIN
(1849–1935) (Fig. 90)

Robert Adkin was born in Lewisham, in 1849. He entered the tobacco business, first as a partner in his family firm Adkin Brothers of Aldgate, London, and, from 1901, as a Board member of the Imperial Tobacco Company. Unlike many wealthy men, he was also extremely generous. His benefactions, though numerous, were never ostentatious. According to his obituarist, W. G. Sheldon, Adkin hated self-advertisement. He was 'one of the simplest and most unassuming of personalities ... always keenly on the look-out for something that was wanted, and when that something was found his usual remark was, "I'll see to that", and see to it he invariably did'.[147]

Over a period of some fifty years, from 1885, when he purchased a brass name-plate for the front door of the 'South London', to the bequest of £200 and 1100 lantern slides on his death, his generosity to that Society was bountiful, and covered, among many other items, printing and engraving costs, furnishings and a reading stand fitted with an electric light.[148] When the Royal Entomological Society was purchasing its present premises at 41 Queen's

Gate, South Kensington, Adkin 'made a handsome donation to the housing fund and subscribed for its debentures, and these latter when drawn for repayment were cancelled by him, as he said, "in the hope that other holders of these debentures might follow his example".'[149] He also supported L. W. Newman's newly established Butterfly Farm by guaranteeing to purchase £100 of stock annually for five years.

Adkin was also noted for his hospitality. He provided an annual supper and conversazione for the officers of the 'South London' and, following his move to Eastbourne in 1915, his house there (a former home of the naturalist Thomas Huxley) was a particularly popular venue for the roving diners of the exclusive Entomological Club.

Fig. 90.
Robert Adkin, tobacco magnate, treasurer of entomological societies and patron of needy entomologists.

Adkin had been interested in entomology from his youth, and built up a magnificent collection of some 47,500 British Lepidoptera, which he left to the BM(NH). He was elected a Fellow of the Royal Entomological Society in 1885 and served on its Council for eight years and, in 1922 and 1928, as its Vice-President. Adkin always modestly refused the honour of that Society's Presidency, though he presided over the 'South London' no fewer than five times and served on its Council almost continuously. He was also an active member of the Ray Society, and on the editorial panel of *The Entomologist.* He was a frequent contributor, mainly of short notes, to this and other entomological magazines. He compiled a complete list of the Lepidoptera of Eastbourne and district, in which he included many species obtained by regular light-trapping.

Late in life, when failing health restricted his mobility, Adkin studied insect pests in his garden and supplied useful information to various institutions studying economic entomology. On his death, the Royal Entomological Society was given the pick of his library, said to be one of the most complete and valuable in private hands. He was a great loss to entomological societies, especially the ever impecunious 'South London' and his local Eastbourne Natural History Society. 'Of them', his obituarist wrote, 'he was known as the father – and a model father he was'.

71. SIR EDWARD BAGNALL POULTON (1856–1943) (Fig. 91)

J. Obadiah Westwood, Poulton's predecessor in the Hope Department, Oxford, had never accepted Darwin's theory of natural selection. He once 'remonstrated with the authorities of Jesus College for allowing such a dangerous volume as *The Origin of Species* to fall into the hands of a young undergraduate'.[150] That undergraduate happened to be Edward Poulton, a promising pupil who would one day be among the foremost exponents of evolution, and succeed Westwood as the Hope Professor of Zoology!

Edward Bagnall Poulton was born in Reading and educated at a succession of boarding schools, of which his main memories were of punishments and the teaching of 'the broad principal that every natural inclination was wrong and dangerous'.[151] He took a first in natural sciences at Oxford, specializing in insects, and embarked on a brilliant career at the University, culminating in his election to the Royal Society at the early age of thirty-four. In 1889, however, Poulton interrupted his university career to devote more time to his own interests and writings, producing in the following year *The Colours of Animals*, the first comprehensive guide to the meaning of the colours of animals, birds and insects, which confirmed him as an ardent champion of the theories of Darwin and Wallace. In 1893, the Hope professorship fell vacant after the death of Westwood. Poulton was elected – he had been associated with the Hope collections and library since his schooldays – and retained the post until his retirement in 1933.

Poulton was a keen collector, so much so that he was nicknamed Edward 'Bag-all' Poulton. Still preserved in the Hope Department are dozens of his biscuit tins (his wife Emily was the daughter of George Palmer of Huntley & Palmer's biscuits) stuffed full of unset butterflies and moths. His colleague and successor, G. D. Hale Carpenter wrote that under Poulton the Hope collections 'grew to an extent embarrassing even to himself, and he could neither cope with them nor refuse accessions'. On the other hand, it was said that Poulton never killed an insect unnecessarily. He was first and foremost a scientist, and 'If he had to kill', argued the Revd L. B. Cross at his funeral, 'it was always in order to make a sacrificial offering, a sweet savour, on the altar of truth'.

Fig. 91. Professor Sir Edward Bagnall Poulton, F.R.S., university academic, who reputedly could forgive his students 'almost anything, except disbelief in the doctrine of evolution'.

Like Westwood, Poulton had a reputation for parsimony. Audrey Smith relates the story of how, when a carpenter came to do a small job in the Department, Poulton would carefully count out the exact number of screws or nails needed, and if one were lost, the carpenter was expected to replace it with one of his own. After one of these episodes, one old cabinet-maker carefully swept up all of his sawdust into a bag and handed it to Poulton, saying solemnly, 'No doubt you will be requiring this, Sir!'[152] Drawers in the Department were full of bits of string that Poulton thought too valuable to throw away, and he continued the Westwood habit of writing his learned articles on odd scraps of paper. Yet despite these thrifty habits he was a most generous benefactor to his college and to the Hope Department.

Though Poulton never produced the *magnum opus* that many expected from him towards the end of his career, he was a highly honoured scientist, a holder of the Darwin Medal and Linnean Medal, and three times President of the Royal Entomological Society, the last on the occasion of its centenary when he was made only its third Honorary President. He was knighted in 1935 for his contributions to science. It was said of him that he lacked ambition, but when such honours came to him he received them graciously, and with deep appreciation. His sympathies and obligations were rooted firmly in the academic and social traditions of Oxford. He would forgive anyone 'almost anything, except disbelief in the doctrine of evolution'.

What one student remembered most about him, many years later, was the way Sir Edward Poulton blew his nose. 'Look at his portrait in the hallway of No 41 [Queen's Gate, home of the Royal Entomological Society],' wrote J. D. Gillett, 'and you will see that a certain obstacle in the shape of an enormous moustache stood in the way of easy operation. The performance started with a more or less almighty blow and was quickly followed by what can best be described as a ritualistic wiping of his nose and moustache and pretty well all his face, first one side then the other and back again. The whole performance was executed deftly and at enormous speed, like some giant dipteran grooming its antennae.'[153]

72. JAMES WILLIAM TUTT (1858–1911) (Fig. 92)

J. W. Tutt was a personality about whom it was impossible to be neutral. His expressed views were always as emphatic as his intellect was prodigious. It was, wrote his friend and admirer, T. A. Chapman, 'simply impossible for him to be idle; he must work away at full steam all the time'.[154] He set himself the highest standards, and expected them of others. His manner, sometimes tart and sarcastic, even rude, offended some of the more gentlemanly entomologists, and South and Frohawk were among those who went so far as to resign their memberships of societies in order to avoid him. 'Yes, I know I am often brutal in the way I put things', Robert Adkin recalled him saying, 'but I can't help it, and you know I am right'.[155] On the other hand, although Tutt had no patience for opinions he considered wrong, he had sufficient intellectual honesty, on finding himself in error, to admit the fact immediately. His friends, at least, thought his heart was in the right place. Chapman probably came close to the nub of the matter in considering him to be 'a born schoolmaster; he must be learning himself or teaching others'. 'The very prominence of his character offended some,' wrote Malcolm Burr. 'Long years devoted to the teaching profession made him didactic, even authoritative and dogmatic, and he made no claim to polished courtliness nor superficial grace.'[156]

Fig. 92. J. W. Tutt, autocratic headmaster, in his prime. 'I know I am brutal in the way I put things', he said, 'but I can't help it, and you know I am right'.

Tutt was born at Strood in Kent, and educated at St Nicholas School in Plumstead and St Mark's Training College at Chelsea. After obtaining the necessary Board of Education certificates and matriculating at London University, though without taking a degree, it was not long before he was appointed headmaster of Snowfields Board School, Bermondsey, followed over the course of his teaching career by the headships of other East London schools at Webb Street and Portman Place and lastly to the higher grade Morpeth Street Central School. He was once offered the post of Principal at a college in Calcutta, but refused it on health grounds: in his youth, Tutt had suffered from a defective heart, and although he seemed to recover in his maturity, it probably shortened his life.

He was, like so many entomologists of the day, an avid collector. Adkin remembered meeting him for the first time: 'I had spent the latter half of a glorious summer day in Chattenden Roughs, had filled perhaps some three score boxes, no doubt was feeling some pride at my success and was preparing to make my way home as the shadows of night were falling, when I met a youth, a stranger. He was in the act of transferring pill-boxes from his pockets to a good sized portmanteau, which was then practically full of them, and which he assured me represented only his day's captures. It was, I suppose, some disparaging remark of mine that drew the reply: "It's no use taking one or two specimens of a species, if you want to know anything about it, and why as to that, I shall be up early tomorrow morning and have them all set by night".'[157]

After he died, everyone had a story to tell about him. H. Smetham recalled one told him by a Mr G. Robinson who was out one night 'sugaring' with Tutt:

they had come to a stile which they daubed with a mixture of treacle, beer and rum before moving on to 'sugar' other prominent sites. On their way back they saw in the gathering gloom a courting couple sitting on that very stile, quite unconscious of its sticky coating. 'As explanations might have been attended with painful possibilities (and could not possibly lend themselves to pleasurable explanations) our friends chose that better part of valour known as discretion, and left the lovers to their transient happiness. What they afterwards thought of the sticky decorations attaching to their garments, history has never yet recorded.'[158]

Tutt often gave public lectures on natural history subjects. One of them, at the Strood Institute, was on 'The Wonders of Insect Life', in which he waxed lyrical on the love of intoxicants of butterflies and moths. It must have attracted some journalist, for the lecture was reported in several daily papers. The editor of *Punch*, spotting an opportunity for a lark, published the following ditty, entitled 'Tut, Tutt'.

'Mr Tutt, who tells no lies,
Tells us that the butterflies
Are alas, what do you think?
Let me whisper, fond of drink!

He has watched them on the flow'rs,
Where they'll sit and suck for hours,
Quite devoid of any motion
Save absorption of the 'lotion.

Thus they spend the summer's day
While the females work away,
For this craving to regale
Is restricted to the male.

Lost illusion of our youth
In a scientific truth,
Tear drops gather in our eyes
When we think of butterflies.'

Tutt was among the most prolific of the Victorian entomologists. A Tutt memorial number of *The Entomologist's Record*, which Tutt had helped to found in 1890 and himself edited for twenty-two years, filled sixteen close-spaced pages with the titles of over 850 contributions he made to nine different entomological journals.[159] He was also a very active member of the 'South London' and the Entomological Society (of which he was president elect at the time of his death), serving on the Council of both bodies, and helping to maintain the high standard of their publications. Of his literary methods, Chapman commented that 'He seems to have explored with meticulous care the magazines, the various systematic works on Lepidoptera, and, in fact, all accessible published matter, and arranged extracts or made indices, so that he had collected a mass of information available in almost every direction. It was in this way that he successfully brought together the many scattered observations which enrich his sadly unfinished *magnum opus*, the 'British Lepidoptera'.

Tutt's two large-scale works, *The British Noctuae and their Varieties* in four volumes (1891–92), which included hundreds of descriptions of new races and aberrations, and the *Natural History of the British Lepidoptera* (1890–1911), were based on his own collection, and represent the culmination of Tutt's entomological career. Nine thick volumes of the latter work, four on butterflies and five on moths and all noted for their encyclopaedic detail and their critical approach to nomenclature and classification, had appeared by the time of his death. Tutt was not a man to do anything by halves, and it must be admitted that readability is not among the strongest points of the great work. The eleventh and last of the planned set of volumes, completed by his old friend the Revd George Wheeler, was published posthumously in 1914, but Volumes VI and VII were never produced. Wheeler had been taken aback when he received the late author's notes, 'delivered into my keeping in four large envelopes ... On opening them I was simply appalled by the task that lay before me. They were filled with pieces of paper of all sorts and sizes, each containing one or more notes (if more than one, generally on points quite unconnected with each other), the beginnings of paragraphs, tiny slips containing a single magazine reference, no subject being specified; small items of information in handwriting quite unknown to me. All this, which Mr Tutt, with his marvellous memory, would have quickly reduced to order, was to me a hopeless chaos.'[160]

Tutt wrote more popular volumes, including *British Butterflies* and *British Moths*, both published in 1896, and a number of other natural history titles which reveal not only his acute powers of observation but also more than a touch of the poet.

Overwork probably contributed to Tutt's early death, still at the height of his powers. He suffered a stroke, and, it is said, his hair turned white. A large number of entomologists from Britain and overseas attended his funeral at Lewisham. In his will, Tutt left no fewer than twenty-four properties, mainly terraced houses in East London and North Kent, and all but one freehold – an interesting legacy from a poorly-paid Victorian schoolmaster![161] Most regrettably, his collection was broken up into lots and sold at auction. As J. M. Chalmers-Hunt remarked, this was a disaster for future work on the variation of Lepidoptera, since Tutt's descriptions were based mainly on his own specimens. The prices realised were surprisingly low, and a great opportunity was missed to acquire the collection for the nation.[162]

Today, while Tutt's reputation remains high, his works are seldom read or cited and in some respects his contribution is probably underrated. The only exception is his still useful *Practical Hints* articles, which have recently been reprinted in facsimile with an Index and cross-referencing giving the original names and the modern names in use today.[163] As Denis Owen has pointed out, Tutt had observed phenomena like industrial melanism years before scientists got to work on it, but his work was not published in weighty science journals but tucked away in periodicals and popular books. Owen suggests he has been ignored for two main reasons: first, Tutt was an amateur and a moth collector, and was not perceived as a scientist, even though he repeatedly advocated a scientific approach to the study of moths. At the same time he did no experiments, and his explanations of the adaptive significance of coloration in various moths were based entirely on observation and supposition – that is, on guess-work.[164] But, as Tutt might have said, 'all the same, you know I was right'.

73. EDWARD MEYRICK (1854–1938) (Fig. 93)

Though an amateur (insofar as you can make any distinction among nineteenth-century entomologists), Edward Meyrick enjoyed a worldwide reputation as an authority on microlepidoptera. Over a long and active life – he was working on micros until a few days before he died – Meyrick described some 20,000 new species from all over the world; indeed, it would be hard to find a country from which he had not described at least one new species. His collection of some 100,000 specimens, including several thousand Types, contained so large a proportion of the world's known microlepidoptera that Meyrick seldom had need to consult the national collection in the BM(NH) (now the Natural History Museum – where his own collection resides).

Fig. 93. Edward Meyrick, F.R.S., classics master and world authority on microlepidoptera.

Meyrick was born in Ramsbury, Wiltshire, the son of the local rector, and educated at Marlborough College, where he was later to spend much of his professional life, and at Trinity College, Cambridge, where he read classics. He spent the early part of his career as a school teacher in Australia and New Zealand, where he found an amazingly rich and largely unworked fauna of micros. The result was a prolific outpouring of papers, which Meyrick later extended to islands of the South Pacific and the Orient. He returned to Britain in 1887 to take up a post at his old school, Marlborough, as an assistant master teaching Latin and Greek, and remained there until his retirement in 1914.

Meyrick's *Handbook of British Lepidoptera*, published in 1895, was the first work to deal with the whole of the Lepidoptera since Stainton's two-volume *Manual*, published nearly forty years earlier. Much had changed since then, and Meyrick's fundamental reclassification raised a storm among the old guard – to which Meyrick made a dignified and reasonable answer in a paper contributed to *The Zoologist* in 1898.[165] His *Handbook* quickly became the standard work especially for the micros, and a revised edition was published in 1928.

In 1912, Meyrick, impatient with the delays in publication of so many foreign journals, started to issue his own '*Exotic Microlepidoptera*', described in his preface as 'a spasmodic entomological magazine on one subject by a single contributor', which he produced at intervals until 1937. It includes much of his later work, which by then was covering the Lepidoptera of India, South Africa and South America, not to mention papers on European pyralids and plumes.

Meyrick was a popular schoolmaster, who did much to encourage the study of natural history in a tradition that lives on at Marlborough College. He was elected a Fellow of the Royal Entomological Society in 1880, of the Zoological Society of London in 1889, and of the Royal Society – by then, an exceptional honour for an amateur – in 1904. In retirement he was president of his local Conservative Association (lepidopterists tend to be conservative by nature, if not always in politics), and his home, 'Thornhanger', received 'microlepidopterous pilgrims' from near and far. Part of Meyrick's prodigious output – some 420 books and papers – can be attributed to his methodical work practices: he made a regular habit of working until an hour after midnight. Like many entomologists, his handwriting was a thing of beauty, all the better for inscribing tiny labels. Personally, he was a modest, retiring man, better known to his Marlborough pupils than to the wider entomological world. Among microlepidopterists, however, he was and is a household word.

74. FREDERICK WILLIAM FROHAWK (1861–1946) (Fig. 94)

With Frohawk we reach one of the greatest names in the history of British butterflies. Frohawk was primarily a zoological artist. He contributed many drawings to *The Field* (in those days classed a newspaper), for the BM(NH), and for scientific periodicals, especially *The Entomologist.* He was also, for many years, natural history editor of *The Field*, and himself contributed many of its nature notes and columns. But for most people, Frohawk will be remembered as the author of the mighty *Natural History of British Butterflies*, published in two folio volumes in 1924 and illustrated throughout with his matchless life-size coloured drawings of the butterflies and their early stages (see Plate 40). The preparatory work for this book took nearly a quarter of a century, from 1890 to 1914. Two other books followed: *The Complete Book of British Butterflies* in 1934 and *Varieties of British Butterflies* in 1938. Together they place Frohawk in the very forefront of twentieth-century butterfly illustrators; in some people's eyes, he was the best of all time.

Fig. 94. F. W. Frohawk, author and artist, wearing his characteristic tweed Norfolk jacket 'cut square at the neck'.

F. W. Frohawk was born at Brisley Hall, East Dereham, Norfolk, in 1861, the youngest child of a gentleman farmer of relatively affluent means. His mother, in particular, encouraged his aptitude for drawing and his interest in natural history. His interest in butterflies probably dated from the moment – still vivid in his memory seventy years later – when, at the age of seven, he saw 'a lovely Pale Clouded Yellow butterfly at rest on a dandelion blossom in our pasture, armed with a very primitive net, the ring of a crinoline-steel, bound to a hazel branch, and the bag made of a muslin curtain. I dropped on my knees and stealthily crept up to the butterfly and suddenly plopped the net over it, greatly to my intense joy. I instinctively seemed to realise the rarity of my capture'.

Soon after this the family moved first to Great Yarmouth and then to Ipswich as Frohawk's father wished to be near the moorings where he kept his yacht. Once again there was plentiful wildlife for the young Frohawk to enjoy, including Large Tortoiseshells which were abundant around Ipswich in 1872–73. Sadly, his father overreached himself and died around this time, leaving his family in very straightened circumstances. His mother moved them south of the Thames, first to Croydon and then to South Norwood. This area was a grave disappointment to Frohawk after East Anglia. To add to his woes, he contracted typhoid while a schoolboy at Norwood College, South London, leading to the almost complete loss of sight in one eye. These vicissitudes make his later achievements all the more remarkable. By using his artistic talents and his interest in natural history, he was able to provide a career for himself though one far removed from his background. In 1880 the family moved to Upper Norwood where Frohawk perfected his illustrative skills before obtaining his first commission from *The Field.*

Frohawk loved to escape from suburbia into the countryside. Of all localities, he was most fond of the New Forest, then in its heyday for butterfly collectors. Of one particular visit in 1888, his first, he recalled:

> 'Insects of all kinds literally swarmed. Butterflies were in profusion, the Silver-washed Fritillary (*Argynnis paphia*) were in hoards [sic] in

every ride and the beautiful var. *valezina* was met with at every few yards, as were both the Dark Green [*Argynnis aglaja*] and High Brown [*A. adippe*] Fritillaries, the elegant White Admiral (*Ladoga camilla*) were sailing about in quantity everywhere. On a bank under a sallow was a large female Purple Emperor [*Apatura iris*] with its wings expanded in the sun, evidently washed out of the sallow by rain. The Large Tortoiseshell (*Nymphalis polychloros*) was a frequent occurrence and the Brimstone (*Gonepteryx rhamni*) abundant in every ride.'[166]

However, to Frohawk collecting was less important than the painstaking task of breeding all the British butterflies in order to draw and describe each stage in minute detail. To realize the magnitude of this self-imposed task, we should remember that in 1890, when he began, most entomologists believed that butterflies would not lay in captivity, or, at least, not readily. As a result, the earliest stages of many species was barely known. Frohawk proved them wrong. The task, though, was formidable:

> 'At every stage of the cycle accurate figures and descriptions were to be attempted.
>
> 'It will be understood that this programme necessitated close and continuous observation from the moment of the deposition of the egg up to the final emergence of the perfect insect. Not infrequently ... specimens had to be watched night and day continuously over considerable intervals of time, and the author will probably be believed when he confesses that in this way he performed many a tedious task.
>
> 'In practice it was found to be impossible to undertake the study of more than three or four species at a time ...The result was that twenty-four years of unintermitting research were absorbed in the course of the author's observations. During this time about 900 drawings were made, none of which represented imagines.'[167]

Frohawk had been publishing detailed descriptions of many butterfly life-histories in *The Entomologist* between 1892 and 1913, and in *The Field* for two years from 1913. Completion of the work was made possible by the patronage of his old friend, Walter, Lord Rothschild, who later purchased all the water-colour drawings for the book and bequeathed them to the nation. The publication of *Natural History of British Butterflies* was nonetheless delayed by a full ten years, first by the outbreak of war in 1914, and later apparently by an insufficiency of subscribers. The expensive colour plates and folio format meant that the work went on sale at six guineas – enough, in 1924, to restrict it to the same limited, well-to-do market as the works of the Aurelians, two centuries earlier. It is unlikely that Frohawk ever made much money out of one of the greatest works of twentieth-century natural history. In 1927 he also sold his collection – noted for its immaculate, reared specimens and almost complete range of varieties – to Lord Rothschild for £1,000, at a time when he was short of money. These butterflies now form part of the huge Rothschild collection in the Natural History Museum.

Frohawk's work was brought before a wider public when *The Complete Book of British Butterflies*, with thirty-two plates, was published in 1934 and became a standard reference book. His last book, *Varieties of British Butterflies*, with forty-eight colour plates, followed in 1938, when Frohawk was seventy-seven years old. No attempt was made to make the latter truly comprehensive; of the several thousand described aberrations, Frohawk figured just 125 particularly striking ones. Both books had an unlucky publishing history. In 1940, the publishing house of Ward Lock was blitzed, and the remaining bound stock went up in flames. They were never reprinted.

Frohawk was probably the first person to rear the Silver-spotted Skipper from new-laid egg to adult insect, and the first to describe fully several others, including the Chequered Skipper, Mazarine Blue and Pale Clouded Yellow. His greatest challenge of all was the Large Blue. As long ago as 1906, Frohawk had proved the association between larvae and ants when he found a full-grown larva on shaking the contents of an ants' nest over a cloth. With the help of E. B. Purefoy (q.v.), he finally succeeded in elucidating the mystery of the final larval instar. For his famous sketch of an ant carrying the larva, Frohawk had knelt in the same position for an hour, a lens in one hand, a pencil in the other (see Fig. 105).

Frohawk was a stocky man of average height. His broad features may have owed something to his Dutch ancestry. It is said that he attended sessions in the gymnasium at Crystal Palace and exercised with dumb-bells. Out of doors he was invariably dressed in a Norfolk jacket, of his own design, cut square at the neck. He was a great walker, who thought nothing of covering twenty miles to visit his invalid mother. He recalled one 42-mile nocturnal walk undertaken in 1884 in characteristic fashion: 'It was a brilliant fine night with a full moon. I met only 7 people in the country districts, two of these were men asleep with a dog between them under a hedge, and two were evidently burglars ... They had bulging sacks on their backs and something soft on their feet ... At 5.30 a.m. I watched a mole hunting amongst a strip of grass beside the road, probably catching the "early worm" '.[168]

Frohawk was married twice. At the age of thirty-three, in June 1895, he married Margaret Grant, who shared his interest in natural history, and their honeymoon in the New Forest doubled as a collecting trip. They enjoyed a number of natural history holidays together but, sadly, Margaret suffered from ill health and died in 1907, leaving her husband to raise two young daughters. He was fortunate to marry again, two years later, Mabel Jane Hart Bowman, a family friend. At the special invitation of Charles Rothschild, they spent their honeymoon at his home at Ashton, Northamptonshire, now the home of Miriam Rothschild. Mabel, who was many years younger than her husband, was the mother of his third daughter, Valezina, named after the New Forest form of the Silver-washed Fritillary (Fig. 95). Mabel died only in 1983, and their daughter in February 2000. After spells of living in Surrey and Essex, in a variety of homes, Frohawk spent his last years in Sutton.

Frohawk was admired by a wide circle of entomologists. 'So much was he for the last fifty years the hub around which amateur lepidopterists gravitated', wrote N. D. Riley, 'especially those interested chiefly in the butterflies, that every bit of news seemed to find its way to him, to be examined, criticized and, if found suitable, published'.[169] He helped to edit *The Entomologist,* to which he contributed practically every year for sixty years. He was an active member of the South London Entomological Society which he had joined in 1886, serving on its Council three times. Like many others, however, Frohawk could not get on with James Tutt and, after enduring a particularly unpleasant outburst from him, resigned his membership, rejoining the Society only after Tutt's

Fig. 95. Frohawk with his wife Mabel, and young daughter Valezina, whom he named after a favourite butterfly.

Plate 10 Valezina, Viscountess Bolingbroke, at the opening of a butterfly walk in the New Forest, named in honour of her father. The moment when a beautiful black, green and silver f. *valesina* butterfly, after having been released from a box and circling once, settled motionless on her dress and was captured on film by a bystander!

death in 1911. He was also a Fellow of the Royal Entomological Society, to which he was elected in 1891, and was made a Special Life Fellow in 1926 – a rare honour.

Frohawk died on the 10th December 1946, and was buried at Headley, Surrey, under a simple wooden cross which bears a carving of the Camberwell Beauty. His friend, Norman Riley paid him a memorable tribute: Frohawk was 'that rare combination, a true naturalist and an accomplished artist, to whom beauty, truth, kindliness and steadfastness of purpose were of greater account than fame'.

As a postscript, in 1996, fifty years after his death, and as a filial tribute, his daughter Valezina, Lady Bolingbroke, unveiled the commemorative sign in the New Forest marking the 'Frohawk Ride', where the Silver-washed Fritillary and its eponymous form are still relatively abundant (Plate 10).

75. HEINRICH ERNST KARL JORDAN (1861–1959) (Fig. 96)

Karl (as he was always known) Jordan, one of the greatest museum entomologists of all time, was brought up on a small farm near Almstedt, Germany. A bright boy, he attended the nearby Hildesheim High School, where he became seriously interested in beetles. Asked much later what he dreamed about, Jordan answered, without hesitation, 'Beetles. Beautiful beetles with tough elytra, crawling near the base of the trees in the woods at Almstedt'. Not the least of his many qualities was steadfastedness. His 'single-mindedness', wrote Miriam Rothschild, 'runs like a steel thread through his life and work'.[170]

Fig. 96. Karl Jordan, F.R.S., the great museum entomologist. He described 2,575 new species and wrote 460 scientific papers, mainly based on the Rothschild collections, as well as major monographs on world butterflies and hawkmoths.

He entered the University of Göttingen and took his degree, and soon afterwards, a doctorate, in zoology and botany. After compulsory military service (Jordan made a surprisingly good soldier) he taught mathematics and science, first at a grammar school, and then at the School of Agriculture in Hildesheim, to which he moved in 1892, the year after his marriage. The turning point in what might have been an ordinary, if moderately distinguished, academic career, came in 1893, when, out of the blue, Jordan was invited to Tring by Walter Rothschild, who had had good reports of him from his newly appointed director, Ernst Hartert, a fellow German. He was offered the job of sorting out Walter's chaotic insect collections on a very modest annual salary of £200. Jordan accepted; his first task was to arrange and classify some 300,000 beetles. He began work, in his methodical way, and had completed it by the end of the following year, describing 400 new species in the process (and Jordan reckoned on taking about sixteen hours to describe a single new species). At first, there was not even a suitable desk for his microscope, and to find sufficient light Jordan had to retire to the landing window-sill. But he realized that Rothschild's long series of specimens and vast numbers of species provided a unique opportunity for descriptive work.

After the beetle marathon, Walter Rothschild needed Jordan's help on swallowtail butterflies. The subsequent *Revision of the Papilios of the Eastern Hemisphere* (1895) was based on the close examination of over 10,000 specimens of 239 species, and was a landmark work on these beautiful butterflies.

There followed a monograph on the African genus *Charaxes*, which took Jordan five years to complete, and then a major revision of the hawkmoths (1903), a sumptuous work of over 1,000 pages and sixty-seven plates. Over his lifetime, Jordan published some 460 scientific monographs and papers, mostly on the Lepidoptera, and described some 2,575 new species himself. Another 851 new species he described in collaboration with Walter or Charles Rothschild.

Jordan was probably as close to Walter Rothschild as any man could be. 'We have terrific arguments', confided Walter, 'but the fellow is always right'.[171] Jordan treated the great collector with tact, good humour and affection. By the 1930s, he was effectively running the whole museum, his microscope and camera lucida buried beneath a never-ending cascade of correspondence, books and incoming specimens. The meticulous labelling of the collections was due mainly to him, as was the labelling of the figures in the Museum's publication, *Novitates Zoologicae*.

The greatest friendship of Jordan's life, however, despite the disparity in their ages, was with Walter's more convivial younger brother, Charles. Charles Rothschild admired him enormously, considering Jordan 'the cleverest man I know'. The two shared a great love for nature and family life, and had similarly liberal views on politics, philosophy, religion, and so forth. Jordan was the more straitlaced of the two, and sometimes seemed not to understand Charles's very English sense of humour. Jordan enjoyed perfect health throughout most of his long life, whereas that of his more intellectually restless, highly-strung friend Charles collapsed in 1916, following a bout of 'Spanish influenza'. Jordan accompanied the ailing Charles to Switzerland, but, as a German by birth, he was not permitted to rejoin his wife and work at the museum until hostilities were over. The deaths, first of Charles Rothschild in 1923 and then of his wife, from renal failure, two years later, had a lasting effect on him. Miriam Rothschild, who knew Jordan well, found that 'it seemed as if he had suddenly grown remote, apart, lost in thought, pondering some insoluble, abstract problem. He often sat for long periods in silence, with his head flung back and his beard tilted at a slight angle, apparently quite unaware of his surroundings, as if a sheet of glass separated him from the outside world'.[172] This sense of isolation was later accentuated by deafness. Though he lived until the 1950s most of Jordan's best work was completed before 1916.

Among Jordan's last great achievements was helping to found the International Congress of Entomology, held in Brussels in 1910, which managed to cut through decades of confusion and national rivalry in the naming and classification of insects. Miriam Rothschild described how, with his brilliant grasp of the principles of systematics, 'he would sweep aside rules, regulations and red tape with tolerant disdain, coming directly to the thing to be done. Precedents that lay in the past did not concern him; it was only the future that mattered'. He remained its permanent Secretary until 1948. The new concepts he introduced, based on his studies of the collections at the Tring Museum, have been of lasting value – indeed, are taken for granted today by many who have never heard of Karl Jordan. Due acclaim was late in coming. In 1932, he received a belated F.R.S., and a 'Festschrift' of papers in his honour was published three years before his death. Part of the reason for this neglect was that his best ideas were often hidden away in obscure journals; another, undoubtedly, was that he was of German birth (with a strong lifelong accent), though he took British nationality in 1911. In Norman Riley's view, it was 'only after his death that fellow systematists understood how much they owed to Jordan's energy and intellectual brilliance'.[173]

Although Jordan is remembered primarily as a museum scientist, he also travelled quite widely, including a strenuous six-month expedition to south-west Africa at the age of seventy-four (the 6,000-odd insects obtained were presented to the BM(NH), as was his private collection of 20,000 beetles, in 1940). After 1918, he would also take his family on an annual collecting holiday in Europe. Like all the best field naturalists, he had an unfathomable instinct for finding 'a good spot'. Miriam Rothschild recalled one instance of this during a wet afternoon's walk at Ashton Wold:

> 'Conditions were depressing, and, after a while, K.J. fell silent and continued to trudge onwards, head bent against the downpour, the rain dripping from the brim of his hat and his eyes fixed upon the ground, apparently lost in thought. After about three-quarters of an hour he suddenly stopped, raised his head hastily as if startled by a loud noise, and remarked to his astounded companion: "This is a very good spot for butterflies". He had halted in a slight bulge in a woodland glade less than fifteen yards square – the only spot where, for some totally inexplicable reason, all the rarities for which Ashton Wold is noted have been found together.'[174]

76. MARGARET ELIZABETH FOUNTAINE (1862–1940) (Fig. 97)

Fig. 97. Margaret Fountaine, intrepid and unorthodox traveller, who nevertheless spoke with 'a voice that Queen Victoria would have recognized'.

'To the reader, maybe yet unborn, I leave this record of the wild and fearless life of one of the "South Acre Children" who never grew up, and who enjoyed greatly and suffered much.' These words close the last letter of Margaret Fountaine, the tireless traveller and lepidopterist. She had been found dying by a roadside in Trinidad, in April 1940, after a heart attack, her butterfly net close at hand. A few years before, she had remarked that she had had 'a lovely life. I should like to pass over quickly and in full harness!' As her obituarist, W. G. Sheldon noted, she had her wish.[175]

Under the terms of her will, Fountaine's magnificent butterfly collection – ten mahogany cabinets housing 22,000 specimens – was left to the Norwich Museum. With it came a mysterious black japanned box in a canvas holdall, which, she instructed, should not be opened until 15th April 1978. No explanation was given, and not until the box was opened (not, in fact, on the appointed day but two days later as April 15th fell on a Saturday that year) was the reason discovered. It contained twelve thick volumes in identical bindings: Margaret Fountaine's journal, which she had maintained almost daily for over sixty years. The day the box should have been opened was the hundredth anniversary of her first entry.

With the journal was a letter explaining the reasons for the embargo. It contained honest accounts of what she termed 'the follies and foibles of youth', and also of her faithful Syrian dragoman or guide, Khalil Neimy, her 'dear companion, the constant and untiring friend' who accompanied her on so many of her travels. They met in Damascus in 1901 where, to her astonishment and initial embarrassment, he developed an undying passion for her despite the difference in their backgrounds and ages – she was just thirty-nine, he was twenty-four. For very many years in several continents they travelled together, always disguising their true feelings from the world at large. They even exchanged rings and planned to marry and settle in America. However, this was never to be. Khalil eventually returned to Syria where he died following a bout of fever in June 1929. On at least one occasion he had saved her life. 'The roving spirit and love of the wilderness drew us closely together in a bond of union', she wrote, 'in spite of our widely different spheres of life, race and individuality, in a way that was often quite inexplicable to most of those who knew us.' Knowing that such things were not likely to be understood, Fountaine had consigned the journals to a long period of oblivion.

Margaret Fountaine's father, the Revd John Fountaine of South Acre, Norwich, had died when she was only sixteen. It was then that she started the journal, recording daily observations, however trivial, in a neat, orderly hand. Her 'greatest passion' was for a drunken Irish Singer, Septimus Hewson, whose desertion left a 'scar upon my heart which no length of time ever quite effaced'. And so, brokenhearted, she turned her back on England and the life of a conventional Victorian young lady and started to travel in search of butterflies, first to Europe, then to Africa, the Middle East, India and the Far East, the New World and the Antipodes. Over her long and extremely unconventional life, Margaret Fountaine became perhaps the most travelled British lepidopterist who has ever lived (Fig. 98).

'She was absolutely fearless, pursuing her beloved butterflies,' Sheldon wrote, 'entirely oblivious to the presence of any inhabitants of tropical jungles, be they lions, tigers, leopards, poisonous snakes, or what not;' and not caring whether 'they were infected with malaria, yellow fever and other tropical diseases.' She was indifferent to normal comforts, and could apparently live quite happily on local food and in primitive living conditions. 'On one occasion, during a sojourn in Asia Minor, she put up with an Armenian family for many weeks, practically the only food available being eggs and rice.'

Miss Fountaine sent reports of her travels and discoveries to the entomological journals. *The Entomologist's Record* for 1902, for example, described her recent adventures in Greece as 'a somewhat grotesque figure, armed with a butterfly-net'. One encounter was with a 'rather evil-disposed person, and in order to disembarras myself of this individual I gave him the slip by pretending to go in hot pursuit after some imaginary butterfly-rarity.'! Even she found the inns 'rough to a degree': 'hemiptera, other than those to be sought for outside by the entomologist engaged in the study of that class of insects, were in some of these inns extremely plentiful.' The amenities of the bathroom were 'a greasy hair-brush and comb, and also with an old clothes brush; to say nothing of a piece of untempting-looking scrubbing soap'.[176] On another journey, this time to Asia Minor, she told how while busy catching some '150 to 200 specimens of the rare Grey Asian Grayling (*Pseudochazara geyeri*)', she encountered a band of Circassian robbers, with whom 'the driver of my yiley (a brute I would like to have kicked many a time if I had been a man) was evidently in league'. She had a ripe vocabulary and found defence in voluble oaths on more than one occasion. In central Turkey, where there were no inns at all, she slept outside on the flat roof 'amongst the storks' nests, with nothing above [her] but a star-strewn sky'.[177] Margaret Fountaine seems to have met a remarkable number of bandits on her travels, like Jacques Bellacoscia, the Corsican brigand, who entertained her in his hide-out in 1893.[178]

Norman Riley first met Margaret Fountaine in 1913. 'Having heard something of Miss Fountaine's exploits', he wrote, 'the announcement [of her arrival] conjured up visions of a well-worn battleaxe to assault me. Instead I met a tall, attractive, rather frail-looking, diffident but determined middle-aged woman. She was pale and looked tired, but the strongest impression she gave me was one of great sadness that seemed to envelop her entirely. It was not long, however, before I discovered that this veil of sadness could be penetrated by self-deprecating flashes of humour that quite transformed her'.[179]

Fig. 98.
Margaret Fountaine, photographed at Palm Springs, California, in 1919. She favoured her own idiosyncratic dress for collecting, often wearing 'a man's checked cotton shirt and striped cotton skirt, both with additional pockets sewn on'.

Our portrait of her was taken in the late 1930s, when Fountaine was over seventy. Barbara Ker-Seymer, the Bond Street photographer, recalled her slipping into the changing room, reappearing some minutes later dressed in the 'somewhat startling costume she favoured for butterfly collecting: ... cotton gloves with the tips of the first finger and thumb cut off. A heavy black chain with a compass on the end was attached to one buttonhole of her shirt; and she was wearing a cork sun helmet ... She sat down bolt upright clasping a large butterfly net and a black tin box for specimens'.[179]

W. F. Cater condensed the twelve volumes of the journal into two books with a linking commentary: *Love among the Butterflies*, which became a best-seller, and *Butterflies and Late Loves*. In his view, though Fountaine 'undoubtedly enjoyed her life of butterfly catching, she failed to catch that one deep happiness she ever sought'.[180] She did most of her work alone, and many of her discoveries of tropical butterflies were never written up. She is one of the strangest lepidopterists Britain has ever produced, in her later years collecting alone on foot or by bicycle, returning to England in the summer to arrange her collection. Few contemporaries knew her well, though whenever she could she attended meetings of the Royal Entomological Society to which she had been elected in 1898. But she was never half so famous during her lifetime as she became forty years after her death, when her literary and artistic talents were at last revealed.

77. PERCY MAY BRIGHT (1863–1941) (Fig. 99)

Percy Bright, the son of a sometime missionary in India, was among the most avid 'chequebook collectors', bidding against the likes of Lord Rothschild, Sir Vauncey Harpur Crewe, Sidney Webb and A. B. Farn for choice specimens. L. Hugh Newman remembered his regular appearances at the Butterfly Farm and in the field at Royston Heath during the Chalkhill Blue season: 'If any freak butterfly took his fancy he would make an offer for it on the spot', his chequebook ready in his pocket together with his collecting boxes and lens.[181] Needless to say, Bright's collection of British butterflies was one of the finest there has ever been, full of stunning aberrations, particularly among the Lycaenidae.

Fig. 99. Percy M. Bright, store-owner and philanthropist, as Mayor of Bournemouth. His 'tremendous' collection was, according to L. Hugh Newman, 'so vast, containing so many rare and unique butterflies, that its sale had to be divided into five separate sessions, spreading over a period of two years ...'.

Bright inherited his father's department store business, Bright & Colson Ltd, as well as his interest in religious and charitable work. He was chairman of the Royal Victoria Hospital at Boscombe, toured missions in India as a delegate of the London Missionary Society, became a J.P., and was so active in municipal affairs that he was elected Mayor of Bournemouth from 1929–31. His interests often contained a strong collecting component. He loved postage stamps, and founded the firm of Bright & Sons in The Strand, the then world capital of philately. His house was full of beautiful antique china, glass and furniture, and his large garden was full of scented flowers, especially those attractive to butterflies and moths.

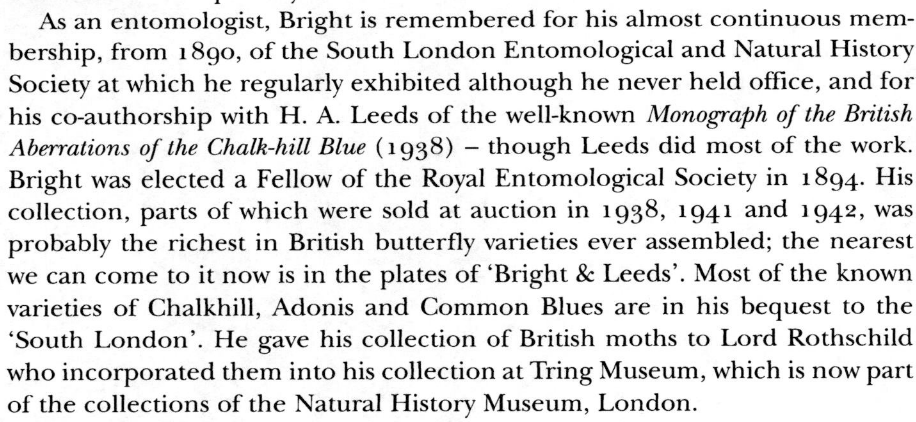

As an entomologist, Bright is remembered for his almost continuous membership, from 1890, of the South London Entomological and Natural History Society at which he regularly exhibited although he never held office, and for his co-authorship with H. A. Leeds of the well-known *Monograph of the British Aberrations of the Chalk-hill Blue* (1938) – though Leeds did most of the work. Bright was elected a Fellow of the Royal Entomological Society in 1894. His collection, parts of which were sold at auction in 1938, 1941 and 1942, was probably the richest in British butterfly varieties ever assembled; the nearest we can come to it now is in the plates of 'Bright & Leeds'. Most of the known varieties of Chalkhill, Adonis and Common Blues are in his bequest to the 'South London'. He gave his collection of British moths to Lord Rothschild who incorporated them into his collection at Tring Museum, which is now part of the collections of the Natural History Museum, London.

Percy Bright was killed, aged 78, in a car accident while returning from a visit to Bournemouth Cemetery to lay flowers on his wife's grave. He was remembered as a surprisingly reserved and unassuming personality, a Christian, a philanthropist, and a connoisseur.

78. SIDNEY GEORGE CASTLE RUSSELL (1866–1955) (Fig. 100)

S. G. Castle Russell, or 'C. R.' as he was usually called, was the king of the butterfly 'variety hunters', and perhaps only Farn could rival him for his knowledge of butterfly life-cycles and their foodplants. His collection, now preserved as part of the Rothschild-Cockayne-Kettlewell collection in the Natural History Museum, contains extreme varieties of every description, including black, white and cream aberrations of all the fritillaries. As a man, Castle Russell was remembered with universal affection. His friend, Colonel S. H. Kershaw (1881–1964), who wrote a long memoir of 'C. R.' in *The Entomologist's Record* on which this account is largely based, claimed that 'he never lost a friend or made an enemy'.[182]

Born at Aldershot, the youngest of eight sons of Major C. J. Russell of the Royal Engineers, Castle Russell spent part of his boyhood in Ceylon, where he first began to collect butterflies. Educated first by army tutors and later at Taplow Grammar School, he joined the Phoenix Fire Office in 1887, and built up a high reputation as an electrical engineer of integrity and great ability, advising companies, factories and schools and colleges on wiring, installation and electricity generation. He retired from the company in 1921, with 35 years still left to him. Much of it was spent collecting and rearing British butterflies.

Fig. 100. S. G. Castle Russell or 'C. R.', electrical engineer and great variety hunter, was said by S. H. Kershaw never to have 'lost a friend or made an enemy'.

Castle Russell's regular haunts lay mainly within a day's length from his home at Fleet in Hampshire. In May and June there would be visits further afield to Pamber Forest, Chiddingfold and the New Forest; in July, to Blean and, again, the New Forest, while August and September would find him on the Downs in search of Blues. In his early years he cycled everywhere, and thought little of riding from Woking to the New Forest, collecting all day, and returning in the dark. On one occasion he nearly blinded himself by painting his forehead with nicotine to keep away the clegs. Although he claimed to know every inch of country around Aldershot and Fleet, Castle Russell had very little sense of direction, and was always getting lost. In his later years, friends gave him a whistle to blow, though he never used it. He had a strong streak of stubbornness. Once he had thought something out and decided what line to take, nothing on earth could make him change his mind.

Castle Russell kept himself remarkably fit throughout his long life. In his youth he boxed and fished, and he could always walk anyone off their feet. His friend P. B. M. Allan told Colonel Kershaw of an occasion when, at the age of eighty-seven, C. R. and he visited a favourite collecting ground in Surrey.

> 'C. R. piloted me along hedgerows and through rough coppiced bottoms to the foot of an escarpment which seemed to me about a hundred feet high, with a gradient of about one in two. "We're not going to climb that, are we?" I asked, apprehensively. "Yes," said C. R., "the common's at the top. Come on." At that time I was still a fairly active man; I could run after a motor-bus in London and jump on it while it was gathering speed; but compared with C. R. that day I was old and decrepit. Never would I have believed that a man of his age could go up that escarpment in the way he did. "Come on", he said, turning round to encourage me as I lagged farther and farther behind him, "the common's just at the top." Up and up he went, his speed never slackening. When he got to the top there was no pausing for a moment's rest, nor was he in the slightest out of breath. He strode straight ahead and at once began using his net. "Come on", he called, waving me on. I sat down to recover my breath, and he disappeared in the distance, walking smartly.'

His luck was legendary. One memorable day in September 1939, Castle Russell, together with his wife, the Revd J. A. Marcon and a few others, were on the South Downs looking for aberrations of the Chalkhill Blue. 'We made quite a good bag at Shoreham of *coridon*' he wrote afterwards. 'A green male, *obsoletas* and *caecas* – the best of the last-named is a male with black upper and brown underside like a female; it may be a *gynandro*, but the body is male. The amusing thing about this bug is that it found its way into a tea-cup and laid down and died, until Mrs Russell spotted it and Burkhardt boxed it. Marcon had remarkable luck here too – a dozen fine vars. and *gynandros* and six green males in *coridon* – and at Eastbourne he netted seven *bellargus Ab. radiata* in an area of 300 square yards; he is as Wells says, "A ruddy Marvel!" I should think his Shoreham vars. are well worth £50'.

On another occasion, Castle Russell was on his way from inspecting the wiring at Eton to Pamber Forest when a sudden instinct made him take a detour to his home to see how his Swallowtail pupae were faring. He arrived just in time to save an all-black Swallowtail from knocking itself to pieces in its efforts to escape. This might have been the specimen described by Allan as really black, 'as black as the inside of a black cow'.[183]

When examining someone else's collection, Castle Russell would study each specimen with great care. As Colonel Kershaw recalled: 'he regarded every specimen as a possible var., and in his mind checked each insect with those in his own cabinets, before he passed on to study the next. He spent 2½ hours over 10 drawers, made some nice comments on the few vars. I possessed and wound up typically with the remark, "I'm glad to see your setting is improving".'

His knowledge of butterfly varieties was such that, when arranging the large Bright collection for sale, his assessments of their value usually came within five percent of the sale price in each case. Castle Russell was not much interested in the genetics of butterfly variation, but he obtained many fine 'vars' by careful breeding, and generally released typical forms in their natural habitats. He also made a practice of 'putting down' eggs or larvae to secure, as he thought, a good crop of butterflies for the coming season. Towards the end of his life he repapered his cabinets with black, finding it isolated each specimen better and emphasized their colours more. Before his death he arranged with Cockayne and Kettlewell that his collection of 6,000 British butterfly varieties be transferred to the Tring Museum with theirs. Characteristically, having done so, he began to build up a second collection, a much smaller one 'of types and vars'.

Sidney Castle Russell died in his eighty-ninth year. Bernard Kettlewell left a moving account of his last hours:

> 'As a doctor, I have seen many people pass on, but this was a truly remarkable last meeting. I was told that he had been unconscious for many hours, but I did want to see the old man once again. I went over to the bed and spoke quietly to him and to my amazement he opened his eyes, recognized me, and slowly sat up and talked to me for about five minutes. He was perfectly lucid, and in those few moments told me that he was about to die and that death was not nearly so unpleasant as he had anticipated. He had no regrets and stated what a wonderful life he had had.
>
> 'I came out of the room completely staggered by the braveness with which he was facing death.'[184]

79. LIONEL WALTER ROTHSCHILD, 2ND BARON ROTHSCHILD OF TRING (1868–1937) (Fig. 101)

Walter Rothschild once showed his mother a battered-looking butterfly he had caught. 'A nice tortoiseshell', she exclaimed, as one does to five-year-old boys. 'No', insisted Walter, 'it's not. It's different.' His mother got down her copy of Morris's *British Butterflies* and looked it up. 'You are quite right, Walter', she agreed after a few minutes. 'It *is* different. It's a Comma Butterfly'.[185] Two years later, Walter decided to make his own museum and told his parents 'Mr Minall [Alfred Minall, a joiner who was working on the house at Tring and a skilled taxidermist] is going to help me look after it'. Walter Rothschild never grew out of that childhood impulse, and devoted much of his life to his private museum at Tring. By the end of it, he had amassed, in his niece Miriam's words, 'the greatest collection of animals ever assembled by one man, ranging from starfish to gorillas. It included 2¼ million set butterflies and moths, 300,000 bird skins, 144 giant tortoises, 200,000 birds' eggs and 30,000 relevant scientific books. He, and his two collaborators, described between them 5,000 new species and published over 1200 books and papers based on the collections'.[186]

Fig. 101. Walter, Lord Rothschild, the greatest collector of them all.

In his youth, at least, Rothschild did have other occupations. For a while, he seemed to take his social duties seriously. He represented Aylesbury in the House of Commons from 1899 to 1910 (though he spent most of his time in the Natural History Museum), and, as Miriam Rothschild relates in her vivid personal memoir of him, *Dear Lord Rothschild,* he played an important behind-the-scenes part in the Balfour Declaration of 1917, which promised a national homeland for the Jews in Palestine. But he was not temperamentally suited to business and, after an unhappy spell in the family bank, withdrew from the world of high finance to devote himself to his all-consuming passion for zoology. By 1892, when the galleries of the museum were first opened to the public, Rothschild's insect collections were already of immense size. There were some 300,000 beetles, for example, piled up in store boxes in his bedroom, on shelves, on chairs, and in sheds and outhouses. Full-time curators were needed, and there Rothschild seemed to have the Midas touch. First, Ernst Hartert was engaged to look after the bird collections, and then, in 1893, Rothschild appointed the exceptionally able Karl Jordan to work on the insect collections. Together Rothschild and Jordan (q.v.) published hundreds of new descriptions, and major revisions of butterflies in America and the Eastern Hemisphere, and of the world's hawkmoths. Rothschild eventually gave up on beetles to concentrate his energies on Lepidoptera, for even he did not have space or resources for both. As Riley wrote, it was his 'considered policy to try to make his collections as complete as he could'.[187] 'I have no duplicates', he would say; in his eye, no two specimens in his endless cabinet drawers were quite alike, and for some of his 100,000 species of butterflies and larger moths, he amassed perhaps the greatest range of variation ever seen, and a uniquely rich mine for scholars of biogeography and genetics. He displayed his treasures 'with almost boyish enthusiasm', seeming to know every detail and scientific name. Walter employed collectors worldwide to find new specimens for the museum, from minute undescribed insects to the bones and skin of the White Rhinoceros. The adventures of Rothschild's collectors is a story in itself. As Miriam Rothschild recalls, 'the occupational hazards of collecting in those pre-penicillin days, and before the development of a host of other useful medicaments and inoculations, were great'. One collector, Alfred Everett died of fever, three others died of yellow fever on the Galapagos expedition, Doherty died of dysentery, and Ockenden of typhoid fever. Webb succumbed to an unspecified illness on his way home, and Stuart Baker had his arm bitten off by a leopard. 'Collecting in the tropics was a risky business.'

In later life, Lord Rothschild (he inherited the title in 1915) became an immense, ponderous man, six feet three inches tall and weighing twenty-two stones. Yet he had relatively tiny feet, so that he seemed to bowl along the marble hall at Tring 'like a grand piano on castors'. All his life he suffered from extreme shyness, which was not helped by an apparent inability to control his voice – when he spoke it was either in a low stammer or a loud bellow. In many photographs he has a downcast look, seeming to stare sadly at his tiny feet. Yet, 'on the rare occasions when he laughed aloud, he raised the rafters with his tremendous roar'. 'Despite his ordered and regular life', wrote his niece, 'despite his shyness, he gave one the feeling of almost dangerous unpredictability, like the proverbial rogue elephant – as if his wealth and size and a certain inner lack of convention and judgement bestowed upon him a special sort of unrestrained freedom'.

His personal foibles were mirrored by his private zoo. Rothschild liked large or bizarre animals. Roaming free in his park at Tring were zebras, kangaroos, ostriches, wild horses and a flock of his beloved cassowaries (which he had had stuffed after their deaths – all sixty-five of them). He harnessed zebras to his carriage (Fig. 102) and once drove them through Piccadilly to Buckingham

Fig. 102.
Lord Rothschild driving the zebras that in the 1890s took him bowling down through Piccadilly to Buckingham Palace! But, according to his niece, 'Walter's heart was in his mouth when Princess Alexandra tried to pat the leading zebra'.

Palace (though, he admitted later, his heart was in his mouth when Princess Alexandra tried to pat the leading animal). He rode his Aldabran Giant Tortoises with the aid of a lettuce dangling from the end of a stick. Even in his university days, at Magdalene College, Cambridge, he had kept a much-loved flock of Brown Kiwis. But, although he had two known mistresses, he never married.

Lord Rothschild died after a long illness in 1937, and was buried at Willesden Jewish Cemetery. His headstone bears the aptest of Biblical quotations: 'Ask of the beasts and they will tell thee, and the birds of the air shall declare unto thee'. He had been honoured for his work on world Lepidoptera by all the leading scientific institutions. He was elected a Fellow of the Royal Society in 1911, and was sometime President of the Zoological Section of the British Association, and in 1921 of the Royal Entomological Society.

Rothschild bequeathed his vast remaining collections and library to the BM(NH) (most of the birds skins had been sold to the American Museum of Natural History in 1931). He had been a regular donor to the Museum during his life, mainly of orders he did not collect, but his collections of Lepidoptera alone were as large as the Museum's general collection put together, and uniting the two was the work of a quarter of a century. 'No scientific collection comparable to this, certainly no collection greater than this has probably ever been given to a nation', said the Archbishop of Canterbury, in his role of Chairman of the Trustees. Eventually it was combined with the much smaller (relatively speaking) scientific collections of Cockayne and Kettlewell as a major source of material on variation and genetics. The wonder and importance of Rothschild's collection lies, as Miriam Rothschild explains, 'in the unfolding and presentation before your eyes of a whole order – in all its variety and complexity, culled from continent to continent ... It made available to the scientific world a vast number of excellent specimens from remote parts of the world – it is rich in type material ... It is also very important for the study of variation within a particular species, and hence as a source for hypotheses about biogeography and genetics. Thirdly, of course, it was the source for the life work of Walter Rothschild and Karl Jordan'.

80. EDWARD BAGWELL PUREFOY (1868–1960) (Fig. 103)

Captain Edward Bagwell Purefoy is best remembered for two great endeavours: his elucidation of the life-cycle of the Large Blue butterfly, and his attempts to reintroduce the extinct Large Copper to Britain. His work on the Large Blue is an entomological legend, and one of the stock butterfly stories. Frohawk gave a full account of Purefoy's rearing experiments, which he carried out in the butterfly garden he established at his home in East Farleigh, Kent (Fig. 104).[188] He raised the larvae with the use of potted plants of thyme up to the third moult. He then placed half an empty walnut shell on a thin layer of soil in a small tin box, making a 3mm hole in the upper part of the shell. At the time when the first *arion* larva was hatching, 'A female ant and about twenty workers of *M. laevinodis*, together with a teaspoonful of brood' were introduced into the tin. The ants took their brood under the shell and after about a week the small colony was found to be established. Following its third moult and the subsequent development of its honey gland, the *arion* larva was placed in the tin and encouraged to enter the shell by means of the 'porthole'. Fresh ant larvae were regularly provided and the growth of the larva monitored until it reached hibernation size in early October when, after the addition of more soil until the shell was almost covered, the tin was transferred to a cool greenhouse. Purefoy checked on his little colony on Christmas

Day and found them all hibernating happily. By mid-April activity had resumed inside the walnut shell. After the addition of some more ant brood, 'the larva finally pupated in the shell, fixed at first to the roof by the cremaster, but later it lay on the soil with the ants constantly over it'. A piece of net was tied over the tin and about three weeks later a 'fine male imago was found hanging to the net' having made its way through the ants' tunnels to the surface. Purefoy's famous walnut shell became the accepted method of rearing the Large Blue; the modern authority, Jeremy Thomas, likens it to a cook-book recipe! Purefoy took on the difficult task to enable his friend, F. W. Frohawk, to illustrate and describe each stage of the life-cycle for his great work, *Natural History of British Butterflies.* With characteristic modesty, Purefoy did not himself publish a full account of the work until 1953.[189] Many entomologists were

Fig. 104. Captain Purefoy's butterfly garden at East Farleigh, the scene of his triumph in rearing the Large Blue for the first time.

Fig. 103. Captain E. Bagwell Purefoy. Soldier and gentleman of independent means. Embarking on his ultimately successful attempts to rear the Large Blue, he told L. Hugh Newman 'he felt like a man setting out on a voyage into the unknown'.

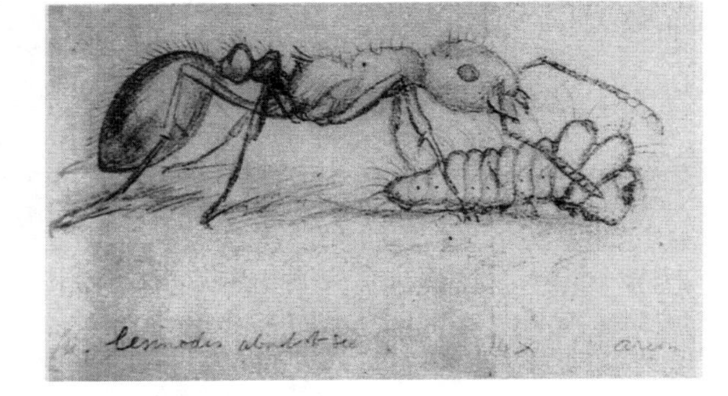

Fig. 105. F. W. Frohawk's most famous drawing – an ant carrying a Large Blue caterpillar - was sketched from life in August 1915 in Captain Purefoy's garden It took him an hour, 'kneeling in the same position ... a lens in one hand, a pencil in the other'.

sceptical about the adoption of the larvae by ants (Fig. 105), suggesting that it was a product of rearing under artificial conditions, but the accuracy of Purefoy's observations were fully confirmed when Jeremy Thomas studied their behaviour in the wild during the 1970s.

Purefoy was the second son of a family of Irish gentry with an estate at Greenfields in County Tipperary. Many of his forebears, as well as his elder brother, were soldiers, and Edward, too, joined the 16th Lancers after completing his education at Tonbridge. He saw active service in South Africa, and was in the forefront of the battle to relieve Kimberley during the Boer War. He seems to have retired from the army to look after his family after the death of his wife in 1903.[190] Purefoy had collected and reared butterflies from boyhood, and became a member of the Royal Entomological Society and Zoological Society. His well-known attempts to reintroduce the Large Copper began in 1913, when Purefoy introduced 700 larvae of the Continental subspecies *rutilus* collected from near Berlin to a ten-acre snipe bog at Greenfields, specially planted with the butterfly's larval foodplant, Great Water Dock (*Rumex hydrolapathum*). There, under the care of James Schofield, an employee of the Purefoys, the butterfly flourished for a time. In 1927, Captain Purefoy and Schofield were able to secure larvae of the subspecies *batavus*, much closer to the extinct British form, from the Friesland marshes in Holland. These were introduced to specially prepared marshy ground at Woodwalton Fen nature reserve known as 'the Copper Fields' (see Chapter 4: 15). During the summer of 1928 Large Coppers jinked and gleamed over the fen dykes for the first time in almost three-quarters of a century (an earlier attempt by G. H. Verrall to establish them at Wicken Fen having failed). The famous colony at Woodwalton Fen survived, helped by periodic topping-up from stock reared in cages, until a severe flood in 1969 wiped out the last survivors. Although the butterfly was reintroduced subsequently, its survival there today is very precarious.

Captain Purefoy was a retiring man, who shunned publicity. He outlived nearly all his peers, and his death at ninety-two, many decades after his most significant achievements in entomology, was hardly noticed by the journals. Contrary to common belief, he was not a wealthy man, the family estate having passed to his elder brother, Wilfred.

81. LEONARD WOODS NEWMAN (1873–1949) (Fig. 106)

Jeremy Thomas remembered the day he met L. W. Newman 'at Folkestone in *Adonis* time'. 'Throughout my boyhood', he wrote 'I had reverently sent the bulk of my weekly pocket money for a few larvae and pupae from the farm at Bexley, and always dreamed of the day when I should meet the man who had become almost a legendary hero. I can still recall vividly my first sight of that tall, spare figure, hatless, the conspicuous silver hair, the cheery voice. From his capacious pocket he drew a small collecting box and showed me therein a pair of immaculately set black Swallowtails. The picture of my boyhood dream was complete. After this came evenings at his summer cottage on the road which skirted Caesar's Camp, the boundless hospitality and good cheer, while the day's captures were discussed over a glass of port.'[191]

Fig. 106. L. W. Newman, pioneering 'butterfly farmer' with tray of part of Sir Beckwith Whitehouse's collection which his son, L. Hugh Newman catalogued for auction in 1944.

It was Newman's achievement to endow commercial entomology with worth and honour, which brought him enduring respect. Born at Singleton in Sussex, Newman joined a tobacco firm after leaving school, but his heart was never in it and Robert Adkin, who at that time was on the Board of the Imperial Tobacco Company, suggested that instead he make entomology his profession. In 1894, Newman established his famous 'butterfly farm' at Bexley, dedicated to supplying livestock of British butterflies and larger moths to entomologists. Adkin helped him by promising to purchase £100 worth of stock each year for the first five years of business.

Newman's was not the first 'butterfly farm' (that honour goes to H. W. Head who had established one near Scarborough during the 1880s). But it soon became *the* Butterfly Farm, where stock was reared in quantities scarcely seen before inside giant sleeves and batteries of cages in a carefully designed garden – part of which was itself enclosed in a wire cage to keep out marauding

Figs 107 & 108. Muslin sleeves, *c.*1939 *(left)* and rearing cages at Newman's butterfly farm at Bexley, Kent, *c.*1900 *(right)*. He used sleeving, then a relatively new technique, to rear large quantities of butterflies and moths for sale.

birds and cats (Figs 107 & 108). At the height of the season, Newman and his assistants were often busy for eighteen hours a day, tending the cages in the daylight hours and collecting moths and their larvae by night. Unlike some of his competitors, however, Newman did not take large quantities of insects from the wild, and he took a keen interest in conservation long before the term became fashionable.[192] The full story of the farm is related by his son L. Hugh Newman in *Living with Butterflies.*[193] With Newman's hard work and talent for rearing difficult insects, the farm prospered, eventually passing on to his son and closing only with the latter's retirement in 1966.

Some of the varieties, bred year after year at the farm, included the beautiful yellow form of the Cinnabar, ab. *lutea*, various attractive forms of the Lime Hawkmoth, and olivaceous forms of the Oak Eggar. However, Newman was unable to rear a regular supply of black Swallowtails, as he had hoped, due to their innate genetic weakness and infertility. Among Newman's patrons were Lord Rothschild and Winston Churchill, who wanted to establish Black-veined White butterflies in his garden at Chartwell. Newman also gave radio broadcasts on 'Butterfly' subjects, which proved very popular, a practice continued with great success by his son.

In 1913, with H. A. Leeds as co-author, Newman published the *Text-book of British Butterflies and Moths*, a popular and still useful work that took seven years to compile. As well as hints on breeding techniques, it listed every British species, with notes on their seasonality, foodplants, distribution, and important aberrations. From the late 1930s, Newman devoted much time to cataloguing collections for sale at auction. He was often also the chief buyer, either on commission or for his business. Of all commercial dealers in entomology, L. W. Newman has perhaps the highest reputation. His obituary in *Entomologist's Record* paid him perhaps the supreme compliment: that he won not only respect but affection, and that his clients were also his friends.[194]

82. HENRY ATTFIELD LEEDS (1873–1958)

H. A. Leeds was one of the great 'variety' collectors, best known for his authorship, with L. W. Newman, of the invaluable *Text-Book of British Butterflies and Moths* (1913) and, with P. M. Bright, of the remarkable *Monograph on the British Aberrations of the Chalk-hill Blue* (1938), based on his and Bright's extensive collections of the species. Leeds coined many convoluted names for butterfly aberrations and was content merely to describe and catalogue them, showing surprisingly little interest in the factors that cause such anomalies. Every specimen captured by Leeds bears one of these names on its label in his exquisite miniature handwriting (Plate 11). The record for length may be held by his ab. *infrasemisyngraphagrisealutescens-albocrenata*, which must have required an unusually long label!

Leeds was born at Somersham, Cambridgeshire, the son of a farmer. In 1887, he joined the staff of the Great Northern Railway, latterly working in the Goods Manager's Office at King's Cross Station where he remained until his retirement in 1933. In his younger days, Leeds had been a talented footballer and sprinter, and was an expert on first-aid, but in retirement he devoted most of his energies to pursuing butterflies and moths. His home, a converted signal box, was conveniently close to Woodwalton Fen, Monks Wood and other fenland localities. He was also a familiar figure on Royston Heath in its heyday. After producing the epic work on the Chalkhill Blue, Leeds started to specialize in the Satyridae, producing an important paper on the aberrations of the Meadow Brown, Gatekeeper and Small Heath butterflies. He was an active member of the South London Entomological and Natural History Society, which he first joined in 1914 and to which he donated many of his celebrated butterfly aberrations. Like so many of the most ardent collectors, he was a bachelor, a man 'of most genial and kindly disposition'.[195]

83. WILLIAM RAIT-SMITH (1875–1958) (Fig. 109)

William Rait-Smith was another collector on the grandest scale. Regarded in his day as the doyen of field collectors, his collection was one of the largest and most complete in the country, and included most of the British microlepidoptera as well as exotic birdwing butterflies. Rait-Smith was especially interested in the Lycaenidae, and each season could be seen wielding his net at Royston Heath and other celebrated downland localities. Baron de Worms recalled his 'seemingly endless forms [of Chalkhill Blue] with the names of each aberration beautifully written in a neat script, the work of an artist'.[196] Rait-Smith was active in the field well into his seventies, and even a severe leg injury could not keep him indoors for long.

Fig. 109. William Rait-Smith, architect, mining surveyor and old-style collector.

Rait-Smith was a qualified architect, who had also worked as a mining surveyor in South Wales, where he published many new records between 1905 and 1914. A man of few words, either spoken or written, Rait-Smith nevertheless welcomed fellow entomologists to his home at Redhill and served for six years on the Council of the Royal Entomological Society of which he had been a Fellow since 1912. He was also a Trustee of the South London Entomological

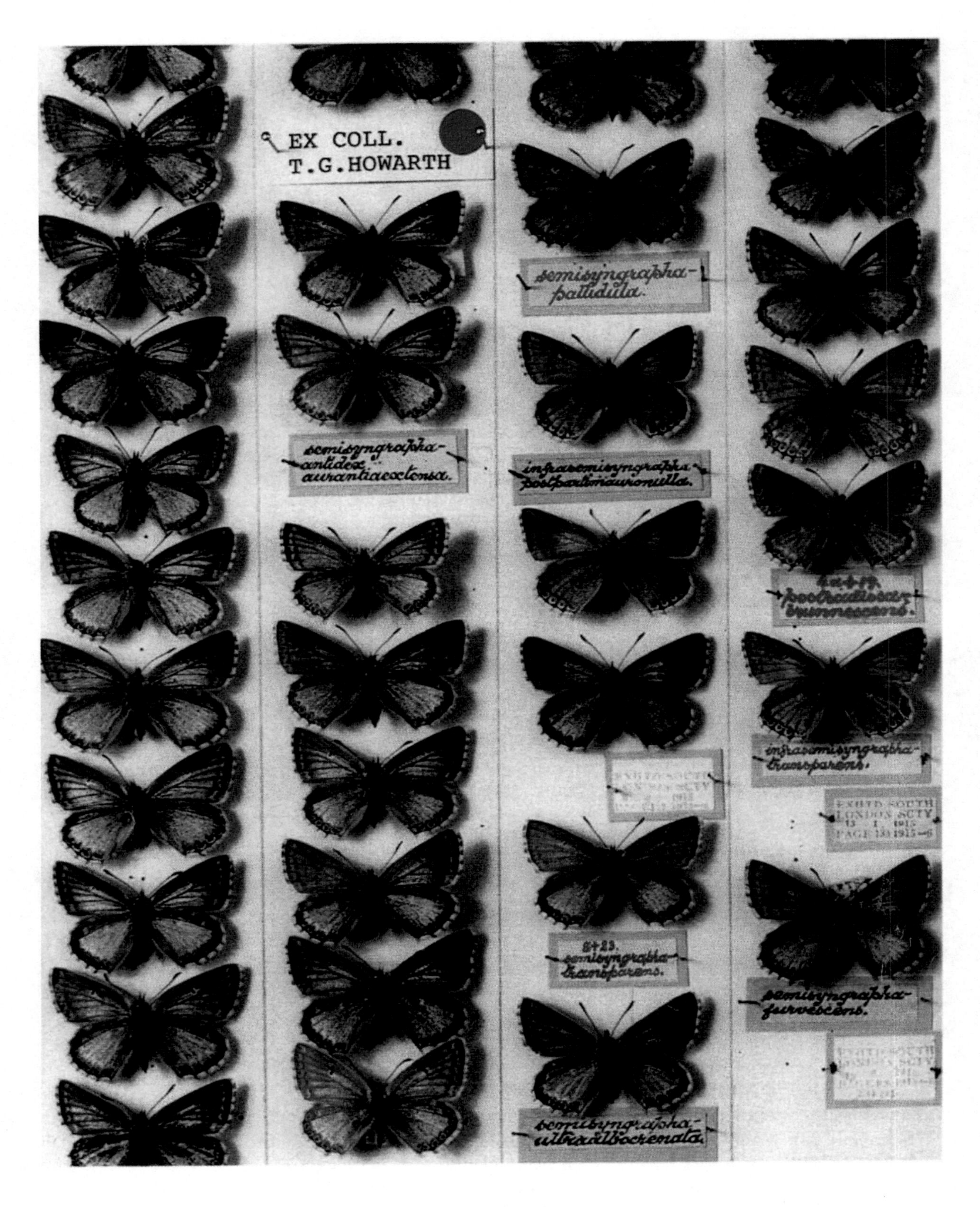

Plate 11
Female Chalkhill Blue varieties collected by H. A. Leeds,
with his characteristic multisyllabic names and beautifully inscribed labels.

and Natural History Society, and regularly attended the meetings of both bodies. He left his magnificent collection of 47,178 specimens to the BM(NH), and a bequest of £250 to the 'South London'.

84. HON. NATHANIEL CHARLES ROTHSCHILD (1877–1923) (Fig. 110)

Like his elder brother Walter (q.v.), Charles Rothschild was a keen field naturalist and a gifted entomologist. His first published entomological work, of which he was co-author with J. L. Bonhote, was *Harrow Butterflies and Moths* (1895, 1897), written when they were both boys at Harrow School. At Trinity College, Cambridge, he was the first undergraduate to own a motor car, mainly, he said, 'so I can get to the "good spots"'. He later became a world authority on fleas. Unlike Walter, though, Charles had a capacity for business, and succeeded his brother as head of N. M. Rothschild & Sons. Hence, his great contributions to natural history and nature conservation were all achieved in whatever hours he could spare outside his long hours in the family bank.

Fig. 110. The Hon. Charles Rothschild, banker, pictured at the age of 29 with his newly-wed wife, Rozsika at his 'holiday house' at Cséhtelek in Hungary, where he collected and studied butterflies and fleas. In the garden beyond, Rothschild discovered a supposedly extinct broom plant *Cytisus horniflorus.*

Charles Rothschild's tastes were less insular than most of his entomological contemporaries. He travelled widely, collecting Lepidoptera and insect parasites in Egypt, Ceylon, Japan, Canada and much of Europe. He met his wife, Rozsika in Hungary (which then included much of modern-day Romania and Czechoslovakia), and built for themselves a fine ranch in Transylvania, with a small laboratory attached. A. von Degen, a Hungarian botanist, remembered how he would search patiently by lantern light for the nocturnal larva of the then little-known Esper's Marbled White (*Melanargia russiae*) 'year after year, night after night, at one particular spot in the woods, scouring the grass stems with a headlamp until at last he found the caterpillar in question'.[197] He was also the first to rear Pallas's Fritillary (*Argynnis laodice*) and describe its life-cycle. His enthusiasm knew no bounds. Once on a train journey, he pulled the communication cord because he had spotted a rare butterfly through the carriage window.

Charles Rothschild was the father of Miriam Rothschild, who remembers him first and foremost as 'an all-round naturalist, a great lover of animals and plants, with an encyclopaedic knowledge of specialized habitats'. His memory was excellent, and, as in his banking world, he could work rapidly but accurately – 'a born sprinter', as Miriam described him. His work on fleas and other blood-sucking insects had begun as a hobby, but improved medical knowledge of transmittable diseases made their study a matter of considerable practical interest. Rothschild's collection of some 100,000 specimens mounted on slides or in glass tubes was the finest in the world. On his travels he discovered many new species, which he described with the help of Karl Jordan, including the tropical rat-flea (*Xenopsylla cheopis*), responsible for carrying bubonic plague from rat to man. He gave this important collection to the BM(NH) in 1913, and, on his death ten years later, made a special bequest for a curator. The entire collection was later catalogued by Miriam Rothschild and G. H. E. Hopkins. Over a period of time Charles Rothschild also gave the museum a total of some 15,000 other insect specimens, mainly European Lepidoptera.

He is especially remembered as the pioneer of nature conservation in Britain. Rothschild was initially worried about the impact of collectors – then perhaps at their most voracious – especially on rare butterflies and moths, and with that in mind acquired part of Wicken Fen and all of Woodwalton Fen as nature reserves. But increasingly he also became concerned at the loss of wildlife habitats to housing schemes and other developments. In 1912, he founded and personally financed the Society for the Promotion of Nature Reserves, the predecessor of today's Wildlife Trusts, and then set about identifying and mapping the best wildlife sites in the country with the intention of saving as many of them as possible. Among those advising him on the best spots for rare insects were friends and entomological colleagues like F. W. Frohawk, W. Parkinson Curtis, William Evans, F. N. Pierce, W. H. Harwood, E. G. B. Meade-Waldo and many others. The 'schedule of areas', which was completed in 1915, contains many of the then classic sites for rarities, especially Lepidoptera (and also for wild orchids – another of Charles Rothschild's passions, though he hunted them with a camera). But although he personally helped to purchase several sites as nature reserves (and his own estate at Ashton, Northamptonshire was a wonderful place for butterflies), the government showed no interest in the project. Britain had to wait another thirty-five years for its first statutory nature reserve – a measure of how far Rothschild was ahead of his time. The full story is told in *Rothschild's Reserves* by Miriam Rothschild and Peter Marren.[198]

Charles Rothschild died in 1923, aged only forty-six. Overwork contributed to a breakdown in his health in 1916 from which he never fully recovered. He was President of the Royal Entomological Society in 1915–16, and a Fellow of the Linnean and Zoological Societies, and he helped to establish the Imperial Bureau of Entomology. Moreover, never one to be discouraged for long, he was busy extending his nature reserves idea to the whole of the British Empire at the time of his death.

85. EDWARD ALFRED COCKAYNE (1880–1956) (Fig. 111)

He was 'an Edwardian bachelor', recalled Richard Bonham-Carter, '... his knowledge and memory were remarkable ... I thought he had the best intellect on the staff of Great Ormond Street. But he was odd!'.[199] There was something birdlike about Cockayne – small, alert and inquisitive, 'but one never knew which bird to expect, the eagle, the raven, or the sparrow'. Another medical colleague, Douglas Gairdner found it impossible to 'get onto his wavelength, or to know how to respond to the curious little giggles which accompanied the remarks he would drop, often after a disconcertingly long silence, in the course of his twice weekly ward-round'.[200] His entomological friends knew him as 'The Doctor'.

E. A. Cockayne was born in Sheffield and educated at Charterhouse School, Balliol College, Oxford, where he gained first-class honours in Natural Sciences, and then at St Bartholomew's Hospital Medical School. He remained in London after qualifying, becoming first medical registrar, and then in 1924 consultant, at The Middlesex Hospital, as well as consultant in 1934 at The Hospital for Sick Children, Great Ormond Street. He retired from medical practice in 1945, and moved to an elegant house, The Oasis, at Tring, conveniently situated only a short walk from Lord Rothschild's famous museum. He died in 1956 from emphysema – said to be a legacy of his lifelong habit of smoking cheap 'Woodbine' cigarettes.

Over a lifetime, Cockayne published over two hundred papers in entomological journals on a vast range of subjects. He was famed for his ability to rear 'refractory' larvae, and through his experiences of breeding a species on through several generations he became interested in heredity. His well-developed scientific mind got to work on the origin of various abnormalities in butterflies and moths, such as gynandromorphs, intersexes and 'mosaics', then regarded by most collectors simply as 'abs'. This led in 1932 to his important theoretical paper on sex-linked inheritance in butterflies, a blueprint for later work on genetics by E. B. Ford, Bernard Kettlewell and others. Cockayne's entomological studies had a parallel in his medical work as, while at The Middlesex Hospital, he ran a clinic to which children with rare genetic disorders were referred, and also wrote a book about inherited skin disorders.

Fig. 111.
Dr E. A. Cockayne, paediatrician. In his obituary in *The Lancet*, it was said of him: 'alert and inquisitive, there was something birdlike about him; but one never knew which bird to expect, the eagle, the raven or the sparrow'.

Cockayne was a keen collector from boyhood, and amassed a collection of butterfly varieties, many of them from breeding experiments, which has rarely if ever been equalled. A friend remembered his bachelor apartment under the rooftops of Westbourne Terrace in West London, furnished 'almost entirely by moth and butterfly cabinets, while every available ledge and cupboard overflowed with receptacles housing feeding caterpillars in various stages of evolution'.[201] In 1947, after his retirement, he and Bernard Kettlewell donated their collections of some 50,000 British Lepidoptera aberrations to the Rothschild museum at Tring, and Cockayne spent most of his last active years amalgamating them with the Rothschild collection on scientifically useful lines. It is now known as the Rothschild-Cockayne-Kettlewell collection and is housed in The Natural History Museum in London. In 1954 he was appointed an O.B.E. for this work, in recognition of its great importance to entomology.

For all his odd behavioural traits, Cockayne was a man with great strength of character. Formidably intelligent and exact, he expected high standards in others, and could be both forthright in his views and acid in his criticisms. But he is also remembered as a generous man, especially to young students. He left £1,000 to the 'South London' of which he had been President, and a collection of about a hundred books for its library. He also served as President of the London Natural History Society and of the Royal Entomological Society of London of which he was elected a Special Life Fellow in 1948.

A little-known aspect of Cockayne's career was his expertise in cipher-breaking, which was put to good use during the First World War. On one occasion, he was himself arrested as a spy, while collecting butterflies on the Wiltshire Downs. It was arranged that he would be taken to Devizes in the morning and charged formally. In the meantime, he was allowed to spend the night at his lodgings in the village. 'How he climbed out of a window at dawn and caught the early morning train to London before the military were on the move,' wrote L. Hugh Newman, 'is one of the minor escape stories of the war!'[202]

86. JOHN WILLIAM HESLOP HARRISON (1881–1967) (Fig. 112)

In 1938, H. M. Edelsten received a startling letter from his friend J. W. Heslop Harrison from the Hebridean island of Rum (in the past often misspelt Rhum). 'The most extraordinary thing here is *Lyc. arion* [the Large Blue butterfly],' he wrote. 'We have seen two but did not catch them.'[203] Later that season, Harrison sent an article to *The Entomologist*, again mentioning the Large Blues, but the editor, N. D. Riley, decided to suppress the news. 'If the species really does occur there', he wrote on 20th October 1938, 'there will almost certainly be a stampede of collectors to try and get it and, if it does not occur there then your reputation will not have been greatly enhanced in the eyes of the same collectors.'[204] Riley plainly suspected a mistake, if not a hoax. All the same, word of the extraordinary and unlikely discovery got around by word of mouth. Few, however, were in a position to verify it, since Rum was, at that time, a private island, forbidden to anyone except Professor Heslop Harrison and his students. When challenged about it by Riley, he seemed to backtrack: he had only meant to record 'the possible occurrence of *M. arion* on Rhum'.[205] However, a letter from Heslop Harrison dated 26th April 1945 survives in the Natural History Museum files claiming that 'captures [of the Large Blue] have been made in the presence of well-known people not connected with us and that such specimens are placed in safe hands'.[206]

Fig. 112. Professor J. W. Heslop Harrison, maverick northern academic. His behaviour was said to have 'something of the traditional, untrammelled spirit of the Borders'.

In 1975, Dr John Lorne Campbell, laird of Rum's neighbouring island, Canna, and a keen entomologist, investigated the whereabouts of these specimens alleged to have been taken on Rum, and concluded that none existed. Perhaps, he suggested, quoting John Raven's letter on 'Alien Plant Introduction on the Isle of Rhum' the Professor's 'ardent and competitive personality may have laid him open to students' practical jokes'.[207] (See also Chapter 4: 22.)

This was a charitable explanation. K. Sabbagh's recent research leaves little room for doubt that Heslop Harrison was himself the guilty party. With the apparent intention of proving his theory that Rum and other Hebridean islands escaped the worst effects of the Ice Age, and that a unique flora and fauna survived there, J. W. Heslop Harrison claimed to have discovered many rare species of plants and insects, some of them new to Britain.[208] Some of them were almost certainly fabricated. At the same time, Heslop Harrison was an exceptionally talented observer and did make many genuine discoveries. The problem today is distinguishing truth from fraud. In the light of what is now known about Heslop Harrison's activities, no unverified record of his can be accepted without independent confirmation.

J. W. Heslop Harrison was, in the words of a friend, admired by many but loved by few. Many of his students and friends, especially in his native north-east England, were devoted to him; for others, he inspired professional and personal antipathy that was warmly returned. Heslop Harrison was a vigorous, forceful man who always spoke his mind. His self-confidence frequently spilled over into dogmatism and aggression. There was about him 'something of the traditional untrammelled spirit of the Borders', thought one friend, E. T.

Burtt.[209] His appearance matched his uncompromising character. Another friend, A. D. Peacock, described him in his obituary as tall, slightly stooping in later life, with black hair and a neat black moustache, steely grey eyes set in a pale face, and, as often as not, clad in a sombre black suit. Heslop Harrison's knowledge of plants and insects was legendary, and his qualities as a teacher were highly praised. His great failing as a scientist, however, as even his friends agreed, was to become so convinced of the rightness of his theories and views that he either ignored reasoned criticism, or perceived it as a personal attack. He was a loner, avoiding as far as possible contact with his professional peers, and conducting much of his experimental work not in the laboratory, where it could be observed, repeated and discussed, but at home at Birtley near Newcastle-upon-Tyne, where his homemade breeding cages and glass-lidded tins would be found 'in the study, bathroom, kitchen, tool-shed and other situations where living conditions were appropriate'.[210]

Unusually for a senior scientist at this period, Heslop Harrison was of humble origins, the son of a foreman pattern-maker at Birtley Ironworks. He gained a place at university by hard work and determination, and later received a doctorate in science at Manchester. Afterwards he taught science at Middlesborough High School before joining the staff at the then Armstrong College, Newcastle, (now Newcastle University), where he taught zoology. His original work in genetics, which utilized his love of breeding butterflies, moths and sawflies, resulted in his election as a Fellow of the Royal Society in 1928. From 1927 until his retirement in 1946 he was Professor of Botany at what was to become Newcastle University.

His best-known work was in the biological exploration of the Hebrides, where Heslop Harrison led parties of students almost annually in the 1930s and '40s, and in the study of the causes of mutation in insects. Both have come under heavy criticism. Heslop Harrison claimed to have experimental proof that physical changes in the life of an individual moth or sawfly could be passed on to its progeny, according to the theory of Lamarck (whose most famous dictum was that the necks of giraffes grew longer as they stretched upwards to the tree-tops). For example, Heslop Harrison thought that melanism resulted from the effect of pollution on individual moths which somehow altered their genes. When others attempted to repeat his experiments, however, they always seemed to come up with different results. According to his obituarist, Peacock, 'to these serious criticisms, Harrison made no comment but held resolutely to his opinion'.

Heslop Harrison's theoretical work never led him far from his roots in field natural history. He was a founder member of the Northern Naturalists' Union in 1924, and none did more to encourage natural history in north-east England. He was a member of the editorial board of *The Entomologist* from 1924 until his death. He also maintained his interest in school biology, and was secretary of the local school examinations board from 1940–50. He is still remembered by his former students as a kind of guru in the field, a man of seemingly inexhaustible knowledge with the ability to awaken interest in others. It was probably as a natural history teacher that Heslop Harrison was at his best. Moreover, his example also inspired his own family. His three children all embarked on biological careers. His son, the late Professor John Heslop-Harrison, ultimately became director of the Royal Botanic Gardens, Kew, and one of the leading plant scientists of his generation.

87. PHILIP BERTRAM MURRAY ALLAN (1884–1973) (Fig. 113)

It was Allan's father, a keen lepidopterist, who, as he put it, 'infected his youngest son with the disorder'. Aged four, he captured his first butterfly, a Clouded Yellow, and was 'an ardent moth-hunter' by his eighth year. Few entomological writers have matched P. B. M. Allan's happy touch for relating the methods and eccentricities of field entomologists, past and present, displayed at its best in his classic book *A Moth-Hunter's Gossip* (1937), and its sequels, *Talking of Moths* (1943) and *Moths and Memories* (1948). They ranged over all aspects of moths and butterflies, with wit and relish, laced with anecdotes, recollections and plain gossip – and a great deal of scholarship too. Many will read these wonderful books over and over again while more turgid works on entomology gather dust in libraries.

Fig. 113. P. B. M. Allan, publisher and author, in the evening of his days, but still full of wisdom and humour, seemingly about to launch into yet another 'mothing' anecdote.

Allan was educated at Charterhouse and Clare College, Cambridge, where he read science and medicine, intending to become a doctor, and also attended The Middlesex Hospital for practical experience. However, following a misdiagnosed illness which led to his being sent on a recuperative cruise, he decided to abandon medicine, and returned to Clare College to read English and mediaeval history. While there, he was a contributor to a dictionary of mediaeval Latin. After service in military intelligence in the First World War, Allan ran the publishing house of Philip Allan & Co., which published a number of his own works, including *The Book-Hunter at Home* (1920), a standard work on rare and antiquarian books. Allan wrote or translated a number of other non-entomological works under such pseudonyms as 'Philip Murray', 'Alban A. Phillip' and 'O. Eliphaz Keat'. In 1928, at the request of the Home Office, he founded *The Police Journal* and, later, *The Journal of Criminal Law*, which he edited for many years, the latter almost until the end of his long life. Though busy with his publishing, between the wars Allan also 'indulged his entomology to the full' from his home on the border of Hertfordshire and Essex.

Allan's writings on entomology are numerous, and usually entertaining as well as informative. Apart from his well-known quartet of books (*Leaves from a Moth-Hunter's Notebooks* was published posthumously in 1980), he contributed regularly to entomological journals in the 1950s and '60s, often under the nom-de-plume of 'An Old Moth-Hunter' or 'O.M.H.'. He took over the publication of *The Entomologist's Record* in 1950 and helped to revitalize it, probably saving it from oblivion. Many of his longer pieces show the benefit of his fine library, and Allan put his unrivalled knowledge of entomological history to good use in articles discussing the extinction of the Large Copper, Black-veined White and Mazarine Blue, as well as other mysteries of British entomology, like the status of the Scarce Swallowtail and Scarce Copper, and the fraudulent activities of past commercial dealers, for example 'the Kentish Buccaneers'.

Allan liked his comforts, as his love of Richebourg wine – now a millionaire's preserve – suggests, and his books today seem set in a lost world of leisured country sportsmen, rectors and gamekeepers. I have a happy memory of my one meeting with him back in the 1950s. On showing him a striking aberration of the Small Copper, caught that morning and still crawling around its pill-box, his enthusiasm knew no bounds. Taking the pill-box solemnly in his hands, he exclaimed with emphasis, 'Surely Solomon in all his glory was not arrayed as one of these'. Later, over sustenance at a nearby inn, he talked about collecting in the past. His greatest regret was the progressive urbanization of the countryside with the loss of so many good butterfly haunts. Ronald Wilkinson recounted another meeting with the 'old moth hunter', by then genuinely old, in the 1970s: 'He met me at the Bishop's Stortford station with an automobile and driver. Although ill and having to walk with the aid of a stick, he entertained me with wit and vigour at a local inn with a fine luncheon with a Chateau Latour of excellent vintage. He recalled with vivid memory and considerable humour his early experiences with both moths and books, which antedated my own by half a century. We were then driven to his ancient and picturesque home at No. 4, Windhill ...'.[211] He was a delightful and learned companion. A fuller account of his life can be found in Ronald Wilkinson's introduction to *Leaves from a Moth-Hunter's Notebooks.*

88. NORMAN DENBIGH RILEY (1890–1979) (Fig. 114)

N. D. Riley was one of the great museum entomologists of the twentieth century. For most of his career he served the BM(NH), having joined the Department of Entomology in 1911 as an assistant in charge of the butterfly collections, and succeeding E. E. Austen as Keeper in 1932, a post Riley retained until his retirement in 1955. During that time, the Museum's insect collections nearly doubled in size (from eight million to fifteen million specimens), and the Department's work and responsibilities increased enormously. The Museum was fortunate to have Riley to oversee them. He was a good administrator as well as a leading entomologist and systematist. In retirement – not that he ever fully retired, for he continued to visit the Museum almost daily – he became half of that famous duo, Higgins & Riley, authors of the most widely-used field guide on the European butterflies. My main memory of him is of a quick, dapper, thoughtful man, dressed in dark grey (including his sober tie), always busy, and moving speedily over the polished Museum floors as though on skates.

Fig. 114.
N. D. Riley, C.B.E., professional entomologist, 'always busy ... moving speedily over the polished Museum floors as though on skates'.

Riley was a Londoner, born in Tooting, and educated at Harlington School, Balham, and Dulwich College. Luckily for him, his boyhood neighbour was none other than Richard South, to whose patience and encouragement Riley probably owed his life-long interest in butterflies and moths. After a spell as a student and demonstrator at the Imperial College of Science, Riley became an assistant to the preoccupied and absent-minded Sir George Hampson in what was soon to become the Natural History Museum's Department of Entomology. Riley might have liked to spend his time there writing monographs on his beloved butterflies, like Hampson or Karl Jordan. 'Hardly one of those drawers', he once remarked, 'does not contain some fascinating problem or other'. But Riley never did produce a great Jordanian revision of a large group of insects. Instead, although he contributed a steady stream of papers, notes, reviews and obituaries to *The Entomologist* – which he also helped edit for nearly forty years – and to many other foreign as well as British journals. Departmental matters, apart from the disruption caused by two world wars, were to take up most of Riley's professional time. In the First World War he served with some distinction in the Army Service Corps and The Queen's Regiment, reaching the rank of Captain (he was known as 'The Captain' in the Department ever afterwards). In the Second, he was responsible for the evacuation and dispersal of the Museum's irreplaceable insect collections to a variety of rural safe-havens, and their reassembly afterwards, following the departure of the Ministry of Health who had taken over the premises! He also served on the International Commission on Zoological Nomenclature and acted as one of its editors. Riley was a very active member of the leading entomological societies. He was an officer and Fellow of the Royal Entomological Society, which he served in one capacity or other for forty-five years, and in 1959 was elected a Special Life Member of the 'Brit. Ent. Soc.' after over fifty years' membership – a rare honour. In his later years he became very interested in nature conservation, and served as chairman of the committee appointed to study and try to save the Large Blue butterfly. He was appointed C.B.E. in 1952 for his services to entomology.[212]

Norman Riley was contemplating writing a handbook on the European butterflies as an entomological swan-song, and mentioned the fact to Lionel Higgins some time in the early 1960s. 'Damn it', exclaimed Higgins. 'So am I!'. The result was their famous collaboration in one of the truly great Field Guides, which opened the eyes of many an insular British lepidopterist to the wonders of Europe. Riley followed it up with another Field Guide, on the *Butterflies of the West Indies* (1975), while his well-written account, *The Department of Entomology of the British Museum (Natural History) 1904–1964. A brief historical sketch*, is an essential source-work for what was happening at the very heart of British entomology between 1904 and 1964.[213] There have been many more-flamboyant characters in British entomology than the reserved, mild-mannered Riley, but few can match him for determination, organization and fluency. 'Riley had a marked distaste for the limelight and was a much more effective administrator than he allowed himself to appear. Men and women of wealth and eminence valued his counsel and young men who worked under him grew in confidence and maturity through his guidance. Few have as much right to claim that they had done their share of the world's work.'[214]

89. LIONEL GEORGE HIGGINS (1891–1985) (Fig. 115)

Lionel Higgins is known to many, even with only a passing interest in butterflies, as the joint author of the Collins *Field Guide to the Butterflies of Britain and Europe* (1970), which has been revised and reprinted many times, and translated into at least ten European languages. 'Higgins & Riley' was the first pocket-sized and affordable book that described and illustrated all the species and main races of the European butterflies, and introduced thousands of naturalists to the still teeming butterflies of the European mainland. However this well-known book represents only a fraction of Higgins's published output. He was a leading butterfly taxonomist, specializing in the smaller fritillaries on which he published five major systematic papers. He would probably have regarded *The Classification of European Butterflies* – published in 1975 and illustrated by 402 text figures of his own line drawings of butterfly genitalia and other structures – and not the Field Guide, as his crowning achievement.

Higgins was born in Bedford, and brought up in rural Hertfordshire, where he became interested in natural history while recovering from rheumatic fever. He studied medicine at Clare College, Cambridge, and went on to a medical career, initially as a family doctor, then as an obstetrician and gynaecologist; he later received a doctorate for his study of anaemia in premature babies. He never fully retired, and was still acting as a locum at the age of ninety.

Fig. 115.
Dr Lionel Higgins, obstetrician and gynaecologist, authority on Palaearctic butterflies, and the other half of 'Higgins & Riley', one of the best field guides of the twentieth century.

We, of course, are more interested in his hobby. Lionel Higgins was among the fortunates who, in C. W. Dale's words, acquired an entomological wife, Nesta, who accompanied him on his travels throughout Europe, North Africa and the Near East, with excursions further afield to Iraq, Iran and Kashmir. Higgins's collection, vast as it was – his batteries of Hill cabinets extended from floor to ceiling on three sides of a large study – was there for a purpose. From 1924, when he published his first important work, a 60-page paper on the skippers of Abyssinia, to the end of his long life, Higgins published some 2,000 pages on the systematics of Palaearctic butterflies. For those few countries he could not visit himself, like Nepal and the outlying parts of the USSR, he nonetheless found people to send him material. His correspondence with entomologists from around the world was enormous and clearly inspirational – perhaps this worldwide influence was his greatest achievement. His splendid library was built up with the same single-minded purpose. Moreover, much of Higgins's work was necessarily done after the age of sixty.

Higgins probably enjoyed the adulation he received late in life – the characteristic twinkle in his eye gave him away. It included awards from the Zoological Society and Linnean Society, and honorary membership of this society and that. He was not without his critics – some thought he relied too heavily on genitalic characters and was too much of a splitter (gone were the days when most of the European fritillaries could be packed into a couple of genera!). But he was universally esteemed as the grand old man of Palaearctic butterflies; during an international congress at Cambridge in 1982, delegates queued to shake his hand, to his embarrassment but secret delight. Tom Tolman, who knew Higgins in his last years, mentions his mild, unassuming manner, coupled with his 'incisive mind, great enthusiasm, gentle humour and generous nature'. To which could be added due scientific humility: 'Our best efforts invariably raised more questions than answers'.[215]

I once had the privilege of examining his magnificent butterfly collection and his library – it took several afternoons – and listening to his marvellous anecdotes. Afternoon tea in the Higgins household was extraordinary for the manner in which he used to manipulate the tea-bags with surgical forceps. And when asked why his butterfly collection contained no aberrations, he retorted: 'Well – they're not perfect, are they?'.

90. HENRY CHARLES HUGGINS (1891–1977)

The elder son of Henry Huggins, J.P., also an entomologist, Henry Charles Huggins ('Harry') started collecting Lepidoptera at the age of eight, after breeding a one-spot aberration of the Lime Hawk-moth. An early mentor was

A. B. Farn, a second cousin of his grandmother, who taught him how to set specimens and tell one moth from another. Most of his earlier collecting was done in Kent, especially along the Medway Valley and around Sittingbourne. Much of his later collecting was in Ireland, Huggins making no fewer than 36 collecting trips there, and describing a number of new Irish subspecies, including ssp. *baynesi* of the Dingy Skipper and ssp. *gravesi* of the Brimstone. His Irish memories are recalled in 'A Naturalist in the Kingdom of Kerry', published in 1960. Although from 1922, Huggins concentrated mainly on microlepidoptera, on which he became a leading authority, he also at one time had a passion for molluscs.

Huggins was a gifted and often amusing writer, contributing over 400 notes and articles, on entomology and other wildlife subjects. In one article, 'They were Irish Gannets', he debunked an alleged discovery of the Large Blue at Dunboy, in West Cork[216] (see Chapter 4: 22). In another he recalled past times at Chattenden Roughs, in Kent, and the conflict in those days between amateur collectors and unscrupulous dealers.[217] He was an entertaining raconteur with a fund of memories of past entomologists, such as J. W. Tutt and Richard South. Huggins's career linked the classic years of butterfly-hunting and our own conservation-conscious age. Having been a founder member of both the South Essex and the Colchester Natural History Societies, it is appropriate that one of his last publications, as a contributing author, was *A Guide to the Butterflies and Larger Moths of Essex* (1975). His fine collections of British and Irish Lepidoptera and of land and freshwater molluscs was presented after his death to the Natural History Museum, London.

91. ARTHUR FRANCIS HEMMING (1893–1964) (Fig. 116)

Francis Hemming was an Oxford-educated civil servant with a brilliant mind that brought him various private secretaryships and the rank of Assistant Secretary in the Ministry of Home Security during the Second World War, and later, as Under-Secretary at the Ministry of Fuel and Power, for which work he was awarded both the C.M.G. and the C.B.E. On his death in 1964, a colleague wrote a long and most appreciative review of his government work in *The Times*.

Fig. 116. Francis Hemming, C.M.G., systematist and civil servant. In the words of Miriam Rothschild, he 'knew more about butterfly nomenclature than anyone before or since but made a terrible mess of non-intervention in the Spanish Civil War'.

We, of course, want to know what he did for butterflies. A crippled left arm, shattered during the First World War, did not seem to affect his netmanship. In the 1920s, Hemming travelled widely and produced a stream of papers on European butterflies, contributing much to our knowledge of their distribution. He also undertook a revision of the Palaearctic Lycaenidae before producing a long paper on the butterflies of Jordan.

His main contribution, however, was to nomenclature – indeed, he was probably the most important nomenclaturist of insects this century. In the 1930s, Hemming assumed the secretaryship of the International Commission on Zoological Nomenclature, then in the doldrums, and helped to revive it by arranging its accumulated papers and launching a new journal, the *Bulletin of Zoological Nomenclature* – which simultaneously ensured the Commission an adequate income. Under the editorship of Hemming, sixteen volumes of the Bulletin were published, containing a vast amount of new work on directions and 'opinions' in nomenclature, with the ultimate, though perhaps impossible, aim of reaching an acceptable finality on names. The eventual Code of Rules, published in 1961, owed more to Hemming's work than that of anyone else.

His second great achievement, the work of many years, was in the systematics of Holarctic butterflies, on which he produced one volume, and was busy completing a second when he died. Among other self-imposed tasks were his post of Treasurer to the Royal Entomological Society between 1929–39, and membership of the editorial panel of *The Entomologist.*

Norman Riley remembered Francis Hemming as 'a man so brilliant, so productive in so many fields, so practical and methodical in affairs and yet at times so wayward and unorthodox'.[218] He had an unusual capacity for concentration, and at the zoological congress he attended seemed to put in long hours of work without much sleep. He was something of an idealist, even a perfectionist. Much of his work in the dry, technical aspects of nomenclature can be fully understood only by specialists, some of whom believed that Hemming over-reached himself. But, as R. V. Melville, who took over from Hemming as Secretary of the ICZN, said of him (quoting from Browning), '... a man's reach should exceed his grasp, or what's a heaven for?'.[219]

92. SIR ROBERT HENRY MAGNUS SPENCER SAUNDBY (1896–1971) (Fig. 117)

Right at the end of the many orders and decorations that follow the name of Air Marshal Sir Robert Saundby are four incongruous letters: F.R.E.S. For as well as being a Knight Commander of both the Orders of the Bath and of the British Empire, and holder of the Military Cross, the Distinguished Flying Cross and the Air Force Cross for gallantries performed during and after the First World War, Saundby was also, unknown to most of his service colleagues, a distinguished lepidopterist, and a Fellow of the Royal Entomological Society.

Fig. 117.
Air Marshal Sir Robert Saundby, one of an elite band of military top-brass who combined an enthusiasm for natural history with their service life.

Robert Saundby was the son of Dr Robert Saundby, professor of medicine at Birmingham University. Educated at King Edward VI School, Birmingham, he entered the London and North Western Railway Company at the age of seventeen, with some idea of commencing a career in transport control. A year later, however, the First World War broke out. Saundby obtained a commission in the Royal Warwickshire Regiment, and in 1916 was seconded to the newly formed Royal Flying Corps. His subsequent career as a remarkably young senior officer in the RAF need not detain us long – it is in the history books. It was due in large measure to Saundby's efforts at the Air Ministry that the RAF was prepared for war when it came in 1939. From 1943 to 1945 he was number two in Bomber Command, under Sir Arthur Harris, and was invalided from the service a year after the war ended – the legacy of a leg injury suffered in a plane crash thirty years before. The end of his military career formed the beginning of his main entomological and conservation efforts, which continued until his death in 1971.

Saundby had retired to Burghclere, near Newbury, where he made regular light-trap records from his garden, and, as far as his injured leg would let him, entomologized in the West Berkshire woods, recording some 44 species of butterfly and 501 of larger moths. He was a stalwart of local natural history societies, notably the Newbury District Field Club and the newly formed Bucks, Berks & Oxon Naturalists' Trust (BBONT) of which he was a very active first president, chairing all its main meetings. He also attended field excursions run by the British Entomological Society, and exhibited regularly at its annual exhibition. He had his share of luck. Once, when cutting some shoots of sallow for some moth larvae he was rearing, he found a Purple Emperor caterpillar sitting there. On another occasion he captured and boxed a striking melanic Pearl-bordered Fritillary, and then lost the box. Three weeks later, he happened to be back in the same area, and there it was, by the tree stump where he had eaten his lunch, the butterfly still inside.

Saundby's fine and near complete collection, described as 'a model of good arrangement and documentation', and particularly strong on butterfly races and varieties, was presented to the Natural History Museum, London.[220] He was himself a rare gift to Thames Valley natural history, being wise, dedicated and always good-humoured. He was also a shining example of the many men in military service – Field Marshall Lord Alanbrooke was another – who were also good field naturalists.

Fig. 118. Professor E. B. Ford, Oxford don and author of 'the best butterfly book ever written' according to Miriam Rothschild, who also said of him, 'the good collector not only has a magic touch but also a sort of second sight which combined to make a work of art rather than a science'.

93. EDMUND BRISCO FORD (1901–1988) (Fig. 118)

E. B. Ford, known to his contemporaries as 'Henry' (after the American founder of the Ford Motor Co.), is best known as the author of *Butterflies* (1945), the first book in the famous and long-running New Naturalist library series – described by Miriam Rothschild as the best butterfly book ever written, together with its companion volume, *Moths*, seven years later. Both books represented a distinct departure from earlier popular books about Lepidoptera. Although *Butterflies* was based on field work, and included a chapter on the history of collecting, it was a product of the modern age in the close attention it gave to the developing sciences of genetics, physiology and evolution. Part of the book's success lay in the challenge to old-style entomologists, inviting them to explore the meaning of such well-known phenomena as aberrations and local variation. It gave a new depth to the study of butterflies, blending amateur field study with mainstream experimental science. It is to the credit of British naturalists that such a book – though admittedly it was exceptionally well written – proved a best seller, and stayed in print for forty years.

Its author was a remarkable man, of cultivated tastes and an equally cultivated eccentricity. Ford was born at Papcastle in Cumberland, the only son of Harry Dodsworth Ford, a local clergyman. He was educated at St Bees School and Wadham College, Oxford, his father's college, where he had originally intended to read classics. However both father and son collected butterflies and had become interested in varieties. E. B. Ford decided, as a result, to study biology, specializing in genetics. He remained at Oxford throughout his professional life, becoming one of the first scientific Fellows of All Souls for nearly three hundred years and was Professor of Ecological Genetics on which he was a world authority. He was elected a Fellow of the Royal Society in 1946. As objects of study he chose butterflies and moths, partly because he liked them and partly because their wing patterns made them ideal subjects for the study of genetically controlled characters. This work had really begun while Ford was still in his teens. By observing a local colony of Marsh Fritillary over several years, he noticed that not only did the butterfly vary greatly in numbers from year to year, but that variation in wing markings was greatest when it was most abundant. He realized that evolution was taking place in front of his very eyes. Here was a field in which butterfly collectors could make important scientific discoveries. At Oxford, Ford developed his ideas by studying the changing proportions of varieties of the Scarlet Tiger, Meadow Brown, Common Blue and other Lepidoptera over the years. Yet, to all appearances at least, he looked like an ordinary butterfly collector, net in hand, hazel switch in the other, a hat on his bald head and with no equipment more technical than a notebook.

At Oxford, 'Henry' Ford is remembered equally for his eccentric behaviour. Miriam Rothschild recalled her first meeting in Ford's office in the Zoology Department, Oxford, in 1956. She knocked and awaited a reply.

> 'After a moment's silence there was rather a plaintive long drawn out cry: "Come in!" I opened the door and found an empty room. I looked round nervously – not a soul to be seen, but an almost frightening neatness pervaded everything. Each single object, from paper knife to *Medical Genetics* was in its right place. Each curtain hung in a predestined fold ... the sight of this distilled essence of neatness and order took my breath away. I stood there, probably with my mouth open, trying to reconcile this vacant room with that ghostly cry – had I dreamed it? – when suddenly Professor Ford appeared from underneath his desk like a graceful fakir emerging from a grave. Apparently he had been sitting crosslegged on the floor in the well of his writing table, lost in thought, but he held out his hand to me in a most affable manner ..."My *dear* Mrs Lane [her married name] – I didn't know it was you".' As she added, 'you never have to ask a great man for an explanation'.[221]

His behaviour in the lecture theatre could be equally individual. He would eye late-comers with a gimlet gaze, and, it is said, refused to teach women students at all. He had deeply old-fashioned tastes, disapproving of, among many other things, radios, newspapers, fish-and-chips, and most of Oxford's attempts to catch up with the twentieth century. Among the modern inventions he could just about bring himself to recognize were what he called the 'motor racing car', 'the photographic camera', and the 'cinematic projector'. When a complex piece of machinery was needed for an experiment, Ford referred to it as 'the engine'. History does not relate whether he appreciated the irony of his nickname! His other, non-entomological, interests were heraldry, archaeology and ecclesiastical architecture, the last two affording more opportunities for field expeditions.

Ford never married; indeed he never really approved of women, though he allowed certain exceptions. He ignored most scientific work outside Oxford, and to read his books you would be forgiven for thinking that ecological genetics was the exclusive preserve of himself and his colleagues. That might be one reason why E. B. Ford is such an exceptionally lucid and readable scientist: most of his writing is based on personal experience. This singular man hoped to live to the age of one hundred, and write a final book about extinction.

Instead he died peacefully in his sleep, aged eighty-seven, after a final dinner at All Souls. His ashes were scattered on a Cotswold hillside where he had spent many happy hours in good company, studying the eternal mysteries of insect variation.

94. BARON CHARLES GEORGE MAURICE DE WORMS (1904–1980) (Fig. 119)

The memorable title of Baron de Worms was inherited from a great-uncle who had been made a hereditary baron of the Austrian Empire in 1871. Permission to use it in Britain was granted in recognition of the family's service in the development of tea-planting in Ceylon.

Fig. 119. Baron Charles de Worms, industrial chemist, and a well-loved enthusiast, who, according to his obituarist, 'usually carried about half a dozen kite nets on collecting expeditions'.

Charles de Worms was born at Egham, Surrey, son of Baron Anthony de Worms, and educated at Eton, where he was a King's Scholar, and at King's College, Cambridge, where he won the Drewitt Prize for agricultural chemistry. After obtaining his doctorate at London University on 'orthodiphenyl ethers', he went on to a career in chemistry, conducting research for the Royal Cancer Hospital, and, during and after the Second World War, at Porton Down, latterly in charge of one of the military research laboratories.

He is remembered by many for the informative and often entertaining articles he contributed to entomological journals, especially the accounts of his regular collecting trips, which he began in the 1930s. Baron de Worms always seemed to know exactly where and how to obtain an elusive species, and, as was by no means always the case, was willing to share this knowledge with others. Typical was his main collecting trip of 1937, 'my main summer holiday', which consisted of 'on the whole, a successful tour of the British Isles'. Starting on 8th July near Portsmouth, 'where *Leucania turca* [The Double Line] was found to be the commonest moth on the sugar', he travelled west to the Dorset coast and heaths, north to the Cotswolds, North Wales and the Lake District, to reach Rannoch and Aviemore late in the month, from whence he returned by way of Witherslack and the Norfolk coast and Broads, where, needless to say, he found plenty of Swallowtail larvae.[222] Such tours made him an authority on the distribution of butterflies and moths, especially in his regular hunting grounds around London and in Wiltshire. He also collected butterflies in Europe and other parts of the world.

De Worms was a warm-hearted and hospitable man. Entomological friends always looked forward to his Christmas parties, where prizes for games were generally unusual or rare entomological specimens. His friend Russell Bretherton described him as 'a kindly man, good with cats, dogs and small children. In short, a "character" '.[223] His friends affectionately referred to him as '*c-nigrum*', a punning reference to the English name of this moth.

My one memory of him was at a collecting excursion to Bagshot Heath, Surrey, in the late 1950s, when we were reduced to laughter at the sight of the Baron fleeing from a swarm of bees. He ran extremely fast for a rather stout, middle-aged man with a gammy hip, not to mention a jacket, at least two old pullovers, and a satchel full of collecting equipment. He always carried a great deal of equipment, much of which was of uncertain use and often caused comment. He would bring anything up to half-a-dozen kite nets on collecting expeditions. In his later years, Charles de Worms found difficulty in boxing specimens owing to failing eyesight. Clumsy attempts at closing a pillbox often meant trapping a moth between the lid and the box – usually with dire consequences. Other lepidopterists who noticed the unfortunate removal of wings or bodies in this way, began to talk of 'charlesing a moth' when they did it themselves.

Charles de Worms was elected President of the 'South London' in 1933, and for almost half a century he attended meetings of both this Society (and its successor, the 'British Ent. Soc.') and the Royal Entomological Society, to which he had been elected in 1926. His collection of Lepidoptera – all macros – filled 350 drawers and contained numerous series of rare or extinct species, all captured or reared by himself. With characteristic generosity, he bequeathed it to the Royal Scottish Museum, Edinburgh, and his fine library to the British Entomological Society.

95. RUSSELL FREDERICK BRETHERTON (1906–1991) (Fig. 120)

Russell Bretherton, the son of a Gloucester solicitor, was educated at Clifton College and Wadham College, Oxford, of which he later became a Fellow. He was an expert on butterfly migration, his interest in which is said to have begun in boyhood when he spotted but could not catch one of the few American Painted Ladies (*Vanessa virginiensis*) ever to be seen in Britain. He was the author of more than 200 articles and publications on butterflies and moths, many of them based on the records of migrant species which he analysed year after year with Michael Chalmers-Hunt. He also contributed an introductory chapter on Migrant Lepidoptera to Volume 10 of the *The Moths and Butterflies of Great Britain and Ireland* series, as well as providing accounts of the adventive and migratory species covered in Volumes 7, 9 and 10. Although listing walking and mountaineering as his other recreations in *Who's Who*, he had some difficulty reading grid references on maps and in his latter years tended to get his eastings and northings mixed, which led to some confusion!

Fig. 120.
R. F. Bretherton, C.B., senior civil servant, who made a special study of the migration of Lepidoptera. The late Robert Mays can be seen in the background.

He joined the British Entomological and Natural History Society in 1947, and served it in various capacities, including Treasurer and President (1969–72), being made an honorary member at the end of his term of office in 1972, its centenary year, along with HRH The Duke of Edinburgh. He was also a Fellow of the Royal Entomological Society. Bretherton's entomologizing had to fit in with a very busy professional career as a civil servant. He reached the rank of Under-Secretary at the Treasury, for his services to which he was appointed a Companion of the Order of the Bath. He left his fine collection to Reading Museum, together with his books and diaries.

96. HENRY BERNARD DAVIS KETTLEWELL (1907–1979) (Fig. 121)

Dr Bernard Kettlewell's name is famous in scientific circles around the world for his classic investigation of industrial melanism in the Peppered Moth (*Biston betularia*). Dark forms of this and other moths had long been known, but it was he who showed, through a series of experiments, that the colour forms of this moth indicated evolution taking place through natural selection. The dark form was less easy to see on the soot-blackened trunks of trees in industrial areas; *ergo* predators found fewer of them, and so more black ones survived to breed than white ones. It is possible to quibble about the details of some of his investigations – few wild Peppered Moths have ever been found sitting on trunks (they seem to prefer shadowy nooks and crannies). But the wider evolutionary principles demonstrated by Kettlewell are the stuff of school and university textbooks the world over.

Fig. 121.
Dr Bernard Kettlewell, general practitioner, geneticist and rugged individualist. R. P. Demuth noted that 'He would consider it a failure if he did not find twice the number of caterpillars that I did'.

His work on ecological genetics was the result of what one contemporary called 'a medical man playing truant'. Born and confirmed a Yorkshireman, the son of a member of the local corn exchange, Kettlewell went to school at Old College, Windermere, and Charterhouse, before studying medicine at Gonville and Caius College, Cambridge (where he was remembered as a particularly boisterous undergraduate, replacing his bicycle with a fast car, useful for reaching the 'good spots' around Cambridge), and then at St Bartholomew's Hospital Medical School.

Despite his many forays to Wicken Fen, Warboys Wood, Bedford Purlieus, and other classic localities, Kettlewell worked hard enough to obtain his qualifications and join a medical practice at Cranleigh in Surrey, as well as to find time to get married. Unlike some entomologists, however, Kettlewell was as fond of wild parties as he was of larva-hunting or butterflying with net and satchel. Even then he never did anything by halves. Demuth recalled that 'when looking for larvae with me, he would consider it a failure if he did not find twice the number that I did'.[224] 'Little did we realise', recalled his college contemporary, Sir Cyril Clarke, 'that he already knew where he was going whereas we did not. Much later as his and my interests converged we saw increasingly more of each other, and our houses and gardens, with their cages and sleeves, bore strong similarities'.[225]

From the 1930s, he began to publish observations and articles in *The Entomologist's Record*, especially on breeding the less well-known moths. An early sign of his experimental bent was a paper on the effect of temperature on the pupation of the Bordered Straw moth (*Heliothis peltiger*), and its effects on the pigmentation of the resulting imago. Such work brought him into contact with E. A. Cockayne (whom he regarded as his mentor) and E. B. Ford, and led to his increasing interest in the genetics of British Lepidoptera.

All this might have come to a premature end in 1946, when, shocked at the inauguration of the National Health Service, Kettlewell resigned as a general practitioner, and two years later, emigrated with his family to South Africa. At the same time, he presented his already fine collection, including the material from his experimental breeding programmes, to the BM(NH), now the Natural History Museum, where – amalgamated first with Cockayne's and then Rothschild's collections – it is studied by researchers the world over for its raw material on the genetic basis of variation.

Kettlewell returned to Britain in 1952 to take up, with Ford's help, a Nuffield Research Fellowship at Oxford, where began his most fruitful work on industrial melanism in Lepidoptera. A stream of scientific papers followed, culminating in 1973 in his (some would say much-delayed) book *The Evolution of Melanism*, the classic work on the subject, which left him exhausted. Along the way Kettlewell also studied pest control, life histories, pheromones, insect migration and the effect of radioactivity on feeding larvae; for example he proved, from its contamination by radioactive dust, that the tiny moth *Nomophila noctuella* in his light-trap was a migrant from the Sahara Desert where a French nuclear test had recently taken place.

Cyril Clarke regarded Bernard Kettlewell, in a sense, as being in the mould of the great nineteenth-century medical entomologists, like Power or Guard Knaggs, though he was very much of his time in his forceful advocacy of his ideas. Kettlewell always said what he thought: 'In character he was touchy, argumentative and often maddening, but he could laugh at himself and was extremely good company'. All his obituarists remembered his 'rugged individualism', robust sense of fun, and delight in his pleasures, from insect varieties to fast cars and good food. The day all the moths escaped at one of his demonstrations and fluttered in a mass around the chandeliers was the sort of thing that often happened when Kettlewell was around. Eric Classey has told me of a typical 'Kettlewell incident'. Early one morning they were driving at some speed through the market square at Wells-next-the-Sea in Kettlewell's powerful Hispano-Suiza when they nearly hit a man carrying a tray of muffins on his head. The muffin-man emitted a loud expletive. Kettlewell stopped and reversed back to the suddenly nervous muffin-man, who explained that the word had really been directed against the only other person in the square, and not at Kettlewell. So Bernard turned his attention to this innocent bystander and bellowed at him across the square 'in the manner of a town-crier' that the muffin-man had called him a so-and-so! They then drove off leaving the two occupants of the square to sort it out.

Kettlewell retired in 1976, after which he did little further work as ill-health took its toll of his once enormous physical and intellectual energy. He died in 1979, giving instructions that his dog's ashes be scattered over his grave.

97. ROBERT WATSON (1916–1984) (Fig. 122)

Robert Watson never knew his father, who was killed in the First World War. He was brought up by his mother and grandmother who encouraged him to study Lepidoptera around their home in Bournemouth. Here he was able to explore the nearby rich coastline and later, his family having moved to Hordle, also The New Forest. However, his early career was colourful and often impecunious. He began with a morning paper round, sold rabbits for a local farmer and even exploited his unusual strength to win wrestling prizes at fairs – though it was said that he often went hungry. Later he became a milkman (quickly rising to branch manager) and, during the war, an ARP (Air Raid Precaution) warden. After these somewhat unpromising beginnings, in 1945 he established a successful accountancy business in Lymington, and became an authority on income tax and other branches of tax law. His marital life was equally eventful, with four marriages and one remarriage. His somewhat porcine appearance led to the nickname 'Porker' which he acknowledged by accepting a challenge that he would not name the house he built in 1957 at Sandy Down 'Porcorum'!

Fig. 122. Robert 'Porker' Watson, tax accountant, whose exceptionally fine collection of British butterfly aberrations is now in the Natural History Museum.

Watson aimed not for a perfect collection, but for a collection of perfect specimens. Damaged ones, however rare, he always released. His setting was described as a miracle of perfection, despite his being virtually blind in one eye. Like many leading field lepidopterists of earlier times, Watson was, despite appearances, a first class shot and fly fisherman. He was also a hospitable and liberal man, holding open days when entomologists could view his collection, and often sending away young visitors with gifts of store boxes or setting boards.

Among entomologists, Watson's name will always be associated with the striking red form of the Cinnabar Moth (*Tyria jacobaeae*), which he bred over some twenty-five years and himself described as ab. *coneyi* (from A. W. Coney, who lent him material). The Watson collection of British butterflies and moths, housed in sixty-five Hill cabinets, was eventually bequeathed to the Natural History Museum, London.[226]

Fig. 123. E. C. ('Teddy') Pelham-Clinton, a meticulous microlepidopterist and all-round entomologist at the Royal Scottish Museum, Edinburgh who, in the last weeks of his life, inherited a historic English dukedom.

98. EDWARD CHARLES PELHAM-CLINTON, 10TH DUKE OF NEWCASTLE (1920–1988) (Fig. 123)

Edward Pelham-Clinton – known as Ted or Teddy by his friends – was, by any standard, among the finest lepidopterists of the twentieth century. When he succeeded to the title of Duke of Newcastle, only five weeks before his sudden death, the popular press were keen to cast him in the role of a retiring and eccentric museum curator and butterfly collector. But he was much more than this caricature made out.

Pelham-Clinton was educated at Eton and Trinity College, Cambridge. After war service in the Royal Artillery, he joined the Agricultural Research Council in 1951 as a parasitologist. His youthful interest in moths and butterflies had by then broadened considerably. He collected caddis flies, sawflies and beetles, and, on a professional level, he studied fleas, blood-sucking flies and dung beetles. He is associated above all with the Royal Scottish Museum in Edinburgh, which he joined as an Assistant Keeper in 1960, later becoming Deputy Keeper of Natural History. Unfortunately, Museum rules forbade private collecting. Pelham-Clinton's solution was to work on the Museum's *exotic* collections of Lepidoptera, enabling him to maintain his large private collection of British species, especially 'micros', on which he was a leading authority. Pelham-Clinton's high curatorial standards and wide influence encouraged many collectors to bequeath their collections to the Museum, and by the time of his retirement in 1981, he had transformed the Museum's insect collections into a major institution.

Few entomologists have been more meticulous record keepers than Teddy Pelham-Clinton, whose sixty-four foolscap volumes of pencil notes (1935–88) span the whole of his entomological career. He used a specially devised standardized system to record the locality, frequency, early stages and collection serial numbers of practically every significant insect he ever captured or reared. His collection of 35,600 specimens was equally methodically referenced: each bore a label in 4-point type from his own press, and was supplemented by genitalia slides and a herbarium of larval workings. During the 1960s he got about in a Morris Minor Traveller with a built-in setting table, and during this time he added immensely to our knowledge of the distribution of microlepidoptera, especially in Scotland.[227]

Pelham-Clinton published many papers in entomological journals, and was an associate editor of *The Moths and Butterflies of Great Britain and Ireland*, contributing the sections on several families of micros. He was working on the Elachistidae and the Gelechiidae at the time of his death. He added several new species to the British list, including a gelechiid new to science named *Scrobipalpa clintoni* in his honour.[228]

Pelham-Clinton was not one to reveal his feelings to strangers, but those who knew him always appreciated his humour and infectious enthusiasm. He never married and, leaving no heir, his title became extinct. With characteristic generosity, he left a large bequest to the British Entomological and Natural History Society to help it fund the permanent home it badly needed, following the expiration of its lease at the Alpine Club. This was eventually found at Dinton Pastures Country Park near Reading where new headquarters were built to house the Society's library and collections, in the appropriately named 'Pelham-Clinton Building'. This was opened by Sir Richard Southwood in the presence of many leading entomologists, on Sunday, 27th June 1993. His collection was left to the Royal Scottish Museum.

99. RICHARD LAWRENCE EDWARD FORD (1913–1996) (Fig. 124)

R. L. E. Ford was born in Bexley, the eldest son of the noted microlepidopterist, L. T. Ford (1881–1961). The son inherited many of his father's interests and talents. Though he left Hurstpierpoint School without formal qualifications, he soon took up a career in entomology, initially at L. W. Newman's Butterfly Farm, and then at the Imperial (later Commonwealth) Institute of Entomology, based at the Natural History Museum. During this time, and for long afterwards, Ford collected and bred insects, including many insect parasites. He was second to none as a field collector – Riley described his work as 'inspired'. Some 50 type specimens of the latter bear R. L. E. Ford's name on their data labels.

Fig. 124. R. L. E. Ford, proprietor of Watkins & Doncaster, an honest and generous entomological dealer, very different from the unprincipled scoundrels of the nineteenth century.

For nearly thirty years, between 1941 and 1969, R. L. E. Ford ran the well-known firm of natural history suppliers, Watkins & Doncaster (see Appendix 2). During his time in charge of Watkins & Doncaster, Richard Ford donated nearly 27,000 specimens of Lepidoptera to the Natural History Museum in London, as well as choice items from his other great passion, British fossils.

As director of Watkins & Doncaster, he bought and sold many book collections. During the war one of these contained some excellent aerial photographs of Berchtesgaden, Hitler's mountain retreat. Ford sent them on to his friend, Sir Robert Saundby, then second-in-command of Bomber Command, with the message: 'Do your stuff!' Saundby's reply was: 'What a turn-up for the book!'. Ford's list of entomological clients was also used to locate people who knew the beaches of what was to become the allied landing-grounds on D-Day. Apparently parts of them were popular venues for entomologists between the wars.[229]

Many young entomologists and naturalists knew of Ford through his authorship of a series of excellent educational books in the 1950s and '60s, including *The Observer's Book of Larger Moths* (1952), *Practical Entomology* (1963) and *Studying Insects* (1973), as well as books on wild animals, birds and pond life in the popular Black's Young Naturalist series. He was responsible for publishing at Watkins & Doncaster the third of P. B. M. Allan's books, *Moths and Memories*, in 1948, and the second edition of Allan's *Moth-Hunter's Gossip* in the previous year. With E. W. Classey, he was co-founder in 1950 of the *Entomologist's Gazette.*

In retirement, Ford spent much of his time studying fossils on the Isle of Wight, which became his home until his death in 1996. There he made a number of important finds, especially of tiny fossil mammals, two of which bear his name. He also took part in several of the 'dinosaur digs' which unearthed some of Britain's most spectacular fossil skeletons. When asked by passers by what he was doing, Ford would usually answer, 'digging for pineapples', after which he was generally left in peace as a harmless lunatic. He kept up his entomology, too, writing papers on insect migration, butterfly and ladybird parasites, and spiral rotation in Large White caterpillars, among other topics.

One of R. L. E. Ford's stories concerned a celebrated royal physician who collected butterflies using his chauffeur, Jones, as his 'gillie'. Each time he caught a specimen, he would say 'The box, Jones', and Jones would duly produce a pill-box from his satchel. 'The lid, Jones', he would then order, if the specimen passed muster. One day, while collecting in the company of L. T. Ford, the latter offered him a moth in a box which he knew the physician wanted. But, 'Oh no', exclaimed the physician, 'I only take specimens I have caught myself'. Fortunately Jones was on hand with a net. The moth was released, and the physician deftly caught it. Having thus established the moth as his, he began the routine again. 'The box, Jones', he said.[230]

100. PETER WILLIAM CRIBB (1920–1993) (Fig. 125)

Much of the credit for the success of the Amateur Entomologists' Society, whose annual exhibition is the liveliest event in the entomological calendar, must be due to the sterling efforts of Peter Cribb. Among the Society's earliest members, Cribb was always in the thick of things, sitting on its Council, organizing many of the annual exhibitions, editing and contributing to the Society's bulletin, and writing some of its excellent pamphlets and label lists. 'He had a predisposition to both mental and physical activity and was always busy', wrote one friend.[231] Cribb's regular articles about his butterflying expeditions around Europe were always a highlight of the AES Bulletin, lively, interesting and full of *joie de vivre.* He was given the rare distinction of Honorary Life Membership of the AES in 1980.

Fig. 125. Peter Cribb, crematorium superintendent and registrar, and gardener, lynchpin of the Amateur Entomologist's Society for decades and entertaining chronicler of his entomological travels in its Bulletin.

Cribb was interested in every aspect of British and European butterflies. He was active in conservation, long before it became a fashionable cause, advocating releases from reared stock and campaigning successfully to save Ditchling Common in Sussex from ploughing. Remarkably skilled at breeding butterflies, he reared many of the then little known European butterflies, and maintained a culture of Marsh Fritillaries in his garden for forty years. Possibly part of his secret was that he was a good gardener, with botanical knowledge (his son, Phillip, became the orchid authority in the Herbarium at the Royal Botanic Gardens, Kew). But he was also something of a craftsman, skilled at pottery, wood-carving, printing and book-binding.

Perhaps Peter Cribb will be remembered best for those articles in the AES Bulletin – some 240 of them – and above all for his descriptions of camping holidays in the wilder parts of France, Spain, Switzerland, Italy, Greece and Yugoslavia. Written between 1960 and 1993, they were an inspiration to so many young and not-so-young entomologists. His enthralling accounts of rare butterflies were always mixed with accounts of local scenery, culture and food, and show him to be a born anecdotist; his style has been much copied, but rarely equalled. These pieces were much in character: Cribb's fellow AES Council member, Brian Gardiner, remembers his jokes and stories at meetings, his readiness to volunteer and his easy way of putting newcomers at their ease. He was one of those rare people who help a small society to flourish by their energy and charm. His sudden death in 1993 was a serious blow to the AES and to all who looked forward to the next account of Cribb's entomological adventures.

101. JOHN HEATH
(1922–1987) (Fig. 126)

Many of those who attended entomological gatherings in the 1970s and '80s will remember John Heath, with his friendly presence and the latest distribution maps of moths and butterflies. His sudden and unexpected death at the age of sixty-five left a gap in British entomology felt by everyone.

John Heath was the son of an Indian Army officer, and educated at King Edward VI School, Southampton. During the war he served with REME (the Royal Electrical and Mechanical Engineers) on isolated radar stations around the English coast. Not only did that often place him in good country for moths and butterflies but the electronic skills he learned then came in useful for computer work later, and enabled him to invent a portable moth trap called the Heath Trap and still used today. After the war, he began his professional entomological career in pest control, before joining the newly-formed Nature Conservancy in 1953. He spent fourteen years at its research station at Merlewood in the Lake District, specializing mainly in the ecology of invertebrates, and his published work included a well-known study of the Netted Carpet moth (*Eustroma reticulatum*).

Fig. 126. John Heath, professional entomologist and Lepidoptera recorder, who was the instigator and founding editor of the series *The Moths and Butterflies of Great Britain and Ireland.*

From 1967 until his retirement in 1982, Heath worked at the Biological Records Centre at Monks Wood Experimental Station, becoming Head in 1979. He was also organizer of the Lepidoptera Recording Scheme from 1967 until 1981. Through the systematic recording encouraged by regular newsletters, Heath built up the first ten-kilometre square distribution maps of our butterflies and larger moths. These maps are being published in Heath's second great contribution to British entomology, the multi-volume series *Moths and Butterflies of Great Britain and Ireland*, yet to be completed. John Heath was the architect and senior editor of this ambitious work, and wrote the systematic section on the Micropterigidae in Volume 1, a family in which he specialized and of which he formed an important collection (it was the only group he did collect), left in his will to the Natural History Museum, London. His wife Joan, whom he married in his Merlewood days, was a great support to him and typed up much of the text for several of the early volumes in the *MBGBI* project.[232]

His expertise was also in demand in Europe, especially for the European Invertebrate Survey, begun in 1969. He was a Vice-President of SEL (Societas Europaea Lepidopterologica) and his work on *Threatened Rhopalocera (butterflies) in Europe* was published by the European Committee for the Conservation of Nature and Natural Resources of the Council of Europe in 1981.

Heath was a stalwart of most of the leading entomological societies, and was elected President of the British Entomological Society in 1982. He was a modest but friendly and gregarious man. Without formal qualifications he built up a reputation as a leading professional ecologist, and his work in biological recording earned him the gratitude of every field entomologist.[233]

Like so many great Aurelians before him, John Heath was both a professional who retained all his amateur zeal and love of insects, and an amateur who developed professional scientific standards. Though he is the last in our choice of one hundred and one Aurelians running over more than three centuries, he will certainly not be the last of that honourable line. For as long as there are butterflies and moths to study in Britain there will be talented Aurelians to discover something new and startling, to produce works of lasting value, or to inspire others by their example.

NOTES TO CHAPTER THREE

1. Bristowe, W. S. (1958). *The world of spiders*, p.12.
2. Raven, C. E. (1942). *John Ray, naturalist: his life and works*, p.407.
3. Raven, *Ibid., passim.*
4. Fitton, M. & Gilbert, P. (1994). Insect Collections. *In* MacGregor, A. (Ed.), *Sir Hans Sloane*, pp.112–122.
5. Foster, W. (1924). *The East India House: its history and associations*, p.124.
6. Petiver, J. (1695). *Musei Petiveriani centuria prima rariora naturae*, p.3.
7. Allen, D. E. (1976). *The naturalist in Britain, a social history*, p.38.
8. Fitton & Gilbert, 'Insect Collections', p.118.
9. Bristowe, W. S. (1967a). The life and work of a great British naturalist, Joseph Dandridge (1664–1746). *Entomologist's Gaz.* **18**: 73–89.
10. Bristowe, W. S. (1967b). More about Joseph Dandridge and his friends James Petiver and Eleazar Albin. *Ibid.* **18**: 197–201.
11. Albin, E. (1720). *The natural history of English insects*, Preface, p.[4].
12. DaCosta, [E.] M. (1812). Notes and anecdotes of literati, collectors, etc. *Gentleman's Mag.* **82**(1): 204–207, 503–517.
13. Fitton & Gilbert, 'Insect Collections', p.119.
14. Wilkes, B. (1747–49). *The English moths and butterflies*, p.[v–vi].
15. Allen, D. E. (1966). Joseph Dandridge and the First Aurelian Society. *Entomologist's Rec.J. Var.* **78**: 89–94.
16. Wilkinson, R .S. (1966c). Elizabeth Glanville, an early English entomologist. *Entomologist's Gaz.* **17**: 89–94.
17. Bristowe, W. S. (1967d). The life of a distinguished woman naturalist, Eleanor Glanville (circa 1654–1709). *Ibid.* **18**: 202–211.
18. Fitton & Gilbert, 'Insect Collections', p.116.
19. Ford, E. B. (1945). *Butterflies*, p.10.
20. Bristowe, 'Joseph Dandridge', pp.197–201.
21. Whalley, P. E. S. (1972). *The English moths*

and butterflies by Benjamin Wilkes [1749], an unpublished contemporary account of its production. *J.Soc.Biblphy.nat.Hist.* **6**: 127.
22. Wilkinson, R. S. (1977c). New Data concerning James Dutfield's *New and Complete Natural History of English Moths and Butterflies* (1748–49). *Entomologist's Rec.J.Var.* **89**: 272–274.
23. Mays, R. (*c*.1965). *James Dutfield*, unpublished MS.
24. Lisney, A. A. (1960). *A bibliography of British Lepidoptera*, pp.127–144.
25. Edwards, G. (1770). *Essays upon natural history, and other miscellaneous subjects*, pp.117–119.
26. Smith, C. H. (1842). 'Memoir of Dru Drury'. In *Introduction to Mammalia*, pp.17–71.
27. Mays, R. (1986). *The Aurelian by Moses Harris* [Facsimile, with modern introduction], p.8.
28. Hough, R. (1994). *Captain James Cook*, p.193.
29. Hoare, M. E. (1976). *The tactless philosopher: Johann Reinhold Forster (1729–1798)*, pp.151–204, 328–329.
30. Jones, W. (1794). Papilios. *Trans.Linn.Soc.Lond.* **2**: 63–69, pl.viii.
31. Poulton, E. B. *et. al.* (1934). English Names regularly used for British Lepidoptera up to the end of the eighteenth century, with a biographical account of William Jones of Chelsea. *Trans.Soc.Br.Ent.* **1**: 139–155, 164–167.
32. Smith, A. Z. (1986). *A history of the Hope entomological collection in the University Museum, Oxford*, pp.80–81.
33. Emmet, A. M. (1989). The vernacular names and early history of British butterflies. *In* Emmet, A. M. & Heath, J. (Eds), *The moths and butterflies of Great Britain & Ireland* **7**(1), p.19.
34. Gilbert, P. (1998). *John Abbot: birds, butterflies and other wonders.*
35. Baker, C. R. B. (1997). Some Bedfordshire Lepidopterists. *In* Arnold, V. W. *et al.*, *The butterflies and moths of Bedfordshire*, pp.10–17.
36. Freeman, J. (1852). *A Life of the Rev. W. Kirby*, passim.
37. Freeman, *Ibid.* p.105.
38. Allan, P. B. M. (1943). *Talking of moths*, pp.87–88.
39. Pickard, H. A. *et al.* (1858). *An accentuated list of the British Lepidoptera*, p.xxv.
40. Gage, A. T. & Stearn, W. T. (1988). *A bicentenary history of the Linnean Society of London*, p.30.
41. Smith, A. Z., *History of Hope collection*, p.37.
42. Chalmers-Hunt. J. M. (1976). *Natural history auctions 1700–1972. A register of sales in the British Isles*, pp.6, 83.
43. Chalmers-Hunt, *Ibid.* pp.5, 77.
44. Pickard *et al.*, *Accentuated list*, pp.xiv–xv.
45. Morris, F O. (1853). *A history of the British butterflies*, p.146.
46. Dickens, C. (1838). *Memoir of Joseph Grimaldi* (rev. edn, 1968), p.49.
47. Bree, W. T. (1832). Notice of some singular varieties of Papilionidae in Mr Weaver's Museum, Birmingham. *Mag.nat.Hist.* **5**: 749–753.
48. Weaver, R. (1845). Occurrence of *Melitaea Dia* near Birmingham. *Zoologist* **3**: 887–888.
49. Allan, *Talking of Moths*, p.232.
50. Thomson, G. (1980). *The butterflies of Scotland*, p.172.
51. Wallace, A. J. (1832). Weaver's Museum of Natural History in Birmingham. *Mag.nat.Hist.* **5**: 546–548.
52. Firmin, J. (1958). Meet Laetitia the bug-hunter. *Essex County Standard*, 14 November 1958.
53. Stearn, W. T. (1981). *The Natural History Museum at South Kensington. A history of the British Museum (Natural History) 1753–1980*, p.206.
54. Anon. (1863). Obituary. John Curtis. *J.Proc.Linn.Soc.Lond.* **7** (1863–64): xxxv–xli.
55. Allan, P. B. M., (1980). *Leaves from a moth-hunter's notebook*, pp.74–75.
56. Dale, C. W. (1887). British Butterflies. *Young Nat.* **12**: cxxi (Supplement).
57. Curtis, W. Parkinson (1984). *In* Thomas, J. [A.] & Webb, N., *Butterflies of Dorset*, p.5.
58. Stephens, J. F. (1853). Reply to Mr. Doubleday's 'Notes on Mr. Stephens' "Catalogue of Lepidopterous Insects in the Cabinet of the British Museum, (Tortrices)" '. *Zoologist* **11**: 3733–3744.
59. Newman, E. (1833). British Entomology, Nos. 111–116. *Ent.Mag.* **1**: 451.
60. Newman, E. (1853). Editorial Postscript to J. F. Stephens's 'Reply to Mr Doubleday'. *Zoologist* **11**: 3744–3745.
61. Anon. (1867–68). Death of Professor Rennie. *Entomologist's mon.Mag.* **4**: 191.
62. Allan, *Moth-hunter's notebook*, pp.116–135.
63. Smith, A. Z., *History of Hope collection*, pp.1–10.
64. Newman, T. P. (1876). Preface. *Entomologist* **9**: v–xxiv.
65. *Ibid.*, p.xiv.
66. Stewart, A. M. (1912). *British butterflies.*
67. Allan, P. B. M. (1937). *A moth-hunter's gossip*, p.58.

68. Thomson, *Butterflies of Scotland*, p.211.
69. Pickard-Cambridge, O. (1893). Some reminiscences of the late Professor Westwood. *Entomologist* **26**: 74–75.
70. McLachlan, R. (1893). Obituary. Professor John Obadiah Westwood. *Entomologist's mon.Mag.* **29**: 49–51.
71. W. L. D[istant]. (1893). Obituary. Professor J. O. Westwood. *Ibid.* **26**: 25–26.
72. Smith, A. Z., *History of Hope collection*, p.44.
73. *Ibid.*, p.45.
74. Doubleday, E. (1832). Singular mode of capturing *Noctuae*. *Ent.Mag.* **1**: 310.
75. Humphreys, H. N. & Westwood, J. O. (1843–45). *British moths and their transformations*, p.198.
76. Mays, R. (1978). *Henry Doubleday. The Epping naturalist*, p.66.
77. *Ibid.*, p.118.
78. Doubleday, H. (1856). Remarks on Mr Burton's Note on *Argynnis Lathonia* and *Pieris Daplidice*. *Zoologist* **14**: 5146–5147.
79. Mays, *Henry Doubleday*, p.58.
80. Dunning, J. W. (1886). Obituary. John Arthur Power. *Entomologist* **19**: 193–200.
81. *Ibid.*, p.195.
82. Dunning, J. W. (1889). *In Memoriam*: Frederick Bond. *Ibid.* **22**: 265–269.
83. Anon. (1889). Obituary. Frederick Bond. *Entomologist's mon.Mag.* **25**: 384.
84. Saunders, E. (1905). *In Memoriam.* J. W. Douglas. *Ibid.* **41**: 221–222.
85. Stainton, H. T. (1857c). Portrait painting. *Entomologist's Wkly Intell.* **2**: 113–114.
86. Hellins, J. (1884). Obituary. William Buckler. *Entomologist's mon.Mag.* **20**: 229–236.
87. Anon. (1899). Obituary. Charles Stuart Gregson. *Ibid.* **35**: 96–97.
88. Bankes, E. R. (1906a). Obituary. Mrs Emma Sarah Hutchinson. *Ibid.* **42**: 43.
89. Hutchinson, E. S. (1879). Entomology and Botany as pursuits for Ladies. *Young Nat.* **1**: 3–4.
90. Bankes, E. R. (1906b). *Eupithecia consignata* Bork.: a correction. *Entomologist's mon.Mag.* **42**: 274.
91. Hutchinson, E. S. (1881). On the supposed extinction of *Vanessa C-album*. *Entomologist* **14**: 250–252.
92. Anon. (1878). Obituary. Thomas Vernon Wollaston. *Entomologist's mon.Mag.* **14**: 213–215.
93. Wollaston, T. V. (1856). *On the variation of species with especial reference to the Insecta, &c.*, p.140.
94. Douglas, J. W. & McLachlan, R. (1893). Henry Tibbats Stainton, F.R.S., &c. *Entomologist's mon.Mag.* **29**: 1–4.
95. Allan, *Moth-hunter's gossip*, p.13.
96. Stainton, H. T. (1855). An Address to young Entomologists. *Entomologist's Annu.* **1855**:13.
97. Longstaff, G. B. (1912). *Butterfly hunting in many lands*, p.7.
98. Weir, J. J. (1845). Capture of Ino Globulariae, Agrotis cinerea and Crambus pygmaeus at Lewes. *Zoologist* **3**: 1084.
99. McLachlan, R. (1894). Obituary. John Jenner Weir, F.L.S., &c. *Entomologist's mon.Mag.* **30**: 116–117.
100. Tutt, J. W. (1894). Obituary. John Jenner Weir, F.Z.S., F.L.S., F.E.S. *Entomologist's Rec.J.Var.* **5**: 103–105.
101. Greene, J. (1857). On pupa digging. *Zoologist* **15**: 5382–5398.
102. Allan, *Moth-hunter's gossip*, p.208.
103. Allan, *Moth-hunter's notebook*, p.79.
104. Greene, J. (1854). A list of Lepidoptera hitherto taken in Ireland as far as the end of the Geometrae. *Nat.Hist.Rev.* **1**: 165, 238.
105. Greene, J. (1851). List of Lepidoptera occurring in the County of Suffolk. *Naturalist (Morris)* **7**: 253–258.
106. Anon. (1906). Obituary. The Rev. Joseph Greene, M.A., F.E.S. *Entomologist's mon.Mag.* **42**: 66–67.
107. J. H. K[eys]. (1910). Obituary. George Carter Bignell. *Ibid.* **46**: 94–95.
108. J. J. W[alker]. (1908). In Memoriam: H. Guard Knaggs, M.D. *Ibid.* **44**: 49–51.
109. Knaggs, H. G. (1870). *A list of the Macro-Lepidoptera occurring in the neighbourhood of Folkestone.*
110. Knaggs, H. G. (1869). *The lepidopterist's guide*, p.68.
111. McLachlan, R. (1897). Obituary. Joseph William Dunning. *Entomologist's mon.Mag.* **33**: 281–283.
112. Stainton, H. T. (1857a). *A manual of British butterflies and moths* **1**: 295.
113. McLachlan, 'Obituary. Dunning', p.282.
114. Pickard-Cambridge, A. W. (1918). *Memoir of the Reverend Octavius Pickard-Cambridge by his son*, p.11.
115. F. D. W[heeler]. (1904–05). Obituary Notice: C. G. Barrett. *Trans.Norfolk Norwich Nat.Soc.* **8**: 152–154.
116. Neave S. A. (1933). *The history of the Entomological Society of London, 1833–1933*, p.49.
117. Turner, H. J. (1920). Obituary. William West (of Greenwich). *Entomologist's Rec.J.Var.* **32**: 175–176.
118. Porritt, G. T. (1917). Obituary. J. Platt

Barrett. *Entomologist's mon.Mag.* **53**: 69–70.
119. Firmin, J. (1992). *Lepidoptera of north east Essex*, pp.9–14.
120. Rothschild, M. & Marren, P. (1997). *Rothschild's reserves: time and fragile nature*, pp.46,145,179,185.
121. Frohawk, F. W. (1938a). *Varieties of British butterflies*, pp.35,100.
122. Kershaw, S. H. (1956). Some recollections of Albert Brydges Farn. *Entomologist's Rec.J.Var.* **68**: 150–155.
123. Jenkinson, F. (1922). Obituary. Albert Brydges Farn. *Entomologist's mon.Mag.* **58**: 20–22.
124. Champion, G. C. (1922). Obituary. Dr Thomas Algernon Chapman, F.R.S. &c. *ibid.* **58**: 40–41 (Portrait).
125. Allan, *Moth-hunter's notebook*, pp.93–101.
126. Miles, B. E. (1981). Past Aurelians and Lost Butterflies. *Trans.Woolhope Nat.Fld Club* **43**: 337–338.
127. Allan, *Moth-hunter's notebook*, p.98.
128. Harvey, J. M. V., Gilbert, P. & Martin, K. (1996). *A catalogue of manuscripts in the Entomology Library of the Natural History Museum, London*, pp.211–214.
129. Anon. (1895). Obituary. Dr F. B. White. *Entomologist* **28**: 25–27.
130. Thomson, *Butterflies of Scotland*, p.212.
131. Riley, N. D. (1964a). *The Department of Entomology of the British Museum (Natural History) 1904–1964. A brief historical sketch*, p.15.
132. Stearn, *Natural History Museum*, p.210.
133. W. E. K[irby]. (1912). Obituary. William Forsell Kirby. *Entomologist's Rec.J.Var.* **24**: 314–317, Portrait.
134. Baker, B. R. (1994). *The butterflies and moths of Berkshire*, pp.xiii–xvi.
135. Allan, *Talking of moths*, p.98.
136. Wallis, H. M. (1930). Obituary. William Holland. *Entomologist's mon.Mag.* **66**: 187.
137. Riley, *History of Department of Entomology*, p.97.
138. Stearn, *Natural History Museum*, p.212.
139. Colvin, H. (1985). *Calke Abbey, Derbyshire. A hidden house revealed*, pp.70–74.
140. Harpur Crewe, H. (1858). Larvae of the Genus *Eupithecia* wanted. *Entomologist's wkly Intell.* **4**: 158.
141. Anon. (1883). Obituary. The Rev. H. Harpur Crewe, M.A. *Entomologist's mon.Mag.* **20**: 118–119.
142. Desmond, R. (1994). *Dictionary of British and Irish botanists and horticulturalists*, p.319.
143. Colvin, *Calke Abbey*, p.70–74.
144. Riley, N. D. (1932). Obituary. Richard South. *Entomologist* **65**: 97–100, portrait.
145. *Ibid.*, p.97
146. South, R. (1885). Address to the South London Entomological and Natural History Society. *Abstr.Proc.S.Lond.ent.nat. Hist.Soc.*, **1885**: 11–25.
147. Sheldon, W. G. (1935). In Memoriam: Robert Adkin. *Entomologist* **68**: 145–147.
148. James, M. J. (1973). *The new Aurelians*, pp.31–32.
149. Sheldon, 'Robert Adkin', p.146.
150. Smith, A. Z., *History of Hope collection*, p.38.
151. *Ibid.*, p.46.
152. *Ibid.*, p.47.
153. Gillett, J. D. (1983). A third of the way back and beyond. *Antenna* **7**: 159–163.
154. Chapman, T. A. (1911). Obituary. J. W. Tutt. *Entomologist* **44**: 77–80.
155. Adkin, R. (1911). "Tutt as I knew him." *Entomologist's Rec.J.Var.* **23**: 115.
156. Burr, M. (1911) "Tutt as I knew him." *Ibid.* **23**: 107.
157. Adkin, R. (1911). "Tutt as I knew him." *Ibid.* **23**: 115.
158. Smetham, H. (1911). Another reminiscence. *Ibid.* **23**: 138.
159. Turner, H. J. & Edwards, S. (1911). Bibliography of J. W. Tutt's works. *Ibid.* **23**: 140–155.
160. Wheeler, G. (Ed.) (1914). *In* Tutt (1910–1914). *A natural history of the British butterflies* **4**: iii.
161. J. W. Tutt's Will. Photocopy in possession of the author (E. W. Classey, pers.comm.).
162. Chalmers-Hunt, J. M. (1976). *Natural history auctions*, pp.11–12.
163. Gardiner, B. O. C. *et al.* (1995). *An index to the modern names, for use with J. W. Tutt's Practical hints for the field lepidopterist.*
164. Owen, D. F. (1997). Natural selection and evolution in moths: homage to J. W. Tutt. *Oikos* **78**(1): 177–181.
165. Meyrick, E. (1898). Moths and their Classification. *Zoologist* (Series 4) **2**: 289–298.
166. Chatfield, J. (1987). *F. W. Frohawk – his life and work*, pp.13–14, 20, 150.
167. Frohawk, F. W. (1924). *Natural history of British butterflies* **1**: ix [hereafter cited as *British butterflies*].
168. Chatfield, *Frohawk's life*, pp.27–28.
169. Riley, N. D. (1947). Obituary. F. W. Frohawk. *Entomologist* **80**: 25–27.
170. Rothschild, M. (1955). Karl Jordan – A biography. *Trans.R.ent.Soc.Lond.* **107**: 1–9.
171. Rothschild, M. (1983). *Dear Lord Rothschild*, p.127.
172. *Ibid.*, p.147.

173. Riley, N. D. (1960). Heinrich Ernst Karl Jordan. *Biogr.Mem.Fellows R.Soc.* **6**: 107–133.
174. Rothschild, 'Karl Jordan', p.8.
175. W. G. S[heldon]. (1940). Obituary. Margaret Elizabeth Fountaine. *Entomologist* **73**: 193–195.
176. Fountaine, M. E. (1902). Butterfly hunting in Greece, in the year 1900. *Entomologist's Rec.J.Var.* **14**: 29–35, 64–67.
177. Fountaine, M. E. (1904). A 'Butterfly Summer' in Asia Minor. *Entomologist* **37**: 79–84, 105–108, 135–137, 157–159, 184–186.
178. Cater, W. F. (Ed.) (1980). *Love among the butterflies*, pp.68–69.
179. Cater, W. F. (Ed.) (1986). *Butterflies and late loves* [quoting N. D. Riley and B. Ker-Seymer], p.134.
180. *Ibid.*, p.136.
181. Newman, L. H. (1967). *Living with butterflies*, p.100.
182. Kershaw, S. H. (1958). Some memories of S. G. Castle Russell. *Entomologist's Rec.J.Var.* **70**: 1–4, 37–41, 94–100, 56–160.
183. Allan, *Moth-hunter's notebook*, p.180.
184. Kershaw, 'S. G. Castle Russell', p.160.
185. Rothschild, *Lord Rothschild*, p.21.
186. *Ibid.*, pp.1–2 *et passim.*
187. Riley, N. D. (1937). Obituary. Lord Rothschild. *Entomologist* **70**: 217–220.
188. Frohawk, *British butterflies* **2**:147–149.
189. Purefoy, E. B. (1953). An unpublished account of experiments carried out at East Farleigh, Kent, in 1915 and subsequent years on the life history of *Maculinea arion*, the Large Blue Butterfly. *Proc.R.ent.Soc.Lond.* (A) **28**: 160–162.
190. Goodden, R. (1999). Unpublished letter from E. B. Purefoy's granddaughter, (pers.comm.).
191. J. A. T[homas]. (1949). Obituary. Leonard Woods Newman. *Entomologist* **82**: 143–144.
192. Gardiner, B. O. C. (1993). Father and Son: The Newmans and their Kent Butterfly Farm. *Entomologist's Rec.J.Var.* **105**: 105–114.
193. Newman, L. H., *Living with butterflies*, pp.1–55.
194. Cockayne, E. A. (1949). Obituary. Leonard Woods Newman. *Entomologist's Rec.J.Var.* **71**: 80–81.
195. 'Q.' (1959). Obituary. Henry Atfield Leeds. *Entomologist* **92**: 22.
196. Worms, C. G. M. de (1959). Obituary. William Rait-Smith. *Ibid.* **92**: 65–66.
197. Rothschild, *Lord Rothschild*, p.174.
198. Rothschild & Marren, *Rothschild's Reserves*, pp.1–51.
199. Gairdner, D. (1988). Great Ormond Street 50 years ago. *Archs Dis.Childhood* **63**: 1272–1275.
200. *Ibid.*, p.1272.
201. Twistington Higgins, P. (1956). Obituary. E. A. Cockayne. *Br.Med.J.*, 8th Dec., p.137.
202. Newman, *Living with butterflies*, p.109.
203. Campbell, J. L. (1975). On the Rumoured Presence of the Large Blue Butterfly (*Maculinea arion* L.) in the Hebrides. *Entomologist's Rec.J.Var.* **87**: 161.
204. *Ibid.*, p.162.
205. *Ibid.*, p.162.
206. Sabbagh, K. (1999). *A Rum affair. How botany's 'Piltdown Man' was unmasked*, p.163.
207. Campbell, 'Large Blue in the Hebrides', p.166.
208. Heslop Harrison, J. W. (1948). The Passing of the Ice Age and its effect upon the Plant and Animal Life of the Scottish Western Isles. *New Nat.*, pt. 2 (Summer), pp.83–90.
209. Burtt, E. T. (1967). Obituary. Prof. J. W. Heslop Harrison, F.R.S. *Entomologist* **100**: 113–114.
210. Peacock, A. D. (1968). John Heslop Harrison. *Biogr.Mem.Fellows R.Soc.* **14**: 248.
211. Wilkinson, R. S. (1975a). P. B. M. Allan: An American's tribute. *Entomologist's Rec.J.Var.* **87**: 49–51.
212. R. F. B[retherton]. (1979). Obituary. N. D. Riley, C.B.E. *Proc.Trans.Br.ent.nat.Hist.Soc.* **12**: 103–112; **13**; 39–40.
213. Riley, *History of Department of Entomology*, pp.1–48.
214. R. V. M[elville]. & I. W. B. N[ye]. (1979). Obituary. N. D. Riley, C.B.E. *Bull. zool.Nomencl.* **36**: 137–138.
215. Tolman, T. (1997). *Butterflies of Britain and Europe*, p.6.
216. Huggins, H. C. (1973). They were Irish gannets. *Entomologist's Rec.J.Var.* **85**: 234–237.
217. Huggins, H. C. (1955). The old days at Chattenden Roughs. *Entomologist's Gaz.* **6**: 56–57.
218. Riley, N. D. (1964b). An appreciation of the late Francis Hemming, C.M.G., C.B.E., for many years Secretary of the International Commission on Zoological Nomenclature. *Bull.zool.Nom.* **21**: 402–404.
219. Riley, N. D. (1964b). [Quoting tribute from R. V. Melville.] *Ibid.* **21**: 404.
220. B. R. B[aker]. (1972). Obituary. Air Marshal Sir Robert Saundby. *Proc.Trans.Br.ent.nat.Hist.Soc.* **5**: 31–32, portrait.

221. Rothschild, M. (1984). Dedication: Henry Ford and Butterflies. In *The biology of butterflies* (Eds: Vane-Wright, R. & Ackery, P. R.). *Symp.R.ent.Soc.Lond.* No. 11, p. xxii–xxiii.
222. Worms, C. G. M. de (1938). British Lepidoptera collecting, 1937. *Entomologist* **71**: 179–183, 201–205.
223. Bretherton, R. F. (1980). Obituary. Baron Charles George Maurice de Worms. *Proc.Trans.Br.ent.nat.Hist.Soc.* **13**: 37–39.
224. Demuth, R. F. (1979). Obituary. H. B. D. Kettlewell, D.Sc., M.A., F.R.E.S., etc. *Entomologist's Rec.J.Var.* **91**: 255–257.
225. Clarke, C. A. (1979). Obituary. Dr. Henry Bernard Davis Kettlewell (1907–1979). *Ibid.*: 253–255.
226. B. J. M[acNulty]. (1984). Obituary. Robert W. Watson. *Proc.Trans.Br.ent.nat.Hist.Soc.* **17**: 51–52.
227. Shaw, M. R. & Agassiz, D. J. L. (1990). Obituary. Edward Charles Pelham-Clinton (1920–1988). *Br.J.ent.nat.Hist.* **3**: 107–114.
228. Emmet, A. M. (1989). Obituary. E. C. Pelham-Clinton. *Nota lepid.* **12**(1): 2–3.
229. Blows, W. T. (1998). Conversations with a naturalist: the life and geological work of Richard Ford 1913–1996. *Geol.Curator* **6**: 329.
230. *Ibid.*, p.328.
231. Marshall, D. & Gardiner, B. (1994). Obituary and Appreciation (with an A.E.S. bibliography) – Peter William Cribb, 1920–1993. *Bull.amat.Ent.Soc.* **53**: 1–9.
232. Harley, B. (1989). John Heath – an Appreciation. *In* Emmet, A. M. & Heath J. (Eds), *The moths and butterflies of Great Britain and Ireland* **7**(1): vii.
233. Anon. (1988). Obituary. John Heath (1922–1987). *Br.J.ent.nat.Hist.* **1**: 113–116.

CHAPTER 4

'Immense Swarmes of Butterflyes as e'en to Darken the Skyes'

SOME SPECIES OF HISTORICAL INTEREST

'You ask what is the use of butterflies?
I reply to adorn the world
and delight the eyes of men ... '
John Ray, *Historia Insectorum*, 1710

THE INDIGENOUS BUTTERFLY SPECIES OF GREAT BRITAIN and Ireland are few in number compared to those found on the Continent, even as close as north-western France. This may not always have been so. The land bridge to the Continent was cut 8,000 years ago and the consequent marine barrier has ensured that Britain has been separated from the continent of Europe since the end of the last Ice Age. Few butterflies, except migrants, can now reach us. Although butterflies are seldom preserved as fossils, we can speculate that in earlier periods, when Britain was attached to present-day France and Holland, our butterfly fauna would have been much richer. As it was, only the hardier and more mobile species reached us in time, before the sea level rose.

Mass butterfly migrations have been witnessed on many occasions in the past. One of the earliest accounts is to be found in the *Chronicles of Calais in the Reigns of Henry VII and Henry VIII to the Year* in which Richard Turpyn recorded the following remarkable event: '1508, the 23rd year of Henry the 7, the 9 of July, being relyke Sonday, there was sene at Calleys [Calais] an innumerable swarme of whit buttarflyes cominge out of the north este and flyinge south-estewards, so thicke as flakes of snowe, that men being a shutynge in St. Petars fields without the towne of Calleys could not see the towne at foure of the clock in the aftarnone, they flew so highe and so thicke'.[1]

This account is almost certainly describing one of the regular migrations of Large White butterflies – on this occasion on a phenomenally large scale – which lead to immigration in most years into the British Isles. There are similar movements of other species, such as the vast invasion of Small White butterflies that invaded Dover from across the Channel on 5th July 1846. *The Canterbury Journal* reported at the time that 'such was the density and extent of the cloud formed by the living mass that it completely obscured the sun from the people on board the continental steamers on their passage for many hundreds of yards'. This was accomplished with 'scarce a puff of wind', but, after their arrival at about twelve noon, a strong SW wind dispersed them up to ten miles from Dover.[2]

Most butterfly movements are far smaller and pass unnoticed, but at least seventeen of the British butterflies and many more moths are migrants.[3] It should be no surprise, therefore, that some species that may survive the British winter only in small numbers, such as the Red Admiral, can be so abundant in some years but not in others; nor that extreme rarities, such as the Bath White, should be recorded in hundreds in 1945 – a great year for immigrants – but be practically never seen in normal years.

Observations made from lightships contain records of normally non-migratory Small Copper, Common Blue, Holly Blue, Small Tortoiseshell, Peacock, Wall and Meadow Brown butterflies,[4] as well as of the known migratory whites and nymphalids. For very small species, under favourable conditions, to cross from the Continent singly or in numbers is therefore not the least improbable. What is almost more surprising is that on arrival and dispersal they should be noted at all. It has been said with some truth that the great increase of rare bird records can be correlated with the increased number of bird watchers and 'twitchers'. Until very recently there have arguably been many fewer butterfly 'twitchers' than there were in the latter part of the 19th century and the first half of this one. In consequence, the odds against the recording of such random rarities increased, but with more popular interest in butterfly recording and conservation today this is changing, though it unfortunately takes place against a background of habitat deterioration on the Continent as well as in Britain, and consequently of smaller populations to provide the reservoir from which migration can take place.

This chapter is about a selection of historically interesting butterflies that have excited lepidopterists and stimulated their imagination. It includes a few species that have been added to the British list in fairly recent times; others that were once native but have become extinct; yet others that come into the category of rare migrants or vagrants. A few, although traditionally on the British list, must be regarded as very doubtfully British, either because they were accidentally introduced or, most probably, were fraudulently passed off as British by unscrupulous dealers who imported specimens from the Continent to sell to avid collectors. The category of a few species remains uncertain although it may be possible in the future, with the advance in DNA techniques, to ascertain the provenance of some specimens.

1. THE CHEQUERED SKIPPER – *Carterocephalus palaemon* (Pallas)

Papilio palaemon Pallas, 1771
Papilio paniscus Fabricius, 1775

In a letter dated 12th August 1798 and still in the Linnean Society's archives, the Revd Dr Charles Abbot informed the Society, of which he had been elected a Fellow five years earlier, that he had discovered *Papilio paniscus* in Clapham-Park Wood, Bedfordshire. His notebook, which was subsequently bought by J. C. Dale, further records that the butterfly was 'Copiose' [copious] and that it had been taken on 8th May 1798. This is the first known record of the Chequered Skipper in Britain. Abbot had taken up the study of Lepidoptera only the season before and was clearly excited about his find. However, he seems to have suffered a rebuff from fellow collectors since he wrote:

> 'I here beg leave to offer a few Remarks on the actions and manners of a Butterfly, which the Aurelians' Company (formerly established in London) passed over in silence, but which I have too high an opinion of our present institution to suppose that the members of the Linnean Society will disregard – The times when this Papilio is most easily taken are May and June when the Lucina [Duke of

> Burgundy Fritillary] is out, though its term of existence must be longer, as I have taken several of them in good condition a full fortnight after the Burgundy has disappeared. They should be sought for from seven to nine o'clock in the morning, indeed I have observed them playing in pairs just after sunrise, or at least as soon as the morning Fog has evaporated – Its flight is extremely short, and the insect is far from being timid, for should the maladroitness of the Aurelian suffer him to escape after capture, he may be easily traced among the herbage by a vigilant Eye and retaken – They fly very near the ground and delight to settle on the blades of very long grass, or the various species of the Carex tribe – The Larva or Chrysalis I do not profess to know.' [1]

In a further letter, sent on 1st November 1798, Abbot asked that the species be called the Duke of York Fritillary, in honour of 'our hero of the House of Brunswick', subsequently familiar to generations of children in the well-known nursery rhyme. Possibly he had in mind the butterfly's similar flight-time and coloration to the Duke of Burgundy. Abbot's proposal was never adopted but his observations were read out at a Society meeting on 6th November and subsequently published in the *Transactions.*[1]

Edward Donovan[2] illustrated the species (Plate 12) but gave it only a Latin name, *Papilio paniscus.* Adrian Haworth[3] was the first to use the name Chequered Skipper and, apart from George Samouelle[4] who called it the Scarce Skipper and the Revd F. O. Morris[5] who endowed it with the name Spotted Skipper, this name was universally adopted.

Four years later, Abbot took the species again, this time at Gamlingay, Cambridgeshire, and a number of other localities were soon identified. In 1823, a Mr Henderson, Lord Milton's gardener, found the Chequered Skipper in great plenty at Castor Hanglands, near Peterborough, and many entomologists of the time (and for long afterwards) regarded this locality as its headquarters, until they realized that the butterfly had probably existed in scattered colonies throughout a number of midland and southern counties for decades. And so it was that, by 1828, J. F. Stephens regarded the Chequered Skipper as not 'a scarce, but merely a very local species', noting it in 'great plenty in several parts of Northamptonshire and Bedfordshire at the end of May'.[6] More than half a century later, the Revd W. W. Fowler, 'In a wood about seven or eight miles from Lincoln, while hunting for Coleoptera, on June 2nd last, saw "*Hesperia Paniscus*" evidently not uncommon in one locality.' He continued, 'On two subsequent occasions I visited the wood, but each time a thunderstorm, followed by heavy rain, came on just as we reached it, and stopped our operations; we, however, took one specimen each time, showing that it was still out, and I have no doubt that the insect was fairly plentiful'.[7]

English populations of the Chequered Skipper were largely confined to counties south of Lincolnshire, with the greatest numbers recorded from Northamptonshire. Although it flourished in few midland counties, there were also reports from Devon and Hampshire.[8] It has declined steadily since 1900, but its sudden disappearance from long-established localities in the late 1960s and early 1970s took lepidopterists by surprise. By 1976, when the last specimen was seen in Rutland, it was considered to be extinct in England. The very rapid and catastrophic decline was attributed to progressive changes in

Plate 12
The Chequered Skipper, as illustrated by Edward Donovan in Volume 8 of his *Natural History of British Insects* (1799), a year after its discovery by the Revd Dr Charles Abbot.

the habitat, principally a decrease in coppicing, and the replanting of broad-leaved woodland with conifer plantations.[9]

In 1942, Lt.-Col. C. W. Mackworth-Praed caused a stir among the British entomological fraternity when he declared in the pages of *The Entomologist* that:

> 'It may be a surprise to British entomologists to know that *Carterocephalus palaemon* occurs in Western Inverness-shire. It certainly was a surprise to me, and I could not believe my eyes when the first one settled in front of me. Since then, however, I have seen a considerable number and have taken a small series. I rather hope it may be fairly widely distributed, as the situations it affects are rides in birch woodland and thinly wooded banks, which are extensive in this part of the country. The specimens I have appear to be pale and sharply marked as compared with a series from eastern England. I saw the first one on May 10th. There is neither *Brachypodium* nor *Bromus* in the places it frequents, and in spite of careful watching I failed to spot its food-plant.'[10]

Seven years after this discovery, Miss C. Ethel Evans made it known that she had found the Chequered Skipper at Loch Lochy a few years earlier but had kept the locality a secret.[11] However, even that sighting may not have been the first. Many years previously, in 1907, two entomologists had reported seeing a strange skipper at Glenshian in Inverness, which they had been unable to identify.[12] In 1974, a search organized by the Scottish Wildlife Trust found that the Chequered Skipper in fact occurred over a large area of Argyll and Inverness-shire. Although most colonies were small, there was one strip of roadside verge five miles in length along which this species was found without a break. Scottish Chequered Skippers, which were probably always isolated from their English counterparts, are darker and greener in colour with the underside lacking the yellow wash that is found in English specimens. The foodplant and behaviour of the two races also differed. English colonies fed mainly on false brome (*Brachypodium sylvaticum*) whereas the foodplant of the Scottish ones is thought to be purple moor-grass (*Molinia caerulea*). Another difference is that English colonies tended to be very sedentary but in Scotland the males are territorial while the females fly longer distances between territories.[13] Dennis has suggested that the Scottish populations acquired their characteristic features during the Boreal and Atlantic periods when dense forest confined them to limited parts near the West Coast.[14] Such conditions may have persisted for up to 7000 years. The Scottish population thrives but there can be no doubt that the English colonies no longer exist.

In 1995, a five-year research project aiming at re-establishing the Chequered Skipper in England was begun by Butterfly Conservation. Stock was obtained from the region of the Meuse in north-eastern France, where conditions and foodplants are the most similar to those of the former English sites, and controlled rearing has been continuing at a site in the East Midlands. Butterflies are again flying in the wild in this limited area and the outlook is promising.[15]

2. THE ESSEX SKIPPER – *Thymelicus lineola* (Ochsenheimer)
Papilio lineola Ochsenheimer, 1808

Until 1890 when F. W. Hawes published his discovery of this species as native to Britain in *The Entomologist*,[1] the Essex Skipper had been completely overlooked

(Fig. 127). Hawes himself, who first collected it in Essex in 1888, thought it no more than a variety of the Small Skipper (*T. sylvestris*) which it so closely resembles. It is a sobering thought that the species must have been present undetected for the last millenium or more.[2]

HESPERIA LINEOLA, OCHSENHEIMER: AN ADDITION TO THE LIST OF BRITISH BUTTERFLIES.

BY F. W. HAWES.

THE specimens,—three in number, all males,—from which the accompanying description is made, were taken by me during the month of July, 1888, in one of the eastern counties, and remained until quite recently in my cabinet, merely as curious varieties of *Hesperia thaumas*. Happening, however, one day last month, to turn over those plates in Dr. Lang's 'Butterflies of Europe,' on which the genus *Hesperia* is figured, I was struck with the great resemblance of my specimens to a species represented at Plate 81, fig. 10. A reference to the description at p. 351 of that work suggested the probability of the so-called varieties being in reality *H. lineola*, the three main points of distinction between *H. lineola* and *H. thaumas* appearing in strongly marked contrast when the specimens under consideration were compared with undoubtedly fresh examples of *H. thaumas*.

Fig. 127. F. W. Hawes's announcement in *The Entomologist* that what he had thought to be a variety of the Small Skipper was in fact a species 'new' to Britain though it had no doubt been present for centuries.

W. S. Coleman, in 1897, was among the first popular writers on butterflies to publish an account of it as 'The "Lineola" Skipper'.[3] 'This little insect', he wrote, 'tried its best to hide away under cover of its near neighbour, *Linea* which it so closely resembles that it is only lately that the difference between the two has been described'. He added that 'many collections have the insect placed with *Linea* still, as it always had been considered to be the same insect till within the last two years'. C. G. Barrett, who was able to include it in his major work on the British Lepidoptera the first volume of which, devoted solely to the butterflies, was published in 1893, gave details of the results of collectors re-examining their specimens of '*Hesperia linea*', the Small Skipper.[4] The Essex Skipper was shown to have been taken in earlier years in Kent, Sussex, Suffolk, Nottinghamshire, and as far west as Taunton in Somerset. It was clearly much more widespread than first thought, though its headquarters at that time were undoubtedly in Essex and Suffolk.

In recent years its range has extended westwards, possibly as the result of the transportation of hay from East Anglia to the West Country when grass is short due to drought, as it was in 1975 and 1976. It 'is easy to imagine the butterfly's eggs being scattered along the grass verges, establishing new colonies along the way' speculated Jeremy Thomas.[5] Abroad the butterfly is widely distributed through Europe to North Africa and eastern Asia. It also became established in North America around 1910, and has become a pest of crops of timothy grass (*Phleum pratense*).[6]

3. THE LULWORTH SKIPPER – *Thymelicus acteon* (Rottemburg)

Papilio acteon Rottemburg, 1775

August 15th, 1832, was one of those days of which every amateur entomologist dreams. On it J. C. Dale, the popular squire of Glanville's Wootton in Dorset, took no fewer than three insect species new to Britain – of which one was the Lulworth Skipper. It is recorded that the squire, after riding almost twenty miles on horseback from Glanville's Wootton, reached Durdle Door, near Lulworth Cove, where he found considerable numbers of the new skipper on the cliff tops, flitting about among long grass and thistles. However, it was not until John Curtis reported Dale's discovery in Volume 10 of his *British Entomology*, published in 1833, and named it The Lulworth Skipper, that the discovery was announced in print (Plate 13). 'We cannot often hope to record the addition of a Butterfly to our British Fauna', Curtis wrote, 'but this species was discovered at Lulworth Cove in Dorsetshire, last August, by J. C. Dale, Esq., through whose liberality it now ornaments most of our cabinets: it was found upon Thistles, and was very local'.[1]

Lepidopterists soon flocked to Lulworth. Some recorded their success in the entomological journals, often in ecstatic terms. In the first volume of his *Manual of British Butterflies and Moths*, Henry Stainton included a report by J. W. Douglas:

> 'In July, 1849, my late friend H. F. Farr was staying at Weymouth for the benefit of his health, then fast declining, by reason of the malady which not long after caused his death. I staid a few days in his company, and made some entomological excursions with him to Portland and other places adjacent; for, although he was weak, his love of insects clung to him still. One bright sunny morning we hired a boat owned by one of the amphibious long-shore dwellers, whom we took with us, and found he was a character, and could turn his hand and his tongue to anything. An hour's sail across Weymouth Bay, during which we amused ourselves by catching mackerel, brought us to the desired spot, "the Burning Cliff" (or Lulworth Cove), where we had been told we should find *Pamphila Actaeon*, and there, sure enough, we saw it in profusion. The spot, close to the sea, is a kind of undercliff, not very level, of no great extent, and covered with thistles and large tufts of a long coarse grass, or *Carex*, about which our prey were skipping briskly. So abundant were they that I often had five or six in my net at one stroke, and in about two hours I caught a hundred, filling my box and my hat; and Mr. Farr had nearly as many. They were accompanied by a few of the common *P. Linea* [Small Skipper], which, in their flight, they greatly resembled. My un-geological eyes detected nothing particular in the soil, and I confess that two hours' hard work in the sun had not disposed me to look if any particular plant which might serve as the food of the larvae of this Skipper grew there; so that I can offer no supposition as to the cause of the species being confined within such narrow limits in this country.'[2]

Some collectors, like Herbert Goss, described the typical habitat of the Lulworth Skipper. He found it

> '... on rough broken ground on the slopes of the cliffs; the insect sparingly distributed over an extent of ground of about three-quarters of a mile in length, but in one or two places it was abundant. The place in which I found it in greatest plenty was a level plateau about three hundred feet in length, and fifty in breadth, apparently formed by a landslip at no very remote period, at the side of a cliff

Plate 13
The first account of J. C. Dale's discovery of the Lulworth Skipper was published with this hand-coloured engraving from Plate 442 of Volume 10 of John Curtis's *British Entomology*, 1833, the year after Dale had found it in Dorset.

> about a hundred and thirty feet above the sea. The ground was extremely rough, with masses of rock lying about in all directions, and collecting there was not without a spice of danger; the vegetation was of a very varied character, including several species of coarse grass, but apparently no *Calamagrostis epigejos* [Wood Small-reed]. The butterflies appeared particularly fond of the flowers of *Ononis arvensis* (Rest-harrow), but I rarely saw them alight on any other flower. I watched hundreds of specimens but was unable to detect a female in the act of oviposition'.[3]

One enthusiastic hunter who clearly enjoyed his collecting trip in 1856 was A. Pretor. 'On Wednesday, the 6th inst., we made an excursion to the "Burning Cliff", the peculiar haunt of this species. Before our arrival at the "hunting-ground" we had already taken *Paphia, Rhamni, Io* and *Aegon* [Silver-washed Fritillary, Brimstone, Peacock and Silver-studded Blue]; but the greatest success was reserved for the object of the excursion; and the capture of about eighty Skippers fully rewarded us for an hour's exertion under a true August sun, and a reasonable quantity of concussions with "Mother Earth".'[4]

Although the capture of some eighty Lulworth Skippers in one hour may have been rewarding for that collector, it greatly alarmed others. In 1881, W. McRae wrote to *The Entomologist*:

> 'Having for a good many years been in the habit of visiting Lulworth, the head-quarters of *H. Actaeon*, I have noticed that this lively little skipper, so prized by collectors, has been becoming scarcer every year. Formerly, in consequence of the inaccessible nature of the place, Lulworth was visited only by the real lover of Nature or the ardent collector; but now steamers from most of the towns of the South Coast carry thousands of visitors thither, and *H. Actaeon* has not a day's respite from the ceaseless inroads made upon its limited domains by the entomologically-inclined portion of these "trippers". Can nothing be done to avert the complete extermination of *H. Actaeon* at its old head-quarters? If these collectors (many of whom are only indiscriminate destroyers) could be induced to retain only the bright fresh specimens, and give all the faded worn specimens their liberty again, the chances are that "Lulworth Skippers" would not become an anachronism.'[5]

Since those early days, the Lulworth Skipper has been found quite commonly along the coast of Dorset from Weymouth to the Isle of Purbeck, as well as around Sidmouth and Torquay, in Devon. There are also records from Freshwater and Sandown on the Isle of Wight,[6] as well as one highly dubious record by Humphreys from Shenstone, near Lichfield, in 1835.[7] So soon after the discovery of this species in Britain the latter was probably a case of wishful thinking. Although the Lulworth Skipper is at the northern limit of its range in England, it is no longer 'becoming scarcer every year'. Quite the contrary: before the last war fewer than forty colonies were known, but today at least ninety colonies have been identified in Dorset alone, principally confined to a coastal strip no wider than five miles or so between the Purbeck Hills and the county border with Devon. The larvae feed on the taller grasses and it is the recent decline in grazing that has resulted in a continued expansion in its range and numbers. Jeremy Thomas has stated that 'nearly 400,000 adults emerge annually on Bindon Hill, with perhaps a million individuals on the ranges as a whole'.[8] Fortunately the thousands of trippers that flock to Lulworth each summer no longer carry nets and collecting equipment and the future for this little butterfly looks healthy, and particularly so if the military land is kept free of development. In central and southern Europe and beyond into Africa and Asia, this species is widespread and not confined to the coast, being more often found inland on any suitable area of grassland.

4. THE MALLOW SKIPPER – *Carcharodus alceae* (Esper)
Papilio (Plebejus urbicola) alceae Esper, 1780

In 1923, F. W. Frohawk reported the capture by Baron J. A. Bouck of a male and female of this species in Surrey.[1] The butterflies appeared to have recently emerged and were found flying near clumps of Common Mallow (*Malva sylvestris*), the larval foodplant. Frohawk described them as the 'Surrey Skipper', and suggested that a small colony had probably existed in the locality, escaping detection until that time. His suggestion is given slender support by the capture of a third specimen at Gomshall, Surrey, in August 1950, by R. E. Ellison. This specimen is in the author's collection (Plate 14). No other examples have been reported from Britain. On the Continent, *C. alceae* is not uncommon in southern and central Europe to about latitude 52° N, but is absent from North Germany, Denmark, the Baltic countries and Fennoscandia. As the skippers are not migratory butterflies, the presence of this species in Surrey is a mystery but it was almost certainly an accidental introduction with little right to be included on the British list (see Plate 36, figure 17).

Plate 14
The specimen of the Mallow Skipper taken at Gomshall, Surrey, in 1950 by R. E. Ellison.

5. THE APOLLO – *Parnassius apollo* (Linnaeus)
Papilio (Heliconius) apollo Linnaeus, 1758

The earliest references to this beautiful butterfly in the British entomological literature would appear to be those by Thomas Moffet in 1634[1] and James Petiver in 1704,[2] although neither suggested that the species actually occurred in Britain. On the contrary, Petiver, who gave it the name *Papilio alpinus*, or Mr Ray's Alpine Butterfly, described it from the continental specimen presented to him by John Ray following his Grand Tour of Europe. It was Haworth who first reported a British specimen. In 1803, he wrote: 'I have recently heard that the *Papilio Apollo* of Linnaeus has been found in Scotland, but [I] have not seen a British specimen of that beautiful species myself'.[3] Following Haworth, Donovan also included the Apollo in his *Natural History of British Insects*,[4] and William Wood illustrated it in a supplement to his *Index Entomologicus* of 'Doubtful British Species' (see Plate 36, figure 3), but remarked that, although Donovan treated it as British, he did so without sufficient authority.[5]

Coleman, too, had misgivings, illustrating it as a 'Reputed British species' (Plate 15, figure 2).

In an addendum to his *Illustrated Natural History of British Butterflies*, Edward Newman listed a number of further records, several of which he attributed to J. C. Dale.[6] Unfortunately many of them are vague and second, third or even fourth hand! According to Newman, Dale had told him that 'the late Mr Haworth informed me that a lady, whom Mr Curtis believed was the Marchioness of Bute, told him that she had received a specimen from some alpine place on the West Coast in the North of Scotland'. Dale had also been informed by Sir William Hooker, the distinguished botanist, that 'in 1812, or about that date, *Apollo* is said to have occurred in the Island of Lewis, and was taken by a tenant of Lord Seaforth's, who had the specimen, but that there being at that time some communication between Norway and the Island of Lewis, the specimen might have come from Norway'. Dale also reported that 'Mr. Curtis was convinced he saw a specimen of *Apollo* flying over the top of a house at the foot of Ben Lawers; and afterwards, on seeing the species on the Continent, he felt assured he was correct'.

While many of these early reports lack detailed evidence, that of the botanist George Wollaston, even though second hand, came complete with particulars. Following an apparent capture of the Apollo at Dover, he wrote to the editor of *The Zoologist*:

> 'As you wish for more particulars about the capture of Parnassius, I have been to-day to see the person who took it, and hear from his own lips all about it. He was lying on the cliffs at Dover, in the end of August or the beginning of September, 1847 or 1848 (he cannot remember which), when the butterfly settled close to him, and not having his nets with him, he captured it by putting his hat over it; and then carried it to his lodgings and shut window and door, and let it go in the room and secured it. He had not the slightest idea what it was till he saw it figured in some work afterwards. The insect has all the appearance of having been taken as he describes; and as he has no object to deceive, and is a person in whom I can place implicit confidence, I have no doubt, in my own mind, that the specimen is a British one. It will probably be in my own collection before this letter reaches you, when I shall be most happy to show it to you at any time you are this way.'[7]

Newman appended a paragraph stating that Wollaston, a botanist of high standing, was of the most scrupulous veracity and accuracy but went on to add: 'I am quite unacquainted with Mr. Wollaston's informant, with whom the *onus probandi* now appears to rest'. Unfortunately the captor did not rise to the bait and no more was heard. His letter prompted a further observation published later that year in the same journal but again it was second hand and lacking in supporting evidence. G. Austin wrote: 'I beg to inform you that I yesterday met a gentleman who assured me that he saw *Apollo* at Hanwell about six years ago. He chased it without success. This gentleman's veracity may be relied on. At a time when *Apollo*'s claim to be a British insect is under discussion, every scrap of information is of value.'[8] A further record was published in *The Entomologist* for 1889 of one seen flying slowly over the cliffs at Dover.[9] This too had the hallmark of a possible immigrant.

W. F. Kirby stated that attempts to introduce the Apollo to localities where

Plate 15
Colour wood-engraving of the Apollo (figure 2) and other 'Reputed British species' in W. S. Coleman's *British Butterflies* of 1860.
The plate also includes (1) the Scarce Swallowtail; (3) the Arran Brown; (4) Weaver's Fritillary; (5) the Purple-edged Copper; and (6) the Long-tailed Blue.

it was not indigenous always failed, even though its foodplant *Sedum telephium* (Orpine) was plentiful.[10] Morley and Chalmers-Hunt, in support of that view, considered that the old reports from Scotland might have followed efforts, possibly on more than one occasion, to introduce the species. Nevertheless, after a critical review of all the records from Britain, they concluded that some of those from England could reasonably be considered natural immigrants from the mountainous areas of western Europe.[11] Thomson also concluded that some at least of the Scottish records were of genuine immigrants from Scandinavia.[12] There can be little doubt that the Folkestone specimen captured and reported by P. Scott in 1955 was an immigrant.[13] Strong easterly winds prevailing at that time probably assisted its flight and it is significant that its capture coincided with that of a Camberwell Beauty (*Nymphalis antiopa*) on the same day.[14] Further research has shown that many of the previous records were also in years of large immigrations, including 1889, when no fewer than eight Camberwell Beauties were taken. There have been several further reports since the 1950s, all of which were probably genuine immigrants. From this limited evidence it is not unreasonable to suppose that *Parnassius apollo* has reached Britain unassisted on several occasions and deserves its place on the British list as a very rare vagrant.

6. THE SWALLOWTAIL – *Papilio machaon* Linnaeus
Papilio (Eques) machaon Linnaeus, 1758
Subspecies *britannicus* Seitz, 1907
Subspecies *gorganus* Fruhstorfer, 1922

Linnaeus applied the names of Greek heroes to the swallowtails. In choosing two of them for The Scarce Swallowtail and The Swallowtail, he turned to Homer who, in *The Iliad* (II: 731), related that 'the two sons of Asclepius, leaders of the men from Tricce, Ithome and Eurytus, were the admirable physicians Podaleirius and Machaon'. However, of the two, only *P. machaon* is native to Britain, its English subspecies *britannicus* being now confined to the Norfolk Broads; the other subspecies *gorganus* (formerly known as *bigeneratus* Verity) is an occasional visitor or temporary resident.

Although the earliest description of the Swallowtail is that by Moffet (Fig. 128) in 1634,[1] it cannot be certain that he was referring to the distinctive British subspecies though this is highly probable. This is because the Latin text gives descriptions of a number of Continental as well as British species and the illustrations, printed from fairly crude but often recognizable woodcuts, do not permit identification down to subspecies level. Apart from Moffet's description, the earliest definite record of a British specimen is that reported by John Ray in a letter dated 17th July 1670 to Martin Lister. He had found a larva at Middleton Hall, the home of Francis Willughby near Tamworth, Warwickshire, about which he wrote: 'This summer we found here the same horned *Eruca* [caterpillar], which you observed about *Montpelier*, feeding on *Foeniculum* [*Seseli*] *tortuosum*. Here it was found on common Fennel: It has already undergone the first change into a *Chrysalis*, and we hope it will come out a *Butterfly* before winter.'[2] In his *Historia Insectorum* Ray also gave Sussex and Essex as localities,[3] and Petiver said he 'caught [it] about London and divers counties in England, but rarely'.[4] Just how rarely can be judged by his later comment that he had seen only one specimen about London, which had been 'caught by my ingenious friend Mr Tilleman Bobart in the Royal Garden at St. James'. Petiver named the butterfly The Royal William. Moses Harris, who illustrated it (see Plate 28) and was the first entomologist to use the name Swallowtail in print,[5] later anachronistically suggested that the earlier name was probably a compliment to 'His Royal Highness, William, Duke of Cumberland, who was popular for his defeat of the rebels in 1745'.[6] However, since

Fig. 128.
This woodcut from *Insectorum Theatrum*, first published in 1634, is clearly the Swallowtail butterfly, though its execution is not accurate enough to be certain whether it was based on a British or a Continental specimen. However the breadth of the black band in the forewing suggests an English origin.

Petiver had died before the Duke was born he was obviously dedicating it to King William III, in whose reign he lived.

Benjamin Wilkes reported capturing Swallowtails in Kent: 'The first brood appears in May, the second towards the end of July. Being in a meadow near Cookham in Kent, on 5th August, 1746, I observed a female hovering over certain flowers, which taking particular notice I found to be the Meadow Saxifrage (*Seseli pratense*) ... It may be taken in meadows and clover fields about Westram [Westerham] in Kent.'[7] Since the British subspecies is confined to fenland, where its larva feeds on the milk-parsley (*Peucedanum palustre*), it is likely that Wilkes took the Continental subspecies. Similarly, John Ray successfully reared his specimens from larvae found on Burnet Saxifrage (*Pimpinella saxifraga*) in Essex and Sussex,[8] and he too was probably dealing with the Continental subspecies. Other early records also thought to be of this subspecies include those from Bristol.[9] This suggests that the two subspecies were resident in Britain at that time.

J. C. Dale saw or caught some forty to fifty Swallowtails between 1808 and 1816, including no fewer than fourteen between 27th July and 3rd August 1808. These specimens were all taken on the chalk downs near his home at Glanville's Wootton, Dorset, and far away from any marshland.[10] From this it may be inferred that these specimens were also of the Continental subspecies. Unfortunately only one has survived, and examination has proved inconclusive on account of its intermediate markings. Russell Bretherton, who examined the specimen, suggested that, as the species was seen annually and in two separate broods, there was probably an established colony in Dorset.[11] Dale reported no more after 1816, and its sudden demise was possibly brought about by the lowest prevailing summer temperatures since 1750, followed by a decade of further very poor summers.[12]

Formerly the subspecies *britannicus* was found in eastern England as far north as Beverley in south-east Yorkshire.[13] There is, however, clear evidence of a decline from around the beginning of the nineteenth century. In 1824, Lætitia Jermyn, author of *The Butterfly Collector's Vade Mecum*, gave its habitat and localities as 'fenny places' in Catton and Acle in Norfolk, and Cherry Hinton, Madingley and Whittlesea in Cambridgeshire.[14] In 1893, C. G. Barrett noted that the Swallowtail was still common in the undrained fens of Norfolk and Cambridgeshire, and that as the 'latter fens are extremely profitable to their owners and the former are practically undrainable there is little probability of their extermination'.[15] How far from the truth he was! A few years later, W. E. Kirby thought the Swallowtail was 'on the verge of extinction',[16] although this view was premature and has caused confusion, particularly as his illustration was of the Continental subspecies. The species died out in its last stronghold in Cambridgeshire at Wicken Fen in the 1950s, and in Norfolk it is more or less confined to fens around the Broads, but it is still widespread there and locally common.

Some nineteenth-century records invite speculation. Throughout the early entomological literature we find villages on the outskirts of London mentioned as localities where the Swallowtail might be taken regularly. Newman noted that in the 1820s when a young man, he had repeatedly found the caterpillar feeding on rue in a garden of some friends on Tottenham Green in Middlesex.[17] Austin recorded finding larvae 'In the osier beds behind Beaufoy's Distillery in Battersea Fields, year after year'.[18] It is possible that these were the Continental subspecies, although it is tempting to speculate that such marshy places as the osier beds of Battersea harboured an isolated colony of the British race.

Reports of single specimens, most probably of immigrants, occurring in unlikely places, appear intermittently in the entomological literature and there is little doubt that these will continue to be reported from time to time. Some make interesting reading. In 1857, one was taken near Balcombe tunnel, Sussex, on the London and Brighton Railway, 'by a man working near the tunnel, with his cap, and of course spoiled'.[19] In the same year the Revd Walter Wilkinson, the entomological rector of Hyde, near Fordingbridge, wrote of a man who, while driving through the New Forest, observed a specimen alight on his wife's dress.[20] In 1975, Miriam Rothschild reported the capture of a specimen of the British subspecies at Polebrook, Northamptonshire, and suggested that it might have been accidentally introduced with thatching reed from Hickling Broad, Norfolk.[21]

The extinction of the Swallowtail at Wicken Fen, where it remained common until the 1940s, has been attributed to the lack of proper management of its fenland habitat, and the drainage of nearby fenland during and after the Second World War. Several attempts to re-establish it have all eventually failed. Any future success will depend on the revival of healthy populations of Milk Parsley at Wicken Fen, which, in turn, require regular cutting of reed beds on a sufficiently large scale.[22]

7. THE SCARCE SWALLOWTAIL – *Iphiclides podalirius* (Linnaeus)
Papilio (*Eques*) *podalirius* Linnaeus, 1758

Whether or not the Scarce Swallowtail (Plate 16; also Plate 15, figure 1, and Plate 36, figure 1) has ever been truly native in these islands will probably

Plate 16
The Scarce Swallowtail illustration on Plate 578 of Volume 13 of John Curtis's *British Entomology* (1836), was 'drawn from the specimen taken by W. H. Rudston Read Esq. when he was at school at Eton College ... in 1822'.

remain forever a matter of speculation, but all the evidence suggests that it has not. Moffet included the species in his *Insectorum Theatrum*,[1] but did not claim it was of British origin (Fig. 129). It is now accepted that Ray was the first

Fig. 129. Woodcut of the Scarce Swallowtail from *Insectorum Theatrum*, presumably based on a foreign specimen. Moffet noted that it differed from the Swallowtail in that the 'outer border of the innermost [i.e. the hind] wing is sky blue or woad in colour, as well as the three '*spintheres*' [spots] on the underside'.

author to consider that it might be a British species.[2] Even so, his comment – '*Prope Liburnum portum in Etruria invenimus, atque etiam, ni male memini, in Anglia*' [we found it near Leghorn harbour in Etruria (now Livorno in Tuscany, Italy) and also, unless my memory fails me, in England] – represents little more than an uncertain recollection of having seen it in Britain. In his *Outlines of Natural History* of 1769, John Berkenhout also included it as a British species, stating that it was 'rare, in woods'.[3] Some twenty-five years later William Lewin illustrated the Scarce Swallowtail in *The Papilios of Great Britain*, commenting that 'This elegant species of butterfly is said to have been caught in England, and therefore I thought it not improper to give a figure of it'.[4] This was not supported by any further evidence, and Lewin was presumably just following his predecessors.

It was not until 1803 that any definite evidence was forthcoming. In that year Haworth published the information that 'My friend the Revd Dr Abbott [sic] of Bedford has informed me that he took in May last, near Clapham Park Wood in Bedfordshire, a specimen of *Papilio Podalirius* in the winged state'.[5] Abbot was one of those collectors possessed not only of phenomenal luck but also the propensity to be duped by dishonest dealers. It may be recalled that it was he who had genuinely taken the first Chequered Skipper in England – also in Clapham Park Wood – but several species he recorded from that locality were undoubtedly 'planted'.

Haworth later learnt of a second Scarce Swallowtail, which the Revd Frederick William Hope told him he had taken on 14th September 1822 at Netley, in Shropshire. Hope also announced his capture to J. C. Dale as follows: 'My own successes have far outran my expectations, & it will be a piece of News to inform you that I have captured the long desired & much doubted Pap.podalirius, since then I have seen another on the wing, but could not obtain it, after toiling half the day'. It seems that the Scarce Swallowtail was to be seen at Netley over a period of about six years, providing Hope with a specimen – probably one of those that still exist in the Oxford University Museum – and, a couple of years later, with two larvae. However, it is also known that a lady living nearby in Longnor painted butterflies, including this species, and she might have been responsible for introducing it. In 1824 Hope was to write again to Dale that 'The caterpillar of Podalirius has just been taken down in Shropshire, near the spot I took the perfect insect on the wing. A Mrs Plimley is feeding it, & if it comes out of the Chrysalis I am to have it after it is painted.'[6] Unfortunately the larva turned out to have been parasitized by an ichneumon wasp. Another letter from Hope, dated the 10th February 1829, informed Dale that '*Podalirius* was again on the wing ... I saw it distinctly settle on a peach, & was in quest of them two days without success'.

In an elegant account of his investigations into the affair of these Scarce Swallowtails, Allan suggested that, as there were clearly two specimens at Netley that summer, this would militate against the probability of immigration. Rather it seemed that they had hatched from pupae very close to Netley and, if this were so, then there must have been a *third* Scarce Swallowtail at Netley that year – namely a gravid female. She would presumably have arrived near Netley in April or May, for her two progeny to be seen and caught by Hope later that summer. Allan reasoned that all this indicated there was a breeding colony at Netley at that time.[7]

In spite of these discoveries, the Scarce Swallowtail caught by Hope was never exhibited at the Entomological Society – a curious omission, as he was a founding member of that society, and one would have expected him to wish to display such an important find. Allan suggested that Hope could not have been entirely satisfied that his specimen was a true native swallowtail and, being a man of integrity, was probably loath to claim that it was. It seems very likely on balance that the Netley Scarce Swallowtails were from imported stock. There are two specimens in the Hope Collection at Oxford, but as these are unlabelled it is not known which, if either, is the Netley specimen.

Only a few other specimens of the Scarce Swallowtail have been taken in this country. As a schoolboy at Eton, also in 1822, the Revd Rudston Read managed to capture in his hat, near a nursery garden between Datchet and Slough, the specimen illustrated by John Curtis.[8] Some years later, in a letter to J. C. Dale, he stated that the specimen had since been damaged by mould and neglect. However, when the Read Collection was sold in 1865, it was said to be in good condition. A single specimen was taken by Dr H. E. Tracey at Willand, South Devon, in May 1895, and two months later another on Wye Downs, Lyminge in Kent by F. Hall.[9] Yet another was taken in 1901 at Woolacombe, North Devon, ending up in Lord Rothschild's collection at Tring, according to Frohawk.[10] Other sightings have been reported from Badminton, Gloucestershire, on 4th June 1963,[11] and Ross-on-Wye, Herefordshire, on 26th August 1984.[12] There have been others but it seems probable that most British records are the result of introduction, whether accidental or intentional. For at least thirty years this species has been a popular one to rear, and livestock is readily available from dealers.

8–10. THE CLOUDED YELLOWS (*Colias* spp.)

The clouded yellows, which, with their relatives the Brimstone and Cleopatra butterflies, were collectively called Redhorns by some nineteenth-century lepidopterists on account of the colour of their antennae, have long been a favourite of entomologists. Their spectacular but intermittent mass migrations to Britain, their bright and contrasting colours, and their powerful flight have always exercised a great pull on the imagination of lepidopterists of all ages.

Three species of clouded yellows visit Britain as migrants, one a regular annual visitor, though in very variable numbers, and the other two more rarely. All three have bred in Britain, but they cannot survive an English winter and eventually die or even make the reverse migration back to the Continent.[1] A fourth species, *Colias palaeno* (Linnaeus), the Moorland Clouded Yellow (see Plate 36, figure 2), has been recorded on only a handful of occasions and probably as an accidental introduction, perhaps with imported shrubs, as it is not usually migratory. The Clouded Yellow and the Pale Clouded Yellow species have been recognized for a long time and descriptions have appeared in the literature since the end of the eighteenth century. However, there was for a long time confusion over the British species which was finally resolved only in 1948 when the discovery of a new species, the New (now Berger's) Clouded Yellow, was published in *The Entomologist*. Their individual histories are given below.[2]

8. THE PALE CLOUDED YELLOW – *Colias hyale* (Linnaeus)
Papilio (Danaus) hyale Linnaeus, 1758

Because of similarities between this species and other clouded yellows referred to above, the earlier literature is both confused and confusing. It is believed that Linnaeus, in naming it *hyale*, was actually including *all* clouded yellows. He cited Petiver and Ray as authorities, but Ray in his *Historia Insectorum* was clearly writing about *The* Clouded Yellow: 'Butterfly, middle sized, upper wings deep yellow with black edges'. He described the red-rimmed spot on the underwings and goes on: 'In the female white takes the place of yellow', an obvious reference not to *hyale* but to the pale form of the female of *croceus* known as *helice*. Ray's own sources were Moffet and Petiver.

The name Pale Clouded Yellow was first used in 1775 by Harris in the *Aurelian's Pocket Companion*[1] but he did not give it a scientific name, having used the scientific name *hyale* for the Clouded Yellow, following Linnaeus. In his classic work on British Papilios, Lewin illustrated both sexes of the Pale Clouded Yellow (Plate 17) but under different English names, the male being called the Clouded Yellow '*Hayale*' [sic] and the female the Pale Clouded Yellow, with no scientific name.[2] The same name '*hyale*' was thus applied by Harris and Lewin to different butterflies. Donovan tried to sort out the muddle,[3] having recognized that Lewin had incorrectly regarded the sexes of the Pale Clouded Yellow as different species, but it was not until the publication of Haworth's *Lepidoptera Britannica* that the Pale Clouded Yellow was given its present-day specific name for the first time.[4] Unfortunately Haworth, as a result of misidentification of a specimen of Pale Clouded Yellow, added another species, *Colias europome* (Esper), the Clouded Sulphur, which has never been found in Britain and so added to the confusion. James Rennie perpetuated this name, regarding the Clouded Sulphur as 'very rare and even doubtful as a native', yet giving its locality as 'East Coast'.[5] Coleman then dropped the scientific name *europome* but kept the English name Clouded Sulphur as his preferred alternative name for the Pale Clouded Yellow.[6] Although the name Pale Clouded Yellow was thenceforth generally adopted, the scientific name continued to include the next of our historic butterflies, the as yet undescribed 'New Clouded Yellow' as it was first called. That event occurred in 1948 when the status of the groups was finally clarified.[7] The name *europome* has not disappeared altogether since it has been used for a subspecies of *C. palaeno*, the Moorland Clouded Yellow,[8] although more recently treated only as a form.[9]

In the great migration years of 1821, 1828, 1835 and 1842 – all interestingly at seven year intervals – Pale Clouded Yellows arrived in considerable numbers.[10] Unfortunately the 'mystical' seven-year cycle, which had been predicted by Thomas Desvignes,[11] thereafter began to fail: there were only about

Plate 17
William Lewin, on Plate 33 of his *Papilios of Great Britain* (1795), illustrated the sexes of the Pale Clouded Yellow as separate species. Of the male, 'Hayale' [*sic*] (figures 1 and 2) he noted: 'This is a very rare species of butterfly', which he met only in the Isle of Sheppey. Of the female (figures 3 and 4), his Pale Clouded Yellow, found 'in a gravelly pasture field in Kent', he wrote: 'This species is likewise very rare'.

twenty specimens reported in 1849, even fewer in 1856 and practically none in 1863.[12] Since then, there were moderately large influxes in 1868, 1872, 1900, and in the years 1945 to 1949. The largest peaks of the twentieth century were in 1900 (2200 recorded), 1945 (318), 1947 (870), 1948 (310) and 1949 (450). In over thirty of the intervening years it was not recorded at all and has been a rare migrant since 1949. The latest report, said to be reliable, is of one from Norfolk on 18th August 1996.[13] The most probable reason for its rarity in recent years is its decline in France and central Europe,[14] though, in view of the continued occasional mass-migrations of Clouded Yellows *Colias croceus*, its decline to the status of one of our rarest migrants is surprising.

9. BERGER'S CLOUDED YELLOW – *Colias alfacariensis* Berger
Colias alfacariensis Berger, 1948
Colias australis Hemming & Berger, 1950
Colias calida Cockayne, 1952

To discover an insect new to science is a gratifying achievement but to find and then describe a new species of butterfly in western Europe is an historic event. In 1945 a Belgian lepidopterist, Lucien Berger (his name is pronounced with a soft 'g'), recognized that the Clouded Yellow which now bears his name was a distinct species. Two years later, after much study of specimens from old collections, L. Berger and M. Fontaine published the first part of their paper entitled 'une espèce méconnue du genre *Colias*' in a Belgian journal (Fig. 130),

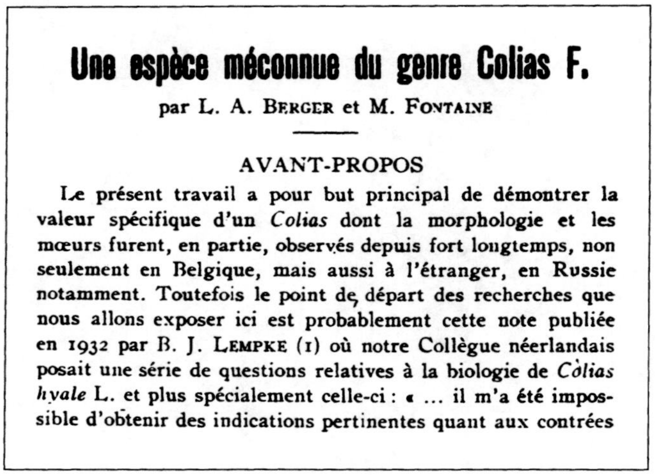

Une espèce méconnue du genre Colias F.

par L. A. Berger et M. Fontaine

AVANT-PROPOS

Le présent travail a pour but principal de démontrer la valeur spécifique d'un *Colias* dont la morphologie et les mœurs furent, en partie, observés depuis fort longtemps, non seulement en Belgique, mais aussi à l'étranger, en Russie notamment. Toutefois le point de départ des recherches que nous allons exposer ici est probablement cette note publiée en 1932 par B. J. Lempke (1) où notre Collègue néerlandais posait une série de questions relatives à la biologie de *Colias hyale* L. et plus spécialement celle-ci : « ... il m'a été impossible d'obtenir des indications pertinentes quant aux contrées

Fig. 130. The original publication of the discovery of a new species of *Colias* by Berger and Fontaine in *Lambillionea*, a Belgian journal.

but they did not then give it a name.[1] The name that Berger had chosen to bestow on it, '*alfacariensis*', had previously been used by Ribbe in 1905 for a form of the Pale Clouded Yellow which may have been this newly-described species. Because the issue of the Belgian journal including the second part of their account, which gave its scientific name, did not come out until late in 1948, the first publication of the scientific name and description of this new butterfly, appeared in an English journal, *The Entomologist*.[2] It would appear that Berger delayed publication of the final part in *Lambillionea* until after his own paper had been published in England in order to establish priority for his authorship of the name. This deprived Dr Maurice Fontaine, 'le veritable découvreur' of this species, of the credit that was his due (J. Hecq, pers.

comm.). But for that, the species might now be known as 'Fontaine's Clouded Yellow'. The English name given it at the time was the 'New Clouded Yellow'.

There was some controversy over the eligibility of the name *alfacariensis* and so, in 1950, Berger, in association with Francis Hemming, substituted *australis* which had been given by Verity in 1911 to a subspecific form of the Pale Clouded Yellow. Two years later, however, Cockayne objected to *australis* on the grounds that the name was unavailable, and proposed *calida*, another Verity name dating from 1916. Nevertheless, for a time the name *australis* prevailed, though different lepidopterists in Britain and on the Continent were by now using all of these names. Finally, in 1982, some order was restored and, under the established rules of nomenclatural priority, *alfacariensis* was readmitted. And so it happened that both the vernacular and the scientific names used by most English lepidopterists became outdated: the New Clouded Yellow *Colias australis* Verity became Berger's Clouded Yellow *C. alfacariensis* Berger. To the average butterfly lover such arcane shenanigans may seem puzzling, but the explanation given above may make them a little clearer. Nomenclature, though fought over fiercely by taxonomists, has to submit to a set of rules if there is to be any stability.

It is very probable that Berger's Clouded Yellow (Plate 18) has long been an occasional visitor to Britain, but had been passed over through its similarity to the Pale Clouded Yellow. It soon became apparent that specimens of Berger's Clouded Yellow had for decades been unwittingly placed by collectors in their series of the Pale Clouded Yellow. According to Bretherton,[3] some fifty specimens have been traced in old museum and private collections. All date from before 1947, the earliest bearing a label that indicated its capture at Folkestone in 1875. There are some twenty-two specimens known to have been caught before 1940 in East Kent, thirteen from the Thames estuary towns of West Kent, and fourteen from Sussex, Hampshire, the Isle of Wight, Dorset, Somerset and Cornwall.

Following the discovery of *Colias australis* [*alfacariensis*] in British collections, Vallins, Dewick & Harbottle pooled their knowledge and resources to present a joint paper.[4] In it Harbottle gave a delightful and graphic account of his experience with this butterfly.

Plate 18
Two historic specimens of the as yet unrecognized species, Berger's Clouded Yellow. Above – from Deal, Kent, taken by Ernest Neal in 1947; below – a male specimen captured at Greenhithe, Kent, by A. B. Farn in June 1900.

> 'A sweltering sun in a cloudless sky was intensified by the reflected brilliance of the cliffs that it bathed. Such were the conditions under which, on Tuesday 26th, July last [1949], whilst on a very brief visit to the Dover-Folkestone area of Kent, I fortuitously came across a small colony of the new British species of butterfly *Colias australis* Verity. At 11.30 am. (B.S.T.) my attention was suddenly drawn to two pale yellow butterflies frisking a few feet to my left. By their flight, which was now swift and reckless, together with the depth of their colour, I suspected their true identity from the first. I made a hasty and ill-judged swipe with the net, missed both, and gave chase to one, cursing my lack of skill. Several times every ounce of energy was exerted to gain on the insect, which eluded every stroke of the net. Eventually, having tired of the sport, it careered from my ken altogether. Periodically this process was repeated, and the steep slopes helped the butterflies to win every time. Later, how-

ever, at about 1.00 pm. I caught a male, which promptly flew out of the only hole in the net! Stung into alacrity I had it replaced by another within a few moments. This insect was less fortunate and, upon examination, proved itself to be a male *C. australis* in immaculate condition. At this juncture my luck seemed to turn, and within the next three hours I had secured four males, making a total of five in all. Their condition varied from good to perfect. I saw only one female, which was very fresh and of the white form. It came with great speed and directness at me, and, before I could regain my balance after the first and only lurch, it had vanished. I was bitterly disappointed, for I had determined from the outset to try to breed the species.

'On the morning of Friday, 12th August [1949], I paid my second visit to the same site. Conditions, although not so ideal as before, were otherwise more pleasant with a slight breeze and considerable cloud. At first, to my dismay, it appeared that all the *C. australis* had either died or migrated elsewhere as not one was to be seen. After strolling about for an hour, however, I saw what I thought was a female *C. australis* flitting lazily along the bank and frequently settling on the purple thyme heads. Having caught it, I boxed it, and thought: "pity, *helice*". Not unnaturally I kept pulling the box out and looking at the creature, and, as it was yellow and heavily bordered, I eventually reached the quite erroneous conclusion that it must be a male *Colias hyale*. However, I refrained from killing it. A little later I netted a male *C. australis*, and in so doing, smashed the ferrule of my umbrella, which I was using as a stick and which I eventually left in the compartment of a railway carriage. This male, which was slightly worn, I boxed. Things were going very well and I was very happy until, in about an hour's time, I was informed in no uncertain manner by a railway official that I was trespassing, and I was told to clear off. This I readily and apologetically consented to do, and perched myself just the other side of the railway's property but within full view of the beloved hunting ground. This forthright devotion to duty on the official's part proved to be a blessing in disguise. Almost immediately I observed a female *C. australis* of the white form. A brief moment passed before it settled, I saw the massive forewing discoidal spot, and I at once knew it for what it was. I slammed the net over it and boxed it. It was in perfect condition. A few minutes later I caught a second male, although I think I had to trespass to do so. I eventually made my way into Folkestone, taking two *C. croceus* ab. *helice* en route.'

Berger's Clouded Yellow migrates to south-eastern Britain from southern Europe and is normally the last of the clouded yellows to arrive, seldom being seen before late summer. The notable immigrant years of 1945 and 1947 produced a crop of records and, following the first account of this new species, some eighty specimens were seen and caught in the area around Folkestone and Dover during 1948 and 1949. Since then there have been relatively few records, but in 1990 there were a probable four: on 3rd August there were records of *alfacariensis/hyale* at Hamstreet, West Kent, and Cooksbridge, East Sussex; on 5th August at Lewes, East Sussex,[5] and on 14th August, a single male was taken near Princes Risborough, Buckinghamshire (Salmon, pers.obs). The following year a remarkable discovery was made. In September 1991, A. S. Harmer found firm evidence that breeding had taken place at Portland, Dorset, having discovered ten males and five females as well as two full and two empty chrysalids. A further sixteen adults were captured by Bernard Skinner and A. J. & C. T. Pickles to give a total from this brood of twenty-two males and nine females. With the further discovery of two more specimens in Honiton, Devon, it was apparent that a significant immigration had taken place the month before.[6]

The only other recent reports are from August 1996, the year of the unprecedented immigration of Painted Lady (*Cynthia cardui*), when there were unconfirmed records of Berger's Clouded Yellow in August and September from Dorset and Hampshire.[7]

10. THE CLOUDED YELLOW – *Colias croceus* (Geoffroy)
Papilio croceus Geoffroy *in* Fourcroy, 1785
Papilio edusa Fabricius, 1787

Although the Clouded Yellow was not formally described until the 1780s, the first published reference to it is found in Moffet's *Insectorum Theatrum*,[1] where it is illustrated by a recognizable woodcut (Fig. 131). Petiver, who also knew it, gave it the name *Papilio croceus*, the Saffron Butterfly,[2] a name adopted by Ray, though he used the feminine *crocea*.[3] Petiver had thought that the dimorphic sexes of this butterfly were different species, so he named them the Saffron and Spotted Saffron butterflies. Ray, however, recognized they were one although he described the female as white, presumably from a specimen of the pale form *helice*. In his *Twelve New Designs of English Butterflies*,[4] published in 1742, Wilkes gave the species its present name for the first time, adding that 'it was taken in the fly state in July'. Harris, in *The Aurelian*,[5] also called it The Clouded Yellow. Berkenhout[6] reverted to Petiver's name, the Saffron Butterfly, but, being one of the first to adopt Linnaean nomenclature, he followed Linnaeus's recently published work which did not include it and wrongly called it by the scientific name *hyale*, as did Harris in an *Index*, issued around 1773 as a supplement to *The Aurelian*,[7] and also in the second edition of the same work, published in about 1775.[8] That two such dedicated lepidopterists should make this mistake may seem extraordinary, but Emmet thought that they should not be blamed 'since Linnaeus himself equated his *hyale* with the unambiguous descriptions given by Petiver and Ray of *C. croceus*'.[9] Their error remained uncorrected for a quarter of a century until Donovan[10] wrote: 'If our observations on the preceding species are satisfactory and conclusive, the Insect before us must be a distinct species, and not the *P. Hyale*, for which it has ever been received'. He thought that as Linnaeus had referred to figures of the Clouded Yellow in other authors' works he must have been acquainted with it. However it is very probable that Linnaeus considered this butterfly to be just a variety of the Pale Clouded Yellow (*hyale*) as he never wrote about it under any other name. Fabricius, who first described the genus *Colias* in 1807, included in it all the known species of clouded yellows including *Papilio edusa*,

Fig. 131.
Moffet illustrated a female Clouded Yellow on page 100 of his *Insectorum Theatrum*. His description was fairly accurate, given the lack of scientific terminology at that time. He called the wings 'a bright yellow ... the utmost parts waxing blackish, beautified with three yellow spots'; and the 'head, feet and horns blood-red'.

by which name he had himself already described the Clouded Yellow in 1787. Donovan continued: '... we are confirmed in our opinion by Mr. Jones of Chelsea (see Chapter 3: 17), who assisted Fabricius with considerable information, and assures us it is certainly the *P. Edusa* of that author.' Donovan therefore adopted the name '*Edusa*' for the Clouded Yellow, disregarding Lewin's scientific name *Papilio electra* but retaining his English name, 'Clouded Orange'.[11]

Until forty years ago, the name *edusa* was universally used by lepidopterists in Britain as well as on much of the Continent and the term 'Edusa years' was applied to their years of abundance by everybody. It was then discovered that, two years earlier than the publication of Fabricius's description, the epithet *croceus* had been applied to it by Geoffroy in France. The names had thus gone full circle since Petiver called it *Papilio croceus* in 1703!

One other ground for confusion is the existence of the very pale female form of this species, f. *helice*, which can be mistaken for other species of Clouded Yellows (Plate 19). The female Clouded Yellow is dimorphic, that is to say it appears in a normal population in two forms. F. *helice*, described by Hübner, was first mentioned as British by Haworth, who considered it a separate species, which he called the White Clouded Yellow. It occurs normally in about 5 to 10 per cent of the female population, and the gene determining it can be carried by either sex. Normal females will produce *helice* offspring.

Moses Harris gave an excellent account of the habits of the Clouded Yellow: 'This beautiful Fly is taken in meadows, in the Month of August; they appear fond of settling on the Yellow Lupins and Thistle. They have been taken flying in Plenty, on Epping Forrest; but as they seldom haunt one Place for many successive Seasons, I can't venture to mention it as a Place where they are to be found. Where there is a Brood, the Times of the Day to find them are at Eight in the Morning and Four in the Afternoon; but never in the Middle of the Day,when they conceal themselves to Rest. They fly very fast, therefore not easily taken; the Male, in particular, flies exceeding swift.'[12]

There is something special about chasing after Clouded Yellows, and the literature on this species is often vivid. For Coleman:

> '*Edusa* has indeed a lively flight, and his pursuer has need of the "seven-league boots", with the hand of Mercury, to insure success in the fair open race, if that can be called a fair race at all, between a heavy biped, struggling and perspiring about a slippery hill-side, such as *Edusa* loves, – and a winged spirit of air, to whom up-hill and down-hill seem all one.
>
> 'In truth the best way to get *Edusa* is to watch and mark him down on a flower, then creep cautiously up till within range, raise the net quietly, and *strike rapidly downwards* over the insect, who usually darts *upward* when struck at; and, in nine cases out of ten, *Edusa* will be fluttering under the net. It is not the most heroic style of sport, this, but it fills the boxes admirably.'[13]

The Clouded Yellow has always brought a thrill of excitement to British entomologists because of its unpredictable appearance and the occurrence of the spectacular 'Edusa Years' – years when massive invasions of migrant Clouded Yellows arrive from the Continent. Frohawk quoted from a letter written to him by the Revd D. Percy Harrison in December, 1933:

Plate 19

Specimens of Clouded Yellow showing (top row) the normal male and female forms (Petiver's 'Saffron' and 'Spotted Saffron' Butterflies); (middle row) the pale female form *helice*, which led John Ray to believe it was yet another species, and ab. *pallida* Tutt, taken in 1909 by J. W. Tutt himself; and (bottom row) a remarkable aberration captured by J. B. Purefoy, son of Capt. E. B. Purefoy, in 1947, alongside another specimen of f. *helice*, taken by E. G. Meek in 1871. Meek 'Naturalist, Plumassier and Furrier' of Brompton Road, London, was alleged to have imported Continental rarities before 'discovering' them!

'My greatest experience was in Cornwall as far back as 1868, when I was only 11, and sat on the cliff near Marazion, and saw a yellow patch out at sea, which as it came nearer showed itself to be composed of thousands of Clouded Yellows, which approached flying close over the water, and rising and falling over every wave till they reached the cliffs, when I was surrounded by clouds of *C.* (=*croceus*) *edusa*, which settled on every flower. I have some of them still, including two *helice*. Whence they came I know not; the nearest coast of France would be Cherbourg. They swarmed in the district for a space of some three weeks and were good specimens when they arrived.'[14]

C. W. Dale, son of J. C. Dale of Glanville's Wootton, gives us a glimpse of the great 'Edusa year' of 1877: 'One of the grandest sights I ever saw in my life, was on a little undercliff to the East of Lulworth Cove, on the 5th of September, 1877. On this undercliff grows a mass of *Inula Crithmoides* [Golden-samphire], then in full bloom; below is the clear blue water of Weymouth Bay, unruffled by a ripple. Every one of its yellow flowers was literally covered with one, two, or more of *Colias edusa*, with its white variety *Helice*, [and also] *Cardui*, *Atalanta*, *Rapae*, *Io*, *Phaelas* [sic], *Janira*, *Corydon*, *Alexis*, *Agestis*, *Sylvanus*, *Linea*, *Actaeon*, and *Galathea*.'[15]

Several other authors have described the way the countryside becomes alive with Clouded Yellows during an 'Edusa Year'; Their reports make nostalgic reading on a grey winter's evening. In 1893, Coleman was amazed to find the cliff front below the Lees at Folkestone a constant flutter of orange specks.[16] L. Hugh Newman recounted an occasion during the Second World War when the Clouded Yellow migrated in such numbers that military observers saw them approach the coast in the form of a great golden ball, which they thought at first to be a cloud of poison gas drifting over the water![17]

Dale published a list of the years during which the Clouded Yellow had been especially common: 1797, 1804, 1808, 1811, 1818, 1822, 1826, 1831, 1835, 1859, 1864, 1865, 1875, 1876, and 1877.[18] The year of 1826 was the finest of all: the '*Annus mirabilis*' for entomologists. Williams added 1857, 1858, 1889, 1892, 1893 and 1899 as years when *croceus* was present in hundreds and even thousands but, unlike Dale, did not include 1864, 1875 and 1876 as 'Edusa years'.[19] Bringing the lists up to date, Bretherton, recorded *C. croceus* as abundant in 1900, 1913, 1928, 1937, and in six of the years 1941 to 1950, with an estimated peak of 36,000 in 1947.[20] More recently it was thought that 'Edusa years' might be a thing of the past, on account of agricultural changes in northern Europe. However, 1983 saw another huge migration of Clouded Yellows, which were recorded from the South Coast to the Orkneys. There were further sizeable invasions to the South Coast in 1996 and 1998 but they did not penetrate so far north.[21, 22] Let us hope there will be many more 'Edusa years' to come!

11. THE CLEOPATRA – *Gonepteryx cleopatra* (Linnaeus)
Papilio cleopatra Linnaeus, 1767

This striking butterfly is superficially similar to the Brimstone (*Gonepteryx rhamni*) but is larger and, in the male, the forewings are largely suffused with bright orange-red. The Cleopatra is a southern European species not occurring in northern France, and is not known to be migratory. Most of the very few specimens of purported British origin have probably been deliberately introduced, or have escaped from breeders or butterfly farms, or may even have been the subject of fraud. Although occasional, unlabelled examples of this species have been found in old British collections, no more than four of these are now thought to be genuine captures. All but one are from the South Coast.

W. S. Coleman wrongly regarded the Cleopatra as a variety of the Brimstone brought about by the higher temperatures prevailing in the Mediterranean regions where it is found. He was, however, puzzled that they were both to be found in the same areas. He tried to explain this as follows: 'It is possible there may be a constitutional difference between individual insects, just as we see that of two Englishmen going to a hot climate, one will brown deeply, while the complexion of the other will hardly alter, though exposed to the very same external influence'.[1] He overlooked their genetic differences and different foodplants!

In 1860 the Revd H. Adair Pickard of Christ Church, Oxford, wrote to the *Entomologist's Weekly Intelligencer*, under the heading 'Singular, considering the Season!' as follows: 'I have to announce an important capture – that of the *Cleopatra* variety of *Gonepteryx Rhamni*. It was taken by my uncle, John Fullerton Esq., in his grounds at Thrybergh Park, near Rotherham, on June 27th, 1860.'[2] One year later, Stainton, editor of the *Entomologist's Annual*, endorsed this discovery, adding the following note: 'The forewings are much more suffused with orange than those of the specimen which Mr. Curtis figured could have been, and the specimen resembles exactly the Italian specimen of Cleopatra in Mr. Hope's Collection'.[3] The whereabouts of the Fullerton specimen are not known, but from Stainton's remarks, and bearing in mind the place of its capture, it was very likely an extremely rare aberration of the Brimstone, with a similar reddish suffusion on the forewings. A specimen of a similar aberration of *rhamni*, taken at Assington, Suffolk, in August 1897, is in the author's collection (Plate 20).

Plate 20
Brimstone and Cleopatra butterflies contrasted. Top – a male Brimstone from Hampshire, 1966; centre – a male Cleopatra from Digne, France, 1974; bottom – Brimstone var. *aureus* Frohawk, taken at Assington, Suffolk in August 1897.

Frohawk recorded two examples of what he described as a new aberration of the Brimstone, ab. *aureus*: one taken at Sandown, Isle of Wight, in August 1873, in the Rothschild Collection; the other taken at Aldeburgh, Suffolk, in 1896 in the Rait-Smith Collection.[4] However, on recent re-examination both, now housed at the Natural History Museum, London, proved to be genuine Cleopatras (D. Carter, pers.comm.). They are among the four British specimens in the Museum listed by Howarth, of which the other two are from Ventnor, Isle of Wight, 1870, and Forfar, Scotland, June 1887.

Another Cleopatra, a male, was captured at Feock, Cornwall, in September 1957 by I. G. M. Reid.[5] Since then, in July 1981, a further specimen was seen and 'positively identified' two miles from Dover near the main road at Temple Ewell, East Kent.[6] It is likely that it had arrived in this country in a car or lorry after crossing the Channel by ferry, but it could possibly have been ab. *aureus* of the Brimstone. In 1984, another male was seen flying south-east along the coast at Pendennis Head, St Mary's, Isles of Scilly. This was considered to be a stray immigrant from the Continent.[7]

12. THE BLACK-VEINED WHITE – *Aporia crataegi* (Linnaeus)
Papilio (Danaus) crataegi Linnaeus, 1758

The Black-veined White was once not uncommon in Britain. It was in fact one of the first British species of butterfly to be described, by Moffet in 1634,[1] though with a barely recognizable figure. Christopher Merrett included it with twenty others in his *Pinax rerum Naturalium Britannicarum*,[2] and Ray described the adult as 'dingy white with black veins', adding that 'the larva feeds on Hawthorn, spins webs and lives gregariously, returning after feeding to its companions'.[3] Petiver[4, 5] and Albin[6] both illustrated it, calling it the 'White Butterfly with Black Veins' (Fig. 132). The name 'Black Veined White' was used by Drury in his journal for 1764, and given it by Moses Harris in 1766.[7] It was subsequently used by all later authors other than Rennie,[8] who characteristically gave it a different name, The Hawthorn, based on its principal foodplant.

Fig. 132. *(facing page)* Petiver's Plate 1 from *Papilionium Britanniae Icones* (1717) containing the engraving of the 'White Butterfly with Black Veins' (figures 1 and 2). His illustration of the Black-veined White is remarkably good with the venation accurately drawn and engraved by H. Terasson.

The early stages of its life history were first illustrated by Albin, and described in detail by Harris:

> 'The Female Fly lays her Eggs on the White Thorn, about the End of *June*; and the young Caterpillars, as soon as hatched from the Eggs, inclose themselves in a slight Web, leaving a Passage to come forth to feed, which they generally do Morning and Evening, retiring within their Web in the Middle of the Day, to avoid the Heat of the Sun: In this Manner they feed the remaining Part of the warm Weather, extending their Web as they increase in Size. At the approach of Winter, they spin a strong web on one of the Twigs of the White Thorn, wherein they remain without eating during the Winter, and come forth again early in the Spring, feeding very greedily on the Buds and young tender Leaves.'[7]

The caterpillar pupated within 'twenty-four hours' after attaching itself to a twig, and remained a chrysalis for twenty-one days.

The Black-veined White butterfly was prized by the early collectors. Drury, in his diary for 1st July 1764, recorded it as 'plentiful and fine' at Enfield Chase (see Chapter 3: 14), and Harris at about the same date noted that: 'They fly in Meadows near Corn-Fields, and as they do not fly very fast, are easily taken in your net'. Haworth stated that 'It is frequently found in gardens'.[9] John Curtis, mindful of the damage that could be inflicted on fruit trees by the larvae, added that 'Fortunately this butterfly is seldom very abundant in England, and from the care taken of our gardens, it seems to become annually more scarce'.[10]

In its heyday the Black-veined White was recorded from thirty-two counties in England and three in Wales, and there were single sightings in Scotland and Ireland. Though common in some areas, like Kent and the New Forest, it seems to have undergone periods of great abundance followed by periods of

Papiliones BRITANNIÆ *albidæ* ENGLISH *White* BUTTERFLIES TAB 1

Fig 1 2 3 4 5 6 7 8 9 10 11 12 13 14 15 16 17 18 19 20 21 22 23 24

scarcity. By 1870, when Newman listed a large number of localities, it was no longer present in many of them. Newman remembered that it was 'Formerly common at Eton Wood, near Leominster [Herefordshire]; I have seen it in cloudy weather settled almost by hundreds on the blossoms of the great moon-daisy (*Chrysanthemum leucanthemum*)'. Another of his informants, H. Ramsay Cox also saw it in large numbers: it was 'most abundant at Herne Bay in 1858: we used, by way of amusement, to see how many we could catch at one stroke of the net; we often took four or five at a time: they appeared particularly fond of fields of broad beans'.[11]

In spite of this apparent abundance, all was not well with the Black-veined White in Britain. From the early years of the nineteenth century, the smaller colonies started to die out and the range of the species to contract. By the latter years of the Victorian era the species had become endangered. Concerned letters discussed the situation. In the *Entomologist's Monthly Magazine*, C. W. Dale asked readers whether the Black-Veined White was still found in the south-eastern counties of England.[12] Herbert Goss replied that: 'During the past ten years it has, to my knowledge, disappeared from all the localities in the New Forest, and in Monmouthshire, where it was formerly found in abundance'. In 1867 he had 'found the species, not uncommonly in grass fields about one and a half miles to the north west of Tintern in the direction of Trelleck. On the 4th of July I discovered another locality one mile west of Tintern, where the species occurred in the greatest profusion. Hundreds of specimens were flying about, or settling on the flowers of ox-eye daisy and on thistles, and I was frequently able to capture five or six specimens at one stroke of the net. Ten years after, on the 30th of June, 1877, I again visited Tintern, and lost no time revisiting the old *Crataegi* haunts, the natural conditions of which were unchanged; but instead of the swarms of this butterfly which occurred there in 1867, only two specimens were seen during the whole of a fine summer's day.'[13]

Other correspondents told a similar story and a variety of theories were propounded for its disappearance. It is now generally considered that the extinction of the Black-veined White was the result of a combination of factors. C. R. Pratt provided an admirable synopsis of the distribution by counties as well as a detailed account of the decline and extinction of this butterfly. Outside the New Forest, which at one time was its headquarters, at least 150 colonies were known.[14] A high September rainfall over a number of years preceded the final demise around 1925, although the sudden decline just after the middle years of the nineteenth century cannot be satisfactorily explained.[15] It has been suggested that birds eating the larvae might have been a secondary cause, while another theory was increased parasitization by the ichneumon, *Apanteles glomeratus* L., coinciding with a significant change in the weather profile over a number of years from a Continental to an Atlantic type, which might have helped the spread of viral diseases infecting the larvae.[16] Subsequent reintroductions of the species to this country have been short-lived. One colony was established for a time from 1974 until the early 1980s at Fife, in Scotland, but it was maintained only by rigorous protection of the larvae from birds.[17]

Being a weak flyer and non-migratory, capable of spreading only by dispersal when populations are at a peak, it is unlikely that the Black-veined White can independently cross the barrier of the English Channel in sufficient numbers to re-establish itself. However a specimen was reported from Sussex on 15th July 1996, the year of the great Painted Lady invasion, which may have been a genuine vagrant.[18]

13. THE BATH WHITE – *Pontia daplidice* (Linnaeus)
Papilio (*Danaus*) *daplidice* Linnaeus, 1758

Although it has been stated that William Vernon captured the first British Bath White in May 1702, a few weeks after Queen Anne ascended the throne,[1, 2] there is some doubt about this date. Vernon's specimen was said to have been taken at Gamlingay, in Cambridgeshire. There is a specimen of the Bath White in the Dale Collection at the Hope Department in Oxford which may be that taken by Vernon but, if so, he captured it several years earlier. In the fourth part of Petiver's *Musei Petiveriani Centuria* published in 1699, Petiver described the Bath White as '*Papilio leucomelanus, subtus viridescens marmoreus*' [black and white butterfly with the underside marbled green] (Fig. 133) and called it 'the greenish marbled half-Mourner'.[3] Petiver knew of only one taken in England. In the first 'decade' of his *Gazophylacium*, written three years later, which referred to it as 'Vernon's half-Mourner', he noted that 'I do not know of any one that has met with this in England but Mr. Will. Vernon about Cambridge, and there very rare'.[4] This would suggest that Vernon had taken his Bath White earlier than 1699, and that the specimen was subsequently acquired by Petiver and was later incorporated into the Dale Collection. It was figured by E. B. Ford.[5] Another specimen taken some time before 1717 at Hampstead was a male, which Petiver figures as 'the slight greenish half-mourner', believing it to be a different species. Further specimens were taken by Vernon and by another early collector, his protegé Robert Antrobus, one being given to John Ray who wrote that he 'had it from Vernon who took it in Cambridgeshire'. Perhaps, then, the Bath White was temporarily established near Cambridge at the end of the seventeenth century.

William Lewin, who first figured the Bath White under that name, derived the latter from 'a piece of needlework executed at Bath by a young lady, from a specimen of this insect – said to have been taken near that place'. 'On my examining the insects purchased by J. J. Swainson Esq., at the sale [in 1786] of the late Duchess Dowager of Portland's subjects in Natural History', he wrote, 'I found this insect mixed with the female Orange Tip; and it then appeared to me that some person collected this box of butterflies, and sent them to the Duchess, and from the great resemblance of this to the female Orange Tip, the difference of this rare species passed without being noticed'.[6] Lewin's comments led Donovan to assume, wrongly, that the butterfly was to be found only in the neighbourhood of Bath.[7] Despite that, Lewin's name has endured, although Haworth's name, the Green Chequered White,[8] was used by several nineteenth-century authorities including Jermyn,[9] Stephens,[10] and Newman.[11] Rennie,[12] always idiosyncratic, named it The Rocket after its supposed foodplant, whilst Morris[13] called it simply the Chequered White.

The Bath White is common in the Mediterranean region and has strong northward migratory tendencies, regularly reaching northern France, the Netherlands and other parts of northern continental Europe. However it is a great rarity in this country. Williams produced a table showing the number recorded in every year from 1850–1955.[14] During the first fifty years there were 123 with a peak of thirty-five in 1872 and, with the exception of two outstanding years in 1906 and 1945, no more than about eighty specimens were reported between 1872 and 1935 (Plate 21).

The first of the two exceptional years was reported by Frohawk.[15] W. W. Collins had seen a 'swarm' of Bath Whites in 1906 on the Dorset coast which

Papiliones *BRITANNIÆ* Luteæ, Albæ &c Mixtæ.

ENGLISH Yellow, White & *other mixt* *BUTTERFLIES.*

Frohawk thought might have been the progeny of an immigrant female in the spring though, in the light of the July 1945 invasion, they may all have been first generation immigrants. Collins's own account ran as follows: 'I captured the Bath Whites in 1906 at the latter end of July, or early in August, on the Dorset cliffs, west of Durdle Door, which is west of Lulworth Cove. I cannot tell the exact day, but I was in camp west of Lulworth, not far from the edge of the cliffs. I went off with my net one morning and saw what I took to be a large hatch out of *P. daplidice*; it is not possible to say how many – the best of my recollection a couple of hundred or probably more, as they were hovering about on the upper part of the cliff.' Frohawk's 'reliable informant', the Revd F. L. Blathwayte, had been shown four of the specimens (two males and two females) taken by Collins.

Fig. 133. *(facing page)* Petiver's descriptions in 1717 of the Bath White show that he considered the two sexes to be different species – the male (figure 8) being 'the *slight* greenish *Half-Mourner*' and the female (figure 9) '*Vernouns* greenish *Half-mourner*'.

The better-attested mass migration of Bath Whites in 1945 was graphically described by H. B. D. Kettlewell, who was on holiday in South Cornwall at the time. Quoting verbatim from his diary, he wrote:

> 'July 14: Fine and sunny. At 9 a.m. (G.M.T.) I went round the hotel gardens to look for larvae ... I noticed an insect fly out of a herb bed which I at first thought was *Abraxas grossulariata* [the Magpie Moth]. It came to rest in some parsnips, where I procured it by picking it up between my finger and thumb. To my amazement it was a newly hatched female *P. daplidice*. I hastily returned to the hotel to get my net, and whilst running through the garden I saw first a *hyale* and then another *daplidice*. On returning with my net I caught the *hyale*, which was a fine male. A quarter of an hour later I caught my second *daplidice*. It was somewhat worn, but the normal green pattern of the underside was entirely replaced with grey-black. It was a female.'[16]

He then crossed the road to some allotments. '*Pontia daplidice* was flying freely here, and I rapidly had two in the net together. Altogether I took 37 during the course of the day, and could have taken several more but for the fact that I spent a great deal of time following the females, endeavouring to find on what they were ovipositing. On four occasions I had two in the net together, including one pair which were flying *in copula* about 1 p.m. (G.M.T.). The majority were newly hatched and in excellent condition ... On at least three occasions I watched a female quivering over low weeds ... but on no occasion could I find an egg.' The weather over the next two days was much less favourable. However, on the 17th July, Kettlewell 'was lucky enough to see at last a female *daplidice* in the act of ovipositing. She lit on a flowerhead of the Common Hedge Mustard (*Sisymbrium officinale*). Her abdomen flexed itself into the unopened flower-buds. The egg was found.' In addition to these rarities, he also caught a fine male Short-tailed Blue. 'At this stage', he wrote, 'an awful feeling of unreality came over me'. He summarized his experience thus: 'In my opinion there must have been an extensive migration in the spring at a date much earlier than usual. This migration must be in fact unique. At the

Plate 21
British specimens of the Bath White from Kent and the Channel Islands.

present moment there must be thousands of ova of *P. daplidice* in this one area of Cornwall alone laid on *Sisymbrium officinale.* The ultimate survival of these will depend entirely on the weather of the next few weeks.' Kettlewell captured a total of 54 specimens between 14th and 17th July.

The same immigration of Bath Whites was reported by C. S. H. Blathwayt,[17] who captured 38 specimens in a field near Looe, on the south Cornish coast, out of an estimated total of some 200 present around midday on 14th July, and a further 5 two days later made a total of 43, of which 25 were males and 18 females.

During that summer when Bath Whites were locally abundant in Cornwall and the Scillies; they were also recorded in sizeable numbers all along the south coast as far as Kent. They even reached Co. Kerry in Ireland. Above average numbers in the following couple of years (13 in 1946, 8 in 1947) would suggest that the butterfly had bred successfully, and that a few survived the winter as chrysalids to emerge the next year.

During the whole of the second half of the twentieth century, it has been among our rarest migrant butterflies: annual numbers have seldom exceeded two and in most years there are none reported. The most recent is of an unconfirmed singleton in Somerset in 1998.[18]

14. THE BLACK HAIRSTREAK – *Satyrium pruni* (Linnaeus)
Papilio (Plebejus) pruni Linnaeus, 1758

Plate 22 (*facing page*) An illustration by H. N. Humphreys of the recently discovered Black Hairstreak (figures 6–10) together with the Brown Hairstreak (figures 1–5), from Plate 25 of his collaborative work with J. O. Westwood, *British Butterflies and their Transformations* (1841).

Most of the native British butterfly species have been known for over two hundred years, only five having been discovered since the beginning of the nineteenth century. One of these is the Black Hairstreak, which was not known in Britain until 1828 (Plate 22). Initially, there was much confusion between this species and the White-letter Hairstreak which was known to Petiver[1] and Ray.[2] However, their descriptions were taken by Linnaeus[3] to refer to his *Papilio (Plebejus) pruni* for which he correctly gave the foodplant as *Prunus domesticus*, the Plum. Linnaeus's *Papilio pruni* seems to be a composite species with the description of the adult butterfly fitting the White-letter Hairstreak, at that time called simply The Hair-streak by English entomologists, and the foodplant matching that of the Black Hairstreak. The foodplant of the White-letter Hairstreak is elm.

Moses Harris, in the second edition of *The Aurelian*,[4] also described the White-letter Hairstreak, giving it a new name, the Dark Hairstreak, while adopting the Linnaean scientific name *pruni.* In 1808 Donovan, while retaining the name *pruni*, changed the vernacular name to Black Hairstreak.[5] Thus a singular situation arose whereby both the scientific and English names for the then unknown Black Hairstreak were being applied to the White-letter Hairstreak! Jermyn, for example, basing her account on Linnaeus, called the latter *Papilio pruni*, The Black Hairstreak,[6] and gave its foodplant as the 'Plumb Tree'; clearly she had no experience of the early stages of the White-letter Hairstreak.

It was Edward Newman who, at the age of twenty-seven, discovered the genuine Black Hairstreak. The following year, in 1829, Curtis published the discovery in his *British Entomology*,[7] but mistakenly transferred to it the English and scientific names then being applied in England to the White-letter Hairstreak, leaving that species without any name at all, and leading to over seventy years of confusion! In 1841, Westwood followed Curtis in calling the new British species the Black Hairstreak, and for our White-letter Hairstreak he

Pl. 25.

proposed a new English name – the 'w-Hairstreak', at the same time adopting its correct scientific name *w-album* under which it had been described in 1782 by A. W. Knoch.[8] Morris likewise called it the Black Hairstreak, dubbing the White-letter the 'White W-Hairstreak' (he meant the 'White-W Hairstreak' but misplaced the hyphen).[9] Wood,[10] Stephens[11] and Newman[12] renamed it the Dark Hairstreak and returned the name Black to the White-letter Hairstreak. It was not until the turn of the twentieth century that Kirby[13] and South[14] restored to the two species the names by which they are still known today. Great caution is clearly needed when determining which species was intended in all of the nineteenth-century records.

The story of his discovery in Britain of the 'true' Black Hairstreak was recounted by Newman himself in his *Illustrated Natural History of British Butterflies*:

> 'In September, 1828, a member of the Entomological Club purchased a number of these butterflies of a Mr Seaman, then a well-known dealer in objects of natural history, and resident at Ipswich. The purchase was made under the impression that the butterflies were the Black Hairstreak [i.e. our White-letter Hairstreak (*Satyrium w-album*)], then a desirable insect to obtain. The purchaser most kindly distributed among the members of the club, then in its infancy, some of the specimens, and I became a recipient. The specimens are still in my possession. On examining them, and comparing them with such specimens of *W-album* as I then possessed, it immediately became manifest to me that the newly-captured species was essentially different, not only on the upper, but also on the under side. Of course, I made it my business to work out the name of the supposed novelty, and soon found, on comparing it with a continental figure, that the new insect was *Thecla Pruni*, and the old one *Thecla W-album*. Like all beginners, I was proud of my discovery, and eager to communicate the intelligence. The late Mr J. F. Stephens then received entomologists every Wednesday evening, and in the most kind and generous manner opened his rich cabinets, and imparted his great entomological knowledge to every one who applied for information. The first Wednesday evening subsequent to my making the discovery found me at the residence of this patriotic entomologist: that night I was his earliest and most enthusiastic visitor. He gave the subject an immediate investigation, and promptly acquiesced in the necessity for a change of name, but at the same time threw a damper over my enthusiasm at the supposed discovery of a new British butterfly, by expressing a doubt whether it was British at all; and represented the intense and praiseworthy desire to do business which was prevalent among dealers in insects as occasionally overcoming the love of truth. Seaman, unconscious of the value of his capture, had given the real and now familiar locality of Monk's Wood as its habitat, and it was so announced when he first disposed of them; but no sooner was it made known that the butterflies were not the [former] Black Hairstreak at all, but a species new to Britain, then the locality became a mine of gold; and Mr Seaman very judiciously concluded to remove the mine to a greater distance, even to *ultima thule* of his geographical knowledge, Yorkshire; and Mr Curtis, who published the insect

under its correct name shortly afterwards, gave Yorkshire as the locality where it had been found.'

Attempts by the dealer, Seaman, to conceal the true habitat soon came to nought when Professor Charles Babington of Cambridge traced the original locality, Monks Wood. Within a few years the *new* Black Hairstreak had been found in a number of other Midland localities. The Revd William Bree found it in blackthorn thickets at Barnwell Wold, Northamptonshire.[15] Herbert Goss, who bred it from stock obtained from there, told friends that 'It was the greatest possible pleasure to see them walking about the table while I was at breakfast'.[16]

This species has always been restricted to the Oxford clay between Oxford and Peterborough, where it inhabits old blackthorn thickets in the vicinity of ancient woodland. According to Thomas,[17] who has made a special study of this species, a number of the Northamptonshire and Buckinghamshire colonies were probably established by Lord Rothschild, who had paid H. A. Leeds to obtain 'large numbers' for release between *c.*1900 and 1917. One of the best-known sites, in Bernwood Forest in Oxfordshire, was not discovered until 1918 when W. F. Burrows found it at Hell Coppice.

The extreme localization of the Black Hairstreak is one of the mysteries of British butterfly distribution, since its foodplant is widespread in woodland and hedgerows. Thomas considers it likely that the limited distribution of the Black Hairstreak in the east Midlands is a reflection of the history of woodland management. Most woods in southern England were cut regularly on a coppice cycle to provide firewood and small-gauge timber. Within the range of the Black Hairstreak, however, many woods were Royal Forests, owned by the Crown and managed less intensively. The growth of dense, economically worthless blackthorn thickets along the rides and wood-edges created ideal conditions for this rare butterfly that have persisted, at least in part, to the present day. Descendants of a colony established at Cranleigh, Surrey, in 1952 by A. E. Collier continued to thrive, despite tree-felling at the original site.[18] Collier apparently released no more than a dozen specimens, but their progeny spread, founding five separate colonies in neighbouring woods, one of which was said by Thomas, to contain more Black Hairstreaks than any other wood in Britain. G. A. Collins reported, however, that it has not been seen in Surrey in recent years,[19] though the species is elusive and is easy to overlook.

The Black Hairstreak, so prized by the early nineteenth-century entomologists, is perhaps one of the least threatened rare butterflies in Britain today and, indeed, 'has fared much better than most butterflies in the past 50 years'.[17]

15–17. THE COPPERS (*Lycaena* spp.)

Of the six species of Copper butterflies that have been recorded in Britain only two have been generally thought to be native, and one of those, the Large Copper, has been extinct for 150 years. However, there are two more, the Scarce Copper (*Lycaena virgaureae*) and the Purple-edged Copper (*L. hippothoe*), which some lepidopterists have considered also to have been indigenous, though now extinct, species. Not all have agreed. South mentioned them in his popular book on butterflies but only to assert that 'there does not appear to be the least reason for considering either of them to be a British butterfly'.[1] On the other hand Kirby, Barrett and others thought it possible that they 'may have existed here at some period and since become extinct'.[2,3] The status of the Scarce and Purple-edged Coppers has been the subject of much speculation and controversy since the early part of the nineteenth century, the picture having been obscured by confusion over nomenclature as well as by the activities of dealers who, aware of the interest in them, imported large numbers of specimens from the Continent. At this distance in time it may never be possible to establish their former status with certainty but, in the light of modern studies and research into the origins of the British Lepidoptera, and by sifting through all the evidence (some of which has emerged only in recent years), it does seem possible that the Scarce Copper, at least, was at one time native to Britain.

The following account of three Copper species describes the taxonomic confusion behind the early records of British Copper butterflies, and some of the evidence for believing that the Scarce and Purple-edged Coppers may once have occurred here. Readers may draw their own conclusions!

15. THE LARGE COPPER – *Lycaena dispar* (Haworth)
Papilio dispar Haworth, 1803

The existence of this species in Britain was known for some years before its first description was published by Haworth in 1803. Up to that time, as it had not been named, its identity puzzled the few naturalists who had seen it, and they naturally tried to relate it to descriptions in the works of Linnaeus, Fabricius and Scopoli. Unsurprisingly they had no success. Lewin, who was the first to figure the Large Copper in print in 1795,[1] called it *hippothoe*, the scientific name Linnaeus had given to the Purple-edged Copper. Lewin's mistake was then repeated by Donovan, who called it the Great Copper.[2] Newman continued to use the Linnaean name *hippothoe*, and included the name *dispar* only as a synonym.[3]

Fig. 134a. Henry Seymer (1745–1800), well-connected landowner and acute observer of Lepidoptera. He knew many leading naturalists of his day and, though never a Fellow, his portrait today hangs in the Council Room of the Linnean Society.

Until comparatively recently it was believed that Lewin's account and illustrations of Large Coppers were the first made of this species but he was, in fact, anticipated by Henry Seymer, nearly twenty years earlier. Lewin had himself noted that 'Some butterflies of this very rare species were met with by a gentleman in Huntingdonshire on a moorish piece of land and afterwards sent to Mr Seymour [*sic*] of Dorsetshire who presented them to the late Duchess Dowager of Portland. They are now in the collection of J. J. Swainson.' This 'Mr Seymour' was in fact Henry Seymer (1714–85) of Hanford House, Dorset (Fig. 134a), a keen collector and talented amateur artist, who was frequently visited by the Duchess, herself one of the outstanding eighteenth-century collectors of natural history items. It is important to be aware that Seymer had a son of the same name with whom he has often been confused. Henry Seymer junior (1745–1800) shared his father's interest in natural history but lacked his father's overriding enthusiasm. He did, however, inherit his father's talent for drawing, and was responsible for colouring the three additional plates of *The Aurelian* published by Harris about 1773, and bound into his father's copy. The year after the death of Henry Seymer senior, his collection of insects was sold, but his books were dispersed only gradually. His copy of Harris's *Aurelian*, which contained extensive annotations as well as excellent original drawings of additional butterflies and moths, expertly fitted in amongst those on the printed plates, came to light only a few years ago.[4] Carefully positioned on the plate with the Small Copper are some fine Seymer drawings of the Large Copper (see Plate 25) with the following note (Fig. 134b): ' ... 'till the year 1776 never seen, or at least taken, in England; this specimen was then taken, with 5

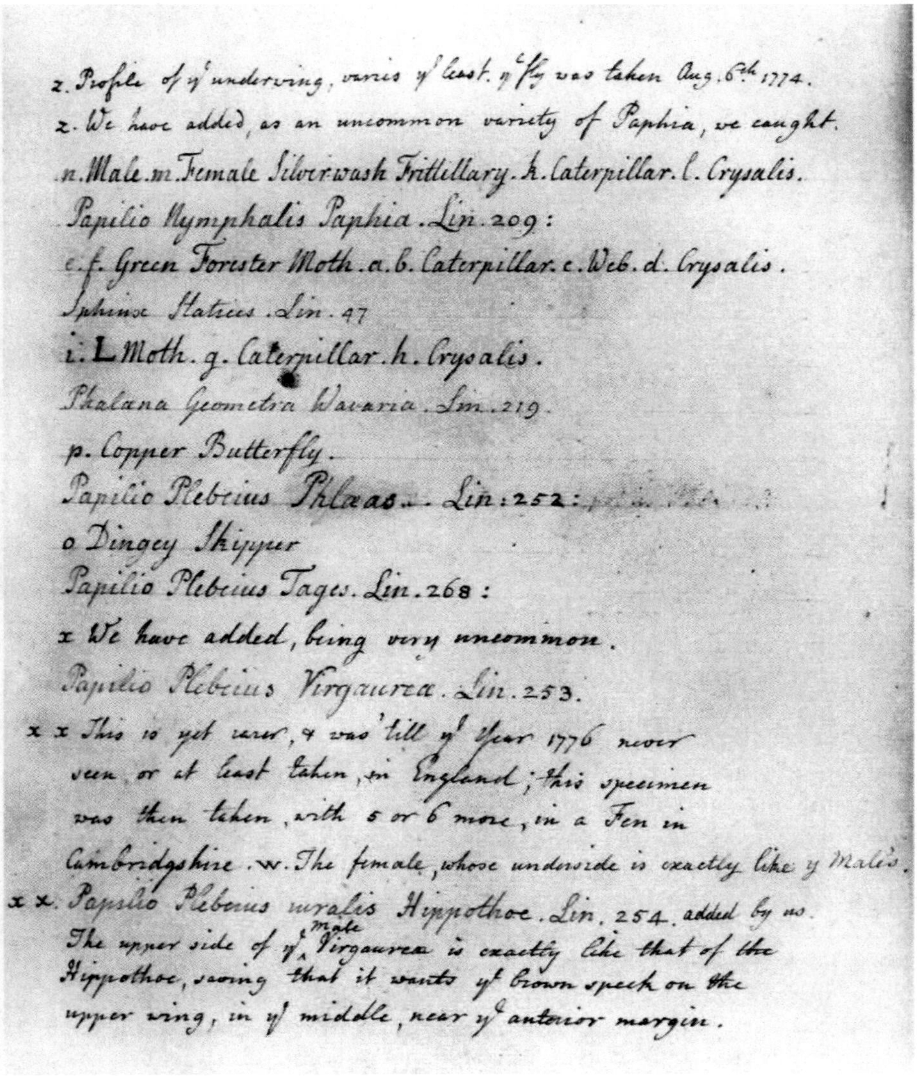

z. Profile of ye underwing, varies ye least. ye fly was taken Aug. 6th 1774.
z. We have added, as an uncommon variety of Paphia, we caught.
n. Male. m. Female Silverwash Fritillary. k. Caterpillar. l. Crysalis.
Papilio Nymphalis Paphia. Lin. 209:
e.f. Green Forester Moth. a.b. Caterpillar. c. Web. d. Crysalis.
Sphinx Statices. Lin. 47
i. L Moth. g. Caterpillar. h. Crysalis.
Phalaena Geometra Wavaria. Lin. 219.
p. Copper Butterfly.
Papilio Plebeius Phlaeas. Lin: 252:
o Dingey Skipper
Papilio Plebeius Tages. Lin. 268:
x We have added, being very uncommon.
Papilio Plebeius Virgaureae. Lin. 253.
x x This is yet rarer, & was 'till ye year 1776 never seen, or at least taken, in England; this specimen was then taken, with 5 or 6 more, in a Fen in Cambridgeshire. w. The female, whose underside is exactly like ye Male's.
x x. Papilio Plebeius ruralis Hippothoe. Lin. 254. added by us.
The upper side of ye male Virgaurea is exactly like that of the Hippothoe, saving that it wants ye brown speck on the upper wing, in ye middle, near ye anterior margin.

Fig. 134b. The key and commentary, in Henry Seymer's hand, for some of the figures of the butterflies and moths which he added to his copy of *The Aurelian* (see Plate 25): '*x*' refers to his drawing of the Scarce Copper, '*xx*' and '*w*' to those of the Large Copper, male and female. Until Haworth described it in 1803 and named it *dispar*, Seymer and others believed the Large Copper to be the Continental *hippothoe.*

or 6 more, in a Fen in Cambridgeshire. *w.* The female whose underside is exactly like ye male's. *xx Papilio Plebeius ruralis Hippothoe.* Lin. 254. added by us', a reference to Linnaeus's catalogue number in the revised 12th edition of his *Systema Naturae* of 1767. Seymer had drawn three figures of this species: male and female upperside and male underside, cross-referred to his notes, like Harris's, by means of key letters. He also commented on the difficulty he experienced in identifying his specimens from the literature to which he had access. It seems from this that Lewin's specimens had not after all been taken in Huntingdon but in Cambridgeshire, and that the year of their capture was 1776.

Amazingly, the Large Copper had, in fact, been noticed, described and illustrated earlier still, though no lepidopterist was aware of it until about twenty years ago. In 1982, E. J. Redshaw, a member of the Spalding Gentlemen's Society, came across an entry for 28th September 1749 in the Society's

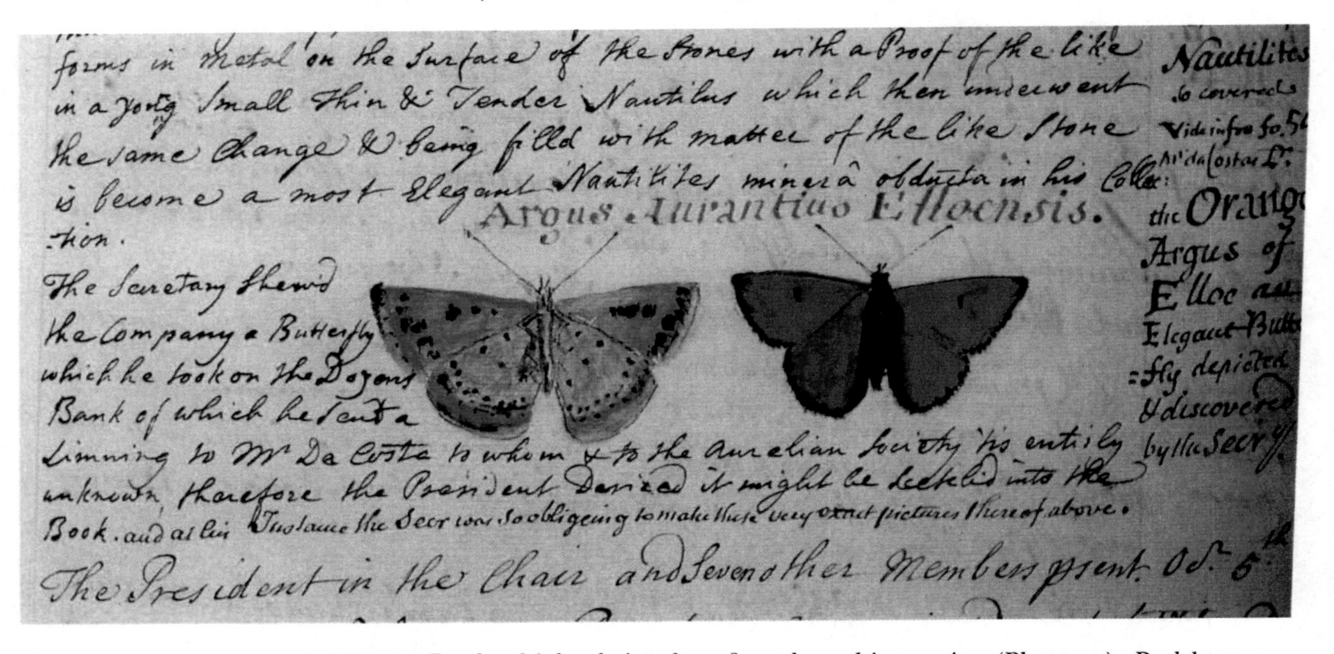
forms in Metal on the Surface of the Stones with a Proof of the like
in a young Small Thin & Tender Nautilus which then underwent
the same Change & being filld with matter of the like Stone
is become a most Elegant Nautilites minerâ obducta in his Colle-
-tion.

Argus Aurantius Elloensis.

The Secretary Shew'd
the Company a Butterfly
which he took on the Dozens
Bank of which he sent a
Limning to Mr Da Costa to whom & to the Aurelian Society 'tis entirly
unknown, therefore the President Desired it might be Scetched into the
Book. and at his Instance the Secr was so obligeing to make these very exact pictures thereof above.
The President in the Chair and Seven other Members present.

Nautilites

the Orange Argus of Elloe an Elegant Butterfly depicted & discovered by the Secr'y.

Plate 23
This, the earliest known illustration of the Large Copper, named the 'Orange Argus of Elloe', was discovered about twenty years ago by E. J. Redshaw in the Minute Book of the Spalding Gentlemen's Society for 1749. The Society, along with its historic Minute Books, survives.

Minute Book which obviously referred to this species (Plate 23). Redshaw reported the discovery to the authors of a forthcoming work on Lincolnshire butterflies,[5] and himself published a short account of his discovery.[6] Then in 1983, John Heath published the entry in full in the *Entomologist's Gazette* with a good reproduction of the watercolour drawing that accompanied it. It reads as follows: '*Argus Aurantius Elloensis.* – The Orange Argus of Elloe, an Elegant Butterfly depicted & discovered by the Secr'y [Dr John Green] – The Secretary Shew'd the Company a Butterfly which he took on the Dozen's Bank of which he sent a limning [watercolour drawing] to Mr Da Costa to whom and to the Aurelian Society 'tis entirly unknown, therefore the President [Mr Maurice Johnson] Desired it might be Scetch'd into the Book and at his Instance the Secr was so obligeing to make these very exact pictures thereof above.'[7] Elloe was the mediaeval adminstrative Wapentake or Hundred which covered what is now known as South Holland in south Lincolnshire. West Pinchbeck, where the Large Copper was found, lies in the western half of the Hundred of Elloe, now known as West Elloe. The centre or meeting point of the ancient Hundred was at the Elloe Stone which still exists on the boundary of the Moulton and Whaplode parishes. The Dozen's Bank, in the parish of Pinchbeck, formed the eastern boundary of the old Counter Drain Washes, which were re-aligned in 1775 when the habitat was presumably destroyed. This account predates the one given by Lewin by some forty-six years and that of Seymer by twenty-seven. E. M. Da Costa, a member of the Spalding Gentlemen's Society, was a prominent naturalist of his day, being a Fellow of the Royal Society and a member of the Aurelian Society. It is, therefore, most surprising that Harris, as Secretary of the latter, was unaware of this discovery.

Adrian Haworth seems to have learnt of the Large Copper from his friends William and Fenwick Skrimshire, who were travelling to Ely in a gig when they chanced to notice a number of unusual 'Copper' butterflies by the roadside.[8] On the return journey they stopped to examine them and confirm their earlier impression that they were an unknown species. There is little doubt that Fenwick Skrimshire would have recognized these butterflies if they had been already known. He was a distinguished naturalist, being President of the Natural History Society of Edinburgh; he was also the author in 1805 of *A Series of Essays introductory to the Study of Natural History.*

The English Large Copper is a larger, more deeply coloured and spectacular butterfly than the two Continental subspecies, *L. dispar batavus* Oberthur (from the Netherlands) and *L. dispar rutilus* Werneburg (from Germany), neither of which were described until much later. In England the Large Copper was locally common and flourished in a number of fens until about the middle of the nineteenth century in Cambridgeshire, Huntingdonshire, Lincolnshire and Norfolk (including the Norfolk Broads) (Plate 24). C. W. Dale recorded that in 1827 Haworth took fifty specimens at Bardolph Fen, Norfolk, in a single day.[9] In the same year Lætitia Jermyn reported it from Benacre in Suffolk,[10] a locality known also to Curtis.[11] An old specimen, set as an underside and simply labelled 'Benacre, Suffolk' with no date, was seen by Bretherton but the label was 'in modern handwriting'.[12] Dale thought that the last five specimens to be taken in Britain were those caught by Mr Stretton at Holme Fen, Huntingdonshire in either 1847 or 1848, but it had also been claimed that the last was taken at Bottisham Fen in Cambridgeshire by a Mr Wagstaff in 1851.[13] However, I have a female specimen, formerly in the Purefoy collection, taken from Holme Fen in 1860, indicating that the butterfly survived there a little longer. The last known specimen from East Anglia was taken from the Norfolk Broads in 1864[14] and that is also almost certainly the last record of a Large Copper taken in England.

Outside East Anglia there have been reports of Large Coppers from near Monmouth, and from the Somerset Levels, although their validity is open to question. The evidence for the Large Coppers in Monmouthshire was recalled by Joseph Merrin, the editor of a local newspaper who, in 1855, had set off with colleagues on their annual 'wayzgoose' or printers' outing:

> 'We called upon Mr. Robert Biddle, of Monmouth, a friend of one of the party. He had a large case of butterflies and moths hanging up, which he had taken, and I was much struck with four specimens of *C.* [*Chrysophanus* (= *Lycaena*)] *dispar* occupying a central position among them. I had then only recently begun to collect. On my drawing his attention to them he said he took them some time previously on the slopes of Doward Hill, bordering the river Wye, not far from Monmouth. He seemed to set no great value upon them. My great admiration of them appeared to interest him, and I was delighted the next day on opening a small packet brought to the *Gloucester Journal* office by the Monmouth coach (there was then no railway) to find two specimens of the *C. dispar* I had admired, which Mr Biddle said, in a short note, he was pleased to present to me. The appearance of these specimens with their "poker pins" and slightly damaged antennae, and the circumstances under which they were given to me, leave no doubt of their British origin.'[15]

A year or two later Merrin went back. 'I reconnoitred the locality as far as I was able, and I saw much marsh land bordering the Wye, but quite

Plate 24 (*facing page*) Specimens of the English Large Copper taken between 1820 and 1860. Those with data are from Holme Fen, Yaxley Fen and Whittlesea Mere in what was then Huntingdonshire, and the one without data was from Cambridgeshire.

unsearchable unless shod with jack boots. In the hopes of getting a better glimpse of the lower slopes of the hill, I rang the bell at the residential gate, but was politely told that as the family was away a stranger could not be allowed to examine the grounds, and I had to leave with regret, 'neath a broiling sun, what seemed classic ground, and sought refuge in the shady streets of old Monmouth.'

The only evidence for Large Coppers on Anglesey is the publication in May 1896, in the weekly *Exchange and Mart*, of an advertisement for 'Butterflies – Three large English Coppers, alive, splendid specimens; also two Camberwell Beauties, alive. What offers, cash?' Later that year the man who had placed the advertisement, J. W. Tattersall, claimed the Large Coppers had been taken 'in the neighbourhood of Beaumaris, North Wales', and that he had sold them 'to a gentleman who refused to give his name or address'.[16] No one else mentions Large Coppers on Anglesey, and these specimens were presumably imported from the Continent, though fenland survives in parts of Anglesey.

The Somerset records are more credible. There are five specimens of the Large Copper from Somerset, together with a single Scarce Copper (*L. virgaureae*), in the Somerset County Museum, Taunton. Though without labels, they were formerly believed to have been taken by E. J. and J. T. Quekett, the famous microscopists, some time during the early part of the nineteenth century. The specimens were presented to the Museum in 1876–77, and are accompanied by the following pencilled label – 'Possibly the specimens of the Large and Purple-edged Copper [the latter misidentified] caught by the Quekett brothers at Langport.' R. G. Sutton,[17] however, has questioned this source and suggested that they were probably caught by a local collector by the name of John Woodland, having found a manuscript note in the hand of a nineteenth century curator stating as follows: 'About the year 1864, Mr. Woodland gave me a small collection of butterflies taken near Langport early in the century; among them were two or three *L. dispar* which he told me were taken by himself. In his early days he had taken care of them, but he got old and neglected them, so that when they came to me they were dilapidated.' Langport lies on the edge of the Somerset levels, then a vast and only partially drained expanse of wet meadows, marshland and peat cuttings in which Great Water Dock (*Rumex hydrolapathum*) is frequent. Tutt expressed the sentiments of all collectors concerning the Langport Large Coppers: 'One would like more authentic information'.[18]

In 1857, an arresting note by W. D. Crotch appeared in the *Entomologist's Weekly Intelligencer* under the title 'Doings in the West',[19] in which he described a collecting trip to Weston-super-Mare and Brean Down, on the opposite side of the Somerset Levels, some 20 miles north of Langport: '*C. dispar* fell ignobly, slain by the hat of a friend, who kindly made the spoil over to me, in utter ignorance of its rarity, and I much regret that my absence from the locality prevents a search, which, if one may trust the aborigines, would have had a fair chance of success'. This discovery, at a time when *L. dispar* was already thought to be extinct, or very nearly so, should have prompted further expeditions. Crotch explained why he did not search for it again. 'We gave up hopes of *C. dispar*, &c., because the time for *T. betulae* [Brown Hairstreak] was drawing near, and we were in many instances pledged to our numerous correspondents of last year to send them, when possible, better specimens.' While Brean Down itself is a dry limestone headland, there is still marshy ground behind the sea walls south of Weston-super-Mare and especially on the nearby Levels.

Ex Newman Jun 1955.
No data.
Newman 5.12.54.
No data.
Cambridgeshire
1814
data incomplete
28.11.51 Lot 57.
One of three for £5.75.
H.H. Colln.
Yaxley 1865.
21.1.53 Lot 52. £1.70
Ex Rev. H. Burney sale
Nov. 1893.
25.1.62 Lot 228. £5.50
Ex Standish colln.
Holme Fen 1860,
23.10.58 Lot 192. £6.25
Holme Fen 1860. Ex Fam collection.
10.4.58 Lot 61. £8
Ex Stevens sale 29.4.1891
Ex Newman 5.12.54.
F.D. Wheeler colln.
28.11.51 Lot 57.
One of three for £5.75.
Ex Nevison colln.
25.1.62 Lot 229. £5.50
From Milford Collection.
Lot 79.
10.2.55 Lot 232. £4.75
Ex colln.
J.A. Clark.
From the Collection of
Charles Felix Palmer, Esq.
Taken about 1840.
Bought at Stevens
10.2.55 Lot 233. £4
Bought at Stevens.

In the same article Mr Crotch also sounded off a diatribe against over-collecting:

> 'And I must here beg pardon for remarking that I do not propose to give localities, within some twenty miles, thus publicly: an advertisement [see Fig. 143] for the purchase of twenty gross of *P.* [*Plebejus*] *artaxerxes* [the Castle Eden or Northern Brown Argus] leads me to this determination: twenty gross! may they haunt the advertiser's slumbers with their incessant wings, and prick the conscience of the pander who may collect them with twenty gross of entomological pins No. 1! For *Thecla W-album* [White-letter Hairstreak] and *Betulae* and I believe and hope *C. dispar*, I possess new localities, which I shall be happy to communicate to every *true* Lepidopterist – every one who would think it a crime, equal at least to the murder of the Innocents, to *clear* a locality, be his wants ever so great.'

There is yet another report of West Country Large Coppers that deserves attention. According to W. S. M. D'Urban, Mr Wentworth Buller had exhibited a specimen of *C. dispar* at a meeting of the Exeter Naturalist's Club.[20] Apparently it had been 'picked up dead among sedges at Slapton Lea'. There are no other details but, as Slapton is one of the few places in Devon where the larval foodplant, the Great Water-Dock, grows, it is just possible that the Large Copper may have existed there.

Further sightings of 'Large Coppers' in the West Country have also surfaced within the last fifty years. However, Allan considered that these were not *dispar* but *virgaureae* (the Scarce Copper) and so are discussed later under that species.[21]

The extinction of the Large Copper was a great loss to British entomology. The possible causes of the decline and extinction were widely debated. Did collecting help to spur *L. dispar* along to its untimely end? P. B. M. Allan certainly thought so: 'Intensive collecting will in time exterminate any animal, no matter how profuse its numbers may be. And the collecting of the Large Copper was intensive indeed.' He concluded 'It was the putting of a price on its head that exterminated *dispar*'.[22] Support for his assertion came in a note to *The Entomologist* which suggested that Benjamin Standish, a London dealer, might have overdone it.[23] Standish had apparently taken a painting of the Large Copper to the fens and shown it to a number of people. An inn-keeper divulged the locality to him and, within a short time, local people realized that they could profit to the tune of two shillings for every Large Copper that they caught. The person that supplied this information many years later, H. J. Harding, was himself a dealer in Coppers. He had raided the Fens a few years later, buying two dozen larvae for ninepence, from which he bred fine specimens which he sold for one shilling each. However, it is very unlikely that collectors could originally have reached all but a portion of their watery habitat before large scale drainage schemes were undertaken.

Drainage of the fens was one of the more frequently advanced explanations for the butterfly's decline and almost certainly the principal cause. In a letter to the Revd William Bree in the first quarter of the nineteenth century, Haworth wrote that, whereas the Large Copper can withstand winter floods and inundations, the destruction of its habitat is fatal to its survival. Referring to a favourite collecting ground which lay under water for a long time during the winter, he commented that the following summer the butterflies were as plentiful as before, despite anxieties expressed by entomologists. However, when the same land was burnt with a view to clearing it for agriculture, the Coppers were no longer to be seen. In 1851, Whittlesey Mere was drained and, as C. W. Dale wrote, 'what was once the home of many a rare bird and insect became first a dry surface of hardened mud, cracked by the sun's heat into multitudinous fissures, and now scarce yields to any land in England, in the weight of its golden harvest'.[24]

No piece of evidence is more telling than that, when the old Fens were successfully drained by steam-driven pumps in the mid-nineteenth century, the Large Copper immediately disappeared, as did other fenland Lepidoptera like the Reed Tussock (*Laelia coenosa*) and the Gypsy Moth (*Lymantria dispar*). Only fragments of the Fens remained, probably too small to hold viable colonies of the butterfly. In the face of such an onslaught there is no need to invoke the spectres of the collector and the dealer. Collecting on a large scale certainly did not help the species but in the face of such wholesale habitat destruction the main reason is depressingly obvious.

There have been a number of attempts at re-establishing the Large Copper in Britain and establishing it in Ireland. In 1909 the subspecies *rutilus*, collected from around Berlin, was unsuccessfully introduced to Wicken Fen, Cambridgeshire by G. H. Verrall.[25] In 1914, specimens from the same area were introduced to a specially prepared ten-acre snipe bog at Greenfields, near Cappawhite, Co. Tipperary, by Captain E. B. Purefoy. In 1926, Purefoy attempted to introduce the butterfly to Woodbastwick marshes in the Norfolk Broads, using stock bred at Greenfields, but he was no more successful than Verrall. By this time Purefoy's Irish colony was also failing (it died out in about 1936) and to obtain new stock he sent his gardener, J. Schofield, to Holland to search for the newly discovered Dutch subspecies *batavus*, much closer in size and markings to the lost English Large Copper. 'Schofield', wrote Lavery,[26] 'searched countless acres of the Friesland marshes for the butterfly's eggs, walking all day long in chin-high sedge and trudging through the difficult marshy terrain'. He brought back over 100 ova from which a colony of the Dutch subspecies was established at Greenfields, which persisted there until about 1949 with the help of reinforcement around 1943–44 by Purefoy of stock from England. In 1926, part of Wood Walton Fen in Huntingdonshire was prepared for an introduction and a colony was established there which survived with much human help until 1969. Lavery commented that: '*Lycaena dispar batavus* owed its existence [in Britain] to the foresight and the excellent work done by E. B. Purefoy and also to J. Schofield, ... flung in at the deep end, jumping from Holland to Greenfields and over to Huntingdonshire, with no experience or training in entomology!' Subsequent releases at Wood Walton Fen have established small colonies for a short period, but the population became extinct again in 1994. Future attempts to re-establish the species in Britain are likely to be centred on the more extensive fenland habitat of the Norfolk Broads.

16. THE SCARCE or MIDDLE COPPER – *Lycaena virgaureae* (Linnaeus)

Papilio (Plebejus) virgaureae Linnaeus, 1758

The first reference in the British literature to *virgaureae* (the name given to the Scarce Copper by Linnaeus) appeared in J. R. Forster's *A Catalogue of British Insects* (see Fig. 45), printed in Warrington for the author in 1770.[1] A Polish

Plate 25 (*facing page*) This excellent coloured drawing of the underside of the Scarce Copper (*x*), painted into his own copy of *The Aurelian* by Henry Seymer in about 1776, offers perhaps the most convincing evidence that this butterfly once occurred in Britain. On the same plate are his equally remarkable watercolours of the male (*xx*) and female (*w*) of the Large Copper, which had yet to be recognized as a distinct species. The butterflies on Harris's original plate are the Small Copper (*p*), the Dingy Skipper (*o*) and the Silver-washed Fritillary (*m, n*) showing both upper- and underside, to which Seymer added the New Forest form *valesina* (*z*), which he 'caught Aug. 6th 1774'. E. J. C. Esper did not name this form until a few years later.

zoologist of English descent, Forster (see Chapter 3: 16) had arrived in England in 1766 (the year Moses Harris published *The Aurelian*),[2] and settled in Warrington. In his native Poland, the Scarce Copper would have been widespread and he certainly ought to have known it well. In his list of British species, he indicated that, following 'three years assiduous collecting in the neighbourhood of *Warrington*' (formerly Lancashire, now Cheshire), he had taken it 'so plentifully as to be enabled to give some to other collectors'. On the other hand, the Small Copper, included in his British list as *Papilio phlaeas* only on Berkenhout's authority,[3] was rather surprisingly shown as one of the species he lacked in his own collection.

All the Copper butterflies ever recorded as native in Britain, or said to be formerly British, have been bedevilled by confusion over names. The Scarce or Middle Copper is no exception. In the Index to his *Aurelian*,[4] published *c.*1775 in what he called his 'new and compleat edition' and also issued separately as a supplement to the first edition, Moses Harris incorporated Linnaeus's names, but mistakenly used *virgaureae* for his figure of the Small Copper. Could not Forster have likewise given the wrong scientific name to the Small Copper? Though perhaps less common in the eighteenth century, the Small Copper was described by Rennie some sixty years later as 'abundant throughout Britain' and it must surely have existed around Warrington in Forster's day.

None of the early naturalists writing on British butterflies – Merrett in 1666,[5] Petiver in 1699 and 1717,[6, 7] and Ray in 1710[8] – listed more than one species of Copper each, and the identity of Merrett's is open to doubt. Linnaeus, in 1758,[9] similarly described just one species, *virgaureae.* He cited Ray's *Historia Insectorum* as one of his authorities, but Ray's small butterfly 'with wings red and shining like Silk, margined and spotted with black', which he found 'in hedgerows at the end of July', was undoubtedly the Small Copper. It is evident that at this period the muddling of *phlaeas* and *virgaureae* was not unique. The Small Copper was not mentioned by Linnaeus until 1761,[10] and therefore, particularly among those unable to refer to the latest books, confusion between these species was bound to persist.

The next known English mention of *virgaureae* was that made around 1776 by Henry Seymer in his annotated copy of Harris's *Aurelian* (see Fig. 134b). In it Seymer corrected Harris's incorrect name *virgaureae* for the figure of *phlaeas,* and added an illustration of the true *virgaureae,* the Scarce Copper (Plate 25), in addition to those of the Large Copper, which he mistakenly called *hippothoe.* He distinguished between the two larger Copper species when he noted of the Scarce Copper that 'The upper side of ye male is exactly like that of the *Hippothoe,* saving that it wants ye brown speck on the upper wing, in ye middle, near ye anterior margin'. 'We have added', he had earlier observed, 'being very uncommon, *Papilio Plebeius Virgaureae*; Lin. 253.' and, of the Large Copper, that 'This is yet rarer'. Both species must have been illustrated from specimens in front of him. Where did he obtain his *virgaureae*? M. J. Perceval,[11] who discovered Seymer's book, suggests that they might have come from his friend the Duchess of Portland, whose collection, according to the catalogue entry for 'lot 3171' of its sale in 1786, contained among a 'fine series of very rare British Papiliones ... two pair of *Virgaureae*'.

In 1786, Dru Drury, replying to a fellow lepidopterist who had asked whether *virgaureae* was native, and if so where he could obtain it, seemed in no doubt: 'English, I can send it'.[12] William Lewin illustrated it with the comment

PL.XXXIV.
H.S.P.

that: 'In the month of August I once met with two of these butterflies settled on a bank in the marshes ... they were exceeding shy and would not suffer me to approach them'.[13] Lewin and Donovan both figured it under its correct name *Papilio virgaureae*, the latter stating that a specimen had been taken at Cambridge.[14] Haworth published the fact that *dispar* (the *hippothoe* of the early English collectors), *virgaureae* and *phlaeas* were three different insects. He gave the habitat for the Large Copper as '*Arundinetis*' ('reedy fens'); that for the Scarce Copper as '*Paludibus*' ('marshy grounds'); and that for the Small Copper as '*Compascius*' ('grassy commons').[15] It had been considered that a reputed British specimen of the Scarce Copper, given by Haworth to Westwood and still in the Hope Entomological Collection at Oxford, might have been given to him by Forster, as Haworth implied that it had not been taken by himself but, in the light of what is written above, that source seems unlikely.

In 1824, John Curtis gave the Isle of Ely, 'near Huntingdon', and Cambridge as localities for the Scarce Copper[16] and, in the same year, Lætitia Jermyn, who included it under the name of 'Middle Copper' in her *Butterfly Collector's Vade Mecum*, stated that it was an inhabitant of 'Marshes' and that the butterfly was to be found on 'Common Golden Rod' (*Solidago virgaurea*) in the 'Isle of Ely, and Huntingdonshire' at the end of August;[17] she did not mention Cambridge. She gave its wingspan as '1 inch 6 lines' [i.e. 1½ inches or 38mm], as compared with 2 inches [50mm] for the Large Copper. However she was inaccurate in giving the larval foodplant as a grass instead of a species of *Rumex*.

Of subsequent authors, Stephens merely repeated what Curtis had written;[18] Rennie, who called it The Golden Rod, believing this to be one of its foodplants, described its status as rare and local from the Isle of Ely and Huntingdonshire,[19] undoubtedly copying earlier writers. Westwood gave the localities of Cambridgeshire and Huntingdonshire, concluding that 'It is therefore by no means impossible that it may have entered here at one period and since become extinct'.[20] Barrett quoted Haworth in giving its habitat as 'marshes',[21] but Coleman did not include it at all.[22] Barrett also mentioned the existence of three further specimens of the Scarce Copper:[23] one taken by Archdeacon Bree and 'still in his cabinet'; one taken by C. H. Capel Cure at Cromer, Norfolk in August 1868[24] and identified as *virgaureae* at the British Museum; and a third, a male, in the Sparshall Collection at the Norwich Museum, where it remains, without a label. It is known, however, that Bree favoured the inclusion of Continental specimens in British collections to make these complete, so the evidence that he had one proves nothing. There is also the unlabelled specimen in the Somerset County Museum, Taunton, mentioned earlier, dating from the last century.

It is thought that Henry Doubleday omitted the Scarce Copper from his *Synonymic List of British Lepidoptera*[25] because of the activity of notorious dealers such as Plastead and Seaman, who supplied the market with falsely labelled specimens of rarities of Continental origin.[26] Stainton also omitted it from his *Manual of British Butterflies and Moths*, but surprisingly did include the Purple-edged Copper (as *L. chryseis*).[27] However, W. F. Kirby wrote of *virgaureae* in his *European Butterflies* that although 'it has been excluded from our British list ... there seems little reason to doubt that it really inhabits this country, although almost extinct',[28] going on to refer to the 'specimen lately taken at Cromer' already mentioned. Tutt was more sceptical, and considered that 'there can be little doubt that *L. dispar* and *L. virgaureae* refer to the same species'[29] and South was equally downright.[30]

It was E. B. Ford, in his classic work *Butterflies*, who reopened the debate and reviewed the evidence,[31] and he was followed by Allan.[32–34] T. G. Howarth, in his revision of South's Butterflies, thought that it 'may well have occurred in Britain long ago and probably became extinct here in the early part of the eighteenth century'.[35] Dennis agreed that it had once occurred here but advanced the date of extinction to the early nineteenth century.[36] Allan went so far as to argue that it might even have survived into the early part of the twentieth century, citing various reported sightings which in his opinion could only be explained as those of the Scarce Copper, mainly from the West Country. He related how S. G. Castle Russell had described to him a strange event that had occurred some forty years previously 'in the middle of June'. While holidaying in Devon, he and his wife had come to a small place in countryside of exceptional beauty somewhat off the beaten track. The following day his wife and a friend, Mr W. G. Mills, hired a conveyance to return to the spot while Castle Russell remained in the hotel to deal with correspondence. Both were experienced collectors and so their claim to have seen numbers of Large Coppers flying was to be taken seriously. 'We tried to knock them down', said Mrs Castle Russell, 'but they flew too fast for us. Flying in the sunshine they looked most beautiful'. Castle Russell was sceptical at the time, but later regretted he never followed it up having 'preferred to go fishing'. Allan strongly believed that these could have been a colony not of Large Coppers but of the Scarce Copper.[37]

Allan reported a further event told him by another friend 'whose words could be implicitly relied on': 'My wife was driving our car down a single track, a remote lane in the West Country, and I was sitting in the back with one of my sons. I had been watching the butterflies on the grass verge, feeding on the flowers, when all at once, and at close range, I saw a butterfly which I was convinced was a Large Copper. I shouted urgently to my wife to stop, but she replied that we were late already for lunch with friends, and drove on; and so, the occasion was lost.'[38] Once again, then, there was no capture and no follow-up and, as a result, no documented records nor authenticated specimens.

In the same review of the status of the Scarce or Middle Copper, Allan summarized the few records reported from northern Britain: the original locality of Warrington in Lancashire by J. R. Forster in the eighteenth century; in Cumberland by J. B. Hodgkinson in 1858; and in Banff in north-east Scotland by Thomas Edward in August 1857. Edward had referred to the butterfly as '*Lycaena Hippothoe*', the name that was still in use for the Large Copper. Edward was a reputable naturalist, and his Banff sighting was made 'while botanizing over a marshy piece of ground'. On biogeographical grounds, *L. dispar* is unlikely to occur so far north, but both *L. virgaureae* and *L. hippothoe* occur in Scandinavia and the Baltic. Could it have been *L. virgaureae*? Allan clearly thought so, but George Thomson[39] had a quite different interpretation, and considered that Edward was using Linnaeus's name for *L. hippothoe* correctly, that is for the Purple-edged Copper. Unfortunately nothing else is known of this colony, if such it was. An additional plausible report of *virgaureae* in the north was provided more recently by R. Rowley, who had inherited a collection made by the Revd W. Robinson, an upstanding Victorian Unitarian minister of Padiham, near Burnley, most of which he had collected in Lancashire.[40] It included three specimens of the Scarce Copper, two males and a female, all set on contemporary pins. They might have been supplied by a dealer, but could equally well have been collected locally. Coincidentally, Burnley is not far from where Forster claimed his English Scarce Coppers.

On the basis of the documented records and known specimens, it seems at least possible that the Scarce Copper may once have been British, and even that it was widespread, if local, from Devon to northern Britain. However, unlike the Large Copper, it is hard to find a reason why it should have died out over so large an area, especially as it is less closely tied to fenland than the Large Copper. On the other hand, the weight of testimony is too great to be attributed entirely to human error or the fraudulent activities of certain dealers. Our verdict must be an open one.

17. THE PURPLE-EDGED COPPER – *Lycaena hippothoe* (Linnaeus)

Papilio (*Heliconius*) *hippothoe* Linnaeus, 1761
Papilio chryseis [Denis & Schiffermüller], 1775

Whether this species has ever been native to Britain is extremely doubtful (see Plate 15, figure 5). Merrett is believed by some entomologists to have intended this species when in 1666 he described one butterfly as having wings '*externis purpurascentibus*' [with purplish edges],[1] a character which E. B. Ford considered was 'applicable to it alone'[2] though it could apply equally to some forms of the Purple-shot Copper (*L. alciphron*) found on the Continent. As already noted, the first use of the present name *hippothoe* in a British publication was by Lewin in 1795, but he was clearly referring to the Large Copper (*dispar*).[3] When, however, Haworth first described *Papilio dispar* in 1803, he also included, as a British species, *Papilio chryseis*, the synonymic name of the Purple-edged Copper. He gave its habitat as 'marshes' (as opposed to fenland for *dispar*) and status as 'very rare', though his only authority was Merrett's description.[4]

Haworth's imprimatur of this species as British gave just the opportunity the dealers needed to establish a profitable trade in allegedly British specimens which had in fact been imported from the Continent and reared on. The notorious Plastead was one who supplied specimens of Purple-edged Copper purporting to have come from Woodside, on the northern fringe of Epping Forest where he then lived, and later, after his removal there, from Ashdown Forest. Among his clients was the easily duped Dr Abbot of Bedford. Others taken in by him included James Sowerby who, in his *British Miscellany*, remarked that 'This new British *Papilio* was caught by Mr Plastead, of Chelsea, in Ashdown Forest, Sussex',[5] and John Curtis who noted that '*Chryseis* was abundant in August and September, 1818, at Woodside, Epping'.[6] Plastead had also been supplying Dr Leach of the British Museum, who 'received fine and recent specimens from the vicinity of Epping, for several successive seasons'.[7] All of these insects probably originated from Plastead's rearing-cages.

The myth that this species existed as a great rarity in south-eastern parts of Britain was sustained by the continued inclusion of *chryseis* in the different works on butterflies published during the first half of the nineteenth century, and by the existence of numerous specimens of dubious provenance in collections. Jermyn, for example, included all four species of copper in her well-known book, but her localities for *chryseis*, the Purple-edged Copper, were from secondary sources and she had accepted unquestioning the Plastead localities of 'marshes' in 'Epping Forest, Essex and Ashdownham, Sussex'.[8] Similarly Rennie, who described no fewer than five supposedly British species of Copper, gave the status of *chryseis*, which he called the 'The Golden Copper', as 'Very rare; near Epping'. He also included *L. hippothoe* as a distinct species under another new vernacular name, 'The Swift Copper', and said it was very rare locally, its localities being given as 'Whittlesea Mere, Norfolk, Suffolk and Kent'.[9] No wonder lepidopterists were confused!

In his *Manual of British Butterflies and Moths* of 1857, Stainton included '*Chryseis*' as a native species: 'Formerly taken near Epping, and in Ashdown Forest, Sussex'.[10] Later however, after corresponding with Henry Doubleday, he changed his mind, realising that the Purple-edged Copper had been introduced into the British list as a dealer's 'dodge'.[11] As a resident of Epping, Doubleday was in a position to make inquiries locally that blew the whistle on Plastead's activities.[12]

Doubleday was appalled by this trade in foreign butterflies misrepresented as British, and in lively correspondence with fellow entomologists protested strongly about the inclusion of Continental specimens in British collections. On visiting the Ipswich-based dealer, Mr Seaman, he saw 'rows of *L. chryseis* and *L. virgaureae*, all of which he assured me were taken by himself in Britain: I purchased three or four specimens for examination, and upon relaxing them, the wings returned to the position in which they had originally been set upon the Continent'.[13] It was for this reason that Doubleday decided to omit both species from his *Synonymic List of British Lepidoptera.* His friend Newman had, in 1842, wryly suggested renaming the Purple-edged Copper after one of the dealers 'who supply us with these delicacies on such liberal terms'.[14]

Though there is a very faint chance that the Purple-edged Copper is an extinct native of Britain (as it is a widespread and polymorphic European species), the absence of any reliable first-hand account of it has led most authorities to believe it is another example of wishful thinking among British collectors, pandered to by unscrupulous dealers.

18. THE LONG-TAILED BLUE – *Lampides boeticus* (Linnaeus)
Papilio (*Plebejus*) *boeticus* Linnaeus, 1767

Ever since the time of its first recorded appearance in Britain in 1859, when three specimens were discovered on the South Coast, this butterfly has been accepted as a rare migrant from the Continent. Delicate though it appears, it is a very strong flier and can travel great distances. To N. McArthur, who caught two on the Downs near Brighton, Sussex, on successive days – the 4th and 5th of August – goes the credit for capturing it first,[1] but another lepidopterist, Captain A. de Latour, also took one on the 4th August, at Christchurch, Hampshire (now in Dorset), and should share the honour of adding this species to the British list. Newman initially dubbed the butterfly 'the Brighton Argus', and it also came to be known as 'the Pea-pod Argus', because of its larval preference for seed-pods of leguminous plants. Other names given it were 'Tailed Blue' by Coleman (see Plate 15, figure 6),[2] a name also used by Furneaux,[3] and 'Large-tailed Blue' by Kirby,[4] but Long-tailed Blue was the most favoured and, following its adoption by South,[5] that name has prevailed.

The Long-tailed Blue is one of the most widespread butterflies in the world, being found throughout southern Europe, Africa, southern Asia and Australia. It has even reached the oceanic islands of St Helena and Hawaii. It is therefore not in the least surprising that this most migratory of the blues should reach the British Isles though, because it is almost continuous-brooded, it cannot live through our winters, requiring a Mediterranean climate to breed successfully. There is evidence of regular northerly migrations, not only in Europe but also in Asia, followed by a reverse migration in the autumn. In Europe it migrates north and south, but in India the movement is altitudinal, one generation of butterflies ascending the Himalayas up to 12,000 feet in the hot summer months, and another descending to the plains in the cooler months of September to November.[6]

In Britain, following the first record in 1859, it was seen only thirty-six times during the next eighty-six years, but in 1945, an outstanding year for rare migrant butterflies, there were records of a further thirty-eight,[6] about half being recorded from Sussex, Hampshire and Dorset, doubling the previous national total. However, the butterfly's inability to survive our climate has inevitably meant that records of sightings soon returned to the previous sparse pattern, though they are more frequent than in the past. In 1952 a female laid twenty-six eggs on everlasting pea (*Lathyrus latifolius*) in a garden at Dorking, Surrey;[7] and in the last week of August 1990 what was possibly the largest-ever invasion took place mainly to the London area,[8] as a result of which eggs were laid and a second generation produced in an urban local nature reserve. This astonishing event was recounted by Brian Wurzell:

> 'For about four weeks, butterflies were observed on most warm days at Gillespie Park [Highbury]. Occasionally an individual was seen amidst bushes in the lower part of the park, but the majority, up to 10–12 adults on the wing at one time, tended to quarter the dry railway land above. Their lively interest in Lucerne flowers was presumed to be for nectar, but I detected several eggshells of unmistakeable Lycaenidae structure on the tube-shaped sepals surrounding the bases of bladder senna flowers (*Colutea arborescens*) [the preferred foodplant for Long-tailed Blues in Southern Europe] ...Two slug-like larvae were also found nestling deep within the keels of flowers whose perforated sides had prompted closer investigation'.[9]

Approximately five miles to the south-east, in Kensal Green Cemetery, another colony had established itself. About a dozen butterflies were counted and a female was seen settling and feeding on the nectar of everlasting pea, raising hopes that eggs would be laid and a second generation produced. These were dashed when the authorities chose that time to strim the area, totally eradicating the foodplants.[10, 11] Together with the other sites where the Long-tailed Blue was seen, which included the Royal Botanic Gardens, Kew; Ranworth Common, Surrey;[12] and Petts Wood, Kent, it was estimated that up to one hundred individuals were present.

Wurzell found it 'tempting to speculate that this continuously-brooded tropical species might permanently colonise the City in response to the progressively warmer "greenhouse" seasons everyone is talking about'. Warmer summers could result in more frequent sightings of Long-tailed Blues in Britain, but it is unlikely ever to establish itself as a resident insect. However, it is one of those species which Dennis believes once occupied these islands some 5,000–10,000 years ago when conditions for it were suitable.[13] Since 1990, annual numbers of immigrants have once again dropped to single figures at most.

19. THE SHORT-TAILED or BLOXWORTH BLUE – *Everes argiades* (Pallas)

Papilio (*Plebejus*) *argiades* Pallas, 1771

The Short-tailed Blue, once also known as the Small-tailed Blue[1] and, more familiarly, as the Bloxworth Blue, is one of our rarest migrant species: discounting two unconfirmed recent records, no more than seventeen have so far been recorded in Britain.[2] It was Richard South who gave it the name by which we know it today.[3] The older name is derived from Bloxworth Heath in Dorset,

THE ENTOMOLOGIST.

VOL. XVIII.] OCTOBER, 1885. [No. 269.

LYCÆNA ARGIADES, PALL.

A BUTTERFLY NEW TO THE BRITISH FAUNA.

BY THE REV. O. PICKARD-CAMBRIDGE.

LYCÆNA ARGIADES.
(From a continental specimen.)

Two specimens (male and female) of this butterfly, which appears to be new to Britain, were taken on Bloxworth Heath, Dorset, on the 18th and 20th of August. The female (which is rather worn) was taken on the 18th by my son Charles Owen, and the male on the 20th, close to the same spot, by my son Arthur; this latter specimen is in good condition. Repeated searches in the neighbourhood since have failed to bring any further success. One of the plants on which I understand the

Fig. 135.
The announcement in the 1885 *Entomologist* of the discovery in Britain of the 'Bloxworth Blue' now usually called the Short-tailed Blue.

a large expanse of formerly open heathland and bog, where in 1885 a pair were netted by the local Rector, Octavius Pickard-Cambridge (see Chapter 3: 55), and his son. Fortunately, Pickard-Cambridge was an observant and expert naturalist and he recognized the butterflies as something new to him (Fig. 135). He described their capture in the following account:

> 'I did not myself see the female before its capture, but my son thought it was *L. aegon* [Silver-studded Blue]; and I do not doubt but I should have thought so myself. The male I saw as it flew up lazily (the sky was cloudy at the moment) from among the grass, less

> than a yard from where I was boxing another insect. I could not distinguish it then from *L. icarus*; but I called to my son Arthur to catch it, which he did at once; and in less than two minutes it was recognized and safely boxed. In our subsequent searches we have captured and examined (and for the most part either pinned for cabinet use or deported to a distance and released) over 500 *icarus* and *aegon*. The spot is one I have gone over constantly for many years, both collecting insects and in ordinary walks, and should never have thought of taking the trouble to catch anything looking so like a worn or dark *icarus* or *aegon*'.[4]

In a postscript to his article, Pickard-Cambridge mentioned a third specimen caught that year:

> 'I have ascertained, beyond a doubt, that an example of *Lycaena argiades* was also taken near Bournemouth, on the 21st of August, by Mr. Philip Tudor. This specimen was named, but doubtfully, for him by Mr. MacRae, of Bournemouth, as a worn example of *L. boetica* [Long-tailed Blue]; but on my yesterday showing Mr. MacRae (who first informed me of the above capture) the description of *L. argiades*, here given, with the woodcut figure above, he at once admitted that Mr. Tudor's specimen was not *L. boetica* but *L. argiades*; and a letter received last night from my eldest son (who is a fellow-pupil of Mr. Tudor's, at Forest School, Walthamstow) informs me that he had just seen a specimen of *L. argiades* in the collection of Mr. Tudor, taken at Bournemouth in August, and thought by Mr. Tudor to be *L. boetica*, but which was identical with our specimens of *L. argiades*. The Bournemouth locality is fourteen miles from that of our captures. I have given the above with perhaps unnecessary particularity; but I think the records of additions to our British fauna cannot be too particular or too accurate.'

Although the Pickard-Cambridge specimens were the first to be recognized in Britain, the Revd J. S. St John found others that predated them in the collection of a Dr Marsh, taken near the Rectory at Frome, Somerset, in 1874:

> 'The collection was fast going to ruin for want of attention, but I selected all those which were worth preserving, cleaned and "doctored" them, and placed the best and most uncommon specimens in my cabinet. Among these I noticed two small blue butterflies which somewhat resembled on the upper surface of the wings the male of *Lycaena icarus*, with the exception of "a small slender, but quite distinct, black, white-fringed tail". I could not quite make them out, and the tail puzzled me. On seeing the woodcut of *L. argiades* in this month's "Entomologist" I at once recognised a strong likeness between it and my two specimens. Comparing them together, and carefully examining the insects with Kirby's description, I found that they were undoubtedly *L. argiades* – both male specimens – and agreed in every detail with the description. On talking over this discovery with my friend, he told me that he took them with several others, eleven years ago, not two miles from this house, close by a small quarry. Thus *L. argiades* would seem to be not quite new to the British fauna.'[5]

These specimens are illustrated in the 'Historic Butterflies' plate in E. B. Ford's classic monograph, *Butterflies*.[6]

Since these early records, others have been reported from Wrington, Somerset (1895 or 1896), New Forest, Hampshire (1921), and Framfield, East Sussex on 1st September 1931. More recent captures include four in 1945 (Falmouth and St Austell, Cornwall; Branksome and Peveril Point, Dorset); Rogate, West Sussex (1977); and Beachy Head, East Sussex (1981). Two recorded from Derbyshire and Warwickshire in 1998, the latter identified after close examination, must be considered of questionable origin.[7]

By remarkable chance, another, even rarer, migrant butterfly turned up at Bloxworth Heath in 1938. This was Lang's Short-tailed Blue (*Leptotes pirithous*) which was netted by M. A. C. Lyell under the impression he was catching a Long-tailed Blue, which it superficially resembled except for its smaller size. Its identification was confirmed by N. D. Riley at the Natural History Museum, London.[8]

20. THE NORTHERN BROWN ARGUS – *Aricia artaxerxes* (Fabricius)

Hesperia artaxerxes Fabricius, 1793

This little butterfly is one of only three species to be first described from specimens collected in Britain, the others being the Large Copper and Berger's Clouded Yellow. It was finally recognized as a distinct species only at the end of the 1960s, over 170 years after it was first taken! Fabricius named it from material that had been obtained in Scotland, possibly by William Skrimshire, the able Edinburgh naturalist, and provided him by William Jones (see Chapter 3: 17). Fabricius, whose knowledge of British geography was clearly limited, gave its type locality as 'Anglia' [England], though the butterfly was known at that time only from the Pentland Hills to the south of Edinburgh and from Arthur's Seat, a hill some 822ft (253m) high to the east of the city. Traditionally this latter locality, where the butterfly flourished for many years, is said to be where Jones's specimens came from, though there is no documentary proof of it.

Its original name, bestowed by Lewin who first figured it in 1795 (Plate 26), was the Brown Whitespot.[1] Haworth called it the Scotch Argus,[2] the present Scotch Argus (*Erebia aethiops*) not yet having been found in Britain. Donovan called it simply '*Artaxerxes*',[3] and Samouelle the White-spot, Brown or Scotch Argus.[4]

When another similar Brown Argus butterfly was discovered at Castle Eden Dene in Co. Durham and described and illustrated by Stephens in 1828 as *Polyommatus salmacis*,[5] though without a vernacular name, there was confusion as to its relationship, not only with the Scottish species, but also with the familiar Brown Argus (*Aricia agestis*). Although Rennie[6] treated them as three distinct species – the Brown Argus Blue, the Durham Argus and the Scotch Argus – Morris still regarded them as one species, though he repeated Rennie's three names.[7] J. C. Dale gave his firm opinion that '*P. Salmacis* or *Titus*, is intermediate between *Agestis* and *Artaxerxes*; in Scotland none of the *Agestis* are to be found, they are all *Artaxerxes*; in the south none of the *Artaxerxes* are to be found, they are all *Agestis*. At Newcastle they appear to be mules, or hybrids between the two species'.[8] Newman went further, concluding that the three were 'nothing more than geographical races of the same species'.[9] However, in 1871, while still retaining that opinion, Newman published separate

Plate 26
William Lewin (1795) was the first artist to depict the 'Brown Whitespot' [the Northern Brown Argus] (figures 8 and 9):
'This new species of butterfly taken in Scotland, is now in the collection of Mr William Jones of Chelsea'.
The other butterflies are the Brown Blue [the Brown Argus] (figures 1 and 2); the Small Blue (figures 3 and 4) and the Silver-studded Blue (figures 5–7).

descriptions of each 'form' – as Brown Argus, Castle Eden Argus and Scotch Brown Argus.[10] In this he was followed by Barrett[11] and even Frohawk.[12] South also recognized only the one species.[13] Although Professor P. C. Zeller, the distinguished German entomologist, was said in 1869 to be 'tolerably well convinced that each must rank as a species distinct from the other',[14] it was not until some hundred years later, and after more detailed research into their life histories and biological differences, that it was conclusively proved in the late 1960s that there were, in fact, two species of *Aricia* in Britain: *Aricia artaxerxes*, the Northern Brown Argus (which includes subspecies *salmacis*, the so-called Castle Eden Argus), and *A. agestis*, the Brown Argus.

The original colony at Arthur's Seat died out before 1870, possibly hastened by over-collecting. On this subject, Stainton quoted a letter written in 1857,[15] from R. F. Logan who collected for him in Scotland, saying that he had not seen the larva, and that he had seen but very few of the imago, adding:

> '*I* have not diminished their numbers, having always a wholesome dread of exterminating species; but I believe a *dealer* has, and a host of small boys who come out of Edinburgh, with *orange*-coloured nets, and bottle them up wholesale, *five or six together, alive*, in the same receptacle, generally a match-box, along with Blues and anything else they can find.
>
> '... Unfortunately, the *Artaxerxes* when at rest is very conspicuous, and becomes an easy prey to these little marauders, whom I would wish to encourage if it were possible; but they are just at the age when their destructive energies are with difficulty to be restrained.
>
> 'In addition to all this, Government has agreed to construct a carriage-road between Edinburgh and Duddingston, much to my disgust, as it is to come along the line of the present footpath, and will destroy all the best localities for *Artaxerxes, Obelisca*, &c.'

Plus ça change!

Despite this, the Northern Brown Argus remains locally frequent in southern and eastern Scotland and northern England. It can still be found at Castle Eden Dene and some of its other localities in Durham after the passage of 150 years.[16]

21. THE MAZARINE BLUE – *Cyaniris semiargus* (Rottemburg)

Papilio semiargus Rottemburg, 1775
Papilio acis [Denis & Schiffermüller], 1775
Papilio cimon Lewin, 1795

The Mazarine Blue butterfly (Plate 27) was once quite widespread in Britain and was occasionally recorded as plentiful. Bretherton found records from at least twenty-four English and Welsh vice-counties.[1] However, it became much rarer in the second half of the nineteenth century, and by its end none were left.

The earliest reference to this species in Britain has been said to be that by Ray in 1710.[2] In his *Historia Insectorum*, he wrote: '*Papilio minor, alis supinis purpureo-caeruleis, pronis ocellis aliquot pictis. An Diurnarum minimarum tertia ?Mouffeti, pag. 105 no. 3.*[3] *Alae supinae ad exortum caerulescunt; inferius a fusco albicant. Ocelli sex septemve in singulis alis. A D.* Dale *capta nobisque ostensa est.*' This translates as 'A small butterfly, with the upperside purplish blue, the undersides marked with a number of eye-spots. Is this the third of Moffet's smallest day fliers, page 105? The uppersides of the wings are bluish towards the base, on the underside they become white from brownish. There are six or seven eye-spots on each wing. A specimen taken by Mr [Samuel] Dale was shown to us.' However, there is disagreement about the species Ray was really describing. P. B. M. Allan made a reasoned case that Ray was actually describing the Small Blue (*Cupido minimus*);[4] E. B. Ford thought he clearly was intending the Mazarine Blue;[5] but Ray's biographer, Canon C. E. Raven, considered that the description best fitted the Silver-studded Blue.[6] The description is insufficiently precise to be certain.

Towards the end of the eighteenth century, Lewin described what was certainly the Mazarine Blue under the name *Papilio cimon*, the Dark Blue. He wrote that it was very rare, but commented: 'The last week in August, 1793, I took two or three of the butterflies, flying in a pasture field at the bottom of a hill near Bath. They were very much wasted in colour and appeared to have been long on the wing; whence we may safely conclude, that they were first out from the chysalides about the middle of July.'[7]

The name Mazarine Blue was first used in 1797 by Donovan for the Large Blue (*Maculinea arion*),[8] but in 1803 Haworth adopted it for Lewin's 'Dark Blue',[9] and his usage has been followed by all subsequent authors. The origin of the name Mazarine is uncertain. The word itself is said to be derived either from the French statesman Cardinal Mazarin (1602–61), credited with the discovery of specially cut precious stones known since then as 'mazarines', or from the Duchesse de Mazarin, who lived in England and died in Chelsea in 1699, and is said to have worn a blue hooded gown, also since known as a 'mazarine'. The word is not to be found in French dictionaries, but in English ones it is defined as a rich blue colour. Perhaps it was the colour of the expensive blue dye used for the duchess's robes.

Many of the early authors considered the Mazarine Blue to be a butterfly of chalk grassland, found more commonly where it was rough and uncultivated, and along the borders of cornfields and meadows. Haworth gave its habitat as 'Chalky places' (*Cretaceis*) and stated that it was a very rare species which had been taken in Norfolk and Yorkshire by two of his friends, J. Burrell and P. W. Watson.[10] Later, it seems to have become more widespread, or perhaps just better known. Bree wrote that '*Acis* was at one time considered to be an insect of very great rarity'.[11] He quoted Haworth as describing it as 'the rarest, perhaps, of our British Blues', adding that since the start of the century 'the species has turned up in a variety of situations. Though by no means common, it appears to be widely distributed; nor is it peculiar to chalk districts; but it seems to delight in woody situations abounding in grass. Probably it may be overlooked on the wing, and passed by for the Common Blue.'

Newman provided a good summary of records from ten of the counties of England and Wales, and cited his own capture, while in the company of Edward Doubleday, of five specimens in June 1832 on his father's meadows at Olden Barn, near Leominster, Herefordshire.[12] At Glanville's Wootton, Dorset, it was 'formerly in plenty', the years of abundance being from 1819 to 1835, whilst the last specimens to be seen alive in Dorset were recorded on the 19th June 1841, indicating a catastrophic decline.[13] The last stronghold in Britain for this butterfly seems to have been at Penarth and Llantrisant in Glamorgan. T. Parry wrote that 'In 1835, '36 and '37 I could take *Polyommatus Acis* in plenty, but have never seen the species since. I fear that both it and *Chrysophanus Dispar* are gone.'[14] (The latter unexpected reference to the Large Copper was

Plate 27
This beautiful plate by H. N. Humphreys, Plate 31 from *British Butterflies and their Transformations* (1841), was made at a time when the now extinct Mazarine Blue (figures 9–11) was still widespread, if very local, in England and Wales. Also figured are the Azure [Holly] Blue (figures 1–3) and the Bedford [Small] Blue (figures 4–8).

thought by Allan to refer to one of the reputed West Country localities of *Lycaena dispar* or possibly even to *L. virgaureae* (the Scarce Copper) (qq.v.).[15]) A. F. Langley, however, reported that he had captured eight males and two females of the Mazarine Blue at Penarth in 1874 and one male in 1875.[16] It was to hang on precariously in South Wales for a few more years. Records from there indicate that the last native specimen may have been that taken at Tenby, Pembrokeshire, in 1883.[17]

In 1871, when the species in England was in terminal decline, A. E. Hudd wrote that 'the cause of its rarity in this country is, I have no doubt, to be found in the fact that the ova and young larvae are destroyed by the haymakers'.[18] Elaborating on this theory, T. A. Chapman suggested in 1909 that its extinction was, in part, due to the widespread sowing of clover (*Trifolium pratense*) as a crop.[19] This practice was likely to have attracted females to these fields, where they laid their eggs on the clover heads, only for the fields to be harvested shortly afterwards and before development would have been completed. Allan disputed this explanation, believing its extinction was more likely to have been due to climatic change during the nineteenth century.[20] Whatever the cause, this rare and pretty little butterfly became extinct as a native and has not bred in Britain for well over one hundred years.

Since 1900 there have been a very few sightings of the Mazarine Blue, all of which are considered to be of immigrants from the Continent. Two specimens were taken in 1900 and three in 1901 at Gorleston, Norfolk. Single specimens have been reported from Beachy Head, East Sussex, 16th July, 1902; Fowey, Cornwall, 1934; Eastbourne, East Sussex in 1917; and Rogate, West Sussex, in 1958.[21]

Plate 28 *(facing page)* The coloured drawings of a male Large Blue (*x*), made by Henry Seymer in his copy of *The Aurelian* in about 1776, predate by nearly twenty years the first published account in Lewin's *Papilios of Great Britain* of this now extinct native butterfly. They fit neatly on Harris's plate of The Swallowtail butterfly.

22. THE LARGE BLUE – *Maculinea arion* (Linnaeus)

Papilio (*Plebejus*) *arion* Linnaeus, 1758

Although William Lewin, in *The Papilios of Great Britain* published in 1795, was the first British author to describe and illustrate the Large Blue as a native species, it had in fact been accurately described and drawn about twenty years earlier by Henry Seymer senior in his personally annotated copy of Harris's *Aurelian* (Plate 28). In it, Seymer wrote that it was 'a species wch is very rarely met with, & in few Collections' (Fig. 136), but he gave no details of its habitat.

f. g. Swallow-tail Fly. a. b. c. Caterpillars. d. e. Crysalis.
Papilio Eques Achivus; Machaon. Lin. 33:
x Papilio Plebeius ruralis Arion. Lin 230.
This we have added, not only as a proper Contrast
to the Swallow tail, but also as being a species wch
is very rarely met with, & in few Collections......
This is a Male, ye black spots, & black border are less,
& paler, as is the whole underside, in the female.
y. a Pyralis found only & but 1 season at Langton near Blanford.
I have as yet met with no figure, or description in any
Author that answers to my satisfaction; but Q. Fab. spt Insect:m
Vol. 2. Page 265. No 135 or 138?

Fig. 136. Henry Seymer's annotations on the Large Blue, then yet to be recognized as British in any published work, remarking that it was 'a species wch is very rarely met with, & in few Collections'. His tasteful positioning of the additional species 'as a proper contrast to the Swallow tail' improved on Harris's layout of the plate.

PL. XXXVI.
To the Rev.d M.r Will.m Ray
This Plate is humbly Dedicated by his most humble Obliged Serv.t
Moses Harris
H.S P.t

Lewin, however, whilst also noting its rarity, stated that 'It is out on the wing the middle of July, on high chalky lands in different parts of the kingdom, having been taken on Dover Cliffs, Marlborough Downs, the hills near Bath, and near Cliefden [Cliveden] in Buckinghamshire'.[1] Donovan too considered it very rare, and added that it did not appear to be any more common in other parts of Europe.[2] Confusingly, he called this species the Mazarine Blue, a name which Haworth in 1803 had adopted for the even rarer *Papilio cimon* (now *Cyaniris semiargus*).[3] Haworth, however, kept to Lewin's English name of Large Blue, by which it has been known almost ever since. By 1832, only Capt. Thomas Brown, author of a popular work on 'Butterflies, Sphinxes, and Moths', continued to call it by Donovan's name[4]. J. F. Stephens was another author to describe the Large Blue as 'an insect of great rarity'.[5] This status increased its desirability in collections and led to its being pursued and captured in very large numbers to the detriment of the more accessible of its colonies.

Plate 29 (*facing page*) Large Blues from Devon, Cornwall and the Cotswolds. The latter, are consistently a deeper shade of blue. Polebrook specimens are also a deep blue.

During the nineteenth century colonies of Large Blues were discovered within six broad areas, mainly in south-western England and usually on limestone (Plate 29). The most northerly site, however, was in the neighbourhood of Barnwell Wold in Northamptonshire. The Revd William Bree, Rector of Polebrook, found six or seven colonies of a distinctive dark race of the Large Blue, centred on ant-hills – though the significance of this association was not appreciated at the time. In 1852 he wrote:

> 'The great prize of all the butterflies of the neighbourhood of Polebrook I hold to be *Lycaena Arion*, which if I mistake not, was discovered here by myself some thirteen or fourteen years since. It is confined entirely, as far as my experience goes, to Barnwell Wold and the adjoining rough fields, with the exception of a single specimen which I once met with in a rough field near Polebrook. Its flight is somewhat peculiar, being different from that of others of the same genus, and more resembling that of *Coenonympha Pamphilus* [the Small Heath] and *Epinephele Tithonus* [the Gatekeeper]. Independently of its manner of flight and size, it is in most cases easily distinguished on the wing from the other blues by its dark and irony appearance. Many entomologists have, of late years, visited Barnwell Wold in search of *Arion*; in short, a summer never passes without meeting in my rambles brother entomologists from different parts of the country; I rejoice, however, to be able to state that its annual occurrence does not appear to be diminished in consequence. Unless my memory fails me, I think Mr. Wolley, of Trinity College, Cambridge, informed me that one year he captured, in a few days, between fifty and sixty specimens in and about Barnwell Wold, though in point of weather, the days were anything but favourable.'[6]

It is thought that these colonies died out in the 1860s.

Discovered in the Cotswolds in 1850, the Large Blue proved quite widespread there, with around thirty colonies, some very small. The typical Cotswold habitats were downland banks or disused quarries. Merrin described the conditions that prevailed on a day's outing on 6th June 1865, when he took eleven specimens:

Devonshire
Cornwall
Cotswolds

> 'There was a strong wind blowing, as there generally is on the exposed places occupied by *Arion*, and doubtless this tends quickly to damage its delicate plumage; the spot most frequented by them was, however, partly sheltered by a stone wall. The same locality subsequently yielded as many as were taken on the first day, while all the district round about, though much of it is of the same character, was perfectly clear of them. This tends to show that the species is very local. On another spot, some miles distant, but of a similar broken character, the species was also found, the area, however, being still more contracted. The ground in both cases consists of deserted quarries, from which broken stone has been taken, the sides of the quarries being left sloping, and thick grass, with the usual herbage of hills, growing near. This herbage includes wild thyme, sun-cistus, wild geranium, wild forget-me-not, milkwort, yellow trefoil, and several species of coarse grass.'[7]

However, twenty years later, Herbert Marsden reflected on the apparent decline of the Large Blue in its Cotswold stronghold and was compelled to contemplate its extinction. The turning point seems to have been about 1877, brought on by a series of wet summers.

> 'It was on June 17th, 1866, that I first saw the species alive, when in the course of a long ramble I captured it in a narrow valley amongst the Cotswold Hills.' During the years 1867 to 1869 the Large Blue was relatively scarce, but 'The year 1870, however, is the one to be marked with a white stone by the lovers of Lycaenidae; and *Arion* appeared much more widely distributed than in any other year I know of, either before or since. It would, I am sure, have been possible for an active collector to have caught a thousand specimens during the season, for in a few visits I secured about an hundred and fifty, not netting half of those seen, and turning many loose again. During the next few years *Arion* continued to appear, but very irregularly as regards numbers. The best seasons since 1870 being those of 1876 and 1877, the latter especially, but on no occasion has it been nearly so abundant as in 1870. Now come the dark days. The latter part of June, 1877, was damp and broken, not at all the bright warm weather which *Arion* loves. In dark, cloudy weather they are always still, and, I believe, they will only deposit their eggs when the sun is warm and bright. In 1878 the weather was worse, there being hardly a fine day in the month, and less than a dozen were seen, mostly worn and weather-beaten, for there were scarcely two consecutive fine days. In 1879 the weather was still worse, and *Arion* scarcer than ever, while in 1880 only two were obtained and two or three more seen. For the four years 1881–4, not one has been seen in the Gloucestershire district that I have been able to trace.'[8]

The Large Blue became locally common again in the Cotswolds during the 1920s and '30s, followed by a further progressive decline in the mid-1940s. Ernest Neal recalled how, as a schoolmaster at Rendcomb College, near Cirencester, in the 1930s he was approached by one of his pupils who told him he had seen a Large Blue:

'It was Founder's Day when parents visited the college and some celebrity was invited and speeches made ... On this occasion, after the parents had collected for tea, Ian [Ian Menzies, one of the boys, later a senior pathologist at a London Hospital] sidled up to me in a conspiratorial manner and excitedly whispered "I've discovered the large blue!" I knew this butterfly was supposed to occur in the Cotswolds, but it was reputed to be extinct. I knew Ian didn't make mistakes over things like that, so I quickly left the gathering. He slipped into my car, and off we went to Withington, a village not far away. He led me over a field to some rough grazing dotted with ant hills, and sure enough there were several of these rare butterflies flying over patches of thyme! It was thrilling! It was not long before he discovered several other colonies in the neighbourhood and was studying them in detail. One thing that became apparent after several years was that emergence was very variable and in one year was as early as the third week in May.'[9]

Neal even attempted to breed the butterfly from egg, and to photograph its life at successive stages of development. He found the caterpillar in the wild, bored inside a thyme flower with 'its back-end filling the hole it had made'. Unfortunately his attempts to breed it further failed – although he successfully established a formicarium, he had chosen the wrong species of ant! By the 1950s the butterfly had become very scarce, although it hung on in small numbers until around 1964.[10]

In Somerset, two colonies existed near Langport between 1833 and 1843. In 1945, a single colony was rediscovered there which persisted until the 1950s.[11]

In mid-Devon, a small inland colony was found on south-eastern Dartmoor near Ashburton in 1870 but, as with colonies elsewhere, it appeared to be heading for extinction towards the end of the century. However, five other small sites were found near Buckfastleigh in the 1930s which survived until the 1970s.[12]

First discovered in 1856, colonies of the Large Blue were found along the coast of South Devon between Bolt Head and Bolt Tail, south-west of Salcombe. Small numbers were also taken at Prawle Point and Beesands to the south-east. For those wishing to collect the Large Blue on this rugged terrain, G. C. Bignell, in a letter to Newman written in 1870 and later published, furnished the following instructions:

'Anyone desiring to take this insect in our neighbourhood must regulate his visit according to the weather during the past spring; he cannot do better than stop at the "King's Arms", Salcombe, for the night. Bolt Head is an out-of-the-way place to get at. The nearest point by rail is Kingsbridge Road; you can take the coach from thence to Kingsbridge, a distance of about ten miles. From thence to Salcombe is about four miles by steamer or boat, and then you have about two miles' walk to Bolt Head; the slopes here are very steep, and in dry seasons you should have spikes or long hob-nails in your boots, to make sure of your footing, for it puts one in mind of walking on ice, it is so excessively slippery.'[13]

Bignell knew what he was talking about for, a few years earlier, on 17th June 1865 he had succeeded in capturing 36 specimens under the cliffs near

Plymouth despite a confrontation with the elements. 'The weather was very boisterous, but I fortunately got into a sheltered nook under some high cliffs, where apparently there had been a land-slip some years before; the ground was very rough, and it was with great difficulty that I could travel over it, or I should have taken more.'[14]

Some years later, in 1884, Bignell wrote pessimistically about the future of this colony: 'I feel quite certain that the haunts of *Lycaena arion* at Bolthead must be looked upon as a thing of the past. I visited the old familiar spots twice this year without seeing a single specimen ... I know *Arion* has been on the wing this year, for I have had the pleasure of seeing nine specimens, taken during the first week in July by a gentleman who had visited Bolthead, but gave it up in disgust.'[15] Writing in 1890, C. W. Dale gave his opinion that these nine specimens, thought to have been taken on a rough piece of ground near a village about ten miles from Kingsbridge, in South Devon, were probably the last of the Large Blue to be collected in England.[16] At the present time the original site on Bolt Head is overgrown with bracken.

However, at a time when the species was in terminal decline or already extinct elsewhere, there was obviously great rejoicing when large numbers of colonies were discovered from 1891 onwards on the north coasts of Devon and Cornwall. The first was found at Millook Common in Cornwall, by E. A. Waterhouse, a friend of Frohawk who, in his old age, committed to his notebook (with a view to publication which never took place) a graphic account of his friend's discovery:

> 'Arriving at a rural spot south of Bude he [Waterhouse] managed to put up at the only cottage there was then available (25/– [£1.25] per week inclusive [of] very good meals). After a good nights rest, when he had a pleasant dream of catching three examples of the Large Blue in his wanderings, he thought no more about it. The following morning being dull with rain at times, he strooled [*sic*] about with the object of doing some botanical work, so took his botanical tin, and a butterfly net in case of it getting finer. To his astonishment, he had not gone far, [when] he saw and caught one of these butterflies, this so surprised him that he kept a sharp look out for more, and succeeded in catching two more. Having caught the three specimens of his previous nights dream, and again turning dull with rain about to fall, he returned to his room at the cottage to medi[t]ate over the queer coincidence. The next day being fine he was rewarded by catching several others, where subsequently they proved to be abundant.'[17]

Following news of this find, large numbers of collectors travelled to the area and hundreds of the butterfly were caught. Within a relatively short time they discovered that the Large Blue was in fact locally common, occurring in scattered colonies in sheltered valleys all the way along the wild and rocky coast from Tintagel to Bude. Nevertheless these colonies too began a slow decline in the 1920s, and the last specimens were taken at the most famous of them at Crackington Haven in the early 1960s. In the 1940s, further large colonies, with up to 10,000 adult butterflies in each, were found in coastal valleys north of Bude as far as Hartland Quay, but there too a decline began in the mid-1950s, with the last specimens being seen in 1973.

Outside its known haunts there have been at least two claimed sightings of the Large Blue. In 1938, J. W. Heslop Harrison made his highly controversial claim that he had seen and others had caught the butterfly on the Island of Rum in the Inner Hebrides,[18] a claim he reasserted in 1948.[19] Despite intensive searching, no such colonies were found and it was tactfully concluded that the Professor's 'ardent and competitive personality may have laid him open to students' practical jokes'.[20] On the other hand, Heslop Harrison had also claimed to have found several species of plants on Rum, otherwise unknown in Britain, and which no one has ever succeeded in refinding.[21] (See also Chapter 3: 86.)

Then, in 1962, a young man called at the National Museum in Dublin and informed the curator that he had discovered the Large Blue at Dunboy, in West Cork, though he had no voucher specimens. In 1963, H. C. Huggins and E. S. A. Baynes explored the locality carefully and found no Large Blues but quantities of the large, very brightly coloured Irish form of the Common Blue. Huggins concluded that the young man's claim, like Heslop Harrison's, was 'very Rhum'![22] Claims of Scottish and Irish Large Blues should be viewed with scepticism.

Until the early years of the twentieth century, the early stages of this butterfly were a mystery. In June 1869, Professor P. C. Zeller, the well-known German entomologist, writing in the *Entomologist's monthly Magazine* on 'Hints for finding eggs and larvae of Lycaena Arion', offered suggestions that might enable English entomologists to 'be the first to unravel the Natural History of the "Large Blue", rare as it is with them'.[23] He recalled his surprise at the abundance of this species both in 'moist open meadows' and also in 'dry fir-forests, on the most barren and sandy ground' until he realized that, in both situations, wild thyme was to be found:

> 'Spending a day (July 28th, 1857) in the Glogau Stadtforst ... I took the opportunity in the morning, before the heat of the day, to watch closely the females of *Arion*, which were flying slowly, and to observe their doings.
>
> 'I saw them sit down on the stems of *Thymus serpyllum*, and, after sipping from a few flowers, bend their abdomens between the flower-stalks, on which they deposited a pale green egg, sometimes not without some apparent pains. I gathered a score or so of twigs, each with a single egg. In the afternoon I noticed them proceeding in the same manner, but as it was then too hot in the sunshine, the oviposition was only performed under the shade of the trees.'

Zeller confessed that he had done nothing more with the eggs and made no attempt to breed from them, concluding that 'as the larvae appear to pass the winter when about half-grown, it will probably be no easy work to rear them to maturity'.

The larva of the Large Blue was one of the few that William Buckler was unable to find to illustrate his classic work, *The Larvae of the British Butterflies and Moths*, published in 1886. In June 1869 he had successfully persuaded captive females to lay eggs on potted thyme. 'Although the eggs hatched both with Mr. Merrin and myself, yet we have failed to detect the young larvae on the plants at present, but we believe they must be very small, hiding away somewhere, and that they will most likely hibernate.' In the following year he repeated the experiment but, although the eggs hatched, he again could find no larvae. However, 'I telegraphed to Mr. Hellins, to whom I had previously sent eleven eggs, and his reply informed me that his were hatched and that he could see one larva feeding. During the following week he reported them to be looking like very small pinkish-brown maggots.' In his preface to Buckler's book, Stainton, commented that: ' ... to this day some mystery seems to prevail as to the proper food of the larva of the rarer *Polyommatus* (*Lycaena*) *Arion*, for though when quite young it eats readily enough the flowers of thyme it seems at a certain stage of its growth to require something else'.[24]

It took nearly half a century longer for the discovery to be made by T. A. Chapman that half-grown larvae of the Large Blue feed on ant larvae rather than wild thyme. Years later, Captain Bagwell Purefoy published his account of the initial breakthrough:

> 'During May of 1915 the late Dr. Chapman went to Cornwall to pursue further investigations into the life history of the Large Blue butterfly (*Maculinea arion*). On pulling up a plant of wild thyme he disclosed a half-grown *arion* larva and damaged it in so doing. The plant was growing on or close to a nest of the ant, *Myrmica scabrinodis*, and the larva was, in fact, among the ants. As it was impossible to try and rear the larva he decided to examine its contents under the microscope and he found that it had been feeding on the small larvae of the ant. That was pretty conclusive but the information was not broadcast at the time and we did not hear about it until the following autumn'.[25]

Until that time, no-one had yet been able to rear the larvae beyond the third instar. It was Frohawk, who first noticed that, during its apparently aimless wanderings, the larva would hunch itself up as a signal to the ant that it was ready to be carried into the nest (see Chapter 3: 74, Fig. 105). The ant would then seize the larva and bodily transport it into the nest, where it would mature, pupate, and finally hatch into the adult butterfly. The ants, in return for providing food and protection, received honey-dew secreted by the larva. With this knowledge, Purefoy was able to rear the Large Blue, using captive ant colonies in artificially constructed nests.

The revelation that the Large Blue was dependent on ants, in particular *Myrmica sabuleti*, for completion of its life-cycle was initially received with scepticism. J. A. Thomas, however, took this research further and was able to supply photographic evidence that confirmed the original findings.[26] He also discovered that the chrysalis produces sounds that stimulate the ants to accompany the newly emerged butterfly as it makes its way to the surface. Although, by 1979, the Large Blue had declined to extinction in Britain, Thomas's research has enabled him to establish small colonies in the west of England from Swedish stock, and the chances of its surviving and increasing on specially managed sites now look promising.

The reasons for the decline and extinction of this fine butterfly have been widely debated and investigated. Over-collecting, leading to the acquisition of thousands of specimens by collectors, undoubtedly contributed. For example, when the collection of Baron J. Bouck came up for auction in 1939, it was found that he had amassed over 900 specimens. The colony at Barnwell Wold, Northamptonshire, was almost certainly destroyed by the activities of dealers and greedy collectors. In 1860 over 200 adults at rest were taken by one dealer, a loss from which it seems never to have recovered.[27] C. G. Barrett told the Essex Field Club of a raid by collectors in 1896 on the newly-found locality in Cornwall: 'It is estimated that more than 2,660 specimens were destroyed: Single collectors in one or two cases taking from 500 to 600 each ...'.[28] Fortunately the rugged ground and inaccessibility of many of the colonies gave the species a degree of protection.

The influence of collecting on the demise of the Large Blue was considered by Spooner as a significant, but by no means the sole, factor. 'The action of collectors probably, if not certainly (i) can reduce an otherwise moderate population to a significantly lower level at which it may be much less able to tide over adverse seasons; (ii) can so cream off a flourishing population in a good year that there are no excess individuals to spread into vacancies in the neighbourhood, or keep some sort of gene-flow operating between adjacent but separated colonies. A degree of spreading of excess members of a colony in good years may be essential for survival in a given district.'[29]

Ploughing or agricultural development led to the destruction of possibly half the original habitats. At the same time the uncontrolled growth of long grass on many sites led to the simultaneous disappearance of the ant *Myrmica sabuleti* and the Large Blue. Grazing by rabbits, which prevented the sward from becoming overgrown, ceased with the spread of myxomatosis in the 1950s, and it grew too tall. A sward-depth of more than four centimetres causes sufficient cooling at ground level to inhibit *M. sabuleti.*[30] Other less congenial species of ant then take over, and this subtle ecological change almost certainly accounted for the extinction of all the remaining sites over a 20-year period. In the end, then, it was habitat loss which brought about the extinction of the Large Blue, and ignorance of its breeding requirements which prevented remedial action being taken in time to save it as a native British butterfly.

23. ALBIN'S HAMPSTEAD EYE – *Junonia villida* (Fabricius)

Papilio villida Fabricius, 1787
Papilio hampstediensis [Jermyn], 1824

At the beginning of the eighteenth century, Hampstead Heath, despite being the haunt of footpads and highwaymen, was a popular hunting ground for naturalists. James Petiver, the apothecary, herbalized there with fellow botanists as well as looking for insects; and Albin collected spiders, butterflies and moths which he then illustrated in his books. The Heath was the place where the 'curious', that is to say the enquiring, Albin was said to have caught the famous and, at that time, unique butterfly illustrated by Petiver[1] alongside the Peacock under the name of 'Albin's Hampstead Eye', grouping them together as 'Brittish Eye-winged Butterflies' (Fig. 137). Petiver described it as '*Papilio oculatus Hampstediensis ex aureo fuscus*' [the eyed butterfly from Hampstead, of a golden fuscous colour], and added that it was 'the only one I have yet seen'. That he did not see another is hardly surprising since this butterfly, which is still preserved among the remains of Petiver's collection in the Natural History Museum in London, is *Junonia villida,* a native of Australia, Tasmania, New Guinea and a few other islands in the Pacific and adjacent Indian Oceans! Apparently it lives for around three weeks as an adult and, where it is indigenous, can be very common.[2]

How Albin's butterfly came to be 'caught' at Hampstead can only be surmised. The most likely explanation is that a specimen had been brought to England, as were so many at the time by equally 'curious' travellers, and that it had become muddled with some British material. In view of the duration of the sea passage from those parts in the early eighteenth century, it is most unlikely that this butterfly could have been brought to England alive at any

Papiliones BRITAN. *Oculatæ*, BRITTISH Eye-wing'd *Butterflies*. TAB. V

stage of its metamorphosis. At that date, few navigators had reached Australia. Tasman was the first European to set foot on Australian soil when, in 1642, he discovered and named Van Diemens Land (now called Tasmania) in honour of the Dutch explorer, Anthony Van Diemen (1593–1645), but it was not until 1770 that Captain Cook discovered the Australian mainland. The clue to its origin lies in the name given for the type locality by Fabricius – 'New Amsterdam'. There were several 'New Amsterdams' at that time. The best known was the Dutch settlement in North America, which the British acquired and renamed New York. It was also, and still is, the name of a town in Guyana, the former Dutch colony in South America. In view of the distribution of this butterfly, neither of these connections seems likely. However, there is an island in the southern Indian Ocean, now called Amsterdam Island, which was originally named by Van Diemen 'New Amsterdam Island'. Could this be the original source of Fabricius's type specimen? If so, we can speculate that Albin's specimen reached him from a Dutch entomologist who had been provided with natural history material brought back from one of their colonies by a Dutch East India Company trading vessel, and that Albin subsequently got it mixed up with another butterfly taken at Hampstead.

Fig. 137. *(facing page)* Petiver's engraved plate of 'Brittish Eye-winged Butterflies' includes 'Albin's Hampstead Eye' (figure 2), based on the specimen said to have been found on Hampstead Heath 'where it was caught by this curious person [i.e. the observant Albin], and is the only one I have yet seen'.

Lætitia Jermyn included Albin's unique butterfly as *Papilio hampstediensis*,[3] giving its locality simply as Hampstead. She described it as 'Rare', adding that it had not been taken since the time of Petiver. Others followed her. Rennie accepted it uncritically but renamed it 'The Hampstead'.[4] The works of Wood (see Plate 36, figure 7),[5] Humphreys & Westwood,[6] and Morris (Plate 30) all illustrate it in colour, the last author stating, inaccurately, that the specimen was 'no longer in existence and cannot speak for itself'.[7] Fortunately it is and

Plate 30
In 1853, F. O. Morris published this illustration of Albin's Hampstead Eye, faithfully copied by Benjamin Fawcett from Petiver's drawing for *A History of British Butterflies*, and printed in colour.

it can, and those writers speculating on its identity who thought it might be an American Painted Lady, or a Peacock, or a variety of Speckled Wood, are conclusively proved wrong. We will never know by what devious route this singular butterfly reached England, but it does confirm that trade in butterflies existed as early as 1717. There is a recent British illustration of the species, but not of the historic specimen itself.[8]

24. THE PAINTED LADY – *Cynthia cardui* Linnaeus
Papilio (Nymphalis) cardui Linnaeus, 1758

Although this graceful butterfly is a visitor to Britain from North Africa, and rarely survives our winters, it was obviously a familiar sight to the early naturalists. Petiver said of it: '*Papilio Bella donna dicta*' ['it is called the beautiful lady butterfly'], and knew it as The Painted Lady,[1–3] as did his contemporaries Adam Buddle, Charles duBois,[4] who described and illustrated it (Fig. 138),

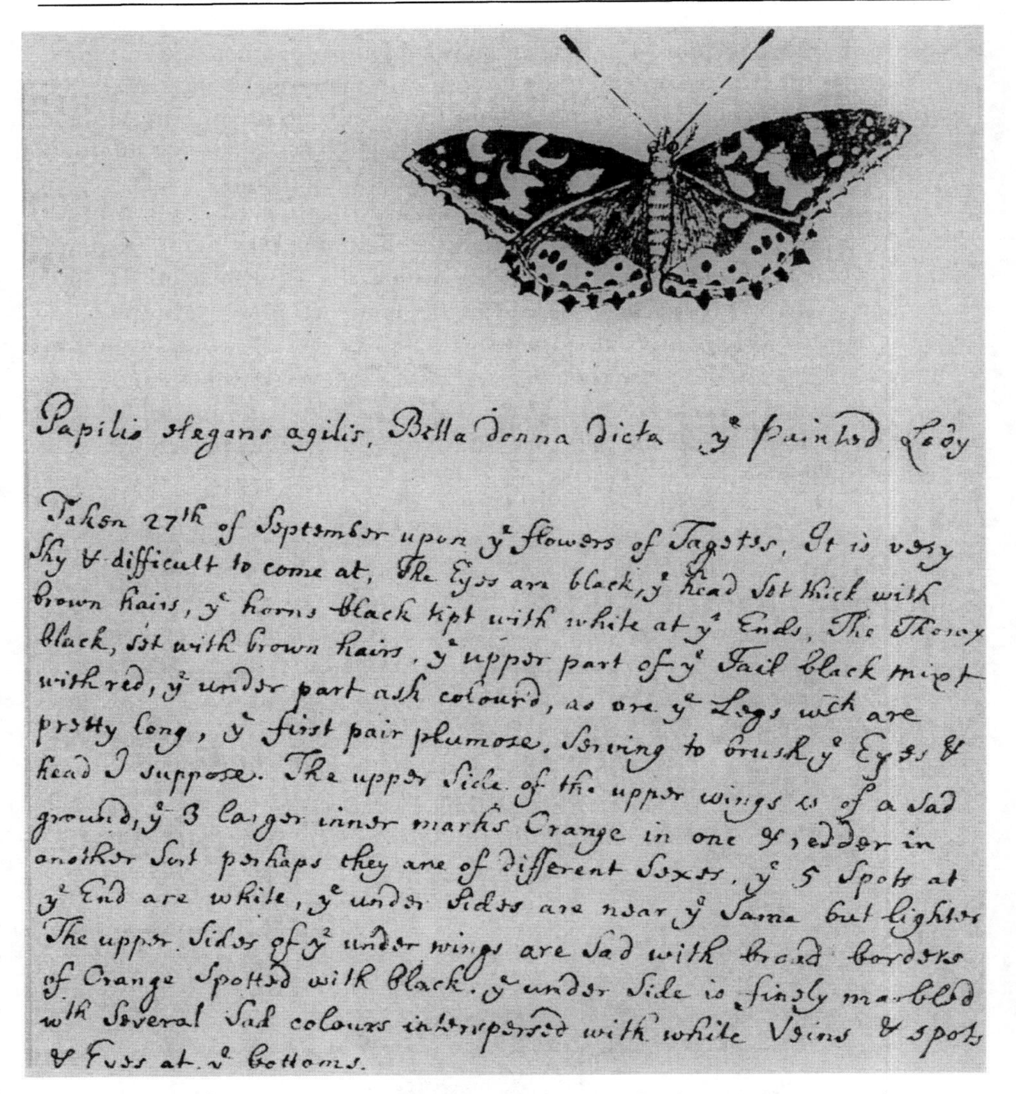

Papilio elegans agilis, Bella donna dicta ye painted Lady

Taken 27th of September upon ye flowers of Tagetes, It is very Shy & difficult to come at, The Eyes are black, ye head set thick with brown hairs, ye horns black tipt with white at ye Ends, The Thorax black, set with brown hairs, ye upper part of ye Tail black mixt with red, ye under part ash colourd, as are ye Legs wch are pretty long, ye first pair plumose, serving to brush ye Eyes & head I suppose. The upper Side of the upper wings is of a Sad ground, ye 3 larger inner marks Orange in one & redder in another Sort perhaps they are of different Sexes, ye 5 Spots at ye End are white, ye under Sides are near ye Same but lighter The upper Sides of ye under wings are Sad with broad borders of Orange Spotted with Black. ye under Side is finely marbled wth Several Sad colours interspersed with white Veins & spots & Eyes at ye Bottoms.

Fig. 138. A drawing of the Painted Lady butterfly from the notebook of Charles duBois. Its name, which was also used by duBois' contemporaries, Petiver and Ray, was probably based on a still older folk-name, 'bella donna'. It refers to the flesh-colours and markings on the butterfly's forewing resembling the painted eyes of seventeenth-century ladies of fashion. The connotation is with beauty, not, as has been suggested, with 'shady ladies'.

Plate 31 *(facing page)* Plate XI from *The Aurelian* displays the larval, pupal and adult stages (including both sexes and their undersides) of the Painted Lady and 'the Marmoress or Marbled White' butterflies, along with their correct foodplants. Characteristically Moses Harris adds a jumble of foreground objects, including broken china and a broken clay pipe to provide depth and scale – and possibly to indicate the habitat of the Painted Lady, which he considered to be 'Waste Grounds, Rubbish Hills, etc.' where thistles abounded.

and John Ray.[5] Some say that 'Bella donna' is an allusion to the cosmetic use of the deadly nightshade plant, which Pierandrea Mattioli (1501–77), a Venetian physician and botanist, referred to as *Herba bella donna*, so called because a distillation of its berries with water was applied by the painted ladies of fashion to their eyes to enlarge the pupils, thus making them more alluring. The name may also be an allusion to the mixture of flesh tints and black markings on the wings, reminiscent of the painted faces of courtesans in the age of Hogarth and Molière.

Ray observed that 'The Painted Lady' was 'common with us around Braintree [Essex] and elsewhere'. In 1795, Lewin dropped the more attractive name used by his predecessors and substituted the 'Thistle Butterfly',[6] after Linnaeus's scientific name *cardui*,[7] Latin for 'of the thistle'. However, all later authors, including even the usually idiosyncratic Rennie,[8] preferred the earlier name of Painted Lady and this has been used right up to the present day.

Because the very idea of butterfly migration was alien to lepidopterists until the present century, all sorts of explanations appear in the literature as to why Painted Ladies were common in some years but not in others. According to Moses Harris, who illustrated it (Plate 31), 'These Flies are not very common:

PL. XI
To the Rt. Hon.ble Lady Charlott Townshend Baroness Ferrers
This Plate is most humbly Dedicated by her Ladyships most obliged & faithful Servts
Moses Harris J. Gretton

the Reason for which is, all Weathers do not agree with them: Yet there are particular Seasons when they are very plentiful, which happens once in about ten or twelve Years. They are then often seen in Town flying in the Streets.'[9] London in Harris's day was a much smaller place than it is today and the countryside was on its doorstep. It is pleasing to picture Painted Lady butterflies patrolling along the Mall and down through Westminster, entering the open windows and laying in untended corners of town gardens. Harris continued: 'They are taken in the Fly state all the month of August and haunt Waste Grounds, Rubbish Hills, &c, settling on the Furzes of the Docks and Blossoms of the Thistle'. There were plenty of rubbish heaps in those days and thistles no doubt grew there in abundance.

It was known very early on that the Painted Lady was to be found widely abroad. One of the type localities given by Linnaeus was 'Africa'. Rennie regarded it as 'common', amusingly giving its localities as 'Edinburgh, Middlesex, Devon, Jersey and in most parts of the globe'! Morris considered it to be 'one of the most universally distributed species of Butterfly in the world'.[10] But although he referred to an immense swarm in 1828 that passed over part of Switzerland in such vast numbers that their transit occupied several hours, he failed to draw the conclusion that the butterfly came to Britain from the European mainland. Similarly, Newman described it as 'truly cosmopolitan but intermittent and irregular', adding that sometimes several years passed without his seeing one: 'Many unsuccessful attempts have been made to discover some law by which to account for the irregular appearance of this insect.'[11] Newman had some years earlier propounded a 'blown-over theory' to account for the appearance of very rare migrants such as the Bath White and the Long-tailed Blue, and suggesting that winds or a migratory tendency in insects accounted for these extremely infrequent sightings. This theory was, in his own words, 'severely ridiculed' by fellow entomologists, to which he responded that 'it is generally safer to investigate a theory based on long experience than to dismiss it as unworthy of consideration'.[12] However Newman was never able to make the leap from his original and partially accurate analysis to the discovery of natural annual migrations of insects, despite incontrovertible evidence that large scale movements took place within continents. It was the very idea of a purposeful flight across seas even as narrow as the English Channel that the nineteenth-century entomologists found incredible. Even as late as 1893, Barrett did not feel able to make the unequivocal statement that our Painted Ladies are immigrants. The closest he would go, after examining several reports of spectacular mass-movements of this species on the Continent in the great Painted Lady year of 1879, was to say that they showed 'the extreme probability, almost amounting to certainty' that 'the sudden appearance of multitudes in this country was a direct result of the great migratory movement noticed in so many parts of Europe'.[13] It is hardly surprising, therefore, that lepidopterists found it so difficult to account for an almost complete absence one year followed by a superabundance the next. They somehow believed that the butterflies successfully 'hibernated' to re-emerge the following spring but that, being very mobile, their presence in large numbers was the result of movement from elsewhere in the kingdom. It was not until the turn of the twentieth century, following Tutt's analysis of insect migration and dispersal,[14] that new light was thrown on a very confused and seemingly intractable problem.

It is now known that the Painted Lady does not regularly overwinter in

Europe and, until 1998, was never recorded as having done so in Britain. The hordes that fly north, breeding as they go in successive broods and even generations, all emanate from North Africa. S. B. J. Skertchly gave the following graphic account of the emergence of the butterflies that produce such a swarm, describing his experience of this extraordinary phenomenon which he had observed while travelling on a camel through the Sudan in 1878:

> 'Our caravan had started for the coast, leaving the mountains shrouded in heavy clouds, soon after daybreak. At the foot of the high country is a stretch of wiry grass, behind which lies the rainless desert as far as the sea. From my camel I noticed that the whole mass of the grass seemed violently agitated, although there was no wind. On dismounting I found that the motion was caused by the contortions of pupae of *V. cardui*, which were so numerous that almost every blade of grass seemed to bear one. The effect of these wrigglings was most peculiar, as if each grass stem was shaken separately – as indeed was the case – instead of bending before a breeze. I called the attention of the late J. K. Lord to the phenomenon, and we awaited the result. Presently the pupae began to burst, and the red fluid that escaped sprinkled the ground like a rain of blood. Myriads of butterflies limp and helpless crawled about. Presently the sun shone forth, and the insects began to dry their wings; and about half-an-hour after the birth of the first, the whole swarm rose as a dense cloud and flew away eastwards towards the sea. I do not know how long the swarm was, but it was certainly more than a mile, and its breadth exceeded a quarter of a mile.'[15]

The destination of this particular swarm would not have been north-western Europe, but this account gives some idea of how such multitudes of butterflies are produced to invade Britain and other parts of northern Europe in phenomenal numbers, as in 1996.

C. B. Williams analysed the numbers recorded in Britain between 1850 and 1955.[16] Peak years, with more than 500 noted, were 1858, 1868, 1879 (with 2,568, the highest to that date), 1883, 1884, 1888, 1892 (2,980), 1894, 1900, 1902, 1903 (3,280), 1906, 1920, 1926, 1928 (2,588), 1931, 1933, 1935, 1936, 1937, 1938, 1939 (4,500), 1940, 1941, 1943, 1945 (6,224), 1949, 1947 (12,000), 1948 (*c.*30,000), 1949, 1950, 1952 (6,700), and 1955. During this period, there were a few years when none was recorded, and all but one of these was in the nineteenth century; 1916 had no record at all and there were seven years in the twentieth century with fewer than twenty sightings. It was again extremely abundant in 1966, 1969, 1980, 1985 and 1988,[17] but by far the greatest invasion ever recorded took place in 1996.[18] The first arrivals were in early spring but the main invasion took place in late spring and early summer (May–June) and this was also the case across Europe.[19] Their numbers were literally countless but estimates of the Painted Lady population in all parts of the British Isles that summer and autumn runs to millions, the total being further boosted by the successful breeding of those that had arrived first. Tens, or even hundreds, of thousands were reported as coming in off the sea at various points along the coast during June,[20] and throughout Britain every garden was likely to have been visited by some of them. In the Garden Butterfly Research counts for 1996 it was by far the commonest recorded butterfly, being much commoner even than the Small White or the Small Tortoiseshell.

The following year proved to be a great anticlimax with very few Painted Ladies in evidence. There had clearly been few if any cases of successful hibernation and those butterflies that were seen were almost certainly new arrivals from the Continent. However, one of the 1997 arrivals in Hayle, Cornwall, was colour-marked in October of that year and seen again in April 1998 – the first proven record of this species having successfully overwintered in Britain.[21] The temperatures, although cold, may never have dropped below freezing in that locality and the butterfly had plenty of suitable evergreen cover in which to hide and receive protection against hard weather.

25. THE AMERICAN PAINTED LADY – *Cynthia virginiensis* (Drury)
Nymphalis cardui virginiensis Drury, 1773
Papilio huntera Fabricius, 1775

The American, Beautiful or Scarce Painted Lady, or Hunter's Butterfly, as it has variously been called, is a very rare visitor from North America, having been recorded no more than twenty times (Plate 32). Although the name

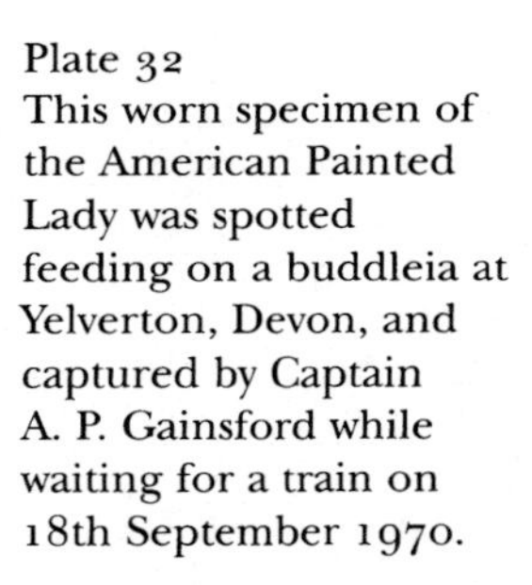
Plate 32
This worn specimen of the American Painted Lady was spotted feeding on a buddleia at Yelverton, Devon, and captured by Captain A. P. Gainsford while waiting for a train on 18th September 1970.

huntera bestowed on it by Fabricius[1] is no longer valid, it has survived in one of the butterfly's alternative vernacular names. Who was this Hunter? According to W. J. Holland, a one-time Director of the Carnegie Museum, he was a remarkable individual.

> 'Captured by the Indians in his infancy, he never knew who his parents were. He was brought up among the savages. Because of his prowess in the chase they called him "The Hunter". Later in life he took the name of John Dunn, a man who had been kind to him. He grew up as an Indian, but after he had taken his first scalp he forsook the red men, no longer able to join them in their bloody schemes. He went to Europe, amassed a competence, became the friend of artists, men of letters, and scientists. He was a prime favorite with the English nobility and with the King of England. He

> interested himself in securing natural history collections from America for certain of his acquaintances, and Fabricius named the beautiful insect shown on our plate in his honour.'[2]

The American Painted Lady is found in the New World from eastern Canada as far south as Venezuela, and has strong migratory tendencies. It has been recorded from the Canary Islands since the turn of the nineteenth century, and from Portugal since about 1958. As with the Monarch butterfly, it was long debated whether the few specimens reported in Britain were the result of migration from the eastern seaboard of North America, or from the Canary Islands or Portugal.

It was, in fact, an Englishman Dru Drury who, in 1773,[3] first described and named the species *virginiensis*, in deference to his great predecessor James Petiver who, having received a specimen from a collector in Virginia, had written '*Papilio Bella Donna dicta, VIRGINIANA, oculis subtus majoribus.* This chiefly differs from our English Painted Lady in having larger Eyes underneath.'[4] However it was not taken in Britain until fifty-five years later.

The announcement of the first British specimen was made by J. C. Dale in a letter to Loudon's *Magazine of Natural History.*

> 'Sir – On the arrival of every new number of your Magazine of Natural History, I am on the look-out for new discoveries in (especially British) entomology, the most extensive branch of natural history, and as such information, I believe, will be acceptable to your readers, I beg to announce (should not Captain Blomer have previously given you the particulars), for the first time, the capture of *Vanessa huntera* in Britain, by Captain Blomer at Withybush, near Haverfordwest, South Wales (about ten miles from a seaport) in July or August 1828; which, till very lately considered by him as a small and odd variety of *V. cardui* (or Painted Lady Butterfly), and which he has very handsomely added to my cabinet.'[5]

Wood (see Plate 36, fig. 8),[6] Westwood[7] and Morris[8] all illustrated the species, and Barrett gave details of the first two records from Britain but discounted it as a genuine immigrant.[9] Bretherton summarized the twenty known British and Irish records up to 1989.[10] Remarkably, the second recorded specimen was found, in August 1871, inside a railway carriage at Wokingham, Berkshire! This is the only inland record to date, all the others being within sight of the coast, and the butterfly had very probably been transported there by the train from some southern or south-western county. Other sightings were reported from southern Ireland, Devon, Cornwall, Isle of Wight, Dorset, Hampshire, Kent, Lancashire, Glamorgan and Sussex. Most recently one was seen by C. G. Lipscomb, who watched and identified it at Penrice Castle, Gower, South Wales, on 28th September 1981.[11] Since then, there have been credible reports of *virginiensis* from Devon and Cornwall in October, 1995, coinciding with an unprecedented invasion of Monarch butterflies *Danaus plexippus,*[12] and of one from the Isles of Scilly in 1998, which was observed on the same day that a Monarch was seen at Wyke Regis in Dorset.[13]

Although there can no longer be any doubt that most if not all of these records are of genuine migrants, the American Painted Lady was ignored by Frohawk in his two illustrated works on British Butterflies and accorded but four lines by South in his *Butterflies of the British Isles*[14] as he regarded its occurrence as accidental.

26. THE LARGE TORTOISESHELL – *Nymphalis polychloros* (Linnaeus)

Papilio (Nymphalis) polychloros Linnaeus, 1758

Plate 33 *(facing page)* This well-observed study of the Large Tortoiseshell by Eleazar Albin (1720), showing its caterpillar and chrysalis on elm, is one of the earliest published colour plates of British butterflies. Only a few of Petiver's, sold coloured, could be said to be earlier.

Although it has been considered[1] that there is a figure of this butterfly in Moffet,[2] neither Ray[3] nor Linnaeus[4] made any reference to it, perhaps because the woodcut is indeterminate and the description imprecise. The first unmistakeable reference to the Large Tortoiseshell was that of Petiver[5] who called it 'The greater Tortoise-shell Butterfly'. Ray described it as a 'Butterfly like that of the nettle [the Small Tortoiseshell] but larger: we call it the elm-butterfly'.[3] The caterpillar too, he added, 'is rather like the preceding' and went on to say that he had found many in 1695 feeding on sallow. In 1720, however, Albin depicted the butterfly with its chrysalis and the larva feeding on elm (Plate 33). Elm and sallow are indeed its two principal foodplants, the butterfly favouring wooded lanes or the borders of woods where these trees occur. Lewin[6] kept Ray's name but most subsequent authors followed Petiver in calling it Great, Greater or Large Tortoiseshell. Rennie,[7] who could be relied on to be independently minded, revived the name Elm Butterfly, but no one else followed suit, and that is the last we hear of it.

Like the Brimstone butterfly, the Large Tortoiseshell lives for up to ten months as an adult but, despite that, it is not seen for much of the year. This is because it is one of the butterflies that hibernates, like its fairly close relatives the Peacock and the Small Tortoiseshell but, unlike them, goes into hibernation shortly after emerging from its chrysalis. In Britain it was on the wing in mid-July but by early August would already have found a place in which to hide, not reappearing until March or early April the following year. In correspondence with C. R. Bree, Henry Doubleday wrote, 'You ask where *Polychloros, &c.* conceal themselves; I reply, in crevices of old trees, sheds, or any convenient places they can find. Last winter some large stacks of beech faggots, which had been loosely stacked up in our Forest [Epping] the preceding spring, with the dead leaves adhering to them, were taken down and carted away and among these were many scores of *Io, Urticae* and *Polychloros* [Peacocks, Small and Large Tortoiseshells]'.[8] Most females are fertilized before hibernation and, soon after they come forth in the spring and have fed, are ready to lay their eggs on sallow or elm twigs, or sometimes rosaceous trees.

Although this fine butterfly has always had a local distribution, it was not uncommon where it occurred. Its history in Britain is one of years of abundance followed by years of great scarcity, but its stronghold has always been the south-eastern quarter of England. It was, however, also known for a long time as far west as Devon and Cornwall and as far north as Yorkshire, with scattered records beyond, including Scotland, but it has never been recorded with certainty from Ireland. Sadly it is now almost certainly extinct as a British resident, its rare sightings being attributable either to releases or to casual immigration.

During the eighteenth century the Large Tortoiseshell seems to have been not uncommon in southern England and was reported as sometimes quite common in the counties around London, though fluctuating from year to year. This pattern continued throughout the nineteenth century, when its main centres seem to have been Essex and east Suffolk, Sussex, the New Forest and Devon, though there were years when it was not commonly found even in these areas. Morris noted that it was 'rather uncertain in its appearance being much more plentiful in some years than others', adding philosophically,

To John Philip Brayne of Dantzick M.D. & F.R.S.

This Plate is humbly dedicated by Eleaz.r Albin.

'but indeed with what insect is this not the case'.[9] At any rate, nobody was alarmed that the sporadic periods of abundance followed by inexplicable scarcity might lead to an ultimate crash.

In 1892, the Revd J. G. Wood recalled his own somewhat unusual first encounter with 'the Great Tortoiseshell', as he called it: 'The first specimen that I ever took I saw in the window of a grocer's shop in Oxford, one of the very last places where one might have expected such a Butterfly to be found. It was quite plentiful in Bagley Wood, where any number could be taken, and had evidently been blown into the streets and then attracted by the sugar in the window'.[10]

This butterfly had one of its periods of comparative abundance at the turn of the century, with 1901 being reported as a wonderful year for this species in north Essex and Suffolk. Harwood wrote that he 'could have taken hundreds of broods had [he] required them',[11] on both sides of the River Stour. The following year was also a bumper one in Sussex, the best for over twenty years,[12] but then in 1903 the population collapsed; it became rare in Sussex and, from 1905, very scarce in East Anglia too. Other counties fared no better, and the next forty years saw this species in an apparently irreversible decline. Then, for no reason obvious at the time, there was a revival in fortunes, and in the mid-1940s, when England enjoyed a run of warm summers, the Large Tortoiseshell again became relatively abundant in many of its former localities. Were there invasions in 1945 and 1947, as some authors believed?[13, 14] By 1949 it had yet again gone into decline and since that year only occasional individuals have been recorded. Was its failure to stage another recovery the result of climate change, or was it due to some other cause, like parasitization? South noted that 'from fifty chrysalids only one butterfly resulted',[15] all the others being found to be filled with parasites. In another case, of 100 caterpillars, some collected when quite small, only one was not ichneumonized. A combination of these factors may have rendered impossible the recovery of this species from one of its characteristic 'lows', and led to its extinction as a breeding resident.

J. A. Thomas reported that 'of its 120 or so sightings made here since 1950' most were either misidentifications or specimens imported from the Continent. He made the arresting comment that the Large Tortoiseshell is now much rarer than the two spectacular migrants, the Camberwell Beauty and the Monarch, the former now being three times and the latter twice as often seen.[16]

The most recent records noted are from Woburn, Bedfordshire, in 1991;[14] Bathpool, Cornwall, in August 1993;[17] St Leonards-on-Sea, East Sussex, in August 1991;[18] Worthing, Sussex on 14th March 1995;[19] Bradwell on Sea, Essex, on 24th April 1994 (R. Dewick, pers.comm.) and 24th August 1995;[19] again from St Leonards on Sea in March 1997;[18] and Brightlingsea, Essex, on 28th August 1999, the eighth Essex record since 1990.[20] Some of these from southern coastal counties might have been genuine immigrants, and those from around the mouths of the rivers Blackwater and Colne in East Essex could even indicate a small breeding colony. However, it is likely that most of these records result from unauthorized releases of Continental specimens. One of the recent Large Tortoiseshells in Britain was found fluttering against the window of a garden shed, where it had apparently been hibernating since the previous year and had been roused prematurely from its sleep by the warmth. The finder therefore placed it in a breeding-cage in a cold, dark room which sent it back into hibernation for just over a fortnight until it showed further signs of movement. It was then fed with diluted honey and released into the garden where it stayed for a while, allowing itself to be photographed before departing.[18]

27. THE CAMBERWELL BEAUTY – *Nymphalis antiopa* (Linnaeus)
Papilio (*Nymphalis*) *antiopa* Linnaeus, 1758

Although the Camberwell Beauty was figured by Moffet[1] and was later described by Ray,[2] it is fairly certain that neither at that time considered it a British species. In *The English Moths and Butterflies* (Plate 34), Wilkes described the first capture in England of what he named 'the Willow Butterfly', as follows: 'About the middle of August, 1748, two of this species of butterfly were taken near Camberwell in Surrey. But in all my Practice I have never seen any of them in the fields; so they must be looked upon as very great rarities.'[3] Some years later, in 1766, Harris wrote a better-known account of its discovery in *The Aurelian*, and included a good description of its life history (Fig. 139). There

The GRAND SURPRIZE. Or CAMBERWELL BEAUTY.

THIS is one of the ſcarceſt Flies of any known in England, nor do we know of above three or four that were ever found here, the firſt two were taken about the middle of Auguſt 1748, in Cool Arbour Lane near Camberwell; the laſt in St. George's Fields, near Newington Butts, the beginning of that Month; but as theſe appeared very much faded and otherways abuſed, I conclude they appear from the Chryſalis, with the Peacock, about the middle of July, and being of that Claſs tis reaſonable to ſuppoſe they live thro' the Winter in the Fly ſtate, and lay the Eggs in Spring that produce Flies in the July following; for in the ſame manner do all the Flies of this Claſs, and as all that have yet been taken were found flying about Willow-Trees tis the common opinion of Aurelians that their Caterpillars feed thereon; but their Caterpillar and Chryſalis is to us intirely unknown, and the food a mere conjecture. I do intend to make a ſtrict ſearch concerning them and ſhould I make any diſcoveries worthy note I ſhall find a proper place and repeat it. The Fly in the plate at *(d)* was drawn and Coloured from a beautiful large Female in the Cabinett of Charles Belliard Eſq; which is the fineſt we have in England; the underſide is ſhewn at *(e)*. The Caterpillar and Chryſalis were drawn from thoſe of M. Roſels who gives the following account of this Fly.
' This Caterpillar which is the largeſt of this Claſs, ſays he, is called the So-
' ciable Caterpillar becauſe never found alone, but always to be met with in com-
' pany with others of its kind, they feed on all kind of Willow-Trees and exiſt

Fig. 139. Moses Harris was the first to give the Camberwell Beauty the name by which it has been known for over a century. However, the text accompanying his Plate also gives his preferred name, reflecting the excitement this spectacular new British butterfly generated.

is no doubt that he was referring to the specimens reported by Wilkes, although he used the names 'Grand Surprize' and 'Camberwell Beauty': 'This is one of the scarcest Flies of any kind in England, nor do we know of above three or four that were ever found here, the first two were taken about the middle of August 1748, in Cool Arbour Lane near Camberwell; the last in St. George's Fields, near Newington Butts, the beginning of that Month.'[4]

The then village of Camberwell has long since been built over. What is now called Coldharbour Lane is a congested highway, and endless rows of houses now lie over once open fields (though there is a plaque showing the Camberwell Beauty above the road sign). In 1860, W. S. Coleman looked back to the years 'when Camberwell was a real village, luxuriating in its willows, [and] the entomologists of the day were delighted by the apparition, in that suburb, of this well-known "Beauty" whose name since then has always been associated

Plate 34 (*facing page*) In this plate from *The English Moths and Butterflies* (1749), Wilkes was the first artist to illustrate the Camberwell Beauty which he named the Willow Butterfly. He very likely drew it from one of the original specimens taken the previous year 'about the middle of August 1748, in Cool Arbour Lane near Camberwell'. The illustrations of the early stages were based on those in a Continental work by A. J. Roesel von Rosenhof, *Insecten-Belustigung*, Volume 1, published in 1746.

with Camberwell'. Facetiously, he added that it was no longer a promising place for a butterfly hunt, for, 'although it has its "Beauties" still, they are not of the lepidopterous order, nor game for any net that the entomologist usually carries'.[5]

After Wilkes's discovery there were no more records of this butterfly for almost forty years until William Lewin happened upon an apparent colony at Faversham, and reported his outstanding luck as follows:

> 'The middle of August, 1789, I was surprised with the sight of two of these elegant flies, near Feversham [*sic*], in Kent; one of which I thought it great good fortune to take; but in the course of that week I was more agreeably surprised with seeing and taking numbers of them in the most perfect condition. One of my sons found an old decoy pond, of large extent, surrounded with willow and sallow trees, and a great number of these butterflies flying about, and at rest on the trees, many of which appearing to be just out of the chrysalis, left no room to doubt, that this was a place where they bred. In March, 1790, a number of these insects were flying and soaring about for the space of twelve or fourteen days; and then, as if with one consent, they migrated from us, and were no more seen.'[6]

Although Camberwell Beauties were not seen again at Faversham, Donovan reported them shortly afterwards from other places in some abundance: 'There have been several instances of this Insect being found in different parts of the country in mild seasons, as plenty as the Peacock, or Admirable [Red Admiral] Butterflies; in the summer of 1793 particularly, they were as numerous in some places as the common White Butterfly is usually near London.'[7] Donovan, though, was critical of Harris's use of names: 'Harris, in his Aurelian, calls it the *Camberwell Beauty*, though in his list of English Butterflies, Hawk-Moths, and Moths, he uses the name *Grand Surprise*; we mention this circumstance, as it appears very inconsistent that the new name he adopts in one work, and the old one he should have discarded in the other, are equally and indiscriminately used in the several editions of both ... from which it might be readily inferred, that he meant two distinct Insects, were it not for the addition of the Linnaean name, *Pap. antiopa*.' Donovan was being rather unfair, since Harris used the name 'Grand Surprise' in both books and, there appears little reason for confusion. However, at the end of the eighteenth century it seems there were at least three, and probably four, English names for this butterfly: The Willow Butterfly, the Camberwell Beauty, the Grand Surprise and the White Border, the last being used by Haworth in 1803.[8] Berkenhout,[9] Lewin[10] and Rennie[11]continued to use Wilkes's original name, but Jermyn[12] and Newman[13] adopted Haworth's name, though modifying it to White-Bordered. Coleman[5] and South[14] reverted to Harris's alternative name, the Camberwell Beauty, which has remained in use ever since. By then it had long ceased to be such a 'Surprise' and so that name fell out of use! On the Continent, the Germans know it as '*Trauermantel*', which the North Americans who also have the species translated as Mourning Cloak, after its sombre colours, below which a pale underskirt seems to protrude. The Swedes, in the same idiom, call it the '*Sorgmantel*' or Cloak of Sorrow.

Throughout the nineteenth century, accounts of captures of the Camberwell Beauty appeared with regularity in the entomological journals. One such

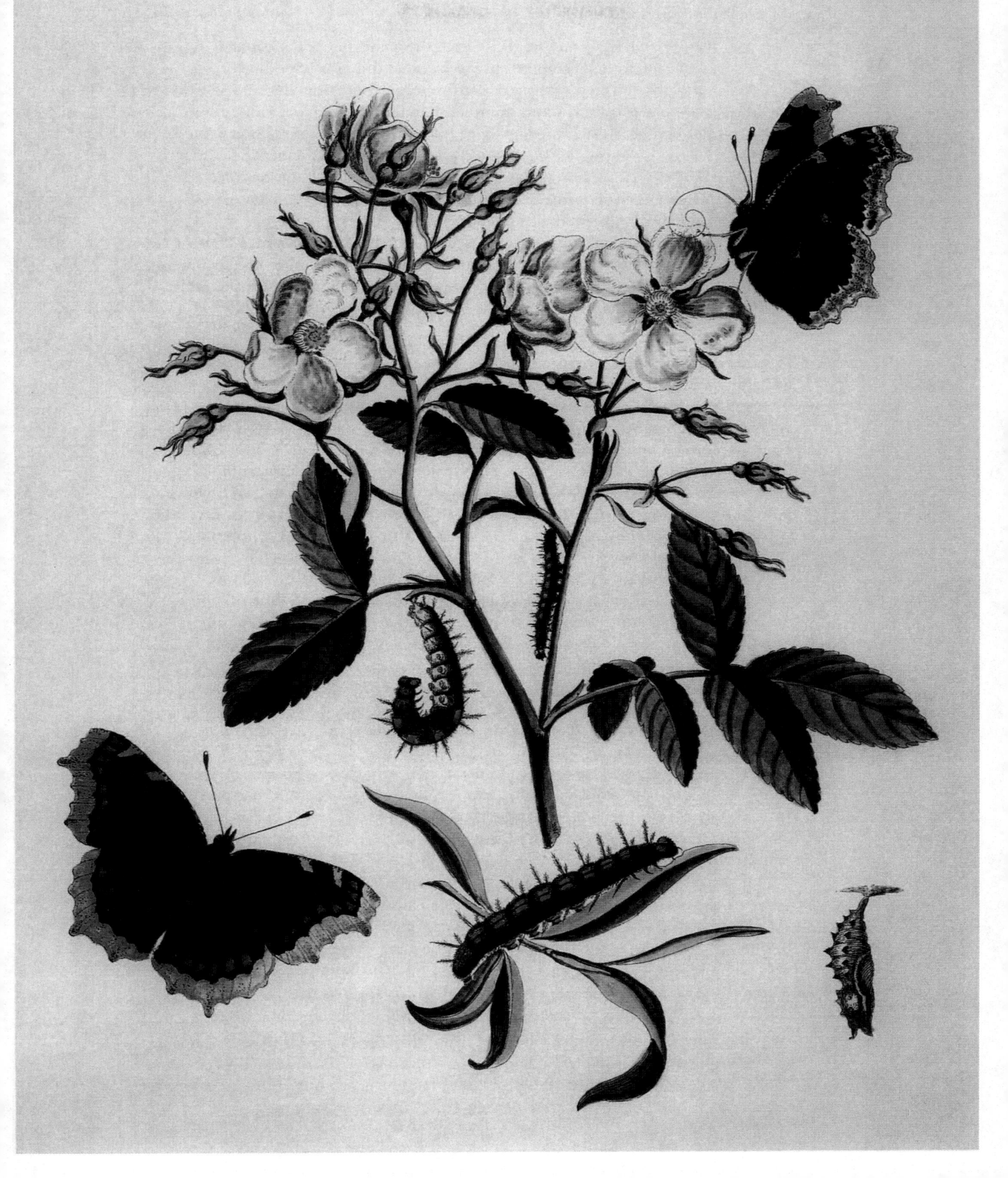

was submitted by the Revd H. Stuart Taylor, and published in the *Entomological Magazine* of 1838:

> 'On the 31st day of August, this year, the specimen of *Vanessa Antiopa*, of which the accompanying is an exact draught [drawing], was taken by my pupil, Edward Pemberton. It is a female; and though much lacerated in both hinder and one fore wing, was a fine strong creature. It was captured about one o'clock, in an angle of a field which is on the west or south-west side of a wood called "Turner's Wood", near to "Caen Wood" [Kenwood], Hampstead, and the property of Lord Mansfield. ... A few days before, I had, twice in the day, observed a Butterfly which I could not make out – large and black to all appearance – flying over the highest part of the wood, as if it were moving in an accustomed haunt. I have now no doubt that it was an *Antiopa*, probably this very specimen.'[15]

Like the Clouded Yellow, famed for its '*edusa* years', the Camberwell Beauty also provides great excitement during the far rarer '*antiopa* years'. It seems to have invaded Britain in large numbers in 1789–90 and 1793. It did so again in the first well-recorded invasion around 1820, when William Backhouse later reported 'vast numbers of this species strewing the sea-shore at Seaton Carew [near the mouth of the Tees], both in a dead and living state'.[16] He added: 'It is surely more reasonable to suppose that these specimens had been blown from the land than that they had crossed a sea at least three hundred miles; and a specimen in Mr Backhouse's collection confirms this opinion, as it has the pale whitish margin to the upperside of the wing so characteristic of our British specimens, which is replaced by yellow in nearly all the Continental and American specimens.'

The '*antiopa* years' were probably 1789 (when Lewin found his 'colony'), 1793, 1820, 1846, 1872, 1976 and 1995, of which 1872 was undoubtedly the greatest (if we exclude Donovan's unsubstantiated statement concerning 1793), with approximately 450 recorded from as far apart as Kent and Roxburghshire, and reports appearing in the daily newspapers, even on their front pages. In terms of magnitude, 1820 and 1872 are followed by 1995 with around 350[17] and then 1976 with about 300.[18] The other years produced smaller peaks. In between these rare occasions, visitations of this beautiful butterfly have been in comparatively small numbers.

Over the past fifty years several attempts have been made to naturalize the Camberwell Beauty, but they have always failed. Between July 1956 and August 1961, 500 marked butterflies of Continental origin were released by L. Hugh Newman at selected sites in Kent and Hertfordshire,[19] but no account was ever published to show whether these releases resulted in even short-lived breeding colonies and it must be presumed that they were unsuccessful.

The origin of our Camberwell Beauties was a cause of controversy from the time they were first recorded. Lewin's discovery of a large number in Kent in 1789 led him to believe he had found a colony, a belief reinforced in March 1790 when a number were seen to be still present. However, it is more probable that those present at the end of 1789 had just arrived, and that those there the following spring had emerged from hibernation, confirmed by his statement that 'they migrated ... and were no more seen'. In the *Entomologist's Annual*, following the great *antiopa* year of 1872, Knaggs posed the question, 'Is *Vanessa Antiopa* a native or an immigrant?' and went on to support Stainton's theory that 'a flight of these beautiful insects took place last August from Sweden and Norway to our east coast'. After enumerating the points in favour of this explanation he continued:

> 'Now, amongst the recorded captures are some, we cannot say how many, yellow-bordered specimens; and Mr. Stainton informs us that both forms occur in Holland, a country situated about one hundred miles from the coast of Norfolk and Suffolk, giving over three and-a-half minutes per mile for a six hours' journey.
>
> 'It seems to me that the migratory hypothesis is the only one which can hold water, for it is inconceivable that the earlier stages of the insect, supposing it to have bred in the country, should have been so entirely overlooked; and again, if our visitors had been true British born, it is only natural to suppose that their appearance in the North would have been considerably later than in the South, whereas such was certainly not the case.'[20]

Despite this strong argument in favour of migration, in the next issue of the *Annual* he reported that a 'Mr. [T. J.] Bold, of Newcastle, one of our most able entomologists, in a recently published pamphlet, declines to adopt the "flown over" theory at any price; he will not have it that they came over to us from Holland, a distance of only one hundred miles, in spite of the fact that the species was unusually common in that country previous to its occurrence in this.'[21]

Two forms of the Camberwell Beauty are to be found. Most British-caught specimens have white or extremely pale yellow borders to the wings, but some appear with wing-borders of a deep yellow (Plate 35). In 1803, Haworth, after dismissing their Continental origin as 'an idle conjecture', went on to claim that 'the English specimens are easily distinguished from all others by the superior whiteness of their borders'. 'Perhaps', he added, 'their eggs in this climate, like the seeds of some vegetables, may occasionally lie dormant for several seasons, and not hatch, until some extraordinary but undiscovered coincidences awake them into active life.'[22] Until the end of the Victorian period, it was always thought that specimens with white borders were of native stock and that those with yellow borders were migrants from the Continent. Some unscrupulous collectors even went so far as to paint the yellow borders of Continental specimens white! However, in 1921 E. A. Cockayne found that the white-bordered British specimens had a wing-scale defect, also present in some Scandinavian examples, which suggested a common origin.[23] Some time later it was shown that butterflies that had overwintered usually lost the fresh yellow coloration of the marginal bands.

The thrill of finding a Camberwell Beauty – the Holy Grail of butterfly collectors – is epitomized in a graphic account by Siegfried Sassoon, the poet and author. Sitting in the attic of his Kentish home, he was disturbed by the flutterings of a butterfly imprisoned between the skylight and the gauze that was tacked over it to soften the glare.

> 'By standing on a chair – which I placed on a table – I could just get my hand between the gauze and the glass. The butterfly was ungratefully elusive, and more than once the chair almost toppled over. Successful at last, I climbed down, and was about to put the butterfly out of the window when I observed between my fingers

that it wasn't the Small Tortoiseshell or Cabbage White that I had assumed it to be. Its dark wings had yellowish borders with blue spots on them. It was more than seven years since I had entomologically squeezed the thorax of a "specimen". Doing so now, I discovered that one of the loftiest ambitions of my childhood had been belatedly realized. I had caught a Camberwell Beauty.'[24]

28. THE EUROPEAN MAP BUTTERFLY – *Araschnia levana* (Linnaeus)
Papilio (Nymphalis) levana Linnaeus, 1758

This pretty little butterfly (Plate 36, figure 9) was first described by Linnaeus as two distinct species, *Levana* and *Prorsa*, to which, while recognizing that they shared the same foodplant, he assigned separate localities. Linnaeus's error was understandable since this butterfly is a quite remarkable example of dimorphism, not between the sexes, as is most commonly the case, but between the spring (May/June) and the summer (July/August) generations, which gives them the appearance of being distinct and dissimilar species. It belongs to a small group of butterflies which are popularly if a little fancifully known as Map butterflies because the pattern on the underside of the wings resembles a map illustrating physical features. It is called the European Map not because of any resemblance to the map of Europe, but because it is the only member of the genus found in Europe.

This butterfly's strongest claim to a place on the British list must rest on the specimen taken in Surrey in 1982. It had been swept from bilberry undergrowth by D. Down at Friday Street, near Dorking, on 21st May at a time when a considerable invasion of Red Admirals and other immigrants was taking place.[1] The fact that it was described as 'in mint condition', far from undermining this conclusion actually supports it since, to cite P. B. M. Allan, 'the fresher the condition, the greater the likelihood of immigration'.[2] Bretherton was of the same opinion as Allan,[3] especially as the species has been spreading in northern Europe and is found in France as near as the Channel coast, and has recently even reached southern Sweden.[4] Nevertheless, G. A. Collins,

Plate 35 *(left)*
The Camberwell Beauty – 'the Holy Grail of British butterfly collectors'. Specimens with white wing-borders have been shown to possess a wing-scale defect in common with some of those from Scandinavia, suggesting a common origin.

Plate 36 *(right)*
This plate from *Index Entomologicus* includes the Map butterfly (figure 9) among eighteen other 'Doubtful British Species' of butterflies, drawn by W. Wood jnr for his father's undeservedly neglected work published in 1854. Other species depicted include (1) the Scarce Swallowtail; (2) the Moorland Clouded Yellow; (3) the Apollo; (6) the Niobe Fritillary; (7) Albin's Hampstead Eye; (8) the American Painted Lady; and (17) the Mallow Skipper. The superb miniature illustrations, here enlarged, are all hand-coloured but not to scale.

writing on Surrey's butterflies, considered it was more likely an escape from captivity.[5]

The earlier history of the Map butterfly in Britain is rather a cloak and dagger affair. It was secretly but successfully introduced to an unknown locality on the borders of Herefordshire, Gloucestershire and Monmouthshire in or around 1912 and two colonies became established. Records of finds in these areas appeared in *The Entomologist* in successive seasons. H. Rowland-Brown exhibited a specimen at the Entomological Society of London on 1st October 1913 that had been captured at the end of May that year by T. Butt Ekins on the outskirts of the Forest of Dean.[6] The following year, A. W. Hughes reported capturing two specimens at Symonds Yat, a few miles to the west in the Wye Valley, between 20th and 24th July, and also mentioned another collector who had taken nearly a dozen examples several miles away.[7] In the same number of *The Entomologist*, G. B. Oliver reported that '*levana* was about in the Forest during the latter half of July last. 8 specimens fell to my share and I heard of five others taken'.[8] However, after the summer of 1914 it disappeared, possibly due to the equally furtive but destructive actions of a then anonymous collector.

Many years later, following correspondence in *The Entomologist* on the subject of 'Liberated Butterflies', Frohawk expressed his long-held view that 'the introduction of foreign species with the object of liberating them in this country is an unwise and disturbing undertaking, as it completely upsets all the work that is being done relating to the study of migration and carefully collected records by scientific societies and naturalists generally'.[9] Referring to the colony of Map butterflies supposedly destroyed by a 'vandal' (as other entomologists had termed the perpetrator of this deed), he continued, '... I happened to be staying shortly afterwards with the "vandal", who was recognized as one of the leading British entomologists, and none other but the late A. B. Farn. He told me that directly he was informed of the liberation of *A. levana* he went to the locality and caught all he could purposely to destroy them, as he was greatly opposed to the introduction of anything foreign to either the fauna or flora of this country' (see Chapter 3: 60). Recently, Thomas has expressed the view that the demise of these colonies was more likely due to the fact that the habitat was not entirely suitable than to the actions of Farn.[10]

In reviewing this controversial but at least partially successful introduction and its subsequent 'destruction' by human agency, E. B. Ford thought the action of the individual an 'arbitrary' and 'improper one', believing that he should at least have consulted a body such as the Entomological Society 'before taking a step which was by no means his personal business alone and on which many other people had as much right to an opinion as he'.[11] Ford himself believed that the introduction of the Map butterfly was less exceptionable than the establishment in a new area of a British species, and would not object to a further attempt to naturalize it in Britain. T. G. Howarth, too, had a sneaking sympathy with the attempts to naturalize it, but recognized the arguments on the other side that subsequent records of natural migration and establishment could be compromised.[12] There is now a strict code on insect re-establishment issued by the Joint Committee for the Conservation of Insects, and a code of practice on Butterfly Releases, published by Butterfly Conservation. These are clearly summarized by David Dunbar in a useful book also published by Butterfly Conservation (see Appendix 2). At present the unlicensed release of further Map butterflies in the countryside would be unlawful under the Wildlife and Countryside Act, 1981.[13]

29. WEAVER'S FRITILLARY – *Boloria dia* (Linnaeus)
Papilio (Nymphalis) dia Linnaeus, 1767

It is improbable that this attractive little fritillary ever reached Britain under its own wing-power. Although it is found as close as France and Belgium, it is not known to migrate. It is almost certain that all the known British specimens of Weaver's Fritillary were accidentally or deliberately introduced. Coleman, however, included it in his book on British butterflies as a 'reputed British species' (see Plate 15, fig. 4), remarking that 'there is little reason to doubt that this insect was really taken by Mr. Richard Weaver in the 1820s at Sutton Park, near Tamworth, [Warwickshire]; also by Mr. [E. J.] Stanley, near Alderley, in Cheshire.'[1] The Revd William Bree, who examined the Weaver Collection at Birmingham, also considered these records were genuine, and went so far as to publish two excellent illustrations of the butterfly (Fig. 140) with a

Retrospective Criticism. 751

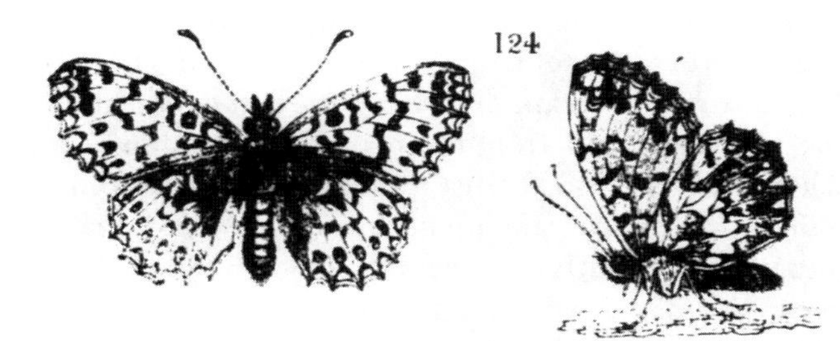

more than five or six. It differs from M. Selène in being rather smaller, and having the black spots and characters on the upper surface of both pair of wings larger and stronger, so that the whole assumes a darker appearance than that insect. But the principal difference consists in the under side of the posterior wings, which are of a brownish purple, interspersed with darker markings of the same colour, and numerous irregular semi-metallic spots; a row of which borders the posterior margin. I would invite the attention of entomologists to this insect, which deserves minute investigation. As Mr. Weaver has taken two examples of it in the same place, in different and distant seasons, it may possibly prove a distinct species; but, for the present, I would refer it to M. Selène, of which it is at least a very strong variety.

Fig. 140. In 1832, Loudon's *Magazine of Natural History* published this excellent engraving of 'Weaver's Fritillary', supposedly caught at Sutton Park by Richard Weaver. The authenticity of this and subsequent records of Weaver's Fritillary has been hotly disputed.

helpful description of the differences between it and the Small Pearl-bordered Fritillary (*Boloria selene*)[2] as, at that date, Weaver's Fritillary was believed by many to be no more than a variety of the latter species.

Of Richard Weaver (see Chapter 3: 27), the first person to collect a specimen in Britain, Allan wrote that it was ill luck that his 'name is destined to be for ever associated with a fritillary that is not, and possibly never has been, a native of this island', adding that he 'was in no wise in the same class with [the unscrupulous dealers] Plastead, Raddon & Co'. He quoted a contemporary entomologist who regarded Weaver as one of the 'best field entomologists ... and a very accurate observer ... a most persevering collector and the first to bring to notice many of the scarcer insects, and his word was never doubted except about *M. dia*'.[3] It seems that a case of insects sold by Weaver somehow contained a specimen of *dia*. 'Weaver, on being questioned, said that if it was amongst his specimens it must have been taken by him ... with other fritillaries, in a wood near Birmingham'.

An animated correspondence over a second apparent record of 'Weaver's Fritillary' appeared in *The Entomologist's Weekly Intelligencer* in 1857. The Revd B. Smith had written concerning the capture of the Fritillary which had been 'knocked down by a village lad with his cap' near his home at Marlow and brought to him by a young collector, the Hon. C. A. Ellis.[4] In his next editorial, H. N. Stainton questioned the probability of this and similar records of non-indigenous species though tactfully not the good faith of those who submitted them.[5] In response, Ellis gave further particulars. 'The specimen in question was taken about the middle of September last, in the Rev. S. Hodson's garden, at Cookham Deane, near Maidenhead. As I killed and set the specimen myself, I am quite certain that it is undoubtedly British; and I trust that this will set the question at rest.'[6] But it did not. John Scott, a distinguished entomologist, was sceptical, and drew attention to two apparently contradictory statements regarding its capture: '"Intelligencer", No.60, says "it had been knocked down by a village lad with his cap, and *was pinned and set in corresponding style*." "Intelligencer", No.64, says, from the pen of *another* individual, "*I killed and set the specimen myself*," &c., and "I trust this will set the question at rest." And, judging from these two distinct statements, my idea is that it makes the thing more doubtful than ever.'[7] Smith responded as follows: 'All we ask is fair play, and I think the capture of *Dia* will have to be admitted by the most sceptical. Mr. Ellis is not in this neighbourhood at present, but, on his return, I will call his attention to the objections of Mr Scott. If I mistake not, there were more witnesses of *Dia*'s last dying speech and confession than one.'[8] Scott, however, was far from pacified, and retorted, under the provocative headline '*Fiat justitia – Dia-bolical outrage*': 'My attention has been drawn still more to the statements in Nos. 60 and 64 of the "Intelligencer", and I perceive that the facts are reconcileable [sic] as there stated, as I presume the boy captured and pinned the insect, which was thus taken to the Hon. C. Ellis, who killed and set it. Is the "*Dire* occurrence" intended for a joke? If so, I should be glad to see it.'[9] Thomas Parry from South Wales joined the fray asking , in the same vein, 'Are the village lad and the Hon. – – one and the same person?'[10] His sarcastic question remained unanswered and, after a few more sallies, the correspondence was brought to an end by the editor.

In 1873, Weaver's Fritillary was back in the entomological headlines when T. Batchelor claimed in *The Entomologist* that: 'On the 23rd [July] I was fortunate enough to take two specimens of *Melitaea Dia* on thistles, in the open spots in a wood, in this neighbourhood [Southborough, Kent]; they were in fine condition, apparently just emerged from the chrysalis.'[11] Of this record, Knaggs sarcastically recalled that: 'A Mr. Batchelor states that he has found a magpie's nest stuffed with about half a hundred cocoons of *C. processionea* [the Oak Processionary], and that he captured *Argynnis dia* in Kent last July. Last year he figured as the discoverer of the South European species *Syntomis phegea* [the Nine-spotted]; next year he will probably treat us to *Saturnia pyri* [the Great Peacock Moth] and *Sphinx Carolina* [the Tobacco Hornworm]. Let us hope that the cocoons of the "Processionary" paid him out'.[12]

Three years later, W. Arnold Lewis claimed another record: 'I have to announce an undoubted British specimen of this fritillary. It is a female, and was taken in 1872 at Worcester Park, Surrey, by a connection of my own, Master Wallace A. Smith. He could not identify his capture, and placed it apart by itself. Very recently, on my looking over his insects, he drew my attention to the specimen as something peculiar; he perfectly recollects making the capture and the exact spot where it was made. I found the specimen pinned and set in beginners fashion. Mr Wallace Smith has never had to do in his life with any dealer or collector; and except things given to him by me, his cabinet contains nothing which he did not catch himself.'[13]

In 1892, C. G. Barrett received, from the Revd E. N. Bloomfield of Hastings, a drawing of a butterfly he identified as Weaver's Fritillary. According to Bloomfield:

> 'It was taken some years ago by Mr. J. C. Arnold, of Hastings. He was collecting a few butterflies and moths about 1876, when he observed two small fritillaries flying together, and secured them both. He took them home and set them, supposing they were "pearl-bordered" (*A. Euphrosyne*); but this year, taking some fresh specimens of the latter, he threw away one of the old ones, and was about to discard the other, when he observed to his surprise that the markings of this specimen differed materially from those of the fresh *A. Euphrosyne* and of *A. Selene*. He says that they were taken on heathy ground in Sussex, somewhere near Tunbridge Wells. ... This would be early in July.' Barrett added that 'There appears to be no reason for doubting the *bona fides* of these statements.'[14]

Of the other records of Weaver's Fritillary, the one taken by N. S. Brodie near Christchurch, South Hampshire [now Dorset], 27th July 1887, is in the Hope Department at Oxford, and another taken by E. W. Platten near Bentley Wood, Ipswich, East Suffolk on 16th May 1899, is in the Ipswich Museum. A specimen taken by Phillip Cribb on the North Downs in Surrey on 24th August 1984 is believed to be one of a number released by a local breeder.[15]

30. THE QUEEN OF SPAIN FRITILLARY – *Argynnis lathonia* (Linnaeus)

Papilio (Nymphalis) lathonia Linnaeus, 1758

It was Moses Harris who coined the name Queen of Spain Fritillary.[1] However, by the time of its first capture in England, at Gamlingay, Cambridgeshire, some sixty years earlier, it had already been described by Petiver under the name of the Riga Fritillary,[2] from a specimen presented to him by the German naturalist Dr David Krieg, which he had collected in Latvia. Ray also included this species as a 'Butterfly from Riga',[3] adding that it had been found around Cambridge by Vernon, Antrobus, and others. There is also a fragment of a specimen in the Buddle herbarium in the Natural History Museum from about the same period but it lacks data. However, it would seem that several specimens of this species may have been taken in England at about that time. Other English names were given it, including the Lesser Silver-spotted Fritillary by Donovan,[4] and the Scallop-winged Fritillary by Lewin.[5] Lætitia Jermyn, however, used Harris's name, and her description of it as 'the most beautiful British Fritillary'[6] is one with which few would disagree; Rennie, that inveterate coiner of names, called it 'The Princess'.[7]

The Queen of Spain Fritillary is known today as a very rare migrant, though one which on a few occasions may have bred here and even produced short-lived colonies. The eighteenth-century Aurelians, however, considered it to be a native of Cambridgeshire, with a well-known locality at White Wood near Gamlingay (now Gamlingay Wood Nature Reserve). This ancient wood on the Lower Greensand and boulder clay of west Cambridgeshire has been noted for rare butterflies since the late seventeenth century. Whether the Queen of Spain Fritillary, Bath White and other desirable butterflies were being bred and released there by some unscrupulous hand will probably never be known. If so, it could not have been the same person, since a full century separates the first capture of a 'Queen of Spain' by William Vernon around 1700 from the equal good luck of Charles Abbot in 1803. Among others who took the Queen of Spain at Gamlingay was 'an old London Aurelian, of the name of Shelfred', who, according to Adrian Haworth, 'was so much attached to Aurelian amusements, and so much enamoured of the beautiful and rare *Lathonia*, that he absolutely determined upon, and accompanied by his daughter, successfully performed (in postchaises) a journey to Gamlingay, in pursuit of that charming *Papilio*, which he had the good fortune to meet with and secure; but his specimens are not now extant'.[8]

Given that butterflies like the Glanville Fritillary and Black-veined White were then so much more widespread, and given the readiness of the Queen of Spain Fritillary to colonize parts of the East Anglian coastline in hot years, it is not impossible that a breeding colony occurred here in the eighteenth century as the records seem to indicate. That, however, some skullduggery was going on for at least part of that time is indicated by the reports of non-British butterflies like the Niobe Fritillary and Scarce Swallowtail from the same locality.

As was the case with other butterfly rarities, the activities of certain notorious dealers in insects, both in selling Continental specimens as British and in claiming to have discovered breeding colonies in out of the way places, muddied the waters to such an extent that verification of many nineteenth-century records of the Queen of Spain Fritillary are made difficult, if not impossible (Plate 37). In consequence, what may have been genuine sightings and accounts of breeding have had to be regarded with suspicion and discounted. For example, Edward Newman, in October 1868, had to publish the following note in *The Entomologist*: 'An account of the capture of thirteen Lathonias near Canterbury has been received but declined'.[9] Unfortunately, the author of the rejected note, George Parry, who reared *lathonia* at his home in Canterbury for local release and to supply his patrons with 'British' specimens, was too fly for Newman and obtained affidavits from local collectors in support of his claims. In consequence, Newman felt obliged in 1872 to publish the following letter from Parry: 'I have sent a specimen of Lathonia alive for you to see, one of four which I took yesterday, August 4th, at Swarling Downs, the same place where I took the species in 1868, which, appeared to have been doubted by many. I took the four specimens off the viper's bugloss, all in about five minutes: it was mizzling rain at the time. There are three very fine ones out of the four.'[10] One month later, Parry was at it again: 'On Friday, the 6th inst., two specimens of Lathonia were taken at Swarling Downs. On Saturday, the 7th, I took eight more, and they are all very fine but two; one is a beautiful female.'[11] Parry's motives were clear enough: he was in effect advertising his services, and his descriptive writing was certainly very imaginative! Henry Doubleday, greatly upset by what he saw as an evil influence pervading entomology at the time, felt compelled to protest, claiming in *The Entomologist* that the majority of

Plate 37
Allegedly British specimens of Queen of Spain Fritillary.
The top one, taken by a 'Mr Gray Jr' at Dover in September 1882, may be genuine.
The others 'netted at Dover by J. Salvage, August 1882', were probably
imported by one of the 'Kentish buccaneers'.

specimens of *lathonia*, and other rarities, 'in collections of professedly British Lepidoptera, are in reality continental'.[12]

Doubleday's outburst echoed that of Edward Newman, writing many years earlier on the subject of 'butterflies questionably British' under the pseudonym 'Inquisitor' in *The Entomological Magazine*. Inquisitor's analysis remains astonishingly topical, as those species he accepted as British and those he did not have scarcely changed. Of the dubious rarities which, supplied by dealers, are to be found in so many collections, he said that: 'it would be far better to forget that such insects have ever been recorded as British, and should they hereafter occur, I would reintroduce them as entire novelties'. Of 65 species of butterflies he considered to be unquestionably British, 'the majority of [specimens of] the rarer ones, as *Daplidice, Lathonia, Antiopa*, although exhibited as British, are decidedly and evidently exotic'. Inquisitor added that he had written under a pseudonym 'in order that I may not offend gentlemen possessing rich series of the questionable species'.[13] The activities of the rogue dealers responsible for this dubious trade have been recounted entertainingly and at some length elsewhere by P. B. M. Allan.[14]

Despite its history of fraudulent releases, the Queen of Spain Fritillary is a genuine migratory butterfly and its occurrence in England, though rare, has continued at intervals since its first reported sighting in the early eighteenth century. During the nineteenth century true immigrants were undoubtedly recorded from various counties, on all but a few instances in the south of England, but rarely in more than single figures. However, interesting records exist from Suffolk, where it may have bred and successfully produced a second generation on more than one occasion. Writing in 1862, William Gaze reported an account by a taxidermist from Sudbury called Barwick who dealt in a minor way in insects which he took locally. Apparently he and his son were visiting Assington Thickets (now known as Assington Thicks) and saw a caterpillar, suspended from a hazel branch, in the process of pupating. They left it undisturbed and returned to it on their next visit by which time it had turned into a chrysalis which they took home. Shortly afterwards, a fine *lathonia* emerged, so his son returned to the site next day and captured two more specimens on the wing. Barwick said they then sold them to a travelling dealer in insects for five shillings [25 pence].[15] Gaze had himself taken a Queen of Spain Fritillary on 3rd September 1858, about two and a half miles from Lavenham and about six from Assington. When Gaze showed Barwick his collection containing this specimen, the 'bird-stuffers' immediately identified it which convinced him of the truth of their story. Barrett,[16] and Allan[17] both accepted this account as genuine which, if so, is the first confirmed report of this species breeding in the wild in this country, possibly following the arrival of a gravid female from the Continent.

In exceptional '*lathonia* years' like 1868, 1872 and 1945, up to 50 were taken in counties from Kent to Devon. From 1850 to 1939, a total of some 300 were recorded, with a further 75 from 1943 to 1950.[18] There are only two Irish records, nearly a hundred years apart, in 1864 and 1960[19] but the species is frequently reported, and occasionally breeds, in the Channel Islands. In recent years there have been increased numbers of sightings, some coinciding with invasions of other migrant Lepidoptera. From 1995 to 1998, the Queen of Spain Fritillary was seen in successive years and in varying numbers at Minsmere RSPB Reserve, on the Suffolk coast.[20] Its larval foodplant grows nearby but, although searched for, no eggs or larvae have yet been found. Two butterflies were recorded in 1995, at least six in 1996, six again in 1997, but in the dreadful summer of 1998 only one. There was justifiable speculation that this species, which is extending its range in France, Belgium and the Netherlands, might have bred on the Reserve and might well be present as a resident species in future years. If so, it would be the first to be added to the British list since the Essex Skipper, one hundred years before.

Seeing the Queen of Spain Fritillary in the British Isles is always a memorable experience. L. Hugh Newman vividly recalled one such moment while collecting on Sark in 1931:

> 'All at once a fluttering butterfly attracted my attention. Its flight was different from anything I had seen before, there was something of the Fritillary about it but it was too large for the Pearl-Bordered Fritillary and not large enough for the High Brown or the Dark Green and, anyhow, it was not the season for them. I crouched down and, making a pair of binoculars with my fingers round my eyes, I followed its erratic flight over the lush herbage in the marsh until it settled and began to fan its wings backwards and forwards in the characteristic way of a Fritillary. Suddenly, with a quickening of the pulse, I realized that I was looking at a Queen of Spain Fritillary, my first glimpse of this insect alive. With trembling fingers I assembled my net and then, drawing on all my knowledge of the best way to catch a butterfly, I approached it, stealthily, on the tips of my toes. But I soon found that my feet and ankles were sinking in the marshy ground and, floundering wildly to keep my balance, I had to stop there, unable to go any further, but still able to watch and admire. Then the sun went in, the Queen closed her wings, and I saw, still out of reach, that glitter of silver on the underside which has gripped the heart of many a collector on those rare occasions when this butterfly has settled on English soil. So I did not capture the Queen of Spain. She remained elusive, protected by the natural defences of her territory where she had chosen to rest awhile.'[21]

31. THE NIOBE FRITILLARY – *Argynnis niobe* (Linnaeus)
Papilio (*Nymphalis*) *niobe* Linnaeus, 1758

The Niobe Fritillary is widespread in Europe and can be found just across the Channel in northern France, yet has never become established in Britain. It is doubtful if any of the records in the literature which have earned it its tenuous place on the British list are genuine (see Plate 36, figure 6), although Stephens included it as an indigenous species,[1] possibly on the basis of a specimen that Dr Abbot, the celebrated entomologist who collected extensively at Clapham Park, Bedfordshire, had in his collection. It is probable that this was one more example of his being deceived by that dealer in rarities, Plastead. Its authentication by Stephens was to provide a golden opportunity for other breeders and purveyors of rare butterflies of Continental origin to pass off specimens purporting to be of British provenance. It was not mentioned by Lætitia Jermyn but Rennie,[2] cautiously following Stephens, included it, under the name of its British 'discoverer', as 'Abbot's Fritillary', adding that it was 'said to be British on doubtful authority'. There are apparently specimens from Kent dating from 1851 and 1856,[3] but the main appearance of *niobe* in that county took place in the 1870s. The story of the controversy stirred up by this butterfly makes interesting reading.

The 'Devil's Kneading Trough' is a steep-sided dry valley about half a mile in length between Wye and Ashford, in Kent. It has been described as 'a huge rent, caused at some remote time by a convulsion of Nature, with almost perpendicular sides, and covered with long grass, amidst which peep out the flowers of the wild Hearts-ease, which grow here in great profusion'. The name of this dramatic ravine is curiously appropriate as it became the scene of much entomological skulduggery. P. B. M. Allan's well-known account of the 'Kentish Buccaneers'[4] includes an investigation into the flush of captures of Niobe Fritillary, involving two dealers, George Parry, of Church Street, and W. Wigan of Wincheap, Canterbury. They (if two persons there were) managed for a time to hoodwink two of the most distinguished lepidopterists of their day, Henry Doubleday and Edward Newman (Fig. 141).

Fig. 141. Niobe Fritillary, drawn by George Willis and engraved by John Kirchner for Edward Newman's *Illustrated Natural History of British Butterflies and Moths*. Newman knew of only one British specimen, but the species afterwards became the subject of entomological fraud.

In 1873, Doubleday wrote to *The Entomologist* concerning Parry's claim to have taken the Niobe Fritillary there: 'Mr. Parry of Canterbury sent me two specimens of an *Argynnis*, to name one of them, which he said he had sent alive to my friend Edward Newman. They are females of Argynnis Niobe, – the typical variety, with the spots on the underside silvery. I know nothing of their history beyond the statement of Mr. Parry, – that he took them twenty miles from Canterbury. Being a sub-alpine species on the Continent, and a native of Sweden, its occurrence in this country might be looked for in the northern counties of England and Scotland rather than in Kent.'[5]

No more Niobe Fritillaries appeared in Kent that year, but next year, W. Wigan sent Doubleday and Newman more specimens, innocently asking what they were. He, too, claimed to have taken them in Kent. Newman published Doubleday's response: 'Mr. Doubleday allows me to publish the following information;– I yesterday received from a Mr. Wigan a pair of Argynnis Niobe, which he says he caught on some hills near Wye, in Kent, flying with Aglaia, of which he took more than one hundred specimens. He states that he took three male Niobe on Monday, two of them being much wasted, and a female on Tuesday; this he sent to me, together with the best male; this was dead, but not stiff, and the female was still alive: they are not more than half the size of my continental specimens.' He added: 'I have no doubt whatever that these two specimens were captured in Kent; they could not have been obtained from the Continent alive in such dry weather as we have had lately'. In a follow-up letter to Newman, also published, Doubleday wrote: 'The butterflies are most certainly Niobe, and quite distinct from Adippe and Aglaia. The moment I opened the box I saw what they were, without looking at the undersides.'[6]

Newman's report which printed Doubleday's response must have delighted Wigan who could hardly have hoped for a better endorsement. He immediately replied: 'Agreeably to your request I beg to inform you that the living specimens of Niobe ... were taken in a hilly district between Wye and Ashford, in Kent.' He then gave the graphic description of the Devil's Kneading Trough quoted earlier, before explaining that 'upon these steepy slopes, where an alpen-stock would not be despised by any but an entomologist, fly Aglaia, but not Niobe. At the bottom of this trough, which is no more than a few yards wide, grows a kind of rush, and there, within a confined area, is to be found Niobe, – and very few, alas, indeed.'[7]

The Niobe Fritillary may have been few in number but it continued to appear in Kent – almost always in association with Messrs Parry and Wigan – sometimes in the company of other rarities like the Queen of Spain Fritillary. Each 'appearance' prompted the publication of further letters, and there is little doubt that the two conspirators were soon in some demand among unsuspecting lepidopterists. Unfortunately they pushed their luck too far. The first seeds of doubt had been sown when Newman declined to publish the reported capture of thirteen Queen of Spain Fritillaries near Canterbury (see earlier). Doubleday had also been having serious misgivings. In a letter to *The Entomologist* in 1873 he revealed that:

> 'Having had the opportunity of examining a considerable number of specimens of some of our rarer Lepidoptera which had been sold as British, nearly all of which proved to be re-set continental specimens, I unhesitatingly say that I believe a very large majority of the specimens of Daplidice, Lathonia ... , which now exist in collections of professedly British Lepidoptera, are in reality continental: they can be purchased at from threepence to sixpence each; and so long as collectors will give as many pounds for them as they cost pence, I am afraid there is no probability of a stop being put to these disreputable proceedings. It is now almost impossible to say what insects are really British, as living pupae of various species are regularly obtained from France and Germany; and the fact of an insect being exhibited alive is no proof of its British origin – living butterflies and moths can be easily obtained from the Continent by post. All interest in collections of Lepidoptera as British is destroyed by the introduction of these continental specimens, which often differ considerably from ours.'[8]

Allan recounts that shortly before Newman died in 1876, he had become convinced of the fraudulent nature of the records to which he, in all innocence, had lent credence.[9] His regret is evident in his editorial response to a sarcastic letter from G. B. Corbin, who wanted to know why a promised illustration of the Kentish Niobe Fritillary had not materialized.[10] Newman replied as follows:

> 'I cannot believe in the Kentish captures of these two species [he referred also to the moth *Cnethocampa* (*Thaumetopoea*) *pityocampa*, the Pine Processionary], in this respect differing from my lamented friend Henry Doubleday, who was so honest and truthful in all his statements that he was ever willing to credit those of others. When I penned the paragraph to which Mr Corbin alludes I certainly intended to figure Argynnis Niobe as British; but the specimens in my possession on further information proved so questionable that I postponed the drawing and engraving *sine die.* I have received records of twenty-six specimens of Daplidice and a round dozen of Podalirius, which I suppress for the same reason.'[11]

Allan ironically concluded that 'lack of *The Entomologist*'s sunshine proved fatal to the remaining broods, and save for their annual appearance in the auction-rooms Niobe & Co are things of the past.'[12]

The Niobe Fritillary superficially resembles the High Brown Fritillary (*Argynnis adippe*), particularly in the male, and may readily be confused with it. Some of the records of *niobe* from other English counties may have resulted from misidentification of the var. *cleodoxa* of the High Brown, which lacks silver spots on the underside. Barrett regarded the specimen from the New Forest, taken by a Mr Gerrard in 1868, as such (*teste* F. Bond and Sydney Webb), though he said that Doubleday considered it to be *niobe.*[13] The Revd W. Hambrough, who then owned the specimen, wrote as follows: 'Being upon an entomological visit to Lyndhurst in last July, I there purchased of a local entomologist an unset variety (as I supposed) of Adippe, which I set to show the underside. On taking it to the British Museum last week, I was informed that it had all the true characters of Niobe; and such it now appears to me undoubtedly to be. I have since written to Mr. Gerrard, of whom I purchased it, for particulars of its capture, and he writes me, "The specimen of Adippe purchased by you was taken by me in 1868: it was flying with Paphia and other Adippe at Lyndhurst, about the middle of July".'[14]

A male specimen of the Niobe Fritillary, collected in or about 1879 from Monks-Park Wood, Bradfield St Clare, Suffolk by a member of the Suffolk Naturalists' Society, is now in the Morley Collection of the Ipswich Museum.[15] Its identity is not in question, and it can fairly claim to be the only confirmed extant specimen to have been taken in the wild in England, though how it got there is not known. It is most unlikely to have been an immigrant.

32. THE MEDITERRANEAN FRITILLARY or THE CARDINAL – *Argynnis pandora* ([Denis & Schiffermüller])
Papilio pandora [Denis & Schiffermüller, 1775]

This fritillary, the largest in Europe, is believed to have reached Britain as a genuine immigrant from the Continent on a few occasions. It superficially resembles the Silver-washed Fritillary and it is therefore possible that its true identity has been overlooked.

E. B. Ford gave an account of its first capture in Britain:

> 'Between August 3rd and 11th, 1911, Mr. A. W. Bennett chanced to walk up a narrow wooded valley near Tintagel in North Cornwall and noticed several very large Fritillaries soaring about and settling on the Purple Loosestrife (*Lythrum salicaria*). Having collected in his youth, though he had long ceased to do so, he realised that these insects differed somewhat from Silver-washed Fritillaries (*Argynnis paphia*) which he had first thought them to be. Accordingly, with great difficulty he managed to catch one in his cap and pinch it. He brought it back to his hotel and set it with an ordinary large-headed pin on a flat piece of wood. On examining his capture later, he realised that the underside did not resemble that of any Fritillary he had seen, the fore-wings being salmon pink and the hind-wings green with silver bars. He showed the specimen to me [Ford was then only ten years old!], but I did not then know the European butterflies and could make nothing of it. Accordingly, I advised him to send it for an opinion to a celebrated entomologist [Richard South] who a few years previously had published one of

> the best-known works on British butterflies and moths. However, this expert did not see fit to reply to Mr. Bennett, merely knocking off one of the antennae of the specimen and pencilling the words "*Argynnis paphia*" on the wrapper when he returned it. Had he given it more than a casual glance he would have realised his mistake.'[1]

Long after this discouraging response, Ford, who by then had it in his own collection, exhibited it at a Royal Entomological Society Meeting in London. He commented that it was remarkable that Bennett saw several specimens and believed their occurrence in numbers could have been the result of the progeny of a previous year's female surviving the winter. He did not believe that anyone would breed and liberate this species in this country, nor that several migrating butterflies would have reached the same wooded Cornish valley together.

In a separate incident, T. G. Howarth recalled that a young visitor to the Department of Entomology at the Natural History Museum in the 1950s brought in a specimen of the Mediterranean Fritillary which he said that he had caught in the New Forest area.[2] Questioning revealed that he had been abroad and the incident was dismissed as a case of confusion on the part of the captor. However, on 13th August 1969, a third specimen was taken, feeding on the pink flowers of thistles, at Durdle Door, Lulworth, Dorset, by C. P. J. Samson, who sent details to the Natural History Museum as he was unsure of its identity. Howarth replied that, from its description, he believed it to be a female *pandora*, and this was confirmed when the specimen itself was presented.[3]

Abroad, this species is found near the Atlantic coast of France as far north as southern Brittany, as well throughout the Iberian Peninsular and the Mediterranean area, including North Africa, and eastwards as far as northern India. It is by no means improbable that this powerful fritillary should occasionally cross the Channel to reach England.

33. THE GLANVILLE FRITILLARY – *Melitaea cinxia* (Linnaeus)
Papilio (*Nymphalis*) *cinxia* Linnaeus, 1758

In a letter dated 28th December 1702, Eleanor Glanville wrote to tell James Petiver that she had sent him 'a 100 several species, such as they are, many comon, & many very smal but some I belive wil prove new, except you have got them lattly, ...'.[1] It is not known what species they were, though we do know he received them safely. A note of acknowledgement was discovered some years ago by P. B. M. Allan which read as follows: 'To this curious Gentlewoman I am obliged for an hundred *Insects* lately sent me (besides others she gave me before) which she has observed in the *West* of England, several of them being altogether new to me'.[2] It is known that around this date Glanville sent Petiver a new butterfly she had collected near Lincoln which the latter illustrated in a part of his *Gazophylacii naturae & artis*, issued in 1703, naming it the Lincolnshire Fritillary.[3]

John Ray also described this 'Fritillary, from Lincoln', having received an 'abundance of it' from Samuel Dale,[4] one of the group of late seventeenth- and early eighteenth-century naturalists who worked closely together and frequently corresponded. It is not known from where Dale's specimens came (he was an Essex neighbour of Ray), but Joseph Dandridge took one 'not far from London', presumably at Dulwich, for, in his larger and more ambitious work on butterflies, published fifteen years later,[5] Petiver was to rename it the White

Plate 38 (*facing page*) One of James Dutfield's regrettably few butterfly portraits is this magnificent plate of 'The Glanvil Fretillary' and its correct foodplant, along with 'The Wood Lady, or Prince of Orange Butterfly'. He obtained the former from a wood near Dulwich, in Surrey, then a well-known locality for this butterfly which is now confined to the Isle of Wight.

Dullidge Fritillary. By then 'Madame Glanville' was dead but her fame clearly lived on among her fellow collectors. Although Benjamin Wilkes gave this butterfly yet another name, the Plantain Fritillary,[6] James Dutfield, in his ill-fated work on English moths and butterflies, named it 'The Glanvil Fretillary' in her honour (Plate 38).[7] This name has been used by most authors since, and is one of only two British butterflies (the other is Berger's Clouded Yellow) to commemorate the name of the person who discovered it.

Dutfield's coloured engraving of the butterfly and its early stages on the food plant is very accurate. He depicted the caterpillars 'which were taken the 5th May, feeding on the long-leaf Plantane near *Dulwich* Wood in *Surrey*'. He went on to state that the caterpillar changed into a chrysalis on 18th May and that the butterfly 'came forth' on 23rd June. Since Dutfield was writing before the Gregorian calendar was introduced in 1752, these dates must be revised and eleven days added to give our present-day dates of 16th May, 29th May and 4th July respectively. Dutfield's butterfly was thus in its chrysalis stage for more than a month, contrasting with the one bred by Moses Harris, which appeared 'fourteen days after'. Harris's colour plate is less well drawn than Dutfield's, but his description of the larval behaviour after hatching from the eggs is both accurate and beautifully written, and merits quoting in full:

> '... the Caterpillars keep sociably together the remaining Part of the summer; during the Winter they cover themselves over with a fine thin Web, which however the Rain is not able to destroy. At the Approach of the Spring, when the Sun gives a kindly Warmth, they come out of their Web, and seek their Food; each Night they retire to their Web, and lie altogether on the Top if it, for they seldom or ever get under it again after their first coming out. They are a very tender Creature, and never are seen to eat, nor scarcely move, but when the Sun shines upon them, and then they commonly feed all the while; but if the Sun should chance to hide his Head by the Interposition of a Cloud, they presently cease. As the warm Weather approaches, and they increase in Size, they separate themselves, wandering in search of their Food, which is the long-leaved Plantain, and such as grow in or near Woods; they are so remarkably timorous, that should you stir the Plant they are on, tho' never so little, or even tread within two Feet of it, they instantly roll themselves up in the Form of a Hedge-hog ... and fall to the Ground, where they lie in that curled-up Form, 'till they think the Danger past.'[8]

The Glanville Fritillary has been recorded from a number of places on the English mainland but it probably died out from many of them during the eighteenth century, and from its last toehold – where it was reported in the *Entomologist's Annual* for 1865 from West Cliff at Folkestone in Kent[9] – in the late 1860s. The isolated record from a wood near Bedford, noted by J. F. Stephens,[10] probably refers to Dr Abbot's notebook entry of '*Cinxia* seen July 5th between Hanger and Aske's Wood'.[11] A record from Yorkshire, also mentioned by Stephens, has been rejected by all subsequent authors on Yorkshire Lepidoptera. H. T. Stainton mentioned Peterborough and Stowmarket[12] but, whereas the Peterborough record could have some foundation, that from Suffolk was quickly rejected by Joseph Greene[13] and should be discounted. Stainton also mentioned Falkland, Fife, as a locality but, according to Thomson,[14] this and another record in 'Argyllshire', both by Duncan,[15] were almost

N°. V

certainly of the Marsh Fritillary *Eurodryas aurinia* – the result of confusion over nomenclature. Newman added Hampshire, near Brockenhurst, where J. C. Dale recorded it, and Wiltshire,[16] and Barrett added other former localities – Cambridgeshire and Leamington, Warwickshire.[17] Some of these, at least, are suspect.

The Glanville Fritillary seems to have become progressively rarer during the eighteenth and early nineteenth centuries. Stephens regarded it as 'very local' by 1828. It had disappeared from most of its former haunts and was by then confined to the coastal region of Kent between Folkestone and Sandgate, and possibly also around Eastbourne in Sussex.[18] Its extinction as a British butterfly seemed to be approaching. Then, in 1824, the 23-year old Edward Newman discovered the now celebrated locality at the Undercliff, Isle of Wight, finding 'the caterpillars, chrysalids and butterflies equally abundant at the same time'. 'With a feeling of triumph that I well recollect' he wrote, 'I recorded the discovery of this beautiful butterfly in the pages of "Loudon's Magazine of Natural History".'[19]

In 1846, the Revd J. F. Dawson visited the site and described what he saw:

> 'As this fritillary is rare in almost every part of the kingdom, some account of its favourite haunts and habits may not prove uninteresting. It cannot be accounted by any means common here, being confined to a few localities only, though where it does occur, it is in general to be found in some abundance. It is not to be expected in cultivated districts, but breeds on steep and broken declivities near the coast, which the scythe or the plough never as yet have invaded, and in such spots it may be met with earlier or later in May, according to the season. Near Sandown, on the side of the cliff, there is one of these broken declivities, occasioned by some former landslip, covered with herbage, which slopes down to the beach. A pathway leads to the base. On the 9th of May, 1844, a hot, sunny day, each side of this pathway was completely carpeted with a profusion of the yellow flowers of Anthyllis vulneraria (var. maritima) [kidney vetch], when I visited the spot; and these flowers were the resort of an abundance of these Fritillaries, which fluttered about them, or rested on their corollas, expanding and sunning their wings, and presenting a most charming picture of entomological loveliness.'[20]

In the British Isles, the Glanville Fritillary is now confined to perhaps a dozen sites on the Isle of Wight and to several more in the Channel Islands.[21] Although attempts have been made to reintroduce it to warm, sheltered sites in the New Forest, Sussex and Gloucestershire, they have all ultimately failed. The New Forest colony, introduced by R. W. Watson in the 1950s, survived in good numbers along a disused railway line at Sway until the site was destroyed some years later.[22] A similar fate befell the colony founded by P. W. Cribb at Ditchling Common in 1965 from thousands of reared larvae of Ventnor stock, when it was largely destroyed by grazing bullocks. The few larvae still present were removed in 1971.[23] From these experiences, it does seem possible that this species can be successfully introduced and will not stray far from any established colonies. Climatic warming may lead to an extension of its range. Stray individuals have been seen in Hampshire and Dorset in recent years, and one day this beautiful butterfly may well colonize the mainland again.

34. THE ARRAN BROWN – *Erebia ligea* (Linnaeus)
Papilio (*Nymphalis*) *ligea* Linnaeus, 1758

The story of the Arran Brown (see Plate 15, figure 3) is among the most intriguing in the history of British butterflies. Whether or not it ever occurred naturally in Britain is still disputed. While there have been many clues to the possibility of its occurring at a low density, especially in the western Highlands and Islands of Scotland, no one has yet managed to find a living colony of the butterfly. If it does occur somewhere, either it is extremely good at concealing the fact or we are being singularly unobservant!

The mystery began in 1804 when James Sowerby published in his *British Miscellany* an account of the discovery of this species by Sir Patrick Walker on the Isle of Arran. He simultaneously reported Sir Patrick's alleged discovery in the same place of the Scotch Argus, and illustrated both species.[1] Unfortunately Sowerby transposed the figures so that the Scotch Argus became *Erebia ligea*, while the Arran Brown was labelled as *Erebia blandina*, at that time the name of the Scotch Argus! It was certainly a strange coincidence that the presence of two species new to Scotland, the Scotch Argus and the Arran Brown, were being published at the same time and this may well have contributed to the consequent confusion. In fact, Dr John Walker (Fig. 142), who was no relation of Sir Patrick, had discovered the Scotch Argus on the Isle of Bute in September ['*in insula Bota septembro*'] some time before 1769. Dr Walker was a good entomologist and Professor of Natural History at Edinburgh University, where his notebooks are still preserved in the Library but his notebook entry of his account of finding this new butterfly, which he called *Papilio amaryllis*, was never published. However, he did show the specimens to Fabricius, who declared them to be 'different from the Ligea and a species not in Linnaeus'.

Fig. 142. Dr John Walker (1731–1804), discoverer in Bute, *c.*1760, of the Scotch Argus which he named 'Papilio Amaryllis' but, as his notebook records were never published, this early name is not acceptable under the Code of Zoological Nomenclature. He was appointed Professor of Natural History at the University of Edinburgh in 1775.

In 1807, Edward Donovan was next to state that *Papilio ligea* had been captured by Sir Patrick Walker at the same time as a number of specimens of *blandina*,[2] and he was followed by Stephens who described and illustrated the Arran Brown in his own work in 1828,[3] affirming that 'the only indigenous specimens which have come to my knowledge were captured in the Isle of Arran by Sir Patrick Walker and A. McLeay Esq. [the then Secretary of the Linnean Society]; but I am not aware of time, locality, or of the period of the year, which is probably about July or August'. As Sir Patrick was said to have been grouse shooting and the season traditionally does not begin until the 12th August, the latter month is more likely.

Sir Patrick Walker was a notable collector who possessed 'an elegant and select collection of Insects, a collection considered at the time the second best in this country [Scotland] – Dr Walker's being superior'.[4] As Sir Patrick's collection was not sold until after his death in 1839, doubt has been cast on whether the specimens illustrated by Stephens ten years before were indeed from the alleged British stock, unless Sir Patrick had loaned or given them to him earlier. The two specimens, from which he prepared his illustrations, still survive as part of the Stephens collection in the Natural History Museum, London.

Since its 'discovery' on Arran, lepidopterists have searched for the Arran

Brown in vain. Morris gave the familiar account of its discovery but added nothing except a new vernacular name, the Arran Argus, while hoping that many more would yet be taken, ending characteristically with the declamation 'Floreat Entomologia'[5] [may Entomology flourish]! Newman included it in his first work on British butterflies (published in 1860)[6] under Stephens's more recent name, the Scarce Scotch Argus,[7] but omitted it from his last work, published in 1871, possibly because he realised that he had been fooled over a number of 'questionably British butterflies'. It was Stainton who, in 1855, posed the question regarding the Arran Brown that must have been in the minds of Weaver and others who had set off in pursuit of the Arran Brown: 'If this be really a British species why does not some enterprising Scotchman rediscover it?'[8]

The by now established account of the discovery of the Arran Brown was repeated much later by Richard Weaver, the indefatigable collector from Birmingham who travelled widely in Britain in search of rare species. It was Weaver who first collected the Small Mountain Ringlet (*Erebia epiphron*) in Scotland in 1844, and who earlier had been the first to collect a female of the English race of the same species. In 1857, he reported in *The Zoologist* on a collecting trip in Scotland during which he visited the Isle of Arran 'in the hopes of obtaining Erebia Ligea: I had seen two specimens of this insect in the cabinet of the late Mr Stephens, which were captured by the late Sir Patrick Walker, on the moors at the back of Brodick Castle, when out one day grouse shooting: I was informed by Mr Stephens that he received that statement from Sir Patrick himself: I have tried that locality three different seasons without finding the insect, still I do not despair. I found E. blandina in profusion ...'.[9]

Despite Barrett's claim that Stephens's specimens were not both of *ligea*,[10] when Pelham-Clinton re-examined them he said that the doubt cast on them, provided they had not been changed, was unwarranted and that both were indeed *ligea*.[11] However, their origin remains uncertain and Pelham-Clinton considered that they might be from the Continent.

Since 1804, several further specimens of the Arran Brown have been discovered in collections of British insects in circumstances that have kept the mystery alive. In 1994, three were found by the author in the G. H. Simpson-Hayward Collection at Malvern College which had all the hallmarks of authenticity (Plate 39). One was labelled '*Ligea* Isle of Mull 1860', and the other two had one label between them stating 'taken on the Isle of Mull 1860 and '62. Wm. Edwards'.[12] All were females. Efforts to identify William Edwards have so far failed, but other butterflies that he collected and labelled were from the Malvern and Worcester district. Simpson-Hayward was born in 1875, so he could have obtained the specimens from Edwards himself. A specimen with no date, said to have been taken by A. E. Gibbs (1859–1917) from 'Galashiels', in the Scottish borders, formerly in the Rothschild Museum at Tring and now in the Natural History Museum, was understood by Pelham-Clinton to have been incorrectly labelled. Another specimen, collected in Scotland at the end of the nineteenth century by the brothers A. B. and J. W. Gillespie, was discovered by Pelham-Clinton in a small collection donated to the Royal Scottish Museum by their nephew, A. C. Gillespie. It was set underside uppermost to show the characteristic white crescentic band of the hindwing, but was unlabelled. Lying loose in the tray, which also included specimens of Scotch Argus, was a label, the only one in the collection, bearing the words: '*Erebia Blandina* / taken on Bute (North End) July 1871'. Pelham-Clinton concluded that it was 'a distinct possibility that *Erebia ligea* once occurred, or may even still occur in damp woods in the south of Scotland '.[13]

South, who seems to have regarded Scottish records of this butterfly as fanciful, commented that its captor 'must have exterminated the species, for, although the island has often been closely explored, no one has been able to detect the "Arran Brown" again'.[14] There can certainly be little doubt that the specimen reported as captured in Margate, Kent, in 1875[15] – 'in the mountain fastnesses of that salubrious watering place', as Allan sarcastically remarked[16] –

Plate 39
These three specimens of the Arran Brown 'taken on the Isle of Mull 1860 and 62', by William Edwards, were discovered in the G. H. Simpson-Heywood collection at Malvern College by the author.

was another of the Continental specimens that emanated from the 'Kentish Buccaneers'.

The discovery of a specimen, exhibited by T. J. Daley, a young member of the British Entomological and Natural History Society, at its annual exhibition in 1977, in a series of three named *Erebia aethiops* – he had assumed it to be an aberration – caused a considerable stir. Bretherton, who investigated the circumstances of its capture discovered that Daley had caught the specimen with others he had not set, in a particular place on the little-known western side of Rannoch Moor, in Argyll. At the date of the capture, the 5th of August 1969, Daley had been a schoolboy, and he confirmed that he had never collected in the countries where this form of *Erebia ligea* occurs, having only twice been abroad and each time to a southern European country.[17] This seems the most credible evidence to date, and raises the possibility that the Arran Brown still awaits rediscovery in some remote glen. The terrain is difficult to search and Bretherton remarked that 'conditions are often such as to make it impossible to be certain that it does not exist there at what perhaps may be a low density'. Moreover, it would be hard to detect if it occurred among large numbers of the superficially similar Scotch Argus.

35. THE MONARCH or THE MILKWEED BUTTERFLY – *Danaus plexippus* (Linnaeus)

Papilio (Danaus) plexippus Linnaeus, 1758
Danaus archippus Fabricius, 1793

Perhaps no other butterfly recorded from Britain conveys the impression of power and majesty evoked by the Monarch, a species unknown to the early entomologists. Its unparalleled size, its powerful, slowly-flapping wings which guide its graceful flight and, above all, its ability to cross oceans must surely make it one of the wonders of the butterfly world. The dynamics of its North American populations cause it to migrate to the far north where it breeds, and then, in the next generation, to return to its traditional wintering quarters far to the south in Mexico (for the eastern population) and in California (for those from the north west). But its strong migratory urge has also led it to colonize places thousands of miles from its native haunts.[1] What change in the mid-nineteenth century brought about this new impulse to spread beyond its existing range is not clear, but the Monarch was first recorded from New Zealand in 1840; from the Canaries in 1860, from the Azores in 1864; and from Australia in 1871.[2] The first European specimen was caught at Neath, South Wales, in 1876, and a year later the butterfly was recorded for the first time on the Continental mainland in La Vendée on the west coast of France. It was first reported from Spain in 1980 and three years later was found to be breeding there.[3]

Plate 40 (*facing page*) In his 2-volume *Natural History of British Butterflies* F. W. Frohawk devoted a full plate to the 'Black-veined Brown', or the Monarch as it is more often known today. After several attempts, Frohawk was able to breed the butterfly from larvae shipped across the Atlantic from Philadelphia, Pennsylvania, by a Dr A. Skinner. Three weeks after their safe arrival, Frohawk was rewarded by 'a fine series of imagines'. The eggs he depicted were laid by butterflies sent the following year by J. H. Gerould from Hanover, New Hampshire, those sent previously by Dr Skinner having hatched on their journey from N. America.

At first it was considered inconceivable that the transatlantic passage of a butterfly even as large and strong as the Monarch could have been achieved without the assistance of a ship. However, most Monarchs have arrived in Britain along with unusually large numbers of immigrant birds from America; there can be little doubt that they had been blown off course from their usual southward migration routes and had flown here under their own power. A few may also have arrived from Spain and the Atlantic islands, where breeding colonies now occur. In recent years, more Monarchs than ever before have been recorded – it was almost a predictable sight on the southern Cornish coast in autumn 1981. Unfortunately, there is little opportunity for the species to breed in Britain since its foodplant, the milkweed (*Asclepias curassavica*) occurs here only as an exotic garden plant. If, however, it were to be grown more widely in Cornwall, who knows what might happen! In 1981, an escaped female from the Butterfly Farm at Syon House homed in on *Asclepias* growing in the Royal Botanic Gardens, Kew, just across the River Thames and laid eggs. From these, larvae hatched in captivity, and adults eventually emerged as our first home-grown Monarch butterflies.[4]

The butterfly is said to have been dubbed 'The Monarch' in honour of the second Prince of Orange.[5] Coleman called it simply by the Fabrician name 'Archippus' and was the first to illustrate it in colour in a work on British butterflies in 1895.[6] It was Richard South who first named it after its foodplant, The Milkweed,[7] replacing another, little-used nineteenth-century name Black-veined Brown, though the latter was retained by Frohawk in his 1924 classic two-volume work on British butterflies (Plate 40).

The Black-veined Brown

Anosia plexippus.

FIG. 1 Egg magd. and nat. size, 2 days old, 10.viii.11
,, 2 Larva, 7th segment magd. directly after emergence, hatched 5.30 a.m., 12.viii.11
,, 3 Larva 28 hours old, showing amount eaten since hatching at 5.30 a.m. 12.viii.11
,, 4 Larva before 1st moult, 7.vi.10
,, 5 Larva before 2nd moult, 10.vi.10
,, 6 Larva after 2nd moult, 5.vi.10
,, 7 Larva after 3rd moult, 14.vi.10
,, 8 Larva, shortly before 4th moult, 17.vi.10

FIG. 9 Larva, after 4th moult, fully grown, 6.vi.10
,, 10 Larva, 7th segment magd., fully grown, 6.vi.10
,, 11-12 Pupa 3 days old, 12.vi.10
,, 13 Dorsal metallic band of pupa, 12.vi.10
,, 14 Neuration ♂
15 ♀ Bred 8.vii.10. F. W. F. coll.
,, 16 ♂ Bred from egg, 10.ix.11, Philadelphia parent. F. W. F. coll.
,, 17 ♀ Captured by Major Chavasse, 20.x.16, co. Cork. First Irish specimen. F. W. F. coll.

Plate 41 (*facing page*) British specimens of the Monarch from Devon and the Isle of Wight, remarkably well preserved after their transatlantic crossings.

About 450 Monarchs were recorded between 1876 and 1988 (Plate 41), more than three-quarters of which arrived in just two years, 1968 and 1981. More recently, there was another large invasion in 1995 when around 200 individuals were seen, and there have been sightings in double figures every year since, almost all of them in the south-western counties of England. The grand total has now passed 700, of which nearly two-thirds have been recorded in the past twenty years, and, for the butterfly-watcher taking a holiday in Cornwall or the Scilly Isles in September or October, the hope of seeing a Monarch is no longer such a vain one. The publicity given by 'Insectline' to rare sightings (similar to 'Birdline' used by ornithological 'twitchers') has given a great fillip to butterfly watchers who can now turn up, armed with camera rather than net, to record the sight of a lifetime.

Details of the very first sighting, in 1876. were carefully recorded for posterity by its fortunate discoverer, J. T. D. Llewellyn:

> 'On the 6th September, my gardener's son, J. Stafford, a lad of 14 years old, on going to a favourite patch of Scabious at once saw this magnificent visitor. It was sitting on a bloom opening and shutting its wings, and flitting in enjoyment of the shelter and the sunshine, from flower to flower. He took the insect to his father, who brought it to me – and I saw it still alive. It is a fine fresh specimen and has not flown much. The locality is two and a half miles from Neath as the crow flies. Ships laden with timber and other merchandise come up to Neath, and the inference is, that the pupa may have come thus imported, although it is marvellous that the insect could have flown so far without more injury to its freshness. The nearest houses, excepting four cottages of mine, are half-a-mile distant, and all enquiry fails to discover anyone who could, by any possibility, have imported the insect.'[8]

Barely a month afterwards, other specimens were taken in Sussex and Dorset. The Revd Thomas Crallan, the butterfly-collecting rector of Hayward's Heath, described how his housemaid brought him 'a collar-box, with the information that a young labourer, living about a quarter of a mile off, had caught a "bug" in a field at the back of his house at dinner-time, and thought I should like to have it. ... I expected a larva of Cossus ligniperda [the Goat Moth], but on applying my ear to the box I heard a rustling of wings; and, opening it very carefully, beheld a fine specimen, almost perfect, of Danais [*sic*] Archippus. We are an entomological household, and the excitement generated amongst us by the sight of so grand an insect ... may be more easily imagined than described.' He added that, 'altogether, considering it had been caught in a hat, and kept seven hours in a box before it came into my hands, it may be considered in very good condition ... I cannot understand this having been an imported specimen, as we are forty miles from London, and sixteen or seventeen from Shoreham, – our nearest seaport, – to which, I think, no American ships come.'[9]

Naturally, neither Llewellyn nor Crallan considered immigration as the means by which their specimens had arrived in Britain, for at that date the idea of insect migration had barely been broached. It was more natural to assume that such exotic specimens had been imported accidentally or by a dealer. In 1886, R. C. R. Jordan made an appeal to 'all true entomologists to give these colonists fair play, and not to attempt to extirpate them', pointing

21.6.1891
Captured
I of Wight
C. King Smith
Date
July 8.1931
Devon.
A.
SIDGWICK
COLL.

out that set specimens could always be purchased for examination if necessary. He had heard that a specimen of the Monarch, captured at The Lizard, Cornwall, had been offered for sale at £5, and commented: 'I never heard of an ornithologist offering £5 for the skin of a grey parrot which had escaped from its cage and been unfortunately shot in England. No one of any sense would believe that the *Archippus* was a true native.' His own view was that, 'as *D. Archippus* is worth sixpence ... the would-be purchaser must regard [the remaining] £4 19s 6d. as the value of the fact that the butterfly may cross the Atlantic in a steamer, and not die of sea-sickness'.[10]

E. B. Ford described his own unforgettable experience of capturing a perfect female specimen of the Monarch:

> '... on the evening of August 30th, 1941, at Kynance Cove, Cornwall, within two miles of the most southerly point in Great Britain. Those who know that exquisite spot, now largely spoiled through having been popularised for tourists, will remember that the steep path up the cliff reaches a short piece of level ground just before the summit. Climbing from the cove, I arrived, net in hand, at this place at 6.20 p.m., double summer time, and glancing to the left saw a Monarch Butterfly about twenty feet away flying inland perhaps fifteen feet from the ground. It was slowly flapping and gliding and looked immense, and the honey-coloured underside of the hind-wings showed clearly. It quickly reached a small rocky hill and disappeared over the top. Now every collector knows that if one loses sight of a butterfly one rarely sees it again. It was with a sinking heart therefore that I gained the top of the hill and, turning to the left in the direction which the insect had taken when last seen, found my way barred by a steep rocky slope. I threw myself over, landing in a heap at the bottom and, on picking myself up, beheld with joy the Monarch about fifty yards away. It was hovering over a path, no more than a foot above the ground, and then slowly rose. By the time I arrived it must have been about two feet above the heather, and I caught it with a single stroke of the net.
>
> 'On this occasion I was much impressed by the resistance of this species to pressure and by its leathery consistency; a well-known characteristic of these protected insects, which allows a bird to peck them sufficiently to realise their disagreeable qualities without killing them. As this specimen was too large to go into my killing bottle or boxes, I kept it in the net and repeatedly pinched it. This would have cracked the thorax of a large Nymphalid and caused its immediate death, but after each pinch the insect would lie still for a few minutes and then revive apparently none the worse. A faint musky odour hung about it, and I was greatly tempted to bite into it to determine if it were unpalatable but, having regard to the interest of the specimen in other ways, I thought it well to restrain my curiosity in this respect.'[11]

In this graphic account of his encounter with the Monarch butterfly, Ford puts into words the mixed emotions felt by all dedicated collectors engaged in the uncertain chase of a butterfly they are so desperate to obtain. They are feelings that all the naturalists included in this book would have experienced when in pursuit of any of the rare and historic butterflies described above.

Today, when for many a camera has taken the place of the net, the thrill and frisson is no less and the resulting pleasure or despair no different. The fact that Britain has so few butterfly species probably increases rather than diminishes the degree of excitement.

In the last century, H. W. Bates, a one-time president of the Entomological Society of London, tantalized British butterfly collectors by noting that in one part of Amazonia there were 'about 700 species of that tribe to be found within one hour's walk of the town, whilst the total number found within the British Isles does not exceed 66, and the whole of Europe supports only 321'.[12] Since he wrote this, many more species have been described and the balance has, if anything, swung even further towards the comparative richness of the South American fauna. The number of recognized native, formerly native, or regular immigrant species of butterfly found in the British Isles now totals around 80. However, this has not deterred British naturalists from rejoicing in and studying them, and it is astonishing how many books have been published on the subject! Indeed more words have probably been written on our impoverished butterfly fauna than on that of any other country. We may have few species but we certainly cherish those that we have.

NOTES TO CHAPTER FOUR

INTRODUCTION

1. Turpyn, R. (1540). *Chronicles of Calais in the Reigns of Henry VII and Henry VIII to the Year 1540.* MSS formerly in the British Museum (now the British Library), reprinted by the Camden Society (1846), **35**: 7.
2. Morris, F. O. (1853). *A history of British butterflies,* p.21.
3. Williams, C. B. (1958). *Insect migration,* p.11.
4. Williams, C. B., Cockbill, G. F., Gibbs, M. E. & Downes, J. A. (1942). Studies in the migration of Lepidoptera. *Trans.R.ent.Soc.Lond.* **92**: 101–280.

1. CHEQUERED SKIPPER

1. Abbot, C. (1798). Manuscript report to the Linnean Society, 12 August 1798. Linnean Society Archive No. 192a; – (1800). XXXI Extracts from Minute Book of the Linnean Society for 6 Nov. 1798. *Trans.Linn.Soc.Lond.* **5**: 276.
2. Donovan, E. (1799). *Natural history of British insects* **8**: 7–8, pl. 254.
3. Haworth, A. H. (1803). *Lepidoptera Britannica,* p.49.
4. Samouelle, G. (1819). *The entomologist's useful compendium,* p.243.
5. Morris, F. O. (1853). *A history of British butterflies,* p.166.
6. Stephens, J. F. ([1827–]28). *Illustrations of British entomology (Haustellata)* **1**: 100.
7. Fowler, W. W. (1882). *Hesperia paniscus* and other Lepidoptera near Lincoln. *Entomologist* **15**: 37–38.
8. Dale, C. W. (1890). *The history of our British butterflies,* pp.209–210.
9. Collier, R. V. (1986). The conservation of the Chequered Skipper in Britain. *Focus on Nature Conservation* No.16, 16 pp.
10. Mackworth-Praed, C.W. (1942). *Carterocephalus palaemon* Pallas in Invernesshire. *Entomologist* **75**: 216.

11. Evans, C. E. (1949a). The Chequered Skipper *C. palaemon* in West Inverness. *Scott.Nat.* **61**: 176.
12. Joicey, J. J. & Noakes, A. (1907). Lepidoptera in Glenshian, Inverness-shire, in July 1907. *Entomologist's mon.Mag.* **43**: 255–256.
13. Collier, R. V. (1989). *In* Emmet, A. M. & Heath, J. (Eds), *The moths and butterflies of Great Britain and Ireland* **7**(1): 53.
14. Dennis, R. L. H. (1977). *The British butterflies*, p.241.
15. Warren, M. S. (1995). The Chequered Skipper returns to England. *Butterfly Cons.News* **60**: 4.

2. ESSEX SKIPPER
1. Hawes, F. W. (1890). *Hesperia lineola* Ochsenheimer: an addition to the list of British butterflies. *Entomologist* **23**: 3–4.
2. Dennis, R. L. H. (1977). *The British butterflies*, pp.205–207.
3. Coleman, W. S. (1895). *British butterflies* (edn 2), p.148.
4. Barrett, C. G. (1893). *The Lepidoptera of the British Islands* **1**: 282–283.
5. Thomas, J. [A.] & Lewington, R. (1991). *The butterflies of Britain and Ireland*, p.16.
6. Burns, J. M. (1966). Expanding distribution and evolutionary potential of *Thymelicus lineola* (Lepidoptera: Hesperiidae), an introduced skipper, with special reference to its appearance in British Columbia. *Can.Ent.* **98**: 859–866.

3. LULWORTH SKIPPER
1. Curtis, J. (1833). *British entomology* **10**: 442.
2. Stainton, H. T. (1857a). *A manual of the British butterflies and moths* **1**: 67–68.
3. Barrett, C. G. (1893). *The Lepidoptera of the British Islands* **1**: 287–288.
4. Pretor, A. (1856). *Pamphila Actaeon. Entomologist's Wkly Intell.* **1**: 154.
5. McRae, W. (1881). Probable extermination of *Hesperia actaeon* at Lulworth. *Entomologist* **14**: 252–253.
6. Goater, B. (1974). *The butterflies and moths of Hampshire and the Isle of Wight*, p.215.
7. Humphreys, H. N. & Westwood, J. O. (1841). *British butterflies and their transformations*, p.129.
8. Thomas, J. [A.] & Lewington, R. (1991). *The butterflies of Britain and Ireland*, p.14.

4. MALLOW SKIPPER
1. Frohawk, F. W. (1923). *Carcharodus alceae* in Surrey. *Entomologist* **56**: 267–268.

5. THE APOLLO
1. Moffet, T. (1634). *Insectorum sive minimorum animalium theatrum*, p.94.
2. Petiver, J. (1702–06). *Gazophylacii naturae & artis: decas prima – decas decima*, p.23.
3. Haworth, A. H. (1803[–28]). *Lepidoptera Britannica*, p.xxix.
4. Donovan, E. (1808). *Natural history of British insects* **13**: 1–2.
5. Wood, W. (1852). *Index entomologicus* (new edn), p.243.
6. Newman, E. (1871). *An illustrated natural history of British butterflies*, p.176.
7. Wollaston, G. B. (1856). Capture of Parnassius Apollo at Dover. *Zoologist* **14**: 5001.
8. Austin, G. (1856b). *Parnassius apollo* at Ealing. *Ibid.* **14**: 5109.
9. Sabine, E. (1889). *Doritis apollo* at Dover. *Entomologist* **22**: 278.
10. Kirby, W. F. (1882). *European butterflies and moths*, p. 4.
11. Morley, A. M. & Chalmers-Hunt, J. M. (1959). Some observations on the crimson-ringed butterfly (*Parnassius apollo* L.) in Britain. *Entomologist's Rec.J.Var.* **71**: 273–276.
12. Thomson, G. (1980). *The butterflies of Scotland*, p.72.
13. Scott, P. (1955). *Parnassius apollo* L. at Folkestone. *Entomologist's Rec.J.Var.* **67**: 273.
14. French, R. A. (1956). Migration Records 1955. *Entomologist* **89**: 174–180.

6. SWALLOWTAIL
1. Moffet, T. (1634). *Insectorum sive minimorum animalium theatrum*, p.99.
2. Ray, J. (1718). *Philosophical letters between the late learned Mr Ray and several of his Ingenious correspondents, &c.*, (Ed. W. Derham), p.69.
3. Ray, J. (1710). *Historia insectorum*, pp.110–111.
4. Petiver, J. (1717). *Papilionum Britanniae icones*, p.1, pl.2, fig.5.
5. Harris, M. (1766). *The Aurelian, &c.*, p.75, pl.36.
6. Emmet, A. M. (1989). The Swallowtail: Vernacular name and early history. *In* Emmet, A. M. & Heath, J. (Eds), *The moths and butterflies of Great Britain and Ireland* **7**(1): 80–81.
7. Wilkes, B. (1747–49). *The English moths and butterflies*, p.47, pl.92.
8. Ray, *Historia insectorum*, p.110.
9. Samouelle, G. (1819). *The entomologist's useful compendium*, p.235.
10. Dale, C. W. (1890). *The history of our British butterflies*, p.5.

11. Bretherton, R. F. (1951b). The early history of the swallow-tail butterfly (*Papilio machaon*) in England. *Entomologist's Rec.J.Var.* **63**: 206–211.
12. Dale, C. W. (1878). *The history of Glanville's Wootton*, pp.145–46.
13. Rimington, W. E. (1987). Historical records of the swallowtail butterfly (*Papilio machaon* L.) in Yorkshire. *Naturalist, Hull* **112**: 81–84.
14. Jermyn, L. (1824). *The butterfly collector's vade mecum*, pp.24–25.
15. Barrett, C. G. (1893). *The Lepidoptera of the British Islands* **1**: 15.
16. Kirby, W. E. ([1906]). *Butterflies and moths of the United Kingdom*, p.2.
17. Newman, E. (1871). *An illustrated natural history of British butterflies*, p.153.
18. Austin, G. (1856a). *Machaon* at Battersea. *Entomologist's Wkly Intell.* **1**: 40.
19. Silvester, E. T. (1857). *Papilio Machaon* in Sussex. *Ibid.* **2**: 197.
20. Wilkinson, W. G. (1857a). *Papilio machaon. Ibid.* **2**: 131.
21. Rothschild, M. (1975). The swallowtail butterfly *Papilio machaon britannicus* Seitz in Northamptonshire. *Entomologist* **87**: 177–179.
22. Friday, L. E. (Ed.) (1997). *Wicken Fen – the making of a wetland nature reserve*, pp.126, 235, 262.

7. SCARCE SWALLOWTAIL

1. Moffet, T. (1634). *Insectorum sive minimorum animalium theatrum*, p.99.
2. Ray, J. (1710). *Historia insectorum*, p.111.
3. Berkenhout, J. (1769). *Outlines of the natural history of Great Britain and Ireland*, p.123.
4. Lewin, W. (1795). *The papilios of Great Britain*, p.74.
5. Haworth, A. H. (1803). *Lepidoptera Britannica*, pp.xxvi–xxvii, 5–6.
6. Wilkinson, R. S. (1975c). The Scarce Swallowtail: *Iphiclides podalirius* (L.) in Britain. I. The evidence before Haworth. *Entomologist's Rec.J.Var.* **87**: 289–293.
7. Allan, P. B. M. (1980). *Leaves from a moth-hunter's notebook*, pp.116–135.
8. Curtis, J. (1836). *British entomology* **13**: 578.
9. Howarth, T. G. (1973). *South's British butterflies*, p.38.
10. Frohawk, F. W. (1935). Letter. *The Field*, 6 July.
11. Harthill, G. G. (1964). *Papilio podalirius* (L.) (Lepidoptera, Papilionidae) in Gloucestershire. *Entomologist* **97**: 19.
12. Bretherton, R. F. & Chalmers-Hunt, J. M. (1985). The immigration of Lepidoptera to the British Isles in 1984. *Entomologist's Rec.J.var.* **97**: 140.

8–10. THE CLOUDED YELLOWS (*COLIAS* SPP.)

1. Williams, C. B. (1958). *Insect migration*, p.27.
2. Cribb, P. W. (1991). The *Colias hyale/alfacariensis* identity problem. *Bull.amat.Ent.Soc.* **50**: 12.

8. PALE CLOUDED YELLOW

1. Harris, M. (1775b). *The English Lepidoptera: or, the Aurelian's pocket companion*, p.7.
2. Lewin, W. (1795). *The papilios of Great Britain*, p.70.
3. Donovan, E. (1798). *Natural history of British insects*, **7**: 57–58, pl. 238.
4. Haworth, A. H. (1803). *Lepidoptera Britannica*, pp.11–14.
5. Rennie, J. (1832). *A conspectus of the butterflies and moths found in Britain*, p.2.
6. Coleman, W. S. (1860). *British butterflies*, p.75.
7. Berger, L. A. (1948). A *Colias* new to Britain (Lep. Pieridae). *Entomologist* **81**: 129–131.
8. Heath, J. (1981). Threatened Rhopalocera (butterflies) in Europe. *Nature & Environment Series* No 23. p.68.
9. Tolman, T. (1997). *Butterflies of Britain and Europe*, p.52.
10. Coleman, *British butterflies*, p.76.
11. Desvignes, T. (1842). Correspondence. *Entomologist* **1**: 388.
12. Newman, E. (1871). *An illustrated natural history of British butterflies*, p.143.
13. Hill, P. M. (1997). Migrant butterflies in 1996. *Atropos* **2**: 13–14.
14. Thomas, J. [A.] & Lewington, R. (1991). *The butterflies of Britain and Ireland*, p.200.

9. BERGER'S CLOUDED YELLOW

1. Berger, L. A. & Fontaine, M. (1947–48). Une espèce méconnue de genre *Colias* F. *Lambillionea* **47**: 91–98; **48**: 12–15, 21–24, 90–110.
2. Berger, L. A. (1948). A *Colias* new to Britain (Lep. Pieridae). *Entomologist* **81**: 129–131.
3. Bretherton, R. F. (1989). *In* Emmet, A. M. & Heath, J. (Eds), *The moths and butterflies of Great Britain and Ireland* **7**(1): 91.
4. Vallins, F. T., Dewick, A. J. & Harbottle, A. H. H. (1950). The name and identification of the New Clouded Yellow. *Entomologist's Gaz.* **1**: 113–125.
5. Chalmers-Hunt, J. M. & Skinner, B. (1992). The immigration of Lepidoptera to the British Isles in 1990. *Entomologist's Rec.J.var.* **104**: 211.
6. Barrington, R. (1992). 1991 Annual Exhibition: British Butterflies. *Br.J.ent.nat.Hist.* **5**: 50, 54, 81, Pl.2.

7. Hill, P. M. (1997). Insects reported in the early part of 1997. *Atropos* **3**: 13–14.

10. CLOUDED YELLOW

1. Moffet, T. (1634). *Insectorum sive minimorum animalium theatrum,* p.100.
2. Petiver, J. (1703). *Gazophylacii naturae & artis: decas prima – decas decima,* p.14.
3. Ray, J. (1710). *Historia insectorum,* p.112
4. Wilkes, B. (1742a). *Twelve new designs of English butterflies,* pl.9.
5. Harris, M. (1766). *The Aurelian, &c.,* p.61, pl.29.
6. Berkenhout, J. (1769). *Outlines of the natural history of Great Britain and Ireland,* p.130.
7. Harris, M. (1773). *The Aurelian, &c.* Supplement, p. 61, pl. 29.
8. Harris, M. (1775a). *The Aurelian, &c.* (edn 2), pp.54–55.
9. Emmet, A. M. (1989). *In* Emmet, A. M. & Heath, J. (Eds), *The moths and butterflies of Great Britain and Ireland* **7**(1): 96.
10. Donovan, E. (1798). *Natural history of British insects* **7**: 61, pl.238.
11. Lewin, W. (1795). *The papilios of Great Britain,* p.68.
12. Harris, *Aurelian* [edn 1], pl.29.
13. Coleman, W. S. (1860). *British butterflies,* pp.71–75.
14. Frohawk, F. W. (1934). *The complete book of British butterflies,* p.330.
15. Dale, C. W. (1890). *The history of our British butterflies,* pp.31–32.
16. Coleman, *British butterflies* (new edn), p.71.
17. Newman, L. H. (1967). *Living with butterflies,* p.160.
18. Dale, C. W., *Our British butterflies,* pp.30–31.
19. Williams, C. B. (1958). *Insect migration,* pp.220–228.
20. Bretherton, R. F. (1989). *In* Emmet, A. M. & Heath, J. (Eds), *The moths and butterflies of Great Britain and Ireland* **7**(1): 95.
21. Hill, P. M. (1998). Insects reported during January – June 1998. *Atropos* **5**: 62–65.
22. Hill, P. M. (1999). Migrant butterflies in 1998. *Atropos* **6**: 68–69.

11. CLEOPATRA

1. Coleman, W. S. (1895). *British butterflies* (edn 2), p.70.
2. Pickard, H. A. (1860). Singular, considering the Season. *Entomologist's Wkly Intell.* **8**: 171–172.
3. Stainton, H. T. (1861). Lepidoptera: Rare British species captured in 1860. *Entomologist's Annu.* **1861**: 93.
4. Frohawk, F. W. (1938a). *Varieties of British butterflies,* p.180.
5. Howarth, T. G. (1973). *South's British butterflies,* p.50.
6. Bretherton, R. F. & Chalmers-Hunt, J. M. (1982). The immigration of Lepidoptera to the British Isles in 1981 including that of the monarch butterfly: *Danaus plexippus* L. *Entomologist's Rec.J. Var.* **94**: 81–87, 141–146.
7. Bretherton, R. F. & Chalmers-Hunt, J. M. (1985). The immigration of Lepidoptera to the British Isles in 1984. *Entomologist's Rec.J. Var.* **97**: 140–145, 179–185, 224–228.

12. BLACK-VEINED WHITE

1. Moffet, T. (1634). *Insectorum sive minimorum animalium theatrum,* p.103.
2. Merrett, C. (1666). *Pinax rerum naturalium Britannicarum,* p.198.
3. Ray, J. (1710). *Historia insectorum,* p.115.
4. Petiver, J. (1699). *Musei Petiveriana centuria prima – decima,* p.33.
5. Petiver, J. (1717). *Papilionum Britanniae icones,* p.1.
6. Albin, E. (1720). *The natural history of English insects,* p.2.
7. Harris, M. (1766). *The Aurelian, &c.,* pl. 9.
8. Rennie, J. (1832). *A conspectus of the butterflies and moths found in Britain,* p.5.
9. Haworth, A. H. (1803). *Lepidoptera Britannica,* pp.6–7.
10. Curtis, J. (1831). *British entomology* **8**: 360.
11. Newman, E. (1871). *An illustrated natural history of British butterflies,* p.168.
12. Dale, C. W. (1887). Historical notes on *Aporia crataegi* in England. *Entomologist's mon.Mag.* **23**: 38–39.
13. Goss, H. (1887). Is *Aporia crataegi* extinct in Britain? *Ibid..* **23**: 217–218.
14. Pratt, C. R. (1989). *In* Emmet, A. M. & Heath, J. (Eds), *The moths and butterflies of Great Britain and Ireland* **7**(1): 99–103.
15. Pratt, C. R. (1983). A modern review of the demise of *Aporia crataegi* L.: the black-veined white. *Entomologist's Rec.J. Var.* **95**: 45–52, 161–166, 244–250; **99**: 21–27, 69–80.
16. Howarth, T. G. (1973). *South's British butterflies,* p.51.
17. Elliot, R. (1982). Further notes on an Introduced "Colony" of the Black-veined white: *Aporia crataegi* L. in Scotland. *Entomologist's Rec.J. Var.* **94**: 245–246.
18. Hill, P. M. (1997). Insects reported in the early part of 1997. *Atropos* **3**: 13–14.

13. BATH WHITE

1. Ford, E. B. (1945). *Butterflies,* p.10.
2. Howarth, T. G. (1973). *South's British butterflies,* p.57.

3. Petiver J. (1699). *Musei Petiveriana centuria prima – decima,* p.304.
4. Petiver, J. (1702). *Gazophylacii naturae & artis: decas prima – decas decima,* p.1.
5. Ford, *Butterflies,* pl.1.
6. Lewin, W. (1795). *The papilios of Great Britain,* p.62.
7. Donovan, E. (1796). *Natural history of British insects* **6**: 47–48, pl.200.
8. Haworth, A. H. (1803). *Lepidoptera Britannica,* pp.xxvii, 10.
9. Jermyn, L. (1824). *The butterfly collector's vade mecum,* p.19.
10. Stephens, J. F. ([1827–]28). *Illustrations of British entomology (Haustellata)* **1**: 17.
11. Newman, E. (1871). *An illustrated natural history of British butterflies,* p.158.
12. Rennie, J. (1832). *A conspectus of the butterflies and moths found in Britain,* p.4.
13. Morris, F. O. (1853). *A history of British butterflies,* p.26.
14. Williams, C. B. (1958). *Insect migration,* pp.220–228.
15. Frohawk, F. W. (1938b). Unrecorded occurrence of *Pontia daplidice* in numbers. *Entomologist* **71**: 66.
16. Kettlewell, H. B. D. (1946). *Pontia daplidice, Everes argiades* and *Colias hyale* in south Cornwall. *Ibid.* **78**: 123–124.
17. Blathwayt, C. S. H. (1945). Immigration of *Pontia daplidice. Ibid.* **78**: 124–125.
18. Hill, P. M. (1999). Migrant butterflies in 1998. *Atropos* **6**: 68–69.

14. BLACK HAIRSTREAK

1. Petiver, J. (1703). *Gazophylacii naturae & artis: decas prima – decas decima,* p.11.
2. Ray, J. (1710). *Historia insectorum,* p.130.
3. Linnaeus, C. (1758). *Systema naturae* (edn 10), p.482.
4. Harris, M. (1775a). *The Aurelian, &c.* (edn 2), pl.23.
5. Donovan, E. (1808). *Natural history of British insects* **13**: 9, pl.437.
6. Jermyn, L. (1824). *The butterfly collector's vade mecum,* pp.34–35.
7. Curtis, J. (1829). *British entomology* **5**: 264.
8. Humphreys, H. N. & Westwood, J. O. (1841). *British butterflies and their transformations,* p.88.
9. Morris, F. O. (1860). *A history of British butterflies,* p.94.
10. Wood, W. (1852). *Index entomologicus* (new edn), p.6.
11. Stephens, J. F. (1856). *List of specimens of British animals in the collection of the British Museum.* **5**: *Lepidoptera,* p.16.
12. Newman, E. (1871). *An illustrated natural history of British butterflies,* p.110.
13. Kirby, W. F. (1896). *A hand-book to the order Lepidoptera 1: Butterflies,* pp.50–54.
14. South, R. (1906). *The butterflies of the British Isles,* p.143.
15. Bree, W. T. (1852). A list of butterflies occurring in the neighbourhood of Polebrook, Northamptonshire; with some remarks. *Zoologist* **10**: 3348–3352.
16. Barrett, C. G. (1893). *The Lepidoptera of the British Islands* [quoting H. Goss], p.51.
17. Thomas, J. [A.] & Lewington, R. (1991). *The butterflies of Britain and Ireland,* p.70.
18. Collier, A. E. (1959). A forgotten discard: the problem of redundancy. *Entomologist's Rec.J.Var.* **71**: 118.
19. Collins, G. A. (1995). *Butterflies of Surrey,* pp.39–40.

15–17. THE COPPERS (*LYCAENA* SPP.)

1. South, R. (1906). *The butterflies of the British Isles,* p.152.
2. Kirby, W. F. (1896). *A hand-book to the order Lepidoptera 1: Butterflies,* pp.115–125.
3. Barrett, C. G. (1893). *The Lepidoptera of the British Islands* **1**: 55.

15. LARGE COPPER

1. Lewin, W. (1795). *The papilios of Great Britain,* p.84.
2. Donovan, E. (1798). *Natural history of British insects* **7**: 3–4, pl.217.
3. Newman, E. (1871). *An illustrated natural history of British butterflies,* p.114.
4. Perceval, M. J. (1983). The Lepidoptera of Henry Seymer (1767–1785). *Entomologist's Gaz.* **34**: 215–227.
5. Duddington, J. & Johnson, R. (1983). *The butterflies and larger moths of Lincolnshire and South Humberside,* pp.104–105.
6. Redshaw, E. J. (1982). An early record of the Large Copper (*Lycaena dispar* Haw.) in Lincolnshire. *Trans.Lincs.Nat.Un.* **20**: 119–120.
7. Heath, J. (1983). Is this the earliest record of *Lycaena dispar* (Haworth) (Lep. Lycaenidae)? *Entomologist's Gaz.* **34**: 228.
8. Haworth, A. H. (1802). *Prodromus lepidopterorum Britannicorum,* p.3.
9. Dale, C. W. (1890). *The history of our British butterflies,* p.47.
10. Jermyn, L. (1827). *The butterfly collector's vade mecum* (edn 2), pp.136–138.
11. Walker, J. J. (1904). Some notes on the Lepidoptera of the "Curtis" Collection of British insects. *Entomologist's mon.Mag.* **40**: 137–194.

12. Bretherton, R. F. (1951a). Our lost butterflies and moths. *Entomologist's Gaz.* **2**: 211–240.
13. Miller, S. H. & Skertchly, S. B. J. (1878). *The fenland: past and present*, p.594.
14. Irwin, A. G. (1984). The large copper, *Lycaena dispar dispar* (Haworth), in the Norfolk Broads. *Entomologist's Rec.J. Var.* **96**: 212–213.
15. Merrin, J. (1899). The 'extinct' *Chrysophanus dispar. Ibid.* **11**: 208–209.
16. Allan, P. B. M. (1958c). French Large Coppers in England. *Ibid.* **70**: 248.
17. Sutton, R. D. (1993). Were the Large Copper *Lycaena dispar* (Haw.) and the Scarce Copper *Lycaena virgaureae* (Linn.) once indigenous species at Langport, Somerset? *Butterfly Conservation Occasional Paper* No. 7.
18. Tutt, J. W. (1906). *Natural history of the British Lepidoptera* **8**: 423.
19. Crotch, W. D. (1857). Doings in the West. *Entomologist's Wkly Intell.* **2**: 165–166.
20. D'Urban, W. S. M. (1865). Exeter Naturalist's Club. *Entomologist's mon.Mag.* **2**: 71.
21. Allan, P. B. M. (1980). *Leaves from a moth-hunter's notebook*, pp.24–34.
22. Allan, P. B. M. (1943). *Talking of moths*, pp.19, 23.
23. 'An Octogenarian', [Harding, H. J.] (1883). Entomological reminiscences. *Entomologist* **16**: 127–132.
24. Dale, C. W. *Our British butterflies*, p.47.
25. Verrall, G. H. (1909). The 'large copper' butterfly (*Chrysophanus dispar*). *Entomologist* **42**: 183.
26. Lavery, T. A. (1990). The history of the Large Copper introductions to Ireland. *Bull.amat.Ent.Soc.* **49**: 33–36.

16. SCARCE COPPER

1. Forster, J. R. (1770). *A catalogue of English insects*, p.10.
2. Harris, M. (1766). *The Aurelian, &c.*, pl.34 (which includes 'the [small] copper')
3. Berkenhout, J. (1769). *Outlines of the natural history of Great Britain and Ireland*, p.130.
4. Harris, M. (1775a). *The Aurelian &c.* (edn 2), p.66, pl.34.
5. Merrett, C. (1666). *Pinax rerum naturalium Britannicarum*, p.199.
6. Petiver, J. (1699). *Musei Petiveriana centuria prima – decima*: Centuria IV & V, p.34.
7. Petiver, J. (1717). *Papilionum Britanniae icones*, p.2, tab. 13 & 14.
8. Ray, J. (1710). *Historia insectorum*, p.130.
9. Linnaeus, C. (1758). *Systema naturae*, p.484.
10. Linnaeus, C. (1761). *Fauna Suecica* ... (edn 2), p.285.
11. Perceval, M. J. (1995). Our lost coppers (Lepidoptera: Lycaenidae). *Entomologist's Gaz.* **46**: 105–118.
12. Allan, P. B. M. (1980). *Leaves from a moth-hunter's notebook*, p.22.
13. Lewin, W. (1795). *The papilios of Great Britain*, p.84.
14. Donovan, E. (1796). *Natural history of British insects* **5**: 93–94, pl.173.
15. Haworth, A. H. (1802). *Prodromus Lepidopterorum Britannicorum*, pp.3–4
16. Curtis J. (1824). *British entomology* **1**: 12.
17. Jermyn, L. (1824). *The butterfly collector's vade mecum*, p.35.
18. Stephens, J. F. ([1827–]28). *Illustrations of British entomology (Haustellata)* **1**: 81.
19. Rennie, J. (1832). *A conspectus of the butterflies and moths found in Britain*, p.17.
20. Humphreys, H. N. & Westwood, J. O. (1841). *British butterflies and their transformations*, pp.97–98.
21. Barrett, C. G. (1893). *The Lepidoptera of the British Islands* **1**: 55.
22. Coleman, W. S. (1860). *British butterflies.*
23. Barrett, *Lepidoptera of British Islands*, p.56.
24. Capel Cure, C. H. (1880). Reoccurrence of *Chrysophanus virgaureae* in England. *Entomologist* **13**: 45.
25. Doubleday, H. (1850). *Synonymic List of British Lepidoptera*, p.2.
26. Doubleday, H. (1856). Remarks on Mr Buxton's Note on *Argynnis Lathonia* and *Pieris Daplidice. Zoologist* **14**: 5146–47.
27. Stainton, H. T. (1857a). *A manual of the British butterflies and moths* **1**: 56.
28. Kirby, W. F (1882). *European butterflies and moths*, p.56.
29. Tutt, J. W. (1906). *A natural history of the British Lepidoptera* **8**: 421.
30. South, R. (1906). *The butterflies of the British Isles*, pp.151–152.
31. Ford, E. B. (1945). *Butterflies*, pp.14–16.
32. Allan, P. B. M. (1956). The Middle Copper. *Entomologist's Rec.J. Var.* **68**: 68–73.
33. Allan, P. B. M. (1966). Copper butterflies in the West Country. *Ibid.* **78**: 161–166, 198–202.
34. Allan, *Moth-hunter's notebook*, pp.21–65.
35. Howarth, T. G. (1973). *South's British butterflies*, p.71.
36. Dennis, R. L. H. (1977). *The British butterflies*, p.112.
37. Allan, *Moth-hunter's notebook*, pp.23–32.
38. *Ibid.*, p.23.
39. Thomson, G. (1980). *The butterflies of Scotland*, p.106.

40. Rowley, R. (1962). *Lycaena virgaureae* (Lepidoptera, Lycaenidae) – Report of possible British specimens. *Entomologist* **95**: 191.

17. PURPLE-EDGED COPPER

1. Merrett, C. (1666). *Pinax rerum naturalium Britannicarum*, p.199.
2. Ford, E. B. (1945). *Butterflies*, p.9.
3. Lewin, W. (1795). *The papilios of Great Britain*, p.84.
4. Haworth, A. H. (1803). *Lepidoptera Britannica*, pp.41–42.
5. Sowerby, J. ([1804]–06). *The British miscellany*, Nos 1 & 2, col. pl., p.27.
6. Curtis, J. (1824). *British entomology* **1**: 81.
7. Stephens, J. F. ([1827–]28). *Illustrations of British entomology (Haustellata)* **1**: 80–81.
8. Jermyn, L. (1824). *The butterfly collector's vade mecum*, pp.34–35.
9. Rennie, J. (1832). *A conspectus of the butterflies and moths found in Britain*, p.17.
10. Stainton, H.T. (1857a). *A manual of the British butterflies and moths* **1**: 56.
11. Stainton, H. T. (1857b). Insect dealers. *Substitute* **15**: 169–170.
12 Allan, P. B. M. (1943). *Talking of moths*, p.226.
13 Doubleday, H. (1856). Remarks on Mr Buxton's Note on *Argynnis Lathonia* and *Pieris Daplidice*. *Zoologist* **14**: 5146–47.
14 Allan, P. B. M. (1980). *Leaves from a moth-hunter's notebook*, p.47.

18. LONG-TAILED BLUE

1 Newman, E. (1876). Editorial note. *Lampides boetica*. *Entomologist* **9**: 92.
2. Coleman, W. S. (1860). *British butterflies*, pp.172–173.
3. Furneaux, W. (1894). *Butterflies and moths*, p.188.
4. Kirby, W. F. (1896). *A hand-book to the order Lepidoptera 2: Butterflies*, 82–85.
5. South, R. (1906). *The butterflies of the British Isles*, p.154.
6. Thomas, J. [A.] & Lewington, R. (1991). *The butterflies of Britain and Ireland*, p.209.
7. Chevallier, L. H. S. (1952). *Lampides boeticus* Linn. in Surrey. *Entomologist's Rec.J.Var.* **64**: 274–277.
8. Chalmers-Hunt, J. M. & Skinner, B. (1992). The immigration of Lepidoptera to the British Isles in 1990. *Ibid.* **104**: 123–127, 209–218, 231–235.
9. Wurzell, B. (1990). The long-tailed blue *Lampides boeticus* breeding in North London. *Bull.amat.Ent.Soc.* **49**: 254.
10. Freed, T. H. (1992). *The butterflies of Kensal Green Cemetery*. (Unpub. M.A. thesis, Royal College of Art).
11. Freed, T. H. (1997). *Butterfly ecology in an urban cemetery*. (Unpub. Ph.D. thesis, Royal College of Art).
12. Smith, J. A. D. (1990). The long-tailed blue on Ranmore Common. *Bull.amat.Ent.Soc.* **49**: 254.
13. Dennis, R. L. H. (1977). *The British butterflies*, p.219.

19. SHORT-TAILED BLUE

1. Kirby, W. F. (1896). *A hand-book to the order Lepidoptera 2: Butterflies*, p.85.
2. Thomas, J. [A.] & Lewington, R. (1991). *The butterflies of Britain and Ireland*, p.210.
3. South, R. (1906). *The butterflies of the British Isles*, p.156.
4. Pickard-Cambridge, O. (1885). *Lycaena argiades* Pallas – A butterfly new to the British fauna. *Entomologist* **18**: 249–252.
5. St John, J. S. (1885). *Lycaena argiades* Pallas in Somerset. *Ibid.* **18**: 292–293.
6. Ford, E. B. (1945). *Butterflies*, p.6, pl.2.
7. Hill, P. M. (1999). Migrant butterflies in 1998. *Atropos* **6**: 68–69.
8. Lyell, M. A. C. (1938). *Tarucus telicanus* (Lep., Lycaenidae) in Dorset. *Entomologist* **71**: 173.

20. NORTHERN BROWN ARGUS

1. Lewin, W. (1795). *The papilios of Great Britain*, p.74.
2. Haworth, A. H. (1803). *Lepidoptera Britannica*, p.47.
3. Donovan, E. (1813). *Natural history of British insects* **16**: 1–3, pl.541.
4. Samouelle, G. (1819). *The entomologist's useful compendium*, p.242.
5. Stephens, J. F. ([1827–]28). *Illustrations of British entomology (Haustellata)* **1**: 95.
6. Rennie, J. (1832). *A conspectus of the butterflies and moths found in Britain*, p.19.
7. Morris, F. O. (1853). *A history of British butterflies*, p.151.
8. Dale, J. C. (1833). Observations on the influence of Locality, Time of Appearance, &c on Species and Varieties of Butterflies. *Ent.Mag.* **1**: 357.
9. Newman, E. (1834). Entomological Notes. *Ibid.* **2**: 515.
10. Newman, E. (1871). *An illustrated natural history of British butterflies*, p.121.
11. Barrett, C. G. (1893). *The Lepidoptera of the British Islands* **1**: 74.
12. Frohawk, F. W. (1934). *The complete book of British butterflies*, p.221.

13. South, R. (1906). *The butterflies of the British Isles*, p.161.
14. Knaggs, H. G. (1869b). *Lycaena Artaxerxes* Fabr. versus *L. Medon* Hufn. *Entomologist's Annu.* **1869**: 132.
15. Stainton, H. T. (1857a). *A manual of the British butterflies and moths* **1**: 62–63.
16. Thomas, J. [A.] & Lewington, R. (1991). *The butterflies of Britain and Ireland*, p.86.

21. MAZARINE BLUE

1. Bretherton, R. F. (1989). *In* Emmet, A. M. & Heath, J. (Eds), *The moths and butterflies of Great Britain and Ireland* **7**(1): 167.
2. Ray, J. (1710). *Historia insectorum*, p.132.
3. Moffet, T. (1634). *Insectorum sive minimorum animalium theatrum*, p.105.
4. Allan, P. B. M. (1980). *Leaves from a moth-hunter's notebook*, p.72.
5. Ford, E. B. (1945). *Butterflies*, p.10.
6. Raven, C. E. (1950). *John Ray, naturalist: his life and works* (edn 2), p.414.
7. Lewin, W. (1795). *The papilios of Great Britain*, p.80.
8. Donovan, E. (1797). *Natural history of British insects* **6**: 11, pl.184.
9. Haworth, A. H. (1803). *Lepidoptera Britannica*, p.48.
10. Haworth, A. H. (1802). *Prodromus Lepidopterorum Britannicorum*, p.48.
11. Bree, W. T. (1833). The Mazarine Blue butterfly. *Mag.nat.Hist.* **6**: 190–191.
12. Newman, E. (1871). *An illustrated natural history of British butterflies*, p.133.
13. Dale, C. W. (1890). *The history of our British butterflies*, p.58.
14. Parry, T. (1859). Observations on Solenobiae of Lancashire, etc. *Entomologist's Wkly Intell.* **6**: 28.
15. Allan, *Moth-hunter's notebook*, p.80.
16. Langley, A. F. (1875). Lycaena Acis near Cardiff. *Entomologist* **8**: 161.
17. Dale, C. W., *Our British butterflies*, p.58.
18. Hudd, A. E. (1871). Notes on the Lepidoptera of South Wales. *Ent.mon.Mag.* **8**: 113.
19. Chapman, T. A. (1909). Why is *Cyaniris semiargus* no longer a British insect? *Entomologist's Rec.J.Var.* **21**: 132–133.
20. Allan, *Moth-hunter's notebook*, pp.85–89.
21. Bretherton, R. F. (1989). *In* Emmet, A. M. & Heath, J. (Eds), *The moths and butterflies of Great Britain and Ireland* **7**(1):167.

22. LARGE BLUE

1. Lewin, W. (1795). *The papilios of Great Britain*, p.78.
2. Donovan, E. (1797). *Natural history of British insects* **6**: 11–12, pl. 184.
3. Haworth, A. H. (1803). *Lepidoptera Britannica*, p.45.
4. Brown, T. (1832). *The book of butterflies, sphinxes and moths* **1**: 164.
5. Stephens, J. F. ([1827–]28). *Illustrations of British entomology (Haustellata)* **1**: 87.
6. Bree, W. T. (1852). A List of Butterflies occurring in the neighbourhood of Polebrook, Northamptonshire; with some remarks. *Zoologist* **10**: 3348–3352.
7. Merrin, J. (1868). Notes on *Lycaena arion*. *Entomologist* **4**: 104–105.
8. Marsden, H. W. (1885). On the probable extinction of *Lycaena arion* in England. *Entomologist's mon.Mag.* **21**: 186–189.
9. Neal, E. (1994). *The badger man, memoirs of a biologist*, pp.78–79.
10. Muggleton, J. (1973). Some aspects of the history and ecology of blue butterflies in the Cotswolds. *Proc.Trans.Br.ent.nat.Hist.Soc.* **6**: 77–84.
11. Thomas, J. A. (1989). *In* Emmet, A. M. & Heath, J. (Eds), *The moths and butterflies of Great Britain and Ireland* **7**(1): 174.
12. *Ibid.*, p.175.
13. Newman, E. (1871). *An illustrated natural history of British butterflies*, pp.138–139, 199.
14. Bignell, G. C. (1865). *Lycaena arion* near Plymouth. *Entomologist* **2**: 295.
15. Bignell, G. C. (1884). *Lycaena arion. Ibid.* **17**: 208–209.
16. Dale, C. W. (1890). *The history of our British butterflies*, p.62.
17. Chatfield, J. (1987). *F. W. Frohawk: his life and work*, p.122.
18. Campbell, J. L. (1975). On the rumoured presence of the Large Blue (*Maculinea arion*) in the Hebrides. *Entomologist's Rec.J.Var.* **87**: 161–166.
19. Heslop Harrison, J. W. (1948). The Passing of the Ice Age & its Effect upon the Plant and Animal Life of the Scottish Western Isles. *New Nat.*, pt. 2 (Summer), p. 89.
20. Campbell, 'Large Blue in Hebrides', p.162.
21. Sabbagh, K. (1999). *A Rum affair. How botany's 'Piltdown man' was unmasked*, pp.65–66.
22. Huggins, H. C. (1973). They were Irish gannets. *Entomologist's Rec.J.Var.* **83**: 234–237.
23. Zeller, P. C. (1869). Hints for finding eggs and larvae of *Lycaena Arion*. *Entomologist's mon.Mag.* **6**: 10–11.
24. Buckler, W. (1886). *The larvae of the British butterflies and moths* **1**: vii, 105, 190.
25. Purefoy, E. B. (1953). An unpublished

account of experiments carried out at East Farleigh, Kent, in 1915 and subsequent years on the life history of *Maculinea arion*, the Large Blue butterfly. *Proc.R.ent.Soc.Lond.* **28**: 160–162.
26. Thomas, J. A. (1987). The return of the large blue. *Br.Butterfly Cons.Soc.News* **38**: 22–26.
27. Thomas, J. [A.] & Lewington, R. (1991). *The butterflies of Britain and Ireland*, p.108.
28. Barrett, C. G. (1897). Address to reception held on occasion of visit to Bishop's Stortford and Hatfield Forest in 1896. *Essex Nat.* **10**: 180.
29. Spooner, G. M. (1963). On causes of the decline of *Maculinea arion* L. in Britain. *Entomologist* **79**: 199–210.
30. Thomas, J. A. (1980). Why did the Large Blue become extinct in Britain? *Oryx* **15**: 243–247.

23. ALBIN'S HAMPSTEAD EYE

1. Petiver, J. (1717). *Papilionum Britanniae icones*, p.2.
2. Common, I. F. & Waterhouse, D. F. (1981). *Butterflies of Australia* (rev. edn), p.402.
3. Jermyn, L. (1824). *The butterfly collector's vade mecum*, pp.28–29.
4. Rennie, J. (1832). *A conspectus of the butterflies and moths found in Britain*, p.10.
5. Wood, W. (1852). *Index entomologicus* (new edn), p.244, pl. 53, fig.7b.
6. Humphreys, H. N. & Westwood, J. O. (1841). *British butterflies and their transformations*, p.58.
7. Morris, F. O. (1853). *A history of British butterflies*, p.80.
8. Emmet, A. M. & Heath, J. (Eds) (1989). *The moths and butterflies of Great Britain and Ireland* **7**(1): pl.23.

24. PAINTED LADY

1. Petiver, J. (1699). *Musei Petiveriana centuria prima – decima*, p.35.
2. Petiver, J. (1704). *Gazophylacii naturae & artis: decas prima – decas decima*, p.4.
3. Petiver, J. (1717). *Papilionum Britanniae icones*, p.2.
4. duBois, C. (*c.*1695). MS notebook, unpublished, in Entomology Library, The Natural History Museum, London.
5. Ray, J. (1710). *Historia insectorum*, p.122.
6. Lewin, W. (1795). *The papilios of Great Britain*, pl.6.
7. Linnaeus, C. (1758). *Systema naturae* (edn 10), pp.475–476.
8. Rennie, J. (1832). *A conspectus of the butterflies and moths found in Britain*, p.10.
9. Harris, M. (1766). *The Aurelian, &c.*, pl.11.
10. Morris, F. O. (1853). *A history of British butterflies*, p.81.
11. Newman, E. (1871). *An illustrated natural history of British butterflies*, p.66.
12. *Ibid.*, p.159.
13. Barrett, C. G. (1893). *The Lepidoptera of the British Islands* **1**: 153.
14. Tutt, J. W. (1898–1902). Migration and dispersal of insects: Lepidoptera. *Entomologist's Rec.J.Var.* **11**: 319–324; **12**: 13–16, 69–72, 127–128, 154–159, 182–186, 206–209, 236–238, 253–257; **13**: 97–102, 124–125, 145–147, 198–201, 233–237.
15. Skertchly, S. B. J. (1879). Butterfly swarms. *Nature* **20**: 266.
16. Williams, C. B. (1958). *Insect migration*, pp.220–228.
17. Bretherton, R. F. (1989). *In* Emmet, A. M. & Heath, J. (Eds), *The moths and butterflies of Great Britain and Ireland* **7**(1): 195.
18. Roper, P. (1996). The Painted Lady Invasion of 1996. *Butterfly Cons.News.* **63**: 2.
19. Davey, P. (1997). The 1996 Insect Immigration. *Atropos* **2**: 2–3.
20. Hill, P. M. (1997a). Migrant butterflies in 1996. *Ibid.* **2**: 14.
21. Wacher, J. (1998). Successful overwintering of Painted Lady *Cynthia cardui* in the U.K. *Ibid.* **5**: 19–20.

25. AMERICAN PAINTED LADY

1. Fabricius, J. C. (1775). *Systema entomologiae*, p.499.
2. Holland, W. J. (1916). *The butterfly guide*, pp.100–101.
3. Drury, D. (1770–73). *Illustrations of natural history*, pp.10–11.
4. Petiver, J. (1702–06). *Gazophylacii naturae & artis: decas prima – decas decima*, p.7.
5. Dale, J. C. (1830). Notice of the capture of *Vanessa Huntera* for the first time in Britain, with a Catalogue of rare Insects captured. *Mag.Nat.Hist.* **3**: 332–334.
6. Wood, W. (1852). *Index entomologicus* (new edn), p.244, pl.53, fig. 8b.
7. Humphreys, H. N. & Westwood, J. O. (1841). *British butterflies and their transformations*, p.57.
8. Morris, F. O. (1853). *A history of British butterflies*, pl.34.
9. Barrett, C. G. (1893). *The Lepidoptera of the British Islands* **1**: 155.
10. Bretherton, R. F. (1989). *In* Emmet, A. M. & Heath, J. (Eds), *The moths and butterflies of Great Britain and Ireland* **7**(1): 197–198.
11. Lipscomb, C. G. (1981). The American Painted Lady: *Cynthia virginiensis* Drury, a

very rare immigrant. *Entomologist's Rec.J.Var.* **93**: 242.
12. Nelson, J. (1996). The Monarch *Danaus plexippus* (L.) influx into Britain and Ireland in October 1995. *Atropos* **1**: 5–10.
13. Hill, P. M. (1999). Migrant butterflies in 1998. *Ibid.* **6**: 68.
14. South, R. (1906). *The butterflies of the British Isles*, p.81.

26. LARGE TORTOISESHELL
1. Emmet, A. M. (1989). *In* Emmet, A. M. & Heath, J. (Eds), *The moths and butterflies of Great Britain and Ireland* **7**(1): 205.
2. Moffet, T. (1634). *Insectorum sive minimorum animalium theatrum*, p.101.
3. Ray, J. (1710). *Historia insectorum*, p.118.
4. Linnaeus, C. (1758). *Systema naturae* (edn 10), p.477.
5. Petiver, J. (1699). *Musei Petiveriana centuria prima – decima*, p.34.
6. Lewin, W. (1795). *The papilios of Great Britain*, pl.2.
7. Rennie, J. (1832). *A conspectus of the butterflies and moths found in Britain*, p.9.
8. Doubleday, H. (1856). The Gonepteryx Rhamni question. *Zoologist*: **14**: 4950–4952.
9. Morris, F. O. (1853). *A history of British butterflies*, p.70.
10. Wood, J. G. (1892). *Insects at home*, p.398.
11. Harwood, W. H. (1906). *Eugonia (Vanessa) polychloros. Entomologist* **39**: 118.
12. Pratt, C. R. (1981). *A history of the butterflies and moths of Sussex*, pp.61–62.
13. Sawford, B. (1987). *The butterflies of Hertfordshire*, p.109.
14. Arnold, V. W., Baker, C. R. B., Manning, D. V. & Woiwod, I. P. (1997). *Butterflies and moths of Bedfordshire*, p.244.
15. South, R. (1906). *The butterflies of the British Isles*, p.65.
16. Thomas, J. [A.] & Lewington, R. (1991). *The butterflies of Britain and Ireland*, p.125.
17. Smith, F. H. N. (1997). *The moths and butterflies of Cornwall and the Isles of Scilly*, pp.202–203.
18. Bruce, E. (1997). Large Tortoiseshell in Sussex. *Butterfly Cons.News* **66**: 22.
19. Anon. (1995). In brief: Large Tortoiseshell sightings. *Ibid.* **60**: 16.
20. Goodey, B. (Ed.) (2000). Round-up of 1999 records. *Essex Moth Group Newsletter* No.15, p. 3.

27. CAMBERWELL BEAUTY
1. Moffet, T. (1634). *Insectorum sive minimorum animalium theatrum*, p.103.
2. Ray, J. (1710). *Historia insectorum*, p.135.
3. Wilkes, B. (1747–49). *The English moths and butterflies*, p.58.
4. Harris, M. (1766). *The Aurelian, &c*, p.26, pl.12.
5. Coleman, W. S. (1860). *British butterflies*, pp.121–123.
6. Lewin, W. (1795). *The papilios of Great Britain*, p.6.
7. Donovan, E. (1794). *Natural history of British insects* **3**: 45, pl.89.
8. Haworth, A. H. (1803). *Lepidoptera Britannica*, p.27.
9. Berkenhout, J. (1769). *Outlines of the natural history of Great Britain and Ireland* **1**: 126.
10. Lewin, *Papilios*, p.6.
11. Rennie, J. (1832). *A conspectus of the butterflies and moths found in Britain*, p.9.
12. Jermyn, L. (1824) *The butterfly collector's vade mecum*, p.21.
13. Newman, E. (1871). *The illustrated natural history of British butterflies*, p.58.
14. South, R. (1906). *The butterflies of the British Isles*, p.73.
15. Taylor, H. (1838). Notice of the capture of *Vanessa antiopa* in the neighbourhood of London. *Ent.Mag.* **5**: 253.
16. Wailes, G. (1858). A catalogue of the Lepidoptera of Northumberland and Durham. *Trans.Tyneside Nat.Fld Cl.* **3**(4): 189–234.
17. Tunmore, M. (1996). The 1995 Camberwell Beauty *Nymphalis antiopa* (L.) influx. *Atropos* **1**: 2–5.
18. Chalmers-Hunt, J. M. (1977). The 1976 invasion of the Camberwell Beauty. *Entomologist's Rec.J.Var.* **89**: 89–105.
19. Newman, L. H. (1967). *Living with butterflies*, pp.207–208.
20. Knaggs, H. G. (1873). Is *Vanessa Antiopa* a native or an immigrant? *Entomologist's Annu.* **1873**: p.37.
21. Knaggs, H. G. (1874). *Vanessa Antiopa* in 1973. *Ibid.* **1874**:153.
22. Haworth, A. H. (1803). *Lepidoptera Britannica*, pp.27–28.
23. Cockayne, E. A. (1921). The white border of *Euvanessa antiopa* L. *Entomologist's Rec.J.Var.* **33**: 205–211.
24. Sassoon, Siegfried (1938). *The old century and seven more years*, p.279.

28. EUROPEAN MAP BUTTERFLY
1. Bretherton, R. F. & Chalmers-Hunt, J. M. (1983). The immigration of Lepidoptera to the British Isles in 1982. *Entomologist's Rec.J.Var.* **95**: 89–94, 141–152.
2. Allan, P. B. M. (1940). British *Calophasia Lunula*: an historical note. *Entomologist* **73**: 203–205.

3. Bretherton, R. F. (1989). *In* Emmet, A. M. & Heath, J. (Eds), *The moths and butterflies of Great Britain and Ireland* **7**(1): 216.
4. Tolman, T. (1997). *Butterflies of Britain and Europe*, p.154.
5. Collins, G. A. (1995). *Butterflies of Surrey.* p.57.
6. Rowland-Brown, H. (1913). *Araschnia levana* in the Forest of Dean. *Trans.ent.Soc.Lond.* **1913**: lxx.
7. Hughes, A. W. (1914). *Araschnia levana* reported from Herefordshire. *Entomologist* **47**: 325.
8. Oliver, G. B. (1914). *Grapta c-album* and *Araschnia levana* from Forest of Dean. *Entomologist* **47**: 325.
9. Frohawk, F. W. (1940). Liberated butterflies. *Entomologist* **73**: 213.
10. Thomas, J. [A.] & Lewington, R. (1991). *The butterflies of Britain and Ireland*, p.197.
11. Ford, E. B. (1945). *Butterflies*, p.170.
12. Howarth, T. G. (1973). *South's British butterflies*, pp.117–118.
13. Dunbar, D. (1993). *Saving butterflies, a practical guide to the conservation of butterflies.*

29. WEAVER'S FRITILLARY
1. Coleman, W. S. (1860). *British butterflies*, p.172.
2. Bree, W. T. (1832). Notice of some singular varieties of Papilionidae in Mr Weaver's museum, Birmingham. *Mag.nat.Hist.* **5**: 749–753.
3. Allan, P. B. M. (1943). *Talking of moths*, pp.231–232.
4. Smith, B. (1857). Capture of *Argynnis dia. Entomologist's Wkly Intell.* **3**: 60.
5. Stainton, H. T. (1857d). Incredulity. *Ibid.* **3**: 65.
6. Ellis, Hon. C. (1857). Communications: Lepidoptera. *Argynnis Dia. Ibid.* **3**: 90.
7. Scott, J. (1857). *Fiat justitia* – Rue IT *Dia. Ibid.* **3**: 107.
8. Smith, B. (1858). *Fiat justitia* – GO IT *Dia! Ibid.* **3**: 114.
9. Scott, J. (1858). *Fiat justitia – Dia-bolical outrage. Ibid.* **3**: 130.
10. Parry, T. (1858). *Dia* again. *Ibid.* **3**: 130–131.
11. Batchelor, T. (1873). *Melitaea dia* in Kent. *Entomologist* **6**: 484.
12. Knaggs, H. G. (1874). *Argynnis Dia* again and *Cnethocampa Processionea. Entomologist's Annu.* **1874**: 155.
13. Lewis, W.A. (1876). Note on *Argynnis Dia. Entomologist* **9**: 69.
14. Barrett, C. G. (1893). *The Lepidoptera of the British Islands* **1**: 184.
15. Hilliard, R. D. (1985). Annual Exhibition 1984. *Bull.amat.Ent.Soc.* **44**: 54.

30. QUEEN OF SPAIN FRITILLARY
1. Harris, M. (1775b). *The English Lepidoptera: or, the Aurelian's pocket companion*, p.21.
2. Petiver, J. (1702). *Gazophylacii naturae & artis: decas prima – decas decima*, p.51.
3. Ray, J. (1710). *Historia insectorum*, p.119.
4. Donovan, E. (1794). *Natural history of British insects* **3**: 1–2, pl.73.
5. Lewin, W. (1795). *The papilios of Great Britain*, pl. 12.
6. Jermyn, L. (1824). *The butterfly collector's vade mecum*, p.21.
7. Rennie, J. (1832). *A conspectus of the butterflies and moths found in Britain*, p.7.
8. Haworth, A. H. (1803). *Lepidoptera Britannica*, pp.xxvi–xxviii.
9. Newman, E. (1868). *Argynnis lathonia* at Canterbury. *Entomologist* **4**: 146.
10. Parry, G. (1872a). *Argynnis Lathonia* at Canterbury. *Ibid.* **6**: 192.
11. Parry, G. (1872b). *Argynnis Lathonia* at Canterbury. *Ibid.* **6**: 212.
12. Doubleday, H. (1873). *Dianthoecia compta. Ibid.* **6**: 563–564.
13. 'Inquisitor' [Edward Newman] (1837). Note on butterflies questionably British. *Ent.Mag.* **4**: 177–179.
14. Allan, P. B. M. (1943). *Talking of moths*, pp.224–259.
15. Gaze, W. (1862). *Argynnis lathonia* in Suffolk. *Zoologist* **20**: 7971.
16. Barrett, C. G. (1893). *The Lepidoptera of the British Islands* **1**: 172.
17. Allan, P. B. M. (1980). *Leaves from a moth-hunter's notebook*, p.214.
18. Bretherton, R. F. (1989). *In* Emmet, A. M. & Heath, J. (Eds), *The moths and butterflies of Great Britain and Ireland* **7**(1): 223–224.
19. Baynes, E. S. A. (1964). *A revised catalogue of Irish Macrolepidoptera*, p.3.
20. Fox, R. (1999). The Queen of Spain Fritillary at Minsmere RSPB Reserve. *Butterfly Cons.News* **71**: 28.
21. Newman, L. H. (1967). *Living with butterflies*, pp.116–117.

31. NIOBE FRITILLARY
1. Stephens, J. F. ([1827–]28). *Illustrations of British entomology (Haustellata)* **1**: 37.
2. Rennie, J. (1832). *A conspectus of the butterflies and moths found in Britain*, p.7.
3. Chalmers-Hunt, J. M. (1960–61). *The butterflies and moths of Kent*, pp.78–79.
4. Allan, P. B. M. (1943). *Talking of moths*, p.245 *et seq.*
5. Doubleday, H. (1873a). *Argynnis Niobe* in Kent. *Entomologist* **6**: 483.

6. Newman, E. (1874). *Argynnis Niobe* in Kent. *Ibid.* **7**: 171–172.
7. Wigan, W. (1874). *Argynnis Niobe* in Kent. *Entomologist* **7**: 172–173.
8. Doubleday, H. (1873b). *Dianthoecia compta. Entomologist* **6**: 563–564.
9. Allan, *Talking of moths*, p.255.
10. Corbin, G. B. (1876). *Cnethocampa pityocampa* and *Argynnis Niobe. Entomologist* **9**: 21–22.
11. Newman, E. (1876). Editorial comment. *Ibid.* **9**: 22.
12. Allan, *Talking of moths*, p.257.
13. Barrett, C. G. (1893). *The Lepidoptera of the British Islands* **1**: 166.
14. Hambrough, W. (1869). *Argynnis Niobe* in the New Forest. *Entomologist* **4**: 357.
15. Cottam, A. (1900). *Argynnis Niobe* var. *Eris* taken in England. *Entomologist's Rec.J. Var.* **36**: 41–42.

32. MEDITERRANEAN FRITILLARY
1. Ford, E. B. (1945). *Butterflies*, pp.160–161.
2. Howarth, T. G. (1973). *South's British butterflies*, p.128.
3. Samson, C. P. J. (1970). Rare migrant specimen. *Bull.amat.Ent.Soc.* **29**: 107–108.

33. GLANVILLE FRITILLARY
1. Glanville, E. (1702). Sloane MS 4063, *f.*188 (renumbered).
2. Petiver, J. (*c.*1703). MS bound at end of *Musei Petiveriana centuria prima – decima.* (See transcription in Allan, P. B. M. (1980), p.269).
3. Petiver, J. (1703). *Gazophylacii naturae & artis: decas prima – decas decima*, p.4.
4. Ray, J. (1710). *Historia Insectorum*, p.121.
5. Petiver, J. (1717). *Papilionum Britanniae icones*, p.2.
6. Wilkes, B. (1747–49). *The English moths and butterflies*, p.58.
7. Dutfield, J. (1748). *A new and complete natural history of English moths and butterflies*, p.[7], pl.5.
8. Harris, M. (1766). *The Aurelian, &c.*, pl.16.
9. Knaggs, H. G. (1866). Notes on British Lepidoptera. Table of local and scarce British Lepidoptera captured in 1865. *Entomologist's Annu.* **1866**: 150.
10. Stephens, J. F. ([1827]–28). *Illustrations of British entomology (Haustellata)* **1**: 33.
11. Arnold, V. W., Baker, C. R. B., Manning, D. V. & Woiwod, I. P. (1997). *Butterflies and moths of Bedfordshire*, p.252.
12. Stainton, H. T. (1857a). *A manual of the British butterflies and moths* **1**: 44.
13. Greene, J. (1857). List of Lepidoptera occurring in the county of Suffolk. *Naturalist (Morris)* **7**: 253–258.
14. Thomson, G. (1980). *The butterflies of Scotland*, p.165.
15. Duncan, J. (1835). *The natural history of British butterflies*, p.144.
16. Newman, E. (1871). *An illustrated natural history of British butterflies*, p.45.
17. Barrett, C. G. (1893). *The Lepidoptera of the British Islands* **1**: 194–195.
18. Pratt, C. R. (1981). *A history of the butterflies and moths of Sussex*, pp.68–69.
19. Newman, E., *Illustrated British butterflies*, p.44.
20. Dawson, J. F. (1846). Habits of Melitaea Cinxia. *Zoologist* **4**: 1271–1272.
21. Thomas, J. [A.] & Lewington, R. (1991). *The butterflies of Britain and Ireland*, p.156.
22. Watson, R. W. (1969). Notes on *Melitaea cinxia* L. 1945–1968. *Entomologist's Rec.J. Var.* **81**: 18–20.
23. Pratt, *Butterflies of Sussex*, p.69.

34. ARRAN BROWN
1. Sowerby, J. ([1804]–06). *The British Miscellany*, Nos 1 & 2, pp.3–4.
2. Donovan, E. (1807). *Natural history of British insects* **12**: 87–88.
3. Stephens, J. F. ([1827]–28). *Illustrations of British entomology (Haustellata)* **1**: 61.
4. Jameson, L. (1797). Unpub. MS journal in the Royal Scottish Museum.
5. Morris, F. O. (1853). *A history of British butterflies*, p.53.
6. Newman, E. (1860). *A natural history of all the British butterflies*, p.19.
7. Stephens, J. F. (1856). *List of specimens of British animals in the collection of the British Museum* **5**: *Lepidoptera*, p.18.
8. Stainton, H. T. (1855). Review of J. O. Westwood's The Butterflies of Great Britain with their transformations. *Entomologist's Annu.* **1855**: 151.
9. Weaver, R. (1857). Note of an Entomological Excursion from Birmingham to Sutherlandshire. *Zoologist:* **15**: 5555–5556.
10. Barrett, C. G. (1893). *The Lepidoptera of the British Islands* **1**: 221.
11. Pelham-Clinton, E. C. (1964). Comments on the supposed occurrence in Scotland of *Erebia ligea* (L.). *Entomologist's Rec.J. Var.* **76**: 121–125, 1 fig.
12. Salmon, M. A. (1985). Further observations on the *Erebia ligea* (Linnaeus) and other controversies. *Ibid.* **107**: 117–126.
13. Pelham-Clinton, 'Occurrence in Scotland'. *Ibid.* **76**: 122.
14. South, R. (1941). *The butterflies of the British Isles* (edn 3), p.117.

15. Mercer, W. J. (1875). *Erebia ligea* at Margate. *Entomologist* **8**: 198.
16. Allan, P. B. M. (1943). *Talking of moths*, p.254.
17. Bretherton, R. F. (1989). *In* Emmet, A. M. & Heath, J. (Eds), *The moths and butterflies of Great Britain and Ireland* **7**(1): 259.

35. THE MONARCH

1. Tolman, T. 1997. *Butterflies of Britain and Europe*, p.140.
2. Miskin, W. H. (1871). Occurrence of *Danais archippus* in Queensland. *Entomologist's mon.Mag.* **8**: 17.
3. Bretherton, R. F. (1984). Monarchs on the move – *Danaus plexippus* (L.) and *D. chrysippus* (L.) *Proc.Trans.Br.ent.nat.Soc.* **17**: 65–66.
4. Thomas, J. [A.] & Lewington, R. (1991). *The butterflies of Britain and Ireland*, p.218.
5. Masó, A. & Pijoan, M. (1977). *Observar mariposas*, p.247.
6. Coleman, W. S. (1895). *British butterflies* (edn 2), p.105.
7. South, R. (1906). *The butterflies of the British Isles*, p.106.
8. Llewellyn, J. T. D. (1876). A foreign visitor (*Danais Archippus*). *Entomologist's mon.Mag.* **13**: 107–108.
9. Crallan, T. E. (1876). *Danais Archippus* in Sussex. *Entomologist* **9**: 265–267.
10. Jordan, R. C. R. (1886). An appeal to entomologists. *Entomologist's mon.Mag.* **22**: 211–212.
11. Ford, E. B. (1945). *Butterflies*, p.159.
12. Bates, H. W. (1864). *The naturalist on the Amazons*, p.62.

CHAPTER 5

'To the Worthy Successors of the Aurelian Society'

CONSERVATION AND COLLECTING

' "This is a new reign", said Egremont, "perhaps it is a new era".
"I think so", said the young stranger.'
Benjamin Disraeli, *Sybil or the Two Nations*

'ENTOMOLOGISTS HAVE LONG CONTENDED THAT THE subject of their concern is the most important animal group in the world', declared Martin Holdgate to the delegates attending a conference on insect conservation in 1991, adding that 'only the prejudice of the mammals that they are addressing prevents this fact from being generally recognized'.[1] This statement probably also sums up the feelings of many butterfly collectors today, as well as those of earlier generations.

Insects constitute without any doubt the most numerous, the most widespread, and the most diverse group of animals on Earth. Their study provides endless opportunities for research and at the same time gives great pleasure. Seen through human eyes, they also include many of the most beautiful creatures ever to have existed and provide examples of some of the most remarkable patterns of behaviour. Until Darwin propounded his revolutionary theory of evolution, this display of variety and beauty was regarded as a clear example of the Wisdom of God manifested in the works of Creation. From John Ray onwards early naturalists wrote at length on this subject. Did not the Holy Bible reveal in the Book of Genesis that 'God created man in his own image' and gave him dominion 'over every living thing that moveth upon the earth'? The idea that collecting and killing animals, for whatever reason, could be in any way morally wrong was not an issue until quite recent times except for a tiny minority. It is worthy of comment that many of the most active collectors – whether of birds and their eggs, or of butterflies and moths, or of many other forms of life – were parsons. The notion that collecting could seriously affect the viability of a species and bring about its extinction, whether locally or nationally, was never even entertained before the nineteenth century. The disappearance of a species was more likely to be attributed to the will of God than to any action by man. As the various eighteenth- and nineteenth-century interpretations of the appearances and disappearances of some migrant species of butterfly demonstrated, there were no credible explanations for these events and it was nearly always assumed that reserves of populations existed somewhere else to fill the gaps in due course. Collecting even on a vast scale rarely seemed to dent populations other than temporarily and, after all, were not all living things created for the utility of man, and birds and butterflies in particular to bring joy to the beholder and 'to adorn the world'?

The Victorian collector certainly made no bones about his enthusiasm, whether it was for big game on the African continent or butterflies in the

English countryside. As with grouse-shooting in Perthshire, it was usually quantity that mattered. One sunny July day in 1857, E. T. Silvester and W. Edwards went collecting in Sussex. They netted 36 Purple Emperors; 3 Large Tortoiseshells; 11 High Brown Fritillaries; 13 Ringlets; 14 Graylings; 21 Silver-studded Blues and 1 Wood White and proudly published their achievement in the pages of *The Entomologist's Weekly Intelligencer*.[2] In the same journal a week later, E. C. Rye announced his recent capture during a two- or three-day collecting trip in Wiltshire of 28 Silver-washed, 17 High Brown and 80 Dark Green Fritillaries, as well as a Large Tortoiseshell, thousands of Ringlets, and an unspecified number of Speckled Woods, Marbled Whites, Purple Hairstreaks, Chalk Hill Blues and Large and Small Skippers. He took the eighty Dark Green Fritillaries 'fresh from the chrysalis ... in about two hours, without moving beyond a yard or two'.[3] Rye was a respected entomologist and author of a successful work on Coleoptera. It is clear from contemporary journals that the numbers of butterflies in the British countryside must have exceeded anything that has been seen since the early 1950s, and that the supply would have seemed to be well-nigh inexhaustible.

Why did these naturalists collect such large and apparently quite unnecessary numbers? Reading on through these journals, we learn that these collectors were not as greedy or as selfish as they might appear at first sight, for many would distribute most of their captures, usually by exchanges of desiderata with friends and other correspondents. This was a frequent practice in those days, when it was no easy matter to reach distant butterfly localities, and most of the species were common, at least locally, at the time. Nonetheless some collectors *did* take extravagant numbers of rare species. For example, among Baron J. Bouck's collection, which was sold in 1938, there were over 900 Large Blues. And even this was relatively modest compared with an advertisement placed in *The Entomologist's Weekly Intelligencer* in 1857, in which a correspondent wanted to purchase 'about Twenty Gross (2,880) of the Scotch Argus [i.e. the Northern Brown Argus *Aricia artaxerxes*], in good condition, either alive or set...' but he would almost certainly have been a dealer rather than a *bona fide* collector (Fig. 143).

WANTED TO PURCHASE, about Twenty Gross of the SCOTCH ARGUS (*P. Artaxerxes*), in good condition, either alive or set. Also any quantity of good second-hand CABINETS, at reasonable prices. By James Gardner, Naturalist, 52, High Holborn, London.

N.B. Carcases with sliding trays not acceptable.

Fig. 143. Wanted: 'Twenty Gross of the Scotch Argus' (i.e. the Northern Brown Argus), an advertisement in *The Entomologist's Weekly Intelligencer* for 25th July 1857. This level of trafficking in butterflies was not uncommon at that time but serious concern was already being expressed in some quarters, notably by A. H. Haworth and H. T. Stainton. Its strong condemnation, by W. D. Crotch, published a month later in the same journal, is quoted in Chapter 4:15.

The price placed on the head of the unfortunate Argus may have resulted in one of the few recorded occasions when a butterfly was actually hunted into oblivion. Stainton quoted a correspondent, R. F. Logan, whose account of its over-collection at Arthur's Seat near Edinburgh revealed his concern[4] (see Chapter 4: 20).

Fear of over-collecting certain species had in fact been voiced as early as 1803 by Adrian Haworth. Large numbers of that 'most lovely of the British Blues', the Adonis Blue, were being taken by 'inferior collectors' as decorations to produce 'pictures of various shapes and sizes'. These collectors made 'annual and distant pedestrian excursions, for the sole purpose of procuring the charming males, to decorate their pictures with; a picture, consisting of numerous and beautiful Lepidoptera, ornamentally and regularly disposed, being the ultimate object of these assiduous people in the science of Entomology'. Apparently there was at that date quite a fashion for set butterflies arranged in circles or geometric patterns, containing up to 500 specimens each.[5]

Similar complaints crop up in the journals throughout the century. Tutt, for example, was incensed by the collection of large numbers of Purple Emperors by private collectors and commercial dealers alike in the 1870s and 1880s, with 'as many as nine amateur collectors standing in a line at three or four yards distance, and netting every specimen as it came up' (see Chapter 2).[6] Frohawk once witnessed a dealer and his companion take ninety-seven Purple Emperors in a single day at Chattenden Woods, Kent, 'as they swooped down from the trees to feed upon putrid flesh hung upon one of the lower branches of an Oak', attributing its apparent extinction there in 1887 to 'the destruction carried on by these two men year after year'.[7]

By the second half of the nineteenth century, some Victorians were beginning to believe, rightly or wrongly, that local extinctions could be the consequence of over-collecting. Stainton was very gloomy about the future of British butterflies. Their extinction seemed to him 'by no means so improbable a contingency as some might be disposed to imagine', the Mazarine Blue seemed to hover on the brink, and the extinction of the Large Blue and Lulworth Skipper 'we may safely prophesy [was] certain to come to pass at no very distant day'. This was before the Large Blue was discovered in Cornwall and Devon. 'Nowadays', Stainton complained, 'species which are at all rare or local are systematically caught and pinned with an unrelenting ardour, such as our butterflies of yore never experienced. The captures of the Purple Emperor this year' he added, 'must pretty nearly have doubled the number of cabinet specimens in the county: insects, it is true, are prolific, but systematic pursuit in all the stages of their existence must eventually thin their numbers; we hope the rising generation will remember this, and not rashly hasten forward the day of "the last British butterfly" '.[8]

Butterfly collectors undoubtedly made themselves unpopular in some quarters. In his *Hampshire Days*, W. H. Hudson revealed his instinctive dislike of the butterfly collector 'with his white, spectacled town face and green butterfly net'. 'Entomologist generally means collector' he went on, 'and his – the entomologist's – admiration has suffered inevitable decay, or rather has been starved by the growth of a more vigorous plant – the desire to possess, and pleasure in the possession of, dead insect cases'.[9]

Hudson's dislike of butterfly collecting pales in comparison with that of John Fowles, who, though a keen collector and country sportsman in his youth, came to anathematize both pursuits with born-again zeal:

> 'I began very young as a butterfly collector, surrounded by setting boards, killing-bottles, caterpillar cages ... One of the reasons I wrote – and named – my novel *The Collector* was to express my hatred of this lethal perversion ... I was trapped by the subtlest temptation of them all: rarity chasing – still a form of destroying, though what is destroyed may be less the rarities themselves than the vain and narrow minded fool who devotes all his time to their pursuit.'[10]

Occasionally members of the public would come to the defence, as they saw it, of the butterflies. Colonel Kershaw recalled one zealot who tried to foil their collecting by himself catching every specimen he could and rubbing their wings to spoil their collectability (in much the same way as rare orchids have been 'conserved' by removing their flowers).[11] Many collectors, however, did practise restraint, and themselves attempted to compensate for specimens captured by releasing bred stock at the locality. L. W. Newman made a regular habit of releasing surplus stock, and once attempted to move an entire colony out of harm's way, as W. S. Berridge described:

> 'At the time in question a large number of Marbled-white butterflies were to be found in a field not many miles from Mr Newman's farm, with the result that no less than seventeen professional collectors soon came down from London to wage war against them. Seeing that the insects were likely to be exterminated in this particular spot if such wholesale slaughter went on month after month, Mr Newman resolved to take matters into his own hands and remove the entire colony himself; but instead of killing them and storing them in boxes, as would otherwise have been their fate, he let them loose in another unknown and private field close by, and had the satisfaction of seeing them settle down permanently in their new home. Although the work of removing them took two years to complete, Mr Newman must have felt considerable satisfaction in having been the means of saving them from annihilation.'[12]

One way of assessing the impact of collecting is to estimate the number of collectors who were active at any given moment. There was a 'depression' in the mid-nineteenth century when Stainton doubted 'much if we could now make out a list of 500 English entomologists.'[13] By analysing entries in entomological journals, Beirne has shown that, like most hobbies, collecting went in and out of fashion. In the last few decades of the century, butterfly collecting came into vogue again, and probably reached its peak of popularity during the long summers of late Victorian and Edwardian England. Numbers dipped again during and after the First World War. Beirne thought that by 1930 there was only one collector where there had been three in 1890.[14]

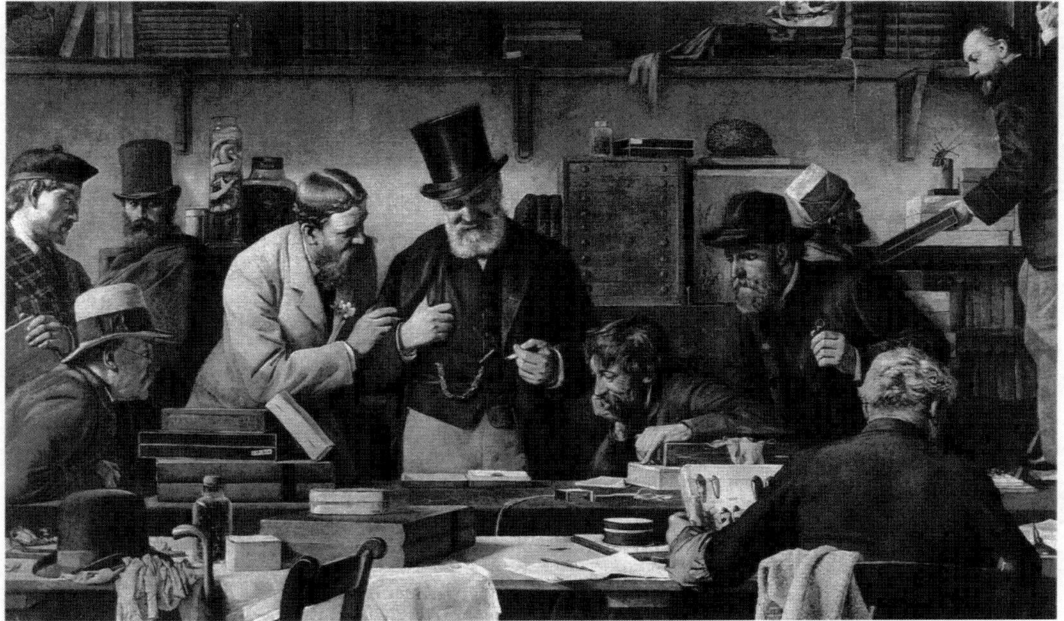

Fig. 144. 'After an Entomological Sale. *Beati possidentes*'. Engraving after a picture by Edward Armitage, R.A., 1878. The figures are presumably all portraits; those on the far left and right are thought to be F. Buchanan White and J. W. Douglas. The central figure may be Edward Newman. Armitage, a Fellow of the Royal Entomological Society, painted himself in the right foreground with back to the viewer.

These trends are to some extent matched by the frequency of entomological auction sales. The principal period, both in sales and prices, was between 1870 and 1930, with the busiest decade of all in the 1880s (Fig. 144).

Over 800 auctions were held during these 60 years, compared with only 70 between 1947 and 1960, and relatively few since. The highest price ever obtained at auction for a pair of British butterflies was in 1943 during the Sir Beckwith Whitehouse sale when £110 was bid for the famous all-white and all-black Marbled Whites from the A. B. Farn collection. Among the most important collections sold, mainly at Stevens' Auction Rooms at Covent Garden (Fig. 145), were those of Samuel Stevens in 1900, J. W. Tutt in 1911 and 1912, and Percy Bright between 1938 and 1942. The latter was said to be 'the richest assemblage of British butterfly varieties ever to have been auctioned'. The sale of Tutt's great collection was a disaster, attracting extremely low prices, with two or more lots often having to be combined in order to find a buyer. The dispersal of this unique collection of races and aberrations was a tragedy for future workers in the field of genetic variation, and it was considered by some that it 'should have (and it undoubtedly could have) been secured for the Nation ...'.[15]

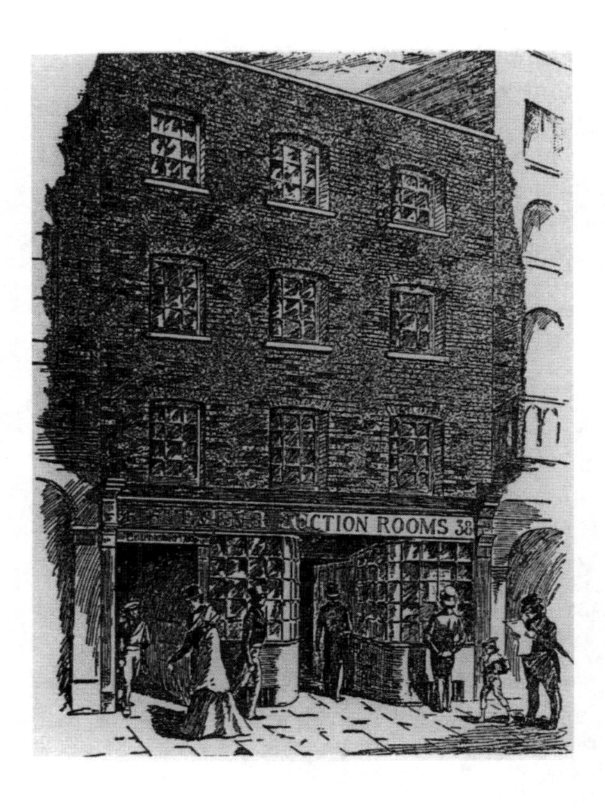

Fig. 145. Stevens's Auction Rooms in Covent Garden, London, as it looked in the 1830s. For over a century many fine collections were sold there under the hammer.

For every serious collector, there must have been dozens of small boys and day-trippers who accumulated a box or two of butterflies as a passing interest. Indeed, collecting butterflies and birds' eggs remained a common pastime among children with access to the countryside up to the middle of the twentieth century. The lifelong interest in natural history of some of our most respected senior zoologists and conservationists was fired in this way! The hobby was popular enough to support a sizeable commercial trade, which seems to have peaked around the 1860s and '70s. An analysis of account books, should they survive, might reveal in more detail the fluctuations in popularity.

Did butterfly collecting do any permanent harm? The answer is a qualified 'No', since the collectors can certainly be exonerated from the charge that they were responsible for any of the half-dozen national extinctions of butterflies, though they did drastically reduce some populations locally. P. B.M. Allan made the claim that 'it was the putting of a price on its head that exterminated *dispar*', the Large Copper, but he had no real evidence that this was so.[16] To explain why the Large Copper died out we need look no further than the steam pumps that drained the Fens from the 1850s onwards. Along with the Large Copper, we lost the Gypsy Moth, the Reed Tussock, the Rosy Marsh Moth, the Orache and the Marsh Dagger in the greatest concentrated extinction of individual species of Lepidoptera in our recorded history. Undoubtedly collectors did contribute to the decline of the Large Blue, but it died out in places where it was not collected as well as those that were. We now know that it was the loss of well-grazed grassland and the all-important ant colonies that spelled doom for the Large Blue. The final blow was a series of wet summers in the late 1970s. No one has seriously accused the collector of wiping out the Black-veined White, the Mazarine Blue, the Large Tortoiseshell or Chequered Skipper in England. R. F. Bretherton considered that over-collecting was a decisive factor in the extermination of the last English colony of the New Forest Burnet (*Zygaena viciae*), but even in that case it was probably no more than a contributory one. The overgrown state of its former habitat today suggests that it could not have survived.[17]

One of the few scientific attempts to assess the potential impact of collecting on a small population was carried out recently on a colony of the Slender Scotch Burnet (*Zygaena loti*) by Nigel Ravenscroft. On the face of it, the colony seemed highly vulnerable to over-collecting, being small, circumscribed and all too visible. However, Ravenscroft found that the average lifespan of the moth was only a single day, so that, given such a rapid turnover, a visiting collector would see only a very small proportion of the total population. Moreover, these thinly-scaled moths deteriorated so rapidly in condition that at any one time only a few would be of interest to the trophy-hunter. His conclusion was that the species is not in fact vulnerable to collection, except in its smallest colonies.[18]

In the past, butterfly numbers were high enough to sustain very heavy collecting over a long period, as at Royston Heath or the New Forest. Yet today, Royston Heath has no collectors, and very few butterflies either. In 1963, I set out to catch aberrations of the Chalkhill Blue, the lure that had attracted Newman, Bright, Leeds and many other great collectors mentioned in these pages. But the area of suitable downland had shrunk to a fraction of its former extent. I searched all morning and half the afternoon but saw no more than a dozen very normal Chalkhill Blues. There were now wardens to stop people collecting butterflies, but there was no one that could stop the habitat destruction.

Today, with our much-improved knowledge of the dynamics of insect populations and their relationship to the environment, we know better than to blame collecting for the overall decline of British butterflies. Except for small, localized populations, it is doubtful whether butterflies have suffered long-term harm from collectors. What has done the damage is habitat change, to which these sun-loving insects, often dependent on practices no longer economic – like coppicing and low-intensity grazing – are vulnerable.

Collecting is no longer the main aim of most entomologists, and even in the past, collecting was often done as a means to an end, and not as an end in itself. The early Aurelians, like Petiver and Sloane, collected butterflies, as they collected everything else, for a purpose – a kind of grand stocktaking of Britain's wildlife. It was through collecting and rearing butterflies that naturalists like Dandridge and Moses Harris discovered the basic facts about their natural history: their caterpillars and food-plants, their sexual and seasonal forms, and their habitat preferences. They corresponded and shared their knowledge and eventually began to publish journals for their observations. Many, like Eleazar Albin, Benjamin Wilkes and Moses Harris, were artists of talent. They were the great all-rounders, observant, dedicated and cultured.

Natural history was more than mere amusement to many Victorians. It contained an extra ingredient of instruction, of moral uplift, shared by Shakespeare in his famous lines from *As You Like It*:

> 'And this our life, exempt from public haunt,
> Finds tongues in trees, books in the running brooks,
> Sermons in stones, and good in everything.'

As Lætitia Jermyn expressed it: 'Is it to be regretted that many of the Spitalfield weavers spend their Saint Monday holidays in search of some of the more splendid lepidoptera, instead of smoking in the ale-house? Or, is it not rather to be wished that they should recreate their leisure hours by breathing the pure air, while in pursuit of this "untaxed and undisputed game"?'[19]

The great Victorian collectors studied their diaphanous subjects with great rigour and seriousness. They created an enormous edifice of scientific literature, recording the most precise observations and the most minute particularities. This was the great museum age, culminating in Tutt's shelf of great books, dignifying every known variation with a Latin name, and describing every pore and bristle. A more readable set of volumes, by C. G. Barrett, formed the basis of nearly every subsequent book on British butterflies and moths for the next eighty years.

The twentieth century, on the other hand, increasingly became the age for the study of living insects. The great practitioner of the first half was Frohawk, who reared every British species from the egg, painting and describing them at each stage. He can be numbered among the greatest butterfly artists, possibly rivalled only by Noel Humphreys, sixty years earlier. Meanwhile butterflies had begun to attract professional scientists as objects of study. E. B. Ford was a combination of Oxford zoologist and amateur collector, and discovered in British butterflies ideal subjects for the study of genetic variation. The 'aberrations' so meticulously described by Tutt, Leeds, and others, suddenly gained a new significance as their role in the continuing evolution of life was revealed. The new natural history of Ford attracted some of its most brilliant exponents in Bernard Kettlewell and Cyril Clarke, who explored fields unknown to the Victorians – such as biochemistry, industrial melanism, and medical genetics. Whether scientists like Ford, Kettlewell and Cockayne would have achieved what they did without also being naturalists and butterfly collectors is open to doubt.

Our own generation's great contribution to butterfly study has been in insect behaviour and population dynamics. Perhaps the most serious accusation that could be levelled at butterfly collecting is that it can blinker the eyes and the mind. Collectors chasing rarities or aberrations have seemed to be more anxious to secure their treasure than to watch them feed, settle and mate. However, one exception among the quarry has always been the Purple Emperor, whose behaviour has fascinated entomologists throughout the centuries, eventually resulting in a book on that species alone by I. R. P. Heslop, G. E. Hyde and R. E. Stockley – three naturalists with the blood of the old Aurelians in their veins.[20]

To understand how an insect lives and relates to its environment requires painstaking work. Few amateurs today can afford the time. This study has, however, become the life work of one professional entomologist, Jeremy Thomas. And, since the 1960s, a number of his protégés have studied individual species of butterfly for postgraduate theses. It is notable that few of these latter-day Aurelians are collectors, though many of them are expert photographers. Thomas's work began in response to the need for more precise knowledge about the requirements of butterflies on nature reserves. This new branch of study is conservation-led as opposed to specimen-led. For the general public, it has produced one of the great butterfly books of our time – *The Butterflies of Britain and Ireland* – which combines Thomas's perceptive descriptions of living butterfly behaviour with the fine artwork of Richard Lewington, and provides a worthy successor to Frohawk.[21]

WHITHER COLLECTING?

Among butterfly enthusiasts today, collecting is a minority pursuit. For most it has been replaced by butterfly watching and photography. Butterflies themselves have never been more popular. The charity Butterfly Conservation has over 8,000 members, and popular works on British and European butterflies enjoy a brisk sale in High Street shops. Quite a high proportion of those who still collect Lepidoptera – an analysis of society membership lists suggests about 1,000 may still do so – have tended, since the 1960s, to turn to microlepidoptera and to Continental butterflies and moths. Many now rear or breed them. Fashion has swung away from amassing specimens for the cabinet. Many conservationists dislike collecting *per se*, and some, it seems, would be happy to ban the collecting of butterflies and moths altogether. Such views are not based on sound reasoning. They are extreme and usually prejudiced. Jeremy Thomas, who is not a collector, noted that 'most detached reviewers have concluded that collectors have had little if any harmful effect on UK butterfly populations, though all believe that heavy collecting might tip the balance against a small colony that was at a low ebb for other reasons ... Nearly all extinctions have occurred on sites where collecting has not occurred.'[22] Unfortunately history shows that if a prejudice is widely shared it has great social force, whether or not it is based on reality.

Most collectors are only too aware of their responsibilities, and of their vulnerability as scapegoats for declines not of their making. In the 1960s, the Amateur Entomologist's Society established a Conservation Committee, and published a regular *Conservation Bulletin*, later renamed *Invertebrate Conservation News*. The Conservation Committee openly discouraged the small number of field naturalists who did overcollect. In 1972, the Joint Committee for the Conservation of British Insects, made up of entomologists and landowners, produced a *Code for Insect Collecting*, which included lists of threatened species and sensible proposals for responsible collecting.[23] All entomologists were asked to observe the code, and demonstrate that they were as concerned as anyone about habitat loss and falling populations. The Committee was concerned that wild places had become so fragmented by ploughing and development that some rare species might be vulnerable to collecting. The opening statement of the Code runs as follows:

> 'This committee believes that with the ever-increasing loss of habitats resulting from forestry, agriculture, and industrial, urban and recreational development, the point has been reached where a code for collecting should be considered in the interests of conservation of the British insect fauna, particularly macrolepidoptera. The Committee considers that in many areas this loss has gone so far that collecting, which at one time would have had a trivial effect, could now affect the survival in them of one or more species if continued without restraint.'

This voluntary code was supplemented in 1975 by the Conservation of Wild Creatures and Wild Plants Act, which protected a number of rare species, including one butterfly, the Large Blue. Rather disobligingly, the Large Blue became nationally extinct four years later, the victim not of collecting but of habitat loss caused by changes in agricultural practice and the near extermination of the rabbit following the introduction of myxomatosis. The Wildlife and Countryside Act of 1981, and its later supplements, extend formal protection to several more rare butterflies and moths, including the Swallowtail, Chequered Skipper (in England only) and Heath Fritillary, joined later by the High-brown and Marsh Fritillaries. The selection was not beyond criticism. The Chequered Skipper had already gone, and the Swallowtail was added not at the request of the Nature Conservancy Council, responsible for drawing up such lists, but at the whim of a politician. Under the terms of the Act, it is illegal to 'kill, take or sell' these species except under licence. Cabinet specimens taken before 1981 can be legally held. More recently, trading was prohibited, except under licence, for a larger number of British butterflies. The law reflects a widely felt concern for the plight of our rare insects, but it ignores the real issues of habitat protection and habitat management.

The amateur entomologist faces more serious restrictions in the ban on collecting on nature reserves or properties owned by the Forestry Commission or National Trust. Until about 1970, the Nature Conservancy Council operated a liberal policy, allowing collecting on its nature reserves, recognizing that it was mainly through the activities of entomologists that rare species were discovered. An early sign of less enlightened views was the blanket banning by the Forestry Commission in about 1960 of unlicensed collecting in the New Forest – a move that was much resented, particularly since it was the policies of the Commission that were held responsible for the decline of butterflies and moths in the Forest. Today, with most of the best remaining butterfly localities on protected land, the collector has a difficult time finding places in which to collect.

The decline in specimen collecting has had unforeseen and unfortunate results. Nature study has largely been replaced in schools by environmental studies, more concerned with international concerns like the ozone layer and global warming, than with naming and observing plants, insects and other wildlife. Examination syllabuses promote physiology and biochemistry at the expense of natural history. In addition the number of school natural history societies – once those fertile nurseries of future naturalists – has greatly diminished, as a result of which there is general ignorance even of the common species of wildlife around us. Where such interest continues, it is increasingly being directed abroad. Several times in the past few years, the Amateur Entomologists' Society junior entomologists' field trip has been held in France.

There has been much debate about whether the protection of certain rare species of butterflies and moths should be extended, perhaps to include all butterflies. In my view, such a step would be counterproductive. Most of the distribution records and knowledge of the habits of rare species that underpins conservation planning comes from 'collectors', most of whom act responsibly and take only small numbers of scarce species or none at all. Penalizing them might damage this essential cooperation without achieving anything positive in return. The *bona fide* amateur lepidopterist, known to respected entomological societies, should not have to suffer the indignity of being pestered by wardens, and should be allowed a small number of voucher specimens in exchange for his expertise. Rupert Barrington, a West Country lepidopterist, has suggested that 'action to restrict trading would serve a far greater purpose than banning the individual collector from his studies'.[24] Field work by collectors will always be scientifically useful – indeed, for some groups, essential. In his opinion, the real threat to rare or local species was excessive trading that lacked all purpose other than monetary gain, but such excesses are largely a thing of the past.

It is unlikely that the small numbers of collectors today make any impression on butterfly numbers. Nevertheless the collector is frequently the object of suspicion. John Moore recalled collecting once on church lands. The inevitable busybody came crashing through the undergrowth to ask breathlessly what the devil he thought he was up to? With an air of great authority Moore told him: ' "I am collecting Noctuid Lepidoptera by permission of the Ecclesiastical Commissioners". The intruder touched his hat, mumbled "Beg pardon, Sir", and took himself off.'[25] Dr Peter Edwards, an active moth-collector, has told me of a highly patrolled and guarded Oxfordshire wood where a number of rare butterflies and moths are known to occur. In spite of wardens, constantly on the look out for collectors, he saw no sign of collectors, but did witness a Forestry Commission vehicle run over a Purple Emperor at rest on a woodland ride!

The majority of collectors now travel overseas, probably no further than France and Spain, but a few travel further afield to Central America and Malaysia, where attitudes to collecting are a great deal more liberal. The increasing interest in foreign butterflies and moths has led a number of travel companies, which previously advertised safari holidays in Africa, to organize holidays specifically for entomologists and ornithologists.

THE FUTURE

The 1990s included two remarkable years for butterfly enthusiasts. In 1995, one of the hottest and driest summers since the mid-eighteenth century occurred, resulting in one of the best 'butterfly-years' for at least two decades. Reports of Camberwell Beauties and other rarities were numerous, and the sight of thousands of Dark Green Fritillaries and Marbled Whites flying on Porton Down, Wiltshire, rivalled the experiences penned by Victorian naturalists. Then in 1996 came the vast invasion of Painted Ladies on an unprecedented scale, leading to very successful breeding which later in the year boosted their numbers to many hundreds of thousands (see Chapter 4: 24). If, as some experts predict, our climate is gradually warming, then many species of butterflies, whether already resident or waiting just across the Channel, are likely to benefit. We might see more frequently the Continental Swallowtail (which used to breed in southern England) and the Queen of Spain Fritillary (which may have bred successfully in Suffolk between 1995 and 1997); and some resident insects currently hanging on by a fore-leg, like the Heath and Glanville Fritillaries, might well become more common.

The future of butterfly collecting is uncertain but it can still inspire the young and enthuse the old. We have surely not become so sophisticated and technological that we cannot experience the same innocent pleasure, the same breathless excitement that the Aurelians of the past described in countless articles to the entomological journals. And when the collecting season is over, perhaps we too will still be permitted to recall golden moments in the open air from the season past, like the Revd B. G. Johns over a century ago: 'All the time we were gradually mounting up to higher and higher ground, and finding out that a west wind could blow very fiercely when it pleased; and then we all came out on the open down, covered with patches of furze, wild thyme, and purple heath, and stopped for a moment to turn our faces to the breeze, and drink in the glorious fresh air that seemed to penetrate to every dusty corner of our lungs, and fill us with new life'.[26]

NOTES TO CHAPTER FIVE

1. Holdgate, M. W. (1991). Foreword. *In* Collins, N. M. & Thomas, J. A. (Eds), *The conservation of insects and their habitats*, p.xiii.
2. Silvester, E. T. & Edwards, W. (1857). Captures in Sussex. *Entomologist's Wkly Intell.* **2**: 139.
3. Rye, E. C. (1857). Doings in Wiltshire. *Ibid.* **2**: 138–139.
4. Stainton, H. T. (1857a). *A manual of British butterflies and moths*, pp.62–63.
5. Haworth, A. H. (1803–28) *Lepidoptera Britannica*, p.44.
6. Tutt, J. W. (1896). *British butterflies*, pp.386–387.
7. Frohawk, F. W. (1934). *The complete book of British butterflies*, pp.179–183.
8. Stainton, H. T. (1858). Lepidoptera. New British Species in 1857. *Entomologist's Annu.* **1858**: 85–86.
9. Hudson, W. H. (1903). *Hampshire days*, p.124.
10. Fowles, J. (1971). *The blinded eye* from *Second nature.*
11. Kershaw, S. H. (1958). Some Memories of S. G. Castle Russell. *Entomologist's Rec.J.Var.* **70**: 40.
12. Berridge, W. S. (1915). *Wonders of animal life*, p. 189.
13. Stainton, H. T. (1861). Regrets. *Entomologist's Wkly Intell.* **10**: 185–186.
14. Beirne, B. P. (1955). Fluctuations in quantity of work on British insects. *Entomologist's Gaz.* **6**: 7–9.
15. Chalmers-Hunt, J. M. (1976). *Natural history auctions 1700–1972. A register of sales in the British Isles*, pp.3–13, 54–180.
16. Allan, P. B. M. (1943). *Talking of moths*, p.23.
17. Tremewan, W. G. (1966). The history of *Zygaena viciae anglica* Reiss (Lep., Zygaenidae) in the New Forest. *Entomologist's Gaz.* **17**: 187–211.
18. Young, M. R. (1997). *The natural history of moths*, p.213.
19. Jermyn, L. (1827). *Butterfly collectors' vade mecum* (edn 2), p.10.
20. Heslop, I. R. P., Hyde, G. E. & Stockley, R. E. (1964). *Notes and views on the Purple Emperor.*
21. Thomas, J. [A.] & Lewington, R. (1991). *The butterflies of Great Britain and Ireland.*
22. Thomas, J. A. (1984). Conservation of Butterflies in Temperate Countries. Butterfly Collectors. *In* Vane-Wright, R. I. & Ackery, P. R. (Eds). *The biology of butterflies*, pp.345–346.
23. Joint Committee for the Conservation of British Insects (1972). *A code for insect collecting.* 4pp.
24. Barrington, R. D. G. (1987). Butterfly collecting. *Proc.Trans.Br.ent.nat.Hist.Soc.* **20**: 123–125.
25. Moore, J. (1954). *The season of the year.*
26. Johns, B. G. (1891). *Among the butterflies*, p.33.

APPENDIX 1

List of the British and Irish butterflies
together with their past and
present common names

APPENDIX 2

Entomological societies,
publications and significant events

BIBLIOGRAPHY
AND
FURTHER READING

INDEX

APPENDIX 1

LIST OF THE BRITISH AND IRISH BUTTERFLIES, TOGETHER WITH THEIR PAST AND PRESENT COMMON NAMES

Appendix 1 Illustrated with woodcuts from Moffet's *Insectorum Theatrum*

Eighty-three species are here listed by their current vernacular names. These are printed in SMALL CAPITALS, and are followed by their scientific names and the names of their authors and the dates of their first description. Then follows the name of the person who first published a record of the species in Britain with the date of its publication. In a few cases, where the date of its first known capture and/or the name of the collector are different, these are given first in square brackets [ ].

Although it has often been said that vernacular names are more stable than scientific ones, which necessarily change from time to time in line with taxonomic revisions, the great range of English names for the different species given below, often with more than one in use at the same date, show that the common names have had even less stability. Some authors give several English names. Their first or preferred name is indicated by a small figure (1), and those in synonymy in order of priority by figures (2), (3), etc. In those cases where a name is the first preference of an author who has published several alternatives, and that name has been adopted by 'all' or 'most subsequent authors', the figure (1) is not shown. The list is by no means comprehensive and not all of the earlier authors have been exhaustively researched, but most names that have been used at some date are included.

Initial square-bracketed numbers indicate a species whose history in Britain is recounted in Chapter 4.

HESPERIIDAE:

1. [1] CHEQUERED SKIPPER (*Carterocephalus palaemon* (Pallas, 1771)): [Abbot, 1798]; Abbot, 1800.
Papilio paniscus, without English name (Donovan, 1799)
The Chequered Skipper (Haworth, 1803; Jermyn 1824; Rennie, 1832; Morris (2), 1853; Coleman, 1860; and all subsequent authors)
The Scarce Skipper (Samouelle, 1819)
The Spotted Skipper (Morris (1), 1853)

2. THE SMALL SKIPPER (*Thymelicus sylvestris* (Poda, 1761)): Petiver, 1704.
The streaked golden Hog: male (Petiver, 1704)
The spotless golden Hog: female (Petiver, 1704)
The Small Skipper (Harris, 1766; and most subsequent authors)
The Great Streak Skipper (Rennie, 1832)
The Common Small Skipper (Heslop, 1959)

3. [2] THE ESSEX SKIPPER (*Thymelicus lineola* (Ochsenheimer, 1808)): [Hawes, 1888]; Hawes, 1890.
The New Small Skipper (Furneaux, 1894; Newman & Leeds (3), 1913; Heslop, 1959)

The Scarce Small Skipper (W. F. Kirby, 1896; W. E. Kirby, 1906; Newman & Leeds (2), 1913)
The Lineola Skipper (Coleman, 1897)
The Essex Skipper (South, 1906; Newman & Leeds (1), 1913; and most subsequent authors)

4. [3] THE LULWORTH SKIPPER (*Thymelicus acteon* (Rottemburg, 1775)): [Dale, 1832]; Curtis, 1833.
The Lulworth Skipper (Curtis, 1833; and all subsequent authors)

5. THE SILVER-SPOTTED SKIPPER (*Hesperia comma* (Linnaeus, 1758)): ?Moffet, 1634; ?Merrett, 1666; [Harris, 1772]; Harris, 1775b.
The Pearl Skipper (Harris, 1775; Donovan, 1800; Samouelle, 1819; Rennie, 1832; Humphreys & Westwood, 1841; Wood, 1852; Morris (2), 1853; Stephens, 1856; Newman, 1860; W. E. Kirby, 1906; Newman & Leeds (2), 1913)
The August Skipper (Lewin, 1795)
The Silver-spotted Skipper (Haworth, 1803; Jermyn, 1824; Morris (1), 1853; Newman, 1871; Furneaux, 1894; South, 1906; and most subsequent authors)

6. THE LARGE SKIPPER (*Ochlodes venata* (Bremer & Grey, 1852)): Petiver, 1704.
The Chequer-like Hog: male (Petiver, 1704)
The Chequered Hog: female (Petiver, 1704)
The streakt Cloudy Hog: male (Petiver, 1717)
The Cloudy Hog: female (Petiver, 1717)
The Large Skipper (Harris, 1766; and most subsequent authors)
The Wood Skipper (Samouelle, 1819)
The Clouded Skipper (Rennie, 1832)

7. THE DINGY SKIPPER (*Erynnis tages* (Linnaeus, 1758)): Merrett, 1666.
Handley's brown Butterfly (Petiver, 1704)
Handley's brown Hog Butterfly (Petiver, 1706)
Handley's small brown Butterfly (Petiver, 1717)
The Dingey Skipper (Harris, 1766)
The Dingy Skipper (Jermyn, 1824; Rennie, 1832; Morris, 1853; Coleman, 1860; and all subsequent authors)

8. [4] THE MALLOW SKIPPER (*Carcharodus alceae* (Esper, 1780)): [Bouck, 1923]; Frohawk, 1923.
Surrey Skipper (Frohawk, 1923)
The Mallow Skipper (Higgins & Riley, 1970) [cf. *Pyrgus malvae*]

9. THE GRIZZLED SKIPPER (*Pyrgus malvae* (Linnaeus, 1758)): Petiver, 1696.
Our Marsh Fritillary (Petiver, 1699)
Mr Dandridge's March Fritillary (Petiver, 1704)
The Grizzled Butterfly (Wilkes, 1747–49)
The Brown March Fritillary (Berkenhout (2), 1769)
The Grizzle; The Gristle (Harris, 1766; 1775a,b; Berkenhout (1), 1769; Rennie, 1832)
The Spotted Skipper (Lewin, 1795)
The Scarce Spotted Skipper (ab. *taras*) (Lewin, 1795)

The Mallow (Donovan, 1813)
The Mallow Skipper (Samouelle, 1819) [cf. *Carcharodus alceae*]
The Grizzled Skipper (Jermyn, 1824; Morris, 1853; Coleman, 1860; and all subsequent authors)

PAPILIONIDAE:

10. [5] THE APOLLO (*Parnassius apollo* (Linnaeus, 1758)): Moffet, 1634 (Fig. 146).

Fig. 146. Apollo

Mr Ray's Alpine Butterfly (Petiver, 1704)
The Crimson-ringed (Rennie, 1832; Wood, 1852)
The Apollo (Brown, 1832; Coleman, 1860)

11. [6] THE SWALLOWTAIL (*Papilio machaon* Linnaeus, 1758): Moffet, 1634 (see Fig. 128).
The Royal William (Petiver, 1699; Newman & Leeds (2), 1913)
The Swallow-tail or Swallowtail (Wilkes 1741–42; Harris, 1766; Jermyn, 1824; South, 1906; Newman & Leeds (1), 1913; and most subsequent authors)
The Queen (Rennie, 1832)
The Swallow-tailed Butterfly (Coleman, 1860)
The Common Swallowtail (Heslop, 1959)

12. [7] THE SCARCE SWALLOWTAIL (*Iphiclides podalirius* (Linnaeus, 1758)): Moffet, 1634 (see Fig. 129); ?Ray, 1710; Berkenhout, 1769.
The Scarce Swallowtail (Lewin, 1795; and all subsequent authors)

PIERIDAE:

13. THE WOOD WHITE (*Leptidea sinapis* (Linnaeus, 1758)): Merrett, 1666.
Papilio albus minor [the small white Butterfly] (Petiver, 1699)
The small white (Ray, 1710) [cf. *Pieris rapae*]
The Small White Wood Butterfly: male (Petiver, 1717)
The White Small Tipped Butterfly: female (Petiver, 1717)
The Wood White (Harris, 1766; Donovan (2), 1799; Jermyn, 1824; Morris, 1853)
The Wood Lady (Donovan (1), 1799) [cf. *Anthocharis cardamines*]
The White Wood (Wood, 1852)

14. [8] THE PALE CLOUDED YELLOW (*Colias hyale* (Linnaeus, 1758)): Harris, 1775b.
The Pale Clouded Yellow (Harris, 1775b); female (Lewin, 1795); both sexes (Jermyn, 1824; Coleman (2), 1860, W. F. Kirby, 1896; W. E. Kirby, 1901; South, 1906; and all subsequent authors)
The Clouded Yellow: male (Lewin, 1795); both sexes (Rennie, 1832) [cf. *Colias croceus*]

The Clouded Sulphur (Haworth, 1803; Coleman (1), 1860)

15. [9] BERGER'S CLOUDED YELLOW (*Colias alfacariensis* Berger, 1948): Berger, 1945.
The New Clouded Yellow (Berger, 1948; South, 1956)
Berger's Clouded Yellow (Heslop, 1959; Howarth, 1973; and all subsequent authors)

16. [10] THE CLOUDED YELLOW (*Colias croceus* (Geoffroy, 1785)): Moffet, 1634 (see Fig. 131).
The Saffron Butterfly: male (Petiver, 1703; Ray, 1710; Berkenhout, 1769; Harris 1775a; Newman & Leeds (3), 1913))
The Spotted Saffron Butterfly: female (Petiver, 1717); both sexes: (Newman & Leeds (3), 1913)
The Clouded Yellow (Wilkes, 1741–42; Harris 1766; Jermyn, 1824; Coleman, (1), 1860; and most subsequent authors)
The Clouded Orange (Lewin, 1795)
The White Clouded Yellow (ab. *helice*) (Haworth, 1803; Jermyn, 1824; Rennie, 1832)
The Clouded Saffron (Rennie, 1832; Coleman (2), 1860)
The Common Clouded Yellow (Heslop, 1959)

17. THE BRIMSTONE (*Gonepteryx rhamni* (Linnaeus, 1758)): Moffet, 1634 (Fig. 147); Merrett, 1666.

Fig. 147. Brimstone

The Pale Brimstone: male (Petiver, 1695)
The Brimstone: female (Petiver, 1695); male and female (Jermyn, 1824; and most subsequent authors)
The Male Straw Butterfly: female (Ray, 1710, in error; Petiver, 1717)
The Primrose (Rennie, 1832)
The Sulphur (Frohawk (2), 1924)

18. [11] THE CLEOPATRA (*Gonepteryx cleopatra* (Linnaeus, 1767)): [J. Fullerton, 1860]; Pickard, 1860.
The Cleopatra (Higgins & Riley, 1970; Howarth, 1973)

19. [12] THE BLACK-VEINED WHITE (*Aporia crataegi* (Linnaeus, 1758)): Moffet, 1634; Merrett, 1666.
The White Butterfly with black Veins (Petiver, 1699; Wilkes, 1747–49; Berkenhout, 1769)

The Black-veined White (Harris, 1766; Jermyn, 1824; Coleman (1), 1860; South; 1906; and all subsequent authors)
The Hawthorn (Rennie, 1832; Coleman (2), 1860)

20. THE LARGE WHITE (*Pieris brassicae* (Linnaeus, 1758)): Moffet, 1634; Merrett, 1666.
The Greater White Cabbage-Butterfly (Petiver, 1703)
The Great Female Cabbage Butterfly (Petiver, 1717)
The Great White Butterfly (Albin, 1720)
The Large White Garden Butterfly or The Large Garden White Butterfly (Wilkes, 1747–49; Lewin, 1795; Donovan, 1808; Humphreys & Westwood, 1841; Wood, 1853; Coleman; 1860; W. E. Kirby (1), 1906; Newman & Leeds (3), 1913; Heslop, 1959)
The Great White Cabbage Butterfly (Berkenhout, 1769; Stephens, 1856)
The Large White (Haworth, 1803; Jermyn, 1824; Morris, 1853; South, 1906; Newman & Leeds (1), 1913)
The Large Cabbage (Samouelle, 1819)
The Large White Cabbage Butterfly (Brown, 1832; Newman & Leeds (2), 1913)
The Cabbage (Rennie, 1832)
The Great White (var.) (Wood, 1852)
The Large Cabbage White Butterfly (W. F. Kirby, 1896; W. E. Kirby (2), 1906)

21. THE SMALL WHITE (*Pieris rapae* (Linnaeus, 1758)): Moffet, 1634 (Fig. 148); Merrett, 1666.

Fig. 148.
Small White

The Lesser White Cabbage Butterfly (Petiver, 1703)
The Smaller Common White Butterfly (Ray, 1710)
The Lesser White Unspotted Butterfly: ?male, spring brood (Petiver, 1717)
The Lesser White Single-spotted Butterfly: ?male, summer brood (Petiver, 1717)
The Lesser White Double-Spotted Butterfly: ?female, spring brood (Petiver, 1717)
The Lesser White Treble-spotted Butterfly: ?female, summer brood (Petiver, 1717)
The Small White Garden Butterfly or The Small Garden White Butterfly (Wilkes, 1747–49; Lewin, 1795; Humphreys & Westwood, 1841; Coleman, 1860; W. E. Kirby (1), 1906; Newman & Leeds (3), 1913; Heslop, 1959)
The Small White (Haworth, 1803; Jermyn, 1824; Morris, 1853; South, 1906; Newman & Leeds (1), 1913)
The Small Cabbage Butterfly (Samouelle, 1819)
The Turnip (Rennie, 1832)
Mr Howard's or Howard's White (var.) (Wood, 1852; Newman & Leeds (4), 1913)

The Small White Cabbage (Stephens, 1856; Newman & Leeds (2), 1913)
The Small Cabbage White (W. F. Kirby, 1896; W. E. Kirby (2), 1906)

22. THE GREEN-VEINED WHITE (*Pieris napi* (Linnaeus, 1758)): Moffet, 1634; Merrett, 1666.
The common white veined-Butterfly (Petiver, 1699)
The lesser, white, veined Butterfly: ?male, spring brood (Petiver, 1717)
The common white veined Butterfly with single spots: ?male, summer brood (Petiver, 1717)
The common white veined Butterfly with double spots: ?female, spring & summer broods (Petiver, 1717)
The Green-veined Butterfly (Albin, 1720)
The White Butterfly with Green Veins (Wilkes, 1747; Berkenhout, 1769)
The Green-veined White (Lewin, 1795; and most subsequent authors)
The Navew (Rennie, 1832)
The Colewort (subsp. *sabellicae*) (Rennie, 1832)
The Early Green-veined White (subsp. *sabellicae*) (Wood, 1852)
The Dusky-veined White (subsp. *sabellicae*) (Newman & Leeds, 1913)

23. [13] THE BATH WHITE (*Pontia daplidice* (Linnaeus, 1758)): Petiver, 1699.
Papilio leucomelanus, subtus viridescens marmoreus [black and white butterfly, with the underside marbled green] (Petiver, 1699)
The Greenish Marbled half-Mourner (Petiver, 1699)
Mr Vernon's half-Mourner (Petiver, 1702)
Vernon's Cambridge half-Mourner (Petiver, 1706)
The slight greenish half-Mourner: male (Petiver, 1717; Morris (3), 1853)
Vernon's greenish half-Mourner: female (Petiver, 1717; Morris (4), 1853)
The Bath White (Lewin, 1795; Morris (2), 1853; Furneaux (1), 1894; South, 1906; Newman & Leeds (1), 1913; Frohawk (1), 1924; and all subsequent authors)
The Green Chequered, Green-chequered or Green-checkered White (Haworth, 1803; Jermyn, 1824; Wood, 1852; Stephens, 1856; Newman, 1871; Furneaux (2), 1894; Newman & Leeds (2), 1913)
The Rocket (Rennie, 1832)
The Chequered or Checkered White (Morris (1), 1853; Frohawk (2), 1924)

24. THE ORANGE-TIP (*Anthocharis cardamines* (Linnaeus, 1758)): Moffet, 1634 (Fig. 149).

Fig. 149.
Orange-tip

The white marbled male Butterfly: male (Petiver, 1699)
The white marbled female Butterfly: female (Petiver, 1699)
The common white marble male Butterfly (Ray, 1710)
The common white marble female Butterfly (Ray, 1710)
The Lady of the Woods (Wilkes, 1741–42; Harris, 1775b)

The Orange-tip Butterfly (Wilkes, 1747–49; Berkenhout, 1769; Lewin (1), 1795; Donovan (1), 1796; Haworth, 1803; and most subsequent authors)
The Wood Lady (Dutfield (1), 1748; Harris, 1766; Lewin (2), 1795; Donovan (2), 1796; Rennie, 1832) [cf. *Leptidea sinapis*]
The Prince of Orange (Dutfield (2), 1748)
The Orange Tipped (Morris, 1853)
The Orange-tip White (Heslop, 1959)

LYCAENIDAE:

25. THE GREEN HAIRSTREAK (*Callophrys rubi* (Linnaeus, 1758)): Merrett, 1666.
The holly under green butterfly (Petiver, 1702)
The holly butterfly (Petiver, 1717)
The Green Butterfly (Wilkes, 1747–49)
The Green Fly (Harris, 1766)
The Green Papillon (Harris, 1775a)
The Bramble or The Green Fly (Harris, 1775b)
The Green Hairstreak or Hair-streak (Lewin, 1795; Samouelle, 1819; Jermyn, 1824; Morris, 1853; and most subsequent authors)
The Green Underside (Samouelle, 1819)

26. THE BROWN HAIRSTREAK (*Thecla betulae* (Linnaeus, 1758)): Petiver, 1703.
The brown double Streak: male (Petiver, 1703)
The Golden brown double Streak: female (Petiver, 1703)
The Brown Hairstreak, Hair Streak or Hair-streak: male (Ray, 1710); both sexes (Wilkes, 1747–49; Harris, 1766; Jermyn, 1824; Rennie, 1832; Humphreys & Westwood, 1841; Wood, 1852; and all subsequent authors)
The Golden Hairstreak: female (Ray, 1710)
The Hairstreak Butterfly (Albin, 1720)

27. THE PURPLE HAIRSTREAK (*Quercusia quercus* (Linnaeus, 1758)): Petiver, 1702b; Ray, 1710.
Mr Ray's purple Streak (Petiver, 1702b)
Mr Ray's blue Hairstreak: male (Petiver, 1717)
'our' blue Hairstreak: female (Petiver, 1717)
The Purple Hairstreak or Hair-streak (Albin, 1720; Wilkes, 1741–42; Harris, 1766; Jermyn, 1824; and all subsequent authors)

28. THE WHITE-LETTER HAIRSTREAK (*Satyrium w-album* (Knoch, 1782)): Petiver, 1703b.
The Hair-streak (Petiver, 1703b)
The Dark Hairstreak (Harris, 1775b; Lewin, 1795; Haworth, 1803)
'w-hairstreak' (Humphreys & Westwood, 1841; W. E. Kirby (2), 1906)
The Black Hairstreak (Donovan, 1808; Samouelle, 1819; Jermyn, 1824; Rennie, 1832; Wood, 1852; Stephens, 1856; Newman, 1871; Furneaux (1), 1894) [cf. *Satyrium pruni*]
The White W-Hairstreak (Morris, 1853;)
The White-W Hairstreak (Morris, 1864)
The White-letter Hairstreak (Furneaux (2), 1894; Coleman, 1860; W. F. Kirby, 1896; W. E. Kirby (1), 1906; South, 1906; and all subsequent authors)

29. [14] THE BLACK HAIRSTREAK (*Satyrium pruni* (Linnaeus, 1758)): [Newman, 1828], Curtis, 1829.
The Black Hairstreak (Curtis, 1829; Humphreys & Westwood, 1841; Morris, 1853; Coleman, 1860; Kirby, 1896; South, 1906; Newman & Leeds (1), 1913; and all subsequent authors) [cf. *Satyrium w-album*]
The Plumb Hairstreak (Rennie, 1832)
The Dark Hairstreak (Wood, 1852; Stephens, 1856; Newman, 1871; Furneaux, 1894; Newman & Leeds (2), 1913)

30. THE SMALL COPPER (*Lycaena phlaeas* (Linnaeus, 1761)): Petiver, 1699.
The small golden black-spotted Meadow Butterfly (Petiver, 1699)
The Small Tortoise-shell (Petiver, 1717)
The Copper (Harris, 1766; Rennie, 1832; Stephens, 1856)
The small golden black-spotted Butterfly (Berkenhout, 1769)
The Small Copper (Lewin, 1795; Samouelle, 1819; Brown, 1832; Morris (1), 1853; and most subsequent authors)
The Common Copper (Haworth, 1803; Donovan, 1808; Jermyn, 1827; Humphreys & Westwood, 1841; Wood, 1852; Morris (2), 1853; Newman, 1871; Newman & Leeds (2), 1913)
The Common Small Copper (Morris (3), 1853)

31. [15] THE LARGE COPPER (*Lycaena dispar* (Haworth, 1803)): [J. Green, 1749; H. Seymer, *c.*1775]; Lewin, 1795.
The Orange Argus of Elloe (Green, 1749 (unpublished)) (see Plate 23) [cf. *Lasiommata megera*]
The Large Copper (Lewin, 1795; Haworth, 1803; and most subsequent authors)
The Great Copper (Donovan, 1798)
The Swift Copper (Rennie, 1832) (as *L. hippothoe*, misidentification)

32. [16] THE SCARCE COPPER (*Lycaena virgaureae* (Linnaeus, 1758)): ?Forster, 1770; [Seymer, *c.*1775].
The Middle Copper (Jermyn, 1824)
The Scarce Copper (Lewin, 1795; Wood, 1852; and all subsequent authors)
The Golden Copper (Newman & Leeds (2), 1913)
The Golden Rod (Rennie, 1832)

33. [17] THE PURPLE-EDGED COPPER (*Lycaena hippothoe* (Linnaeus, 1761): ?Merrett, 1666.
The Purple-edged Copper (Jermyn, 1824; and most subsequent authors)
The Swift Copper (Rennie, 1832)

34. [18] THE LONG-TAILED BLUE (*Lampides boeticus* (Linnaeus, 1767)): Newman, 1859.
The Brighton Argus (Newman, 1860)
The Long-tailed Blue (Coleman (1), 1860; South, 1906; Newman & Leeds (1), 1913; and all subsequent authors)
The Tailed Blue (Coleman (2), 1860; Furneaux, 1894)
The Pea-pod Argus (Newman, 1871; Newman & Leeds (3), 1913)
The Large Tailed Blue (Newman & Leeds (2), 1913; W. F. Kirby, 1896)

35. THE SMALL BLUE (*Cupido minimus* (Fuessly, 1775)): Lewin, 1795.
The Small Blue (Lewin, 1795; Jermyn, 1824; Furneaux, 1894; South, 1906; Frohawk, 1924; Emmet & Heath, 1989; Thomas (1), 1991)
The Bedford Blue (Samouelle, 1819; Rennie, 1832; Humphreys & Westwood, 1841; Morris (2), 1853; Coleman (1), 1860; W. F. Kirby, 1896; W. E. Kirby, 1906; Newman & Leeds (2), 1913)
The Little Blue (Morris (1), 1853; Coleman (2), 1860; Newman & Leeds (3), 1913; Thomas (2), 1991)

36. [19] THE SHORT-TAILED OR BLOXWORTH BLUE (*Everes argiades* (Pallas, 1771)): [J. S. St John, 1874]; Pickard-Cambridge, 1885.
The Small Tailed or Small-tailed Blue (Kirby, 1896; Newman & Leeds (2), 1913; Emmet & Heath (2), 1989)
The Bloxworth Blue (South, 1906; Newman & Leeds (3), 1913; Frohawk (2), 1924,1934; Heslop, 1959; Thomas (1), 1991)
The Short-tailed Blue (Newman & Leeds (1), 1913; Frohawk (1), 1924, 1934; Emmet & Heath (1), 1989; Thomas (2), 1991)

37. THE SILVER-STUDDED BLUE (*Plebejus argus* (Linnaeus, 1758)): ?Moffet, 1634; Harris, 1775b.
?The Small Lead Argus (?Petiver, 1717; Stephens, 1829)
The Silver-studded Blue (Harris, 1775b; Lewin, 1795; Haworth, 1803; and most subsequent authors)
The Lead Blue (Rennie, 1832)
The Lead Argus (Newman & Leeds (2), 1913) [cf. *Lysandra bellargus*]

38. THE BROWN ARGUS (*Aricia agestis* ([Denis & Schiffermuller], 1775)): Petiver, 1704.
The edg'd brown Argus (Petiver, 1704, 1717)
The brown edg'd Argus (Petiver, 1706)
?The Argus Blue (Harris, 1775b)
The Brown Blue (Lewin, 1795)
The Brown Argus (Haworth, 1803; Rennie, 1832; Wood, 1852; Coleman, 1860)
The Black-spot Brown (Samouelle, 1819)
The Brown Argus Blue (Morris (1) [which includes the next species], 1853; Heslop, 1959)

39. [20] THE NORTHERN BROWN ARGUS (*Aricia artaxerxes* (Fabricius, 1793)): Fabricius, 1793, [from material obtained earlier by W. Jones]; Lewin, 1795.
The Brown Whitespot (Lewin, 1795)
The Scotch Argus (Haworth, 1803; Samouelle, 1819; Jermyn, 1824; Rennie, 1832; Wood, 1852; Morris, 1853) [cf. *Erebia aethiops*]
The Artaxerxes (Donovan, 1813; Brown, 1843; Coleman, 1860, 1897)
The White-spot Brown (Samouelle, 1819)
The Durham Argus (subsp. *salmacis*) (Rennie, 1832; Wood, 1852; (both subspp.) Morris (2), 1853)
The Brown Argus Blue (both subspp.) (Morris (1), 1853)
The Dark Argus (subsp. *salmacis*) (Stephens, 1856)
The Castle Eden Argus (subsp. *salmacis*) (Newman, 1871)
The Scotch Brown Argus (Newman, 1871; W. E. Kirby, 1906; Newman & Leeds (1), 1913; Heslop, 1959; Thomas (2), 1991)

The Scotch White Spot (South, 1906; Newman & Leeds (2), 1913)
The Scotch Brown Blue (Heslop, 1959)
The Northern Brown Argus (Howarth, 1973; Thomas (1), 1991)

Fig. 150.
Common Blue

40. THE COMMON BLUE (*Polyommatus icarus* (Rottemburg, 1775)): ? Moffet, 1634 (Fig. 150); Merrett, 1666.
The little Blew Argus (Petiver, 1699)
The Blue Argus: male (Petiver, 1704); both sexes (Wilkes, 1747–49; Berkenhout, 1769)
The Mixed Argus: female (Petiver, 1704)
The Selvedg'd Argus (Petiver, 1717)
The Ultramarine Blue (Wilkes, 1741–42)
The Common Blue (Harris, 1775a,b; and most subsequent authors)
The Caerulean Butterfly (Brown, 1832)
The Alexis (Rennie, 1832)

41. THE CHALK HILL BLUE (*Lysandra coridon* (Poda, 1761)): Petiver, 1704.
The pale blue Argus (Petiver, 1704; 1717)
The Chalkhill, Chalk Hill, or Chalk-hill Blue (Harris, 1775b; and most subsequent authors).

42. THE ADONIS BLUE (*Lysandra bellargus* (Rottemburg, 1775)): ? Petiver, 1717; Harris, 1775.
?The Lead Argus (Petiver, 1717) [cf. *Plebejus argus*]
The Clifden Blue (Harris, 1775b; Jermyn, 1824; Wood, 1852; Morris (1), 1853; Furneaux, 1894; W. E. Kirby, 1906; Newman & Leeds (2), 1913)
The Dartford Blue (Morris (2), 1853)
The Adonis Blue (Coleman, 1860; South, 1906; and most subsequent authors)
The Clifton Blue (Newman & Leeds (3), 1913)

43. [21] THE MAZARINE BLUE (*Cyaniris semiargus* (Rottemburg, 1775)): ?Moffet, 1634; ?Ray, 1710.
The Dark Blue (Lewin, 1795)
The Mazarine Blue (Haworth, 1803; and subsequent authors) [cf. *M. arion*]

44. THE HOLLY BLUE (*Celastrina argiolus* (Linnaeus, 1758)): Ray, 1710.
The Blue Speckt Butterfly: male (Petiver, 1717; both sexes: Newman & Leeds (3), 1913)
The Blue Speckt Butterfly with black tips: female (Petiver, 1717)
The Azure Blue (Harris, 1775b; Haworth, 1803; Donovan, 1810; Samouelle, 1819; Jermyn, 1824; Rennie, 1832, Brown, 1832; Humphreys & Westwood, 1841; Wood, 1852; Morris (2), 1853; Stephens, 1856; Coleman, 1860; Newman, 1871; W. F. Kirby, 1896; W. E. Kirby, 1906; Newman & Leeds (2), 1913; Frohawk (2), 1924)
The Wood Blue (Lewin, 1795)
The Holly Blue (Morris (1), 1853; Kane, 1885; Furneaux, 1894; South, 1906; Newman & Leeds (1), 1913; Frohawk (1), 1924; and all subsequent authors)

45. [22] THE LARGE BLUE (*Maculinea arion* (Linnaeus, 1758)): [H. Seymer, *c.*1775]; Lewin, 1795.
The Large Blue (Lewin, 1795; Haworth, 1803; and most subsequent authors)
The Mazarine Blue (Donovan, 1797; Brown, 1832) [cf. *C. semiargus*]
The Arion (Rennie, 1832)

46. THE DUKE OF BURGUNDY FRITILLARY (*Hamearis lucina* (Linnaeus, 1758)): [?Vernon, 1696]; Petiver, 1699.
Mr Vernon's Small Fritillary (Petiver, 1699, 1717; Ray, 1710)
The Duke of Burgundy Fritillaria or 'The Burgundy' (Harris, 1766)
The Duke (Rennie, 1832)
The Duke of Burgundy Fritillary (Morris, 1853; South, 1906; and most subsequent authors)
The Duke of Burgundy (Heslop, 1959)

NYMPHALIDAE:

47. THE WHITE ADMIRAL (*Ladoga camilla* (Linnaeus, 1764)): Petiver, 1703.
The White Leghorn Admiral (Petiver, 1703)
The Leghorn white Admiral (Petiver, 1706)
The white or White Admiral (Petiver, 1717; Berkenhout, 1769; Haworth, 1803; and most subsequent authors)
The White Admirable (Wilkes 1741–42; Harris, 1766; Brown, 1843; Newman & Leeds (2), 1913)
The Honeysuckle (Rennie, 1832)

48. THE PURPLE EMPEROR (*Apatura iris* (Linnaeus, 1758)): Petiver, 1704.
Mr Dale's Purple Eye (Petiver, 1704)
The Purple Emperour or Purple Emperor (Wilkes, 1741–42; Harris, 1766; Haworth, 1803; Jermyn, 1824; Morris (1), 1853; and all subsequent authors)
The Purple Highflyer (Wilkes (1), 1747–49; Berkenhout (2), 1769; Brown (2), 1832)
The Emperor of the Woods (Wilkes (2), 1747–49; Berkenhout (1), 1769; Brown (1), 1832)
The Purple Shades (Lewin, 1795)
The Emperor (Rennie, 1832)
The Emperor of Morocco (Morris (2), 1853)

49. [23] ALBIN'S HAMPSTEAD EYE (*Junonia villida* (Fabricius, 1787)): Albin, 1717.
Albin's Hampstead Eye (Petiver, 1717; and most subsequent authors)
Papilio hampstediensis (Jermyn, 1824)
The Hampstead (Rennie, 1832)

50. THE RED ADMIRAL (*Vanessa atalanta* (Linnaeus, 1758)): Moffet, 1634 (Fig. 151); Merrett, 1666.

Fig. 151. Red Admiral

The Admiral (Petiver, 1699; Ray, 1710; Albin, 1720; Dutfield (1), 1748; Berkenhout, 1769; Haworth, 1803)
The Admirable (Wilkes, 1747–49; Harris, 1766; Samouelle, 1819)
The Scarlet Admiral (Harris, 1775b; Lewin, 1795)
The Alderman (Dutfield (2), 1748; Rennie 1832; Morris (2), 1853; Newman & Leeds (2), 1913)
The Red Admiral (Donovan, 1799; and most subsequent authors)

51. [24] THE PAINTED LADY (*Cynthia cardui* (Linnaeus, 1758)): Moffet, 1634 (Fig. 152).

Fig. 152. Painted Lady

Papilio Bella Donna (Petiver, 1699; Ray, 1710; Linnaeus, 1746)
The Painted Lady, (Petiver, 1699, 1704, 1717; Buddle, *c.*1700; Ray, 1710; Wilkes, 1741–42; Harris, 1766; and most subsequent authors)
The Thistle Butterfly (Lewin, 1795)

52. [25] THE AMERICAN PAINTED LADY (*Cynthia virginiensis* (Drury, 1773)): [Capt. T. Blomer, 1828]; Dale, 1830.
Virginiana (Petiver, 1704) (from American specimen)
The Scarce Painted Lady (Morris (1), 1853; W. F. Kirby, 1896; Newman & Leeds (2), 1913; Emmet & Heath (2), 1989)
Hunter's Cynthia (Morris (2), 1853)
The Beautiful Painted Lady (Newman, 1860; Emmet & Heath (4), 1989))
Hunter's Butterfly (Holland, 1898; Emmet & Heath (3), 1989)
The American Painted Lady (Newman & Leeds (1), 1913; Frohawk, 1924; Higgins & Riley, 1970; Emmet & Heath (1), 1989; Thomas (1), 1991)
The Scarce Lady (Heslop, 1959)
Hunter's Painted Lady (Thomas (2), 1991)

53. THE SMALL TORTOISESHELL (*Aglais urticae* (Linnaeus, 1758)): Moffet, 1634 (Fig. 153); Merrett, 1666.

Fig. 153. Small Tortoiseshell

The Lesser or common Tortoise-shell Butterfly (Petiver, 1699; Ray, 1710; Albin, 1720)
The Small Tortoiseshell (Wilkes, 1741–42; Jermyn, 1824; Morris, 1853; and most subsequent authors) [cf. *Lycaena phlaeas*]
The Tortoise-shell or Tortoiseshell (Harris, 1766; Rennie, 1832; Stephens, 1856)
The Nettle Tortoiseshell (Lewin, 1795; Brown, 1832)

54. [26] THE LARGE TORTOISESHELL (*Nymphalis polychloros* (Linnaeus, 1758)): Moffet, 1634.
The Greater Tortoiseshell (Petiver, 1699)
The Elm Tortoiseshell (Lewin, 1795)
The Elm (Rennie, 1832)
The Large Tortoiseshell or Tortoise-shell (Wilkes, 1741–42; Jermyn, 1824; Morris, 1853; and all subsequent authors)

55. [27] THE CAMBERWELL BEAUTY (*Nymphalis antiopa* (Linnaeus, 1758)): Moffet, 1634.
The Willow Butterfly (Wilkes, 1747–49; Berkenhout, 1769; Lewin, 1795; Rennie, 1832)
The Grand Surprize (Harris (1), 1766; Newman & Leeds (4), 1913)
The Camberwell Beauty (Harris (2), 1766; Morris, 1853; Newman (2), 1871; Brown, 1832; Wood, 1852; Coleman, 1860; and all subsequent British authors)
The White Border (Haworth, 1803)
The White-bordered or White Bordered (Jermyn, 1824; Newman (1), 1871; Newman & Leeds (2), 1913; Frohawk (2), 1924)
The Willow Beauty (Newman & Leeds (3), 1913; Howarth, (2), 1973)
The White Petticoat (Newman & Leeds (4), 1913; Howarth (3), 1973)
The Mourning Cloak (Frohawk (2), 1924) [N. American name]

56. THE PEACOCK (*Inachis io* (Linnaeus, 1758)): Moffet, 1634 (Fig. 154).
Omnium Regina [The Queen of All] (Moffet, 1634)
The Peacock's Eye (Petiver, 1699; Buddle, *c.*1700; Ray, 1710; Albin, 1720; Linnaeus, 1746 [as *Oculis pavonis*])
The Peacock Butterfly (Wilkes, 1741–42; Harris, 1766; and most subsequent authors)

Fig. 154. *(left)* Peacock

Fig. 155. *(right)* Comma

57. THE COMMA (*Polygonia c-album* (Linnaeus, 1758)): ?Moffet, 1634 (Fig. 155); Ray, 1710.
The silver, pale, jagged-wing and small Commas (Petiver, 1717)
The Comma Butterfly (Wilkes, 1741–42; Harris, 1766; and all subsequent authors)

58. [28] THE EUROPEAN MAP BUTTERFLY (*Araschnia levana* (Linnaeus, 1758)).
The Map Butterfly (Higgins & Riley, 1970)
The European Map (Emmet & Heath, 1989; Thomas, 1991)

59. THE SMALL PEARL-BORDERED FRITILLARY (*Boloria selene* ([Denis & Schiffermuller], 1775)): ?Merrett, 1666; Ray, 1710.
The April Fritillary (Ray, 1710; Petiver, 1717; Morris (2), 1853) [cf. *B. euphrosyne*]
The Small Pearl Border Fritillary (Wilkes, 1741–42)
The Small Pearl Border Fritillaria (Harris, 1766)
The May Fritillary (Lewin, 1795) [cf. *Melitaea athalia*]
The Pearl-bordered Fritillary (Haworth, 1803)
The Pearly Border Likeness (Samouelle, 1819 [cf. *M. athalia*])
The Small Pearl-bordered Fritillary (Jermyn, 1824; Morris, 1853; and all subsequent authors)
The Silver Spot Fritillary (Rennie, 1832)

60. THE PEARL-BORDERED FRITILLARY (*Boloria euphrosyne* (Linnaeus, 1758)): Petiver, 1699.
The April Fritillary (Petiver, 1699; Berkenhout, 1769; Lewin, 1795; Newman & Leeds (2), 1913) [cf. *B. selene*]
The April Fritillary with few Spots (Petiver, 1717)
The Pearl Border Frittillaria (Harris, 1766)
The Pearl-bordered Fritillary (Jermyn, 1824; Morris, 1853; and most subsequent authors)
The Prince (Rennie, 1832)
The Large Pearl-bordered Fritillary (Heslop, 1959)

61. [29] WEAVER'S FRITILLARY (*Boloria dia* (Linnaeus, 1767): [R. Weaver, 1832]; Coleman, 1860.
The Goddess (Rennie, 1832)
Weaver's Fritillary (Coleman, 1860; Howarth (1), 1973; Heslop, 1959)
The Violet Fritillary (Higgins & Riley, 1970; Howarth (2), 1973)

62. [30] THE QUEEN OF SPAIN FRITILLARY (*Argynnis lathonia* (Linnaeus, 1758): [W. Vernon or R. Antrobus, early 1700s]; Ray, 1710.
The Riga Fritillary (Petiver, 1702a; Ray, 1710)
The Lesser Silver-spotted (Ray, 1710; Petiver, 1717; Donovan, (1), 1794; Newman & Leeds (2), 1913)
The Queen of Spain Fritillary (Harris, 1775b; Donovan, (2), 1794; Haworth, 1803; and all subsequent authors)
The Scalloped-winged Fritillary (Lewin, 1795)
The Princess (Rennie, 1832)

63. [31] THE NIOBE FRITILLARY (*Argynnis niobe* (Linnaeus, 1758): [?Abbot, *c.*1800]; Waller, 1879.
Abbot's Fritillary (Rennie, 1832)
The Niobe Fritillary (Newman & Leeds, 1913; Higgins & Riley, 1970)

64. THE HIGH BROWN FRITILLARY (*Argynnis adippe* ([Denis & Schiffermüller], 1775)): Petiver, ?1699, 1717.
?The greater silver-spotted Fritillary (Petiver, 1699) [cf. *A. aglaja*]

The High Brown or High-brown Fritillary (Wilkes 1741–42; Haworth, 1803; Jermyn, 1824; Rennie, 1832; and all subsequent authors)
The High Brown Fritillaria (Harris, 1766; 1775b)
The Violet Silver-spotted Fritillary (Lewin, 1795)

65. THE DARK GREEN FRITILLARY (*Argynnis aglaja* (Linnaeus, 1758)): Moffet, 1634 (Fig. 156).

Fig. 156.
Dark Green Fritillary

The Great Sylver Spotted Fritillary (Buddle, *c.*1700) [cf. *A. adippe*]
Darkned Green Fritillary (Wilkes, 1741–42)
Dark Green Fritillaria (Harris, 1766)
Silver-spotted Fritillary (Lewin, 1795)
Dark Green Fritillary (Haworth, 1803; Jermyn, 1824; Rennie, 1832; and most subsequent authors)
Queen of England Fritillary (ab. *charlotta*) (Haworth, 1803; Jermyn, 1824; Newman & Leeds, 1913)
The Charlotte Butterfly (ab. *charlotta*) (Brown, 1832)

66. THE SILVER-WASHED FRITILLARY (*Argynnis paphia* (Linnaeus, 1758)): Petiver, 1699.
The Greater Silver-streaked Fritillary (Petiver, 1699; Newman & Leeds (2), 1913)
The Silver-Stroaked Fritillary (Ray, 1710)
The Greater Silverstreakt Orange Fritillary (male) (Petiver, 1717)
The Greater Silverstreakt Golden Fritillary (female) (Petiver, 1717)
The Great Fritillary (Wilkes, 1741–42; Berkenhout, 1769)
The Silver-washed Fretillaria (Harris, 1766)
The Silver Wash Fritillary (Harris, 1775)
The Silver Streak Fritillary (Lewin, 1795; Rennie, 1832)
The Silver Stripe Fritillary (Donovan, 1798; Brown, 1832)
The Silver-washed Fritillary (Haworth, 1803; Samouelle, 1819; Humphreys & Westwood, 1841; and all subsequent authors)
The Greenish Silver-washed Fritillary (f. *valesina*) (Howarth, 1973)

67. [32] THE MEDITERRANEAN FRITILLARY or THE CARDINAL (*Argynnis pandora* ([Denis & Schiffermüller], 1775)): [A. W. Bennett, 1911]; Ford, 1945.
The Cardinal (Higgins & Riley, 1970; Emmet & Heath (2), 1989)
The Mediterranean Fritillary (Emmet & Heath (1), 1989)

68. THE MARSH FRITILLARY (*Eurodryas aurinia* (Rottemburg, 1775)): Ray, 1710.
The Dandridge's Midling Black Fritillary (Petiver, 1717)
The Small Black Fritillary (Petiver, 1717)
The Small Fritillary Butterfly (Wilkes, 1747–49)
The Dishclout or Greasey Fritillaria (Harris, 1766)
The Marsh Fritillary (Lewin, 1795; Morris (2), 1853; Coleman (2), 1860; South, 1906; and most subsequent authors)
The Greasy Fritillary (Haworth, 1803; Samouelle, 1819; Jermyn, 1824; Humphreys & Westwood, 1841; Wood, 1852; Morris (1), 1853; Stephens, 1856; Coleman (1), 1860; Newman, 1871; Furneaux, 1894; W. F. Kirby, 1896; W. E. Kirby, 1901; Newman & Leeds (2), 1913)
The Scabious (Rennie, 1832)

69. [33] THE GLANVILLE FRITILLARY (*Melitaea cinxia* (Linnaeus, 1758)): Petiver, 1703.
The Lincolnshire Fritillary (Petiver, 1703b; Ray, 1710)
The Dullidge Fritillary (Petiver, 1717; Newman & Leeds (3), 1913)
The Plantain Fritillary (Wilkes, 1747–49; Berkenhout, 1769; Lewin, 1795; Donovan, 1798; Brown, 1832; Newman & Leeds (2), 1913)
The Glanvil Fretillary (Dutfield, 1748)
The Glanvil Fritillaria (Harris, 1766)
The Glanville Fritillary (Haworth, 1803; Samouelle, 1819; Jermyn, 1824; Rennie, 1832; Humphreys & Westwood, 1841; and most subsequent authors)

70. THE HEATH FRITILLARY (*Mellicta athalia* (Rottemburg, 1775)): Petiver, 1699.
The May Fritillary (Petiver, 1699; Ray, 1710) [cf. *Boloria selene*]
The Straw May Fritillary (Petiver, 1717; Jermyn, 1824)
The White May Fritillary (?ab. *latonigena*) (Petiver, 1717; Humphreys & Westwood, 1841; Morris (2), 1853)
The Heath Fritillary (Wilkes, 1747–49; Berkenhout, 1869; Lewin, 1795; Samouelle, 1819; Wood (1), 1852; Morris (3), 1853; Furneaux, 1894; Coleman, 1901; South, 1906; Newman & Leeds (1), 1913; and all subsequent authors)
The Pearl Border or Pearl-bordered Likeness Fritillary (Harris, 1766; Haworth, 1803; Jermyn 1824; Humphreys & Westwood, 1841; Wood (2), 1852; Morris, 1853; Coleman, 1860; W. F. Kirby, 1896; W. E. Kirby, 1901; Newman & Leeds (2), 1913) [cf. *B. selene*]
The Yellow Crescent (Rennie (1), 1832)
The Morning Crescent (Rennie (2), 1832)
The Black Crescent (Rennie (3), 1832)

71. THE SPECKLED WOOD (*Pararge aegeria* (Linnaeus, 1758)): ? Moffet, 1634 (Fig. 157); ?Merrett, 1666; Petiver, 1704.

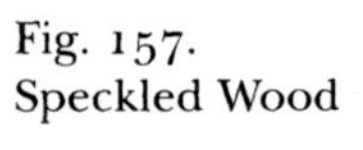
Fig. 157.
Speckled Wood

The Enfield Eye (Petiver, 1704; Newman & Leeds (3), 1913)
The Wood Argus (Wilkes, 1747–49; Lewin, 1795; Morris (1), 1853; Coleman, 1860; Furneaux (2), 1894; W. F. Kirby (2), 1896; W. E. Kirby (2), 1906; Newman & Leeds (2), 1913)
The Speckled Wood (Harris, 1766; Morris (3), 1853; Furneaux (1), 1894; W. F. Kirby (2), 1896; Coleman, 1901; W. E. Kirby (1), 1906; South, 1906; Newman & Leeds (1), 1913; and all subsequent authors)
The Wood Lady (Morris (2), 1853) [cf. *Leptidea sinapis*]

72. THE WALL (*Lasiommata megera* (Linnaeus, 1767)): Moffet, 1634 (Fig. 158).

Fig. 158. Wall

The golden marbled Butterfly, with black Eyes (Petiver, 1699)
The London Eye, with a black List: male (Petiver, 1717)
The London Eye: female (Petiver, 1717)
The Great Argus Butterfly (Wilkes, 1747–49; Berkenhout, 1769)
The Wall (Harris, 1766; Haworth, 1803; Humphreys & Westwood, 1841; Jermyn, 1824; Morris (3), 1853; South, 1906; and most subsequent authors)
The Orange Argus (Lewin, 1795) (cf. *Lycaena dispar*)
The Gatekeeper (Samouelle, 1819; Morris (1), 1853) [cf. *Pyronia tithonus*]
The Speckled Wall (Morris (2), 1853)
The Wall Brown (W. F. Kirby, 1896; W. E. Kirby, 1906; Newman & Leeds (2), 1913; Heslop, 1959)

73. THE SMALL MOUNTAIN RINGLET (*Erebia epiphron* (Knoch, 1783)): [T. Stothard, 1808]; Haworth, 1812.
The Small Mountain Ringlet (Haworth (2), 1812; Newman & Leeds (1), 1913; Frohawk, 1924; South, 1941; Emmet & Heath, 1989)
The Mountain Ringlet (Haworth (1), 1812; Rennie, 1832; Wood, 1852; Coleman, 1860; W. F. Kirby, 1896; W. E. Kirby, 1906; Newman & Leeds (3), 1913; Ford, 1945; Heslop, 1959; Thomas, 1991)
The Small Ringlet (Humphreys & Westwood, 1841; Morris, 1853; Stephens, 1856; Newman, 1871; Newman & Leeds (2), 1913)

74. THE SCOTCH ARGUS (*Erebia aethiops* (Esper, 1777)): [Dr J. Walker, *c.*1760–1769]; Sowerby, 1804–05.
Papilio amaryllis, without English name (Walker, *c.*1769)
The Scotch Argus (Donovan, 1807; and most subsequent authors)
The Scotch Ringlet (Rennie, 1832) [cf. *Aricia artaxerxes*]
The Northern Brown (Newman, 1871; Furneaux, 1894; Heslop, 1959)

75. [34] THE ARRAN BROWN (*Erebia ligea* (Linnaeus, 1758)): [Sir Patrick Walker, 1803]; Sowerby, 1804–05.

The Arran Brown (Sowerby, 1804–05; Samouelle, 1819; Humphreys & Westwood, 1841; Wood, 1852; Morris (2), 1853, Coleman, 1860; and subsequent authors)
The Arran Argus (Morris (1), 1853)
The Scarce Scotch Argus (Stephens, 1856)

76. THE MARBLED WHITE (*Melanargia galathea* (Linnaeus, 1758)): Merrett, 1666.
Our half-Mourner (Petiver, 1695; Newman & Leeds (2), 1913)
The Half-mourner (Ray, 1710; W. E. Kirby (2), 1901)
The common half-Mourner (Petiver, 1717)
The Marmoris (Wilkes, 1741–42; Newman & Leeds (3), 1913)
The Marble Butterfly (Wilkes, 1747–49; Berkenhout, 1769; Donovan, 1799)
The Marmoress (Harris (1), 1766; Humphreys & Westwood, 1841; Newman & Leeds (4), 1913)
The Marbled White (Harris (2), 1766; Jermyn, 1824; Morris, 1853; and most subsequent authors)
The Marbled Argus (Lewin, 1795)
The Marbled Butterfly (Samouelle, 1819; Brown, 1832)
The Marbled White Half-mourner (Humphreys & Westwood, 1841)

77. THE GRAYLING (*Hipparchia semele* (Linnaeus, 1758)): Petiver, 1699.
The black-eyed marble Butterfly (Petiver, 1699)
The Brown Tunbridge Grayling: male (Petiver, 1703, 1717)
The Tunbridge Grayling: female (Petiver, 1703, 1717); both sexes (Newman & Leeds (3), 1913)
The Rock Underwing (Wilkes, 1741–42; Samouelle, 1819; Newman & Leeds (2), 1913)
The Grailing (Harris, 1766)
The Grayline (Harris, 1775b)
The Black-eyed Marbled Butterfly (Berkenhout, 1769; Donovan, 1799)
The Great Argus (Lewin, 1795)
The Grayling (Haworth, 1803; Samouelle, 1819; Jermyn, 1824; Morris (2), 1853; and most subsequent authors)
The Rock-eyed Underwing (Morris (1), 1853)

78. THE GATEKEEPER (*Pyronia tithonus* (Linnaeus, 1771)): Merrett, 1666.
The lesser double-eyed Butterfly (Petiver, 1699)
The Hedge Eye with double specks (Petiver, 1717)
The Orange Field Butterfly: both sexes (Wilkes, 1741–42); female (Harris, 1766)
The Gate Keeper or Gatekeeper (Harris, 1766; Rennie, 1832; Humphreys & Westwood, 1841; W. E. Kirby (2), (1906); South, 1906; Newman & Leeds (1), 1913; Heslop, 1959; Emmet & Heath (1), 1989; Thomas, 1991)
The Large Gatekeeper (Harris, 1775a)
The Clouded Argus (Lewin, 1795)
The Large Heath (Haworth, 1803; Jermyn, 1824; Wood, 1852; Morris (2), 1853; Coleman, 1860; Newman, 1871; W. F. Kirby, 1896); W. E. Kirby (1), 1906) [cf. *Coenonympha tullia*]
The Small Meadow Brown (Samouelle, 1819; Morris, 1853; Newman & Leeds (2), 1913)

The Hedge Brown (Newman & Leeds (3), 1913; Frohawk, 1924; Ford, 1945; Emmet & Heath (2), 1989)
The Hedge Eye (Newman & Leeds (4), 1913)

79. THE MEADOW BROWN (*Maniola jurtina* (Linnaeus, 1758)): ?Merrett, 1666; Petiver, 1699.
The brown Meadow Ey'd Butterfly: male (Petiver, 1699)
The golden Meadow Ey'd Butterfly: female (Petiver, 1699)
The Brown Meadow-Eye: male (Petiver, 1717)
The Golden Meadow-Eye: female (Petiver, 1717)
The Meadow Brown or Meadow-Brown (Albin, 1720; Wilkes, 1747–49; Harris, 1766; and most subsequent authors)
The Meadow Brown Argus (Lewin, 1795)
The Large Meadow Brown (Morris, 1853)

80. THE SMALL HEATH (*Coenonympha pamphilus* (Linnaeus, 1758)): Merrett, 1666.
The Small Heath Butterfly (Petiver, 1699; Haworth, 1803; and most subsequent authors)
The golden Heath Eye (Petiver, 1717; Newman & Leeds (2), 1913)
The selvedg'd Heath Eye (Petiver, 1717; Newman & Leeds (3), 1913)
The Little or Small Gatekeeper (Harris, 1766)
The Small Argus (Lewin, 1795)
The Golden Eye (Rennie, 1832)
The Least Meadow Brown (Morris, 1853)

81. THE LARGE HEATH (*Coenonympha tullia* (Müller, 1764)): Lewin, 1795.
The Manchester Argus (Lewin (1), 1795)
The Manchester Ringlet (Lewin (2), 1795)
The Scarce Meadow Brown (Donovan, 1797)
The Small Ringlet (subsp. *davus*) (Haworth, 1803; Jermyn, 1824; Wood, 1852); (all subspp.): (Coleman (2), 1860)
The Marsh Ringlet (subsp. *polydama*) (Haworth, 1803; Rennie, 1832; Wood, 1852); (all subspp.): (Coleman (1), 1860; Newman, 1871; Furneaux, 1894; Newman & Leeds (2), 1913)
Large Heath (subsp. *tiphon*) (Haworth (2), 1803)
The Scarce Heath (subsp. *tiphon*) (Haworth (1), 1803; Jermyn, 1824)
The July Ringlet (subsp. *davus*) (Rennie, 1832)
The Heath Butterfly (Morris, 1853)
The Scarce Marsh Ringlet (subsp. *polydama*) (Newman & Leeds, 1913)
Rothleib's Marsh Ringlet (subsp. *davus*) (Newman & Leeds, 1913)

82. THE RINGLET (*Aphantopus hyperantus* (Linnaeus, 1758)): Merrett, 1666.
The brown eyed Butterfly with yellow Circles (Petiver, 1699)
The Brown and Eyes (Petiver, 1717)
The Brown seven Eyes (Petiver, 1717)
The Ringlet (Harris, 1766; Jermyn, 1824; Wood, 1852; Coleman, 1860; and most subsequent authors)
The Brown-eyed Butterfly (Berkenhout, 1769)
The Brown Argus (Lewin, 1795) [cf. *Aricia agestis*]
The Wood Ringlet (Morris, 1853)
The Common Ringlet (Heslop, 1959)

83. [35] THE MONARCH (*Danaus plexippus* (Linnaeus, 1758)): [J. Stafford, 1876]; Llewelyn, 1876.
The Archippus (Brown, 1832; Coleman, 1897)
The Milkweed or Milk-weed (South, 1906; Newman & Leeds (1), 1913; Frohawk (2), 1924; Frohawk (1), 1934; Heslop, 1959; Howarth (2), 1973; Emmet & Heath (2), 1989; Thomas (2), 1991)
The Monarch (W. E. Kirby, 1901; Newman & Leeds (2), 1913; Frohawk (3), 1924, 1934; Howarth (1), 1973; Emmet & Heath (1), 1989; Thomas (1), 1991)
The Black-veined Brown (Newman & Leeds (3), 1913; Frohawk (1), 1924; Frohawk (2), 1934; Howarth (3), 1973)

APPENDIX 2

ENTOMOLOGICAL SOCIETIES, PUBLICATIONS AND SIGNIFICANT EVENTS

Most societies produced journals or similar publications. Many are referred to in the text, and these and others are detailed here. The publication dates of the first issue of each journal are given as well as the dates during which each journal continued as a serial. Journals of local natural history societies and field clubs which have devoted substantial space to entomological articles and notes are also included. A number of journals have changed their titles over the years and all such changes are given.

The following chronological account, spanning over three hundred and seventy-five years, includes the principal societies and publications as well as certain other events and organizations of special interest to entomologists, but makes no claim to be comprehensive.

1620: The Society of Apothecaries founded. It organized field excursions, primarily for the purpose of collecting plants of medicinal value.

1660: On 5th December, inaugural meeting of a 'Society for promoting Experimental learning' held. Christopher Merrett and Francis Willughby were among the initial signatories who also included the diarist John Evelyn, Sir Christopher Wren and Samuel Pepys among many of the great and the good of the time. In 1662 it became the Royal Society, by Royal Charter of King Charles II. The Royal Society is the oldest scientific society in Great Britain and one of the oldest in Europe. Other early fellows included John Ray, Charles duBois and James Petiver. Many other eminent entomologists have been elected fellows over the centuries. Scientific papers were (and still are) read at meetings and subsequently published in the *Philosophical Transactions* which began in 1665, or in the *Proceedings* which date back to 1800.

c.1698: The Temple Coffee House Botanic Club founded. It had a number of entomologists and butterfly collectors among its membership, including Adam Buddle, Samuel Dale, Charles duBois, Martin Lister, James Petiver, Leonard Plukenet, Sir Hans Sloane and William Vernon. The Club met on Friday evenings and is said by D. E. Allen (1976) to have been the earliest natural history society in Britain and probably the world.

1710: The Spalding Gentlemen's Society, modelled on the Royal Society, founded by the antiquary and barrister of the Inner Temple, Maurice Johnson (1688–1755). Its aims were to promote the arts and science and to exclude nothing from conversation but politics and religion. For a society in a small provincial town in the Fens of Lincolnshire, the quality of its membership was astonishing. Among them were Sir Isaac Newton and the Swedish astronomer and inventor of the centigrade thermometer, Anders Celsius; the naturalists Mark Catesby, Sir Hans Sloane, Emanuel Mendes da Costa, Thomas Pennant, Sir Joseph Banks, and the apothecary John Hill; and representing the Arts were John Gay, the playwright of *Beggars' Opera* fame, Alexander Pope, the poet, and George Vertue, engraver and antiquary. The Society, one of the oldest of its kind, survives to this day with a fine historic library and a membership of about 350. It is particularly noteworthy to Aurelians for the first record of the Large Copper in 1749 (see Chapter 4: 15).

*c.*1738: The Society of Aurelians (The Aurelian Society) founded. The precise date of foundation is not known but there is good evidence that the founder was Joseph Dandridge. According to Allen the membership was thought to include: Benjamin Wilkes, Stephen Austin, Henry Baker, Ephraim Bell, Elias Brownsword, Walter Blackett, Peter Collinson, Philip Constable Jnr., Thomas Grace, Samuel Hartley, Thomas Knowlton, James Lemon, Samuel Lee, Daniel Marshal, Edmund Overall, William Wells, and Moses Harris Snr. The Society met on Saturday evenings at the Swan Tavern, Exchange Street, Cornhill, London.

1748: On 25th March, The Great Cornhill Fire destroyed the Aurelian Society's meeting room, collections and regalia, following which it seems to have ceased to function.

1762: The second Aurelian Society founded with Moses Harris Jnr. as Secretary. Membership included Dru Drury, Emanuel Mendes Da Costa, Daniel C. Solander, Henry Smeathman and James Lee. The Society met at the King's Arms, Cornhill, London, but foundered early in 1767 – the year after Moses Harris published *The Aurelian*, dedicated to the Members of the Society (Fig. 159) – owing to dissension among its members exacerbated by the irrascible behaviour of Da Costa.

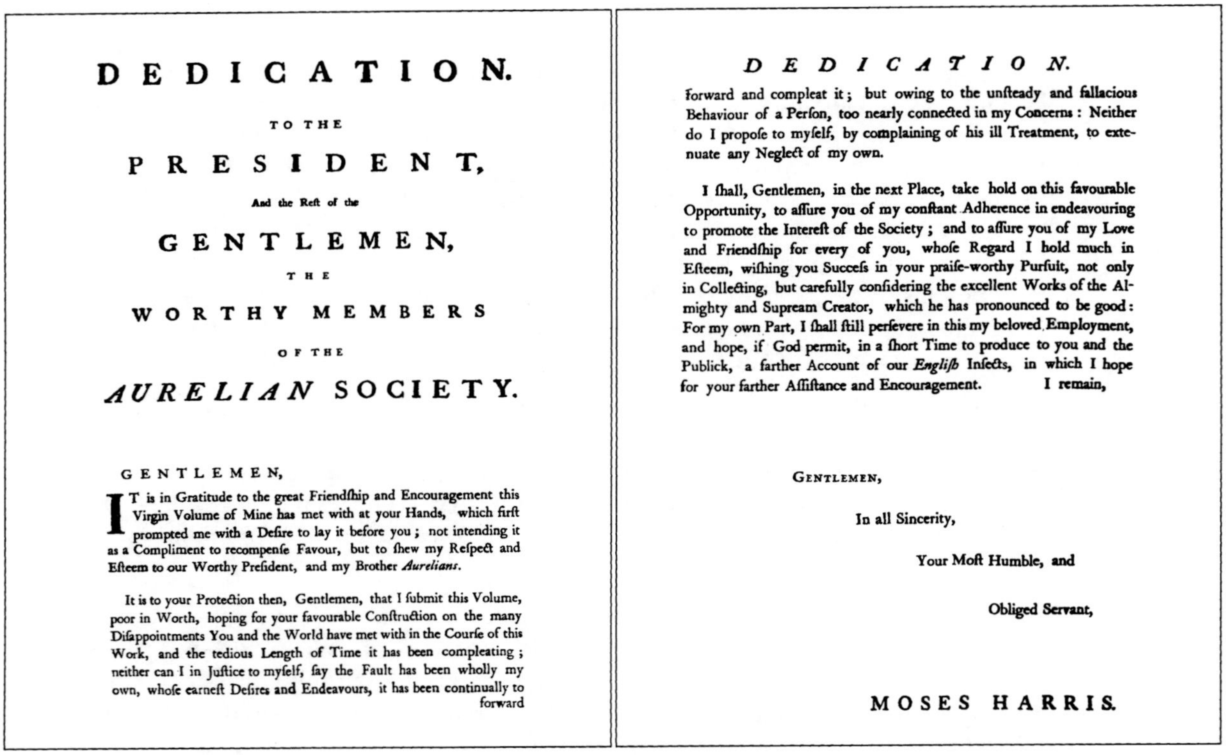

DEDICATION.

TO THE

PRESIDENT,

And the Reſt of the

GENTLEMEN,

THE

WORTHY MEMBERS

OF THE

AURELIAN SOCIETY.

GENTLEMEN,

IT is in Gratitude to the great Friendſhip and Encouragement this Virgin Volume of Mine has met with at your Hands, which firſt prompted me with a Deſire to lay it before you; not intending it as a Compliment to recompenſe Favour, but to ſhew my Reſpect and Eſteem to our Worthy Preſident, and my Brother *Aurelians*.

It is to your Protection then, Gentlemen, that I ſubmit this Volume, poor in Worth, hoping for your favourable Conſtruction on the many Diſappointments You and the World have met with in the Courſe of this Work, and the tedious Length of Time it has been compleating; neither can I in Juſtice to myſelf, ſay the Fault has been wholly my own, whoſe earneſt Deſires and Endeavours, it has been continually to forward

DEDICATION.

forward and compleat it; but owing to the unſteady and fallacious Behaviour of a Perſon, too nearly connected in my Concerns: Neither do I propoſe to myſelf, by complaining of his ill Treatment, to extenuate any Neglect of my own.

I ſhall, Gentlemen, in the next Place, take hold on this favourable Opportunity, to aſſure you of my conſtant Adherence in endeavouring to promote the Intereſt of the Society; and to aſſure you of my Love and Friendſhip for every of you, whoſe Regard I hold much in Eſteem, wiſhing you Succeſs in your praiſe-worthy Purſuit, not only in Collecting, but carefully conſidering the excellent Works of the Almighty and Supream Creator, which he has pronounced to be good: For my own Part, I ſhall ſtill perſevere in this my beloved Employment, and hope, if God permit, in a ſhort Time to produce to you and the Publick, a farther Account of our *Engliſh* Inſects, in which I hope for your farther Aſſiſtance and Encouragement. I remain,

GENTLEMEN,

In all Sincerity,

Your Moſt Humble, and

Obliged Servant,

MOSES HARRIS.

Fig. 159. Moses Harris's dedication to members of the Aurelian Society.

1780: The Society of Entomologists of London founded but survived only until August 1782.

1788: The Linnean Society founded by James Edward Smith who became its first President. In 1802 the Society was incorporated by Royal Charter as The Linnean Society of London. Its objectives were 'the cultivation of the science of natural history in all its branches and more especially of the Natural History of Great Britain and Ireland'. Still active, it is now the world's oldest surviving natural history society.

1801: The third Aurelian Society founded by Adrian H. Haworth. It was dissolved in April 1806 as members were unwilling to donate their much treasured specimens to the Society's collections.

1807: The Entomological Society founded, becoming the first Entomological Society of London later the same year. It published three parts of the *Transactions of the Entomological Society of London* (1807–12), which constitute the first known publication of any Entomological Society. This society survived until about 1822, after which it was re-formed as the Entomological Society of Great Britain (see below).

1810: The Norwich Entomological Society founded. Its establishment owed much to John Curtis, at whose home a number of meetings were held. It appears to have been a very informal society, with meetings and field trips usually arranged on an ad hoc basis, and is thought to be the oldest provincial entomological society. It is not known how long it lasted, but correspondence in the County Records Office, Norwich, suggests that, in 1823, following its demise, the Revd John Burrell, Henry Denny, Thomas Skrimshire, George Sothern, Joseph Sparshall and several others formed an 'Entomological Correspondence Club' which was still active in 1825. The correspondence originated in the desire expressed by a number of entomologists living in Norwich and the surrounding villages to get together. It was suggested that a quarterly meeting might be held at Swaffham, and an annual 'conversation' at Dereham. In the end the members settled for a monthly correspondence, in which entomological notes were sent to the registrar on the first of each month, the registrar replying on the 13th or 14th day following, the letters being preserved in a book for mutual reference. Records show that during the first two-and-three-quarter years, eight of the nine signatories, who had been expected to write 33 letters each, managed to write a total of only 98 letters, while John Burrell, the registrar, wrote over 180.

1822: The first Entomological Society of Great Britain evolved from the Entomological Society of London (founded, 1807; foundered 1822). It came to an end after some of its members had joined Fellows of the Linnean Society who the year before had formed a Zoological Club within the framework of the Linnean Society. In 1825 Sir Stamford Raffles, F.L.S., proposed the founding of a zoological society on parallel lines to the Linnean.

1826: The Zoological Society of London established and held its first general meeting.

1826: The Entomological Club formed. Edward Newman and George Samouelle were among the founding members. This Club still exists today – the small but exclusive membership of eight meeting periodically to dine at each other's invitation. From 1833–38 the Club published a quarterly journal, *The Entomological Magazine*, with Edward Newman as Editor. Since 1887 it has hosted the annual Verrall Supper. Recently, Miriam Rothschild was elected the Club's first Honorary Life President.

1828: Ashmolean Natural History Society, Oxfordshire, founded. The *Abstracts of Proceedings of the Ashmolean Society of Natural History* were published from 1832–58, continuing as *The Proceedings of the Ashmolean Society, Oxford* from 1866–68, and then as the *Journal of the Proceedings of the Ashmolean Society* from 1879–1881. In 1900, it published the *Report of the Ashmolean Natural History Society and Field Club of Oxfordshire*, which continued from 1901 to 1907 as the *Report of the Ashmolean Natural History Society of Oxfordshire* and from 1908 until the mid-1960s as *The Proceedings & Report of the Ashmolean Natural History Society of Oxfordshire.* The Society still exists under the same name and in 1990 launched an 'occasional' publication, *Fritillary*, of which the second issue is due in 2000.

1828: *Magazine of Natural History and Journal of Zoology* launched, and produced by J. C. Loudon until 1836 (Vols. 1–9). It continued from 1837 until 1840 (as *New Series*, Vols. 1–4) and then as the *Annals and Magazine of Natural History* under Edward Charlesworth.

1829: The Natural History Society of Northumberland, Durham and Newcastle-upon-Tyne founded, with the Bishop of Durham as its first President. The Society published its *Transactions* as two volumes in 1831 and 1838, but publication then ceased, although it continued to produce *Annual Reports*. In 1846, the Tyneside Naturalists' Field Club published its *Transactions* and these were continued until 1864, when the Field Club and the Natural History Society combined to produce the *Natural History Transactions of Northumberland, Durham and Newcastle-upon-Tyne*. These were continued from 1864 to 1903. In 1904 the Tyneside Naturalists' Field Club amalgamated with the Natural History Society and a new series of *Transactions of the Natural History Society of Northumberland, Durham and Newcastle-upon-Tyne* was published until 1973. In 1973 the Society became the Natural History Society of Northumbria and since then has published *Annual Reports*.

1831: Berwickshire Naturalists' Club founded, with the Revd A. Baird as President. During its first year the Club consisted of 22 members. It has published the *History of the Berwickshire Naturalists' Club* from the year of its foundation to the present day.

1833: The second Entomological Society of London was founded eleven years after the first Entomological Society of London had ceased to be active, although there is no evidence that that Society had ever formally ceased to exist – see 1807. The headquarters of the second Society was at 17 Old Bond Street, Piccadilly. At the inaugural meeting of 'gentlemen, friends of the science of Entomology', held at the British Museum on May 3rd, N. A. Vigors, M.P. was in the Chair, and J. G. Children, G. R. Gray, J. E. Gray, Dr T. Horsfield, Revd F. W. Hope, Dr G. Rudd, J. F. Stephens and W. Yarrell were present. Shortly afterwards, at the first General Meeting, J. G. Children was elected President and the Revd W. Kirby was elected Honorary Life President. This society was the direct forerunner of the present Royal Entomological Society of London – see also 1852, 1873 and 1885.

1833: *The Entomological Magazine* founded in London, with Edward Newman as Editor, but lasted only until 1838.

1834: The Swaffham Prior Natural History Society, Cambridgeshire, founded by the Revd George Bitton Jermyn, naturalist and antiquary, under the patronage of Dr J. A. Power. Included in its membership were the Revd W. Kirby, John Curtis, J. F. Stephens and J. O. Westwood. The Society ceased to exist in 1838, and its library and collection were sold at auction.

1835: First Lady Fellow of the Entomological Society of London, Mrs E. H. M. Hope, elected.

1835: The Shropshire Natural History and Antiquarian Society established, with the Earl of Bradford as its first President. In 1877, it combined with the newly formed Shropshire Archaeological Society to publish the *Transactions of the Shropshire Archaeological and Natural History Society*. In 1936 the natural history element was dropped and the Society changed its name to the Shropshire Archaeological Society, devoting its publication solely to archaeology.

1840: *The Entomologist* founded by Edward Newman, its first editor. After two years it broadened its compass to become, in 1843, *The Zoologist*, publishing articles on all aspects of zoology and natural history. However, in 1864 the number of entomological contributions caused Newman to re-establish *The Entomologist*, leaving *The Zoologist* to continue independently until its closure in 1916. Newman maintained the editorship till his death in 1876, since when *The Entomologist* continued under different editors without a break until 1973 – see 1960, 1978.

1844: The Ray Society, named after John Ray (1627–1705), the most renowned of early English naturalists, founded by George Johnston with the express purpose of 'promoting Natural History by the printing of original works in Zoology and Botany', chiefly relating to the British fauna and flora. The Society has largely maintained those aims to the present day. Among the 164 works published to date is William Buckler's *The Larvae of the British Butterflies and Moths* in nine volumes (1886–1901).

1846: The Cotteswold Naturalists' Field Club was founded expressly so that members might 'study and record the natural history, geology, and antiquities of Gloucestershire and the adjacent districts'. *The Proceedings of the Cotteswold Naturalists' Field Club* have been published from 1848 to the present day.

1849: The Somersetshire Archaeological & Natural History Society was founded at Taunton with Sir Walter Calverley Trevelyan as its first President. No fewer than 350 persons attended the inaugural meeting. From 1851 until 1968 the Society published *Proceedings*, and subsequently *Somerset Archaeology and Natural History* to the present.

1851: The Woolhope Naturalists' Field Club founded for the study of the natural and archaeological history of Herefordshire. The first president was R. M. Lingwood, and Honorary Members included Sir Roderick Murchison, F.R.S., Sir Charles Lyell, F.R.S., and Professor A. Sedgwick, F.R.S. The *Transactions of the Woolhope Naturalists' Field Club* have been published from 1852 until the present day.

1852: The Entomological Society of London moved to 12 Bedford Row, London, but between 1866 and 1875, because of lack of adequate accommodation, held its meetings in the rooms of the Linnean Society, at Burlington House, Piccadilly, while keeping its library at 12 Bedford Row.

1853: The Malvern Naturalists' Field Club established, with the Revd W. G. Symonds, F.G.S., as its first President. The *Transactions of the Malvern Naturalists' Field Club* were published from 1855–79.

1855: *Entomologist's Annual*, edited by H. T. Stainton, was published in London from 1855–74.

1855: The Holmesdale Natural History Club established at Reigate, Surrey. By 1866 the membership exceeded 160. The *Proceedings and Annual Report of the Holmesdale Natural History Club* were published from 1865–67, and again from 1871–1919. The Society is still in existence and publishes *Bulletins* and *Annual Reports*.

1856: *The Entomologist's Weekly Intelligencer* launched in London, with Henry Stainton as Editor. The first volume was dedicated to William Spence, F.R.S., and William Wilson Saunders, the then President of the Entomological Society of London. Following publication of its 260th issue in 1861, Stainton closed it down.

1856: *The Substitute; or Entomological Exchange Facilitator, and Entomologists's Fire-side Companion*, edited by J. W. Douglas, was published, by request, by Edward Newman on Saturday, 25th October, and continued in twenty weekly parts, price two pence, until Saturday, 7th March 1857. It was designed to take the place of The *Intelligencer* over that winter season when the latter journal did not appear, and contained lists of duplicates and desiderata.

1856: The Oxford University Entomological Society founded under the Presidency of the Revd H. Adair Pickard. The Society continues to this day.

1858: The Haggerstone Entomological Society founded in North London by C. Healy and 'two or three friends'. By 1859 the membership had reached 60. Members paid one shilling entrance fee, two pence for a copy of the rules, and a subscription of one penny per week. The inaugural meeting was held at the 'Carpenters Arms', Queens Road, Dalston, but after a few years the Society removed to the 'Brownlow Arms', Haggerstone. A number of eminent lepidopterists were attracted to the meetings, including Henry Doubleday, Edward Newman, Samuel Stevens, Dr Henry Guard Knaggs, and Henry Stainton. (See also 1888 and 1914).

1860: The Bowden and Altrincham Entomological Society founded in Cheshire by a number of youthful entomologists including the brothers T. and J. B. Blackburn and E. M. Geldart. Date of demise unknown.

1862: *The Weekly Entomologist* founded by the Blackburn brothers and their friend E. M. Geldart and published by the above Society, to fill the gap left by the cessation of the *Weekly Intelligencer.* Three volumes only were published of what was commonly referred to as '*The Entomologist*' and it closed in 1863. It is interesting to note that, when Newman recommenced publication of his own journal *The Entomologist* in 1864, the early volumes bore the title *Newman's Entomologist* – see 1840.

1862: East London Entomological Club formed. Date of demise unknown.

1866: The Bristol Naturalists' Society established, and has continued to publish its *Proceedings* and a *Bulletin* to the present day.

1864: *Entomologist's Monthly Magazine* first published, but there appears to have been no single editor initially, the magazine being 'conducted' by T. Blackburn (formerly of *The Weekly Entomologist*), Dr Henry Guard Knaggs, R. McLachlan, E. C. Rye, and Henry T. Stainton. Familiarly referred to as the *E.M.M.*, it continues still.

1865: The North Staffordshire Naturalists' Field Club and Archaeological Society established at Hanley, issuing *Annual Reports* until 1886. From 1887 until 1915 it published the *Report and Transactions of the North Staffordshire Naturalists' Field Club*, and from 1916 until 1960 the slightly amended title, *Transactions and Annual Report of the North Staffordshire Field Club.* In 1961, Keele University took over publication under the title, *North Staffordshire Journal of Field Studies* (incorporating the *Transactions of the North Staffordshire Field Club*), and this journal continues today.

1868: The Folkestone Natural History & Microscopical Society founded and published a *Quarterly Journal* from 1868 until 1870. *Annual Reports* were then produced from 1871 and 1872, after which the Society published its *Proceedings* from 1883 until 1901. Publication was then suspended owing to lack of funds though occasional newsletters were issued. The Society finally collapsed for want of new members in the mid-1980s.

1868: The West London Entomological Society founded. Date of demise unknown.

1869: The Norfolk and Norwich Naturalists' Society founded. It was unconnected to the Norwich Entomological Society (1810 to *c.*1822). The *Transactions of the Norfolk and Norwich Naturalists' Society* have been published from 1869 to the present day.

1869: I Zingari Entomological Society founded, but changed its name later to the North London Entomological Society. Date of demise unknown.

1872: The South London Entomological Society founded, with J. R. Wellman as the first President. In 1884 its name was changed to the South London Entomological and Natural History Society. The *Proceedings and Transactions of the South London Entomological and Natural History Society* were published under various titles from 1885 but, in 1968, the Society became the British Entomological and Natural History Society (see 1968).

1873: The Entomological Society of London applied unsuccessfully to the Government for rooms at Burlington House and, after a repeat request in 1874 was turned down, plans to affiliate with the Linnean Society were discussed. These plans failed also and, in 1874, the Society moved to rooms owned by the Medical Society of London at 11 Chandos Street, London W1, where it remained for the next 50 years.

1874: William Watkins, Entomologist, founded a business to supply equipment and specimens to collectors from his home, Villa Sphinx, Eastbourne (Fig. 160). He established a butterfly farm in his grounds, three-quarters of an acre being devoted to rearing Lepidoptera collected from all over the world. One of his collectors, in the Solomon Islands, was at risk of being eaten by cannibals while waiting for a passing ship to pick him up! His establishment was visited by the Prince of Wales, later Edward VII, and among his customers was the Hon. Walter, later Lord Rothschild. He also founded the Insect House in the Royal Zoological Society Gardens, Regent's Park, London – see 1879.

Fig. 160. William Watkins, co-founder of Watkins & Doncaster, with butterfly cages in his garden.

1875: The Dorset Natural History and Antiquarian Society founded. The *Proceedings of the Dorset Natural History and Antiquarian Field Club* were published from 1877–1928, since when they have continued as the *Proceedings of the Dorset Natural History and Archaeological Society.*

1877: The Lancashire and Cheshire Entomological Society founded with Samuel Capper as President. This society, originally established with eleven members, is still thriving. From the start it published an *Annual Report* which, in 1905, became its *Annual Report and Proceedings.*

1878: First Grand National Entomological Exhibition organized by South London Entomological Society and others, at the Royal Aquarium, Westminster.

1879: William Watkins formed a partnership with Arthur Doncaster, formerly of Sheffield, in a joint venture to establish the famous firm of Watkins & Doncaster at 36 The Strand, London (see Fig. 20) – see 1874. Although Arthur Doncaster was deaf and dumb, he communicated skilfully by means of a slate hung around his neck. After one year, William Watkins left the partnership to continue independently. In 1884, he wrote *Directions* for collectors of British and foreign Lepidoptera for the Crystal Palace Insectarium, and it seems that he continued to run his butterfly farm in Eastbourne as it was still in existence in 1899. At that date he was also supplying Lepidoptera, cabinets and equipment from an address in Croydon as well as the West

End of London (Fig. 161). The later history of his firm is not known but Watkins & Doncaster continues under their joint names to this day. Doncaster was later joined

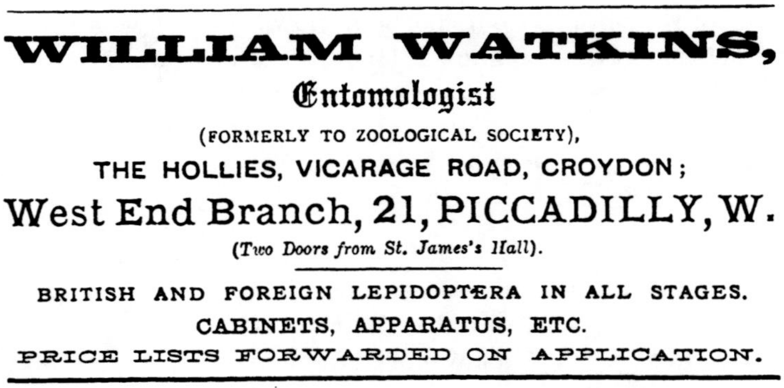

WILLIAM WATKINS,
Entomologist
(FORMERLY TO ZOOLOGICAL SOCIETY),
THE HOLLIES, VICARAGE ROAD, CROYDON;
West End Branch, 21, PICCADILLY, W.
(Two Doors from St. James's Hall).
BRITISH AND FOREIGN LEPIDOPTERA IN ALL STAGES.
CABINETS, APPARATUS, ETC.
PRICE LISTS FORWARDED ON APPLICATION.

Fig. 161. Watkins' advertisement, *c.*1892. After leaving Watkins & Doncaster, he again traded independently.

by Frederick Mette, an expert on oology, and at one time the firm employed no fewer than five full-time taxidermists to stuff birds and mount them under glass domes. When Doncaster retired, Mette took over the day-to-day running of the business and on his death, shortly before the Second World War, R. L. E. Ford became the new owner. Ford re-established its entomological expertise, assisted by E. W. Classey and L. Christie. Watkins & Doncaster occupied the upper four storeys of the five-storey Strand building (see Fig. 27). Its famous signboard depicting a large Swallowtail butterfly was a beacon for entomologists and other naturalists until the mid-1950s – see 1956.

1879: *The Young Naturalist: an illustrated magazine on natural history* commenced in November and published eleven volumes, edited by John E. Robson, the first five with S. L. Mosley as co-editor. In 1891 it changed its name to *The British Naturalist*, still under J. E. Robson, but only three volumes were published before it foundered in 1894 – see 1892.

1880: The Epping Forest and County of Essex Naturalist's Field Club founded, with Raphael Meldola as President. The membership at the end of the first year was 140, and by 1881 had reached 240. The Club was renamed the Essex Field Club in 1884, and published the *Transactions of the Essex Field Club* until 1887, when its publication was renamed the *Essex Naturalist.* The Club is still active.

1884: The South London Entomological Society became the South London Entomological and Natural History Society – see 1872 and 1968.

1885: The Entomological Society of London became the Royal Entomological Society of London by Royal Charter – see 1833.

1885: The Hampshire Field Club established at Southampton. There were 51 members at the inaugural meeting, and by 1889 the membership had increased to 250. *Papers and Proceedings of the Hampshire Field Club* were published from 1887 and *Proceedings of the Hampshire Field Club and Archaeological Society* from 1958 until the present day.

1887: Verrall Supper founded by G. H. Verrall, M.P., as an annual entertainment for entomologists, under the auspices of the Entomological Club – see 1826. It is still held annually.

1888: The Haggerstone Entomological Society (see 1858) changed its name to the City of London Entomological and Natural History Society, which, from 1891, held meetings at 33 Finsbury Square, London, EC, with J. A. Clark as the first President and

J. W. Tutt as Vice-President. The first by-laws stated that the Society was established 'for the study and enjoyment of natural history in the London area, especially within 20 miles of St. Paul's Cathedral, London'. The original annual subscription was five shillings (25p.). The *Transactions of the City of London Entomological and Natural History Society* were published from 1891 until 1913 when the Society once again underwent a change of name – see 1914.

1890: *The Entomologist's Record and Journal of Variation* founded by J. W. Tutt. It has been published without a break until the present time.

1892: *The Naturalists' Journal*, was launched in July and ran for a total of eleven volumes under a variety of titles, the last of which was *Nature Study*, before closing in 1903. It also had a number of editors, one of whom was S. L. Mosley who became involved after the closure of *The Young Naturalist* – see 1879.

1902: The Manchester Entomological Society founded with W. E. Hoyle, D.Sc., as its first President. There were 41 members during the first year, and the annual subscription was five shillings (25p). The Society issued *Annual Reports* from 1903 until 1909, after which it published the *Annual Report and Transactions of the Manchester Entomological Society*. This continued from 1910 until 1960, and was followed by the *Report, Proceedings and Transactions of the Manchester Entomological Society* from 1961 up to the present.

1914: The City of London Entomological and Natural History Society (see 1888) became the London Natural History Society. Its *Transactions* became the *Transactions of the London Natural History Society* which in 1921 were renamed *The London Naturalist*, which is still published annually.

1920: The Isle of Wight Natural History and Archaeological Society founded at Newport with George W. Colenutt, F.G.S., as its first President. There were 50 members at the inaugural meeting. The *Proceedings* of the Society have been published from 1920 until the present day.

1920: Entomological Society of Hampshire founded. Published the *Transactions of the Hampshire Entomological Society* from 1924 until 1928, after which publication continued for the year 1929 as the *Transactions of the Entomological Society of Hampshire and the South of England*, but in 1930 changed once more to *Transactions of the Entomological Society of the South of England*. In 1932 the publication became *Journal of the Entomological Society of the South of England* – see 1934.

1923: The Royal Entomological Society of London moved to 41 Queen's Gate, London SW7, its present headquarters.

1929: The Suffolk Naturalists' Society founded at Ipswich with Dr C. H. S. Vinter, M.A., as its first President. The Society attracted 79 members during its first year. The *Transactions of the Suffolk Naturalists' Society* were produced from 1929 until 1969, since when publication has continued as *Suffolk Natural History*. It is one of the most active natural history societies in Britain, producing occasional publications on Suffolk's fauna and flora and organizing an annual conference on an important theme, attracting speakers of national importance.

1934: The Society for British Entomology established at Southampton with W. Parkinson Curtis as its first President. The Society evolved from the Hampshire Entomological Society (see 1920) and in addition to a *Journal* it published *Transactions of the Society for British Entomology*.

1935: The Entomological Exchange and Correspondence Club founded by L. R. Tesch specifically to cater for the novice collector or junior entomologist. Its journal was published for two years from 1935–36, after which it continued as *The Entomologist's Bulletin* 1937–38. However, in January 1937, the Club changed its name to The Amateur Entomologist's Society – see 1937.

1937: The Amateur Entomologist's Society formed – see 1935. It published the '*Bulletin*' as *The Amateur Entomologist – the Journal of the Amateur Entomologist's Society* in 1939 and 1941. The war curtailed further production but, from 1940–44, the Society produced *The Amateur Entomologist's Society Wartime Exchange Sheet.* In 1945 publication was resumed as *The Bulletin of the Amateur Entomologist's Society* which continues to the present, containing observations by amateur entomologists and Lists of Exchanges and Wants. It also publishes a series of handbooks including *Breeding the British Butterflies, A Lepidopterist's Handbook* and *Killing, Setting and Storing Butterflies and Moths,* and has recently reprinted Tutt's invaluable *Practical Hints for the Field Lepidopterist.* The AES is reputedly one of the largest and most vigorous entomological societies in the world with an active junior branch – The Bug Club – and a website. It has no headquarters, library, collections or calendar of meetings, but it organizes an increasingly well-attended Annual Exhibition Meeting in early October in West London, currently at Kempton Park Racecourse, which is open to the public. Members can meet and exchange experiences and purchase a wide selection of entomological equipment, livestock, set specimens and books. Membership of the Society currently stands at around 1,700, with that of the junior 'Bug Club' at between 250–300.

Fig. 162. Eric Classey, founder of E.W. Classey Ltd. and co-founder of *Entomologist's Gazette.*

1949: E. W. Classey (Fig. 162) set up an entomological bookselling business at his home in Feltham, Middlesex, later moving to nearby Hanworth where, in 1959, it became E. W. Classey Ltd. Eric Classey's entomological career had begun in 1934 at the Natural History Museum but after service in the Second World War he joined Watkins & Doncaster in the Strand. With the assistance of his wife Ivy, Classey built up a worldwide business dealing in new, antiquarian and second-hand books and periodicals, mainly in entomology, and republishing, in facsimile, very rare or unobtainable works of reference – see 1973.

1950: *Entomologist's Gazette* founded and edited by E. W. Classey and R. L. E. Ford, with the first part of volume 1 in facsimile typescript. In 1952 E. W. Classey became sole publisher and editor, being joined in 1956 as co-editor by J. D. Bradley, who took over as sole editor in 1960. In 1964 W. G. Tremewan was appointed co-editor, becoming sole editor in 1965. The journal, taken over by Gem Publishing in 1991, continues to flourish, with Eric Classey still on the editorial board.

1956: Watkins & Doncaster left 36 The Strand after a tenancy of seventy-seven years and moved to Welling, Kent. Ford was later joined by his son, Robin, who took over the running of the business in 1969 – see 1973.

1960: Worldwide Butterflies founded by Robert Goodden at the age of twenty in the garden of his parents' home in Charmouth, Dorset – see 1966.

1960: The British Trust for Entomology Ltd, which incorporated The Society for British Entomology, took over *The Entomologist,* merging it with the *Transactions* of the Society – see 1934. Publication ceased in 1973 – but see 1978.

1966: Worldwide Butterflies transferred to the grounds of the historic but somewhat neglected Goodden family home, Compton House, near Sherborne, and in the same year opened up a showroom in Brighton – see 1976, 1978.

1968: The South London Entomological and Natural History Society – see 1884 – became The British Entomological and Natural History Society, continuing publication of its *Proceedings and Transactions* until 1987. The Society holds an Annual Exhibition in October at Imperial College, London SW7. The library and extensive entomological collections are housed at its headquarters at Dinton Pastures, near Reading, and are open to members on the second and fourth Sundays of each month. Meetings of the Society are held regularly at the rooms of the Royal Entomological Society, and field meetings are held at weekends throughout the summer. The Society flourishes today – see 1988.

1968: British Butterfly Conservation Society founded by Thomas Frankland and Julian Gibbs, keen amateurs who funded its establishment. Its first President was Peter Scott, the ornithologist but also a keen lepidopterist since boyhood. Known as Butterfly Conservation, its aims are to conserve existing species of Lepidoptera (including moths) and to re-establish certain species that have become extinct. It has a considerable national profile and also sponsors scientific research into conservation and education, with junior groups in many schools. It now publishes *Butterfly Conservation News* three times a year.

1973: Watkins & Doncaster moved to their present premises at Hawkhurst, Kent, where they have showrooms, workshops for the manufacture of equipment and a worldwide mail-order service.

1973: E. W. Classey Ltd moved to Faringdon, Oxfordshire, where they now have a shop in addition to their mail-order business. Following the death of Ivy Classey, their son Peter, who had joined the company in 1976, took over the running of the new-book business whilst Eric Classey has continued to run the antiquarian side.

1976: Worldwide Butterflies sold their shop in Brighton to Watkins & Doncaster to finance major restoration of Compton House – see 1978 – but it was closed after some five years.

1977: The Royal Entomological Society of London launched *Antenna*, replacing its *Proceedings*.

1978: Robert Goodden and his wife Rosemary (granddaughter of Captain Edward Bagwell Purefoy (see Chapter 3) extended Worldwide Butterflies – see 1966, 1976 – into the restored house itself where it remains. It is among the longest-established businesses of its kind, combining educational displays of livestock with fine butterfly houses displaying British as well as exotic species. It sells livestock, set specimens, entomological equipment and many other items of interest to the entomologist, and produces colour catalogues. It also houses the Lullingstone Silk Farm, where, incidentally, the silk for Princess Diana's wedding dress was made.

1978: The Royal Entomological Society relaunched *The Entomologist*, but publication ceased once more in 1997.

1988: *British Journal of Entomology and Natural History* founded by the British Entomological and Natural History Society. This replaced its *Proceedings and Transactions* – see 1968.

1989: *British Wildlife*, subtitled '*the magazine for the modern naturalist*', founded by Andrew Branson. Published bi-monthly, each issue includes, in addition to a wide-ranging selection of articles, Habitat Management News, Wildlife Reports (arranged by interest groups), and Conservation News.

1996: *Atropos* founded by its present editor, Mark Tunmore. It is a lively publication, designed for the active field lepidopterist and odonatist, and carries many interesting observations and records. Published twice a year, in July and December.

Addresses of national Societies, Journals and Suppliers mentioned above:

The Amateur Entomologists' Society,
PO Box 8774, London SW7 5ZG
(website www.theaes.org)

British Entomological and Natural History Society,
Dinton Pastures Country Park, Davis Street, Hurst, Reading, Berkshire RG10 0TH

Butterfly Conservation,
Manor Yard, East Lulworth, Wareham, Dorset BH20 5QP
(website: www.butterfly-conservation.org)

The Royal Entomological Society,
41 Queen's Gate, London SW7 5HR
(website: www.royensoc.demon.co.uk)

Atropos,
c/o Mark Tunmore, 36 Tinker Lane, Meltham, Huddersfield, West Yorkshire HD7 3EX

British Wildlife, British Wildlife Publishing,
Lower Barn, Rooks Farm, Rotherwick, Hook, Hants RG27 9BG
(website: www.brit.wildlife.clara.net)

The Entomologist's Gazette,
Gem Publishing Company, Brightwood, Brightwell, Wallingford, Oxon. OX10 0QD

The Entomologist's Monthly Magazine,
Gem Publishing Company, as above.

The Entomologist's Record,
c/o The Editor, 14 West Road, Bishops Stortford, Herts. CM23 3QP

E. W. Classey Ltd.,
Oxford House, 6 Marlborough Street, Faringdon, Oxon. SN7 7JP
or PO Box 93, Faringdon, Oxon. SN7 7DR
(website: www.abebooks.com/home/bugbooks)

Watkins & Doncaster,
Conghurst Lane, Four Throws, Hawkhurst, Cranbrook, Tonbridge, Kent TN18 5ED
or PO Box 5, Cranbrook, Kent TN18 5EZ
(website: www.watdon.com)

Worldwide Butterflies Ltd.,
Compton House, Sherborne, Dorset DT9 4QN
(website: www.wwb.co.uk)

BIBLIOGRAPHY AND FURTHER READING

All books referred to in the text and the footnotes are listed, in addition to a small selection of the very many consulted in the preparation of this work. Abbreviations of journals cited in the footnotes and elsewhere are appended to this Bibliography. Papers from journals cited in the footnotes to Chapters 1–5 are not included below unless they are also referred to in the Appendixes.

Abbot, C. (1798a). *Flora Bedfordiensis*, xii, 351 pp., 6 col. pls. Bedford, Smith.

—- (1798b). Manuscript report to the Linnean Society of London, 12 August 1798. Linnean Society Archive, No. 192a.

—- (1800). XXXI Extracts from Minute Book of the Linnean Society for 6 Nov. 1798. *Trans.Linn.Soc.Lond.* **5**: 276.

Abbot, J. & Smith, J. E. (1797). *The natural history of the rarer lepidopterous insects of Georgia*, 2 vols, xv, 214 pp., 104 col. pls. London.

Albin, E. (1720). *The natural history of English insects*, [12] pp., 100 pls. London, Innys.

—- (1736). *A natural history of spiders and other curious insects*, viii, 76 pp., 54 col. pls. London, Montagu.

Aldrovandi, U. (Aldrovandrus) (1602). *De animalibus insectis libri septem, &c.*, [x], 767, [43] pp., illust., portr. Bologna.

Allan, P. B. M. (1937); (revd edn 1947). *A moth-hunter's gossip*, 269 pp., 1 pl. London, Watkins & Doncaster.

--- (1943). *Talking of moths*, 340 pp. Newtown, Montgomery, privately published.

--- (1948). *Moths and memories*, 316 pp., 1 pl. London, Watkins & Doncaster.

--- (1980). *Leaves from a moth-hunter's notebook*, 281 pp. Faringdon, Classey.

Allen, D. E. (1976). *The naturalist in Britain*, 292 pp., 13 pls. Harmondsworth, Allen Lane.

Arnold, V. W., Baker, C. R. B., Manning, D. V. & Woiwod, I. P. (1997). *Butterflies and moths of Bedfordshire*, 408 pp., 104 col. pls, 44 text figs, 8 tables. Bedford, Bedfordshire Natural History Society.

Baker, B. R. (1994). *The butterflies and moths of Berkshire*, xxxii, 367 pp. Uffington, Hedera Press.

Barber, L. (1980). *The heyday of natural history*, 320 pp., 31 pls. London, Jonathan Cape.

Barrett, C. G. (1893). *The Lepidoptera of the British Islands* **1**: viii, 313 pp., 40 col. pls. London, Reeve.

Bates, H. W. (1863). *The naturalist on the river Amazons* (2 vols), **1**: viii, 351 pp., 4 pls, 1 map; **2**: iv, 423 pp., 4 pls. London, John Murray.

Baynes, E. S. A. (1964). *A revised catalogue of Irish macrolepidoptera (butterflies and moths)*, iii, 110 pp. Hampton, Classey.

Berger, L. A. (1948). A *Colias* new to Britain (Lep. Pieridae). *Entomologist* **81**: 129–131.

Berridge, W. S. (1915). *Wonders of animal life*, 269 pp., 29 pls. London, Simpkin, Marshall Hamilton, Kent & Co.

Berkenhout, J. (1769). *Outlines of the natural history of Great Britain and Ireland* **1**, ix, 233 pp. London, Elmsley.

Bonhote, J. C. & Rothschild, N. C. (1895–97). *Harrow butterflies and moths*, 2 vols, 1 pl. Harrow.

Bremer, O. & Grey, W. (1852). Diagnoses de Lepidoptères nouveaux trouvés par Mss. Tatarinoff et Gaschkewitsch aux environs de Pekin. *In* Motschulsky, V. von, *Études entomologiques*, Teil 1, 80 pp. Helsingfors.

Bright, P. M. & Leeds, H. A. (1938). *A monograph of the British aberrations of the Chalk-hill Blue butterfly* Lysandra coridon *(Poda, 1761)*, ix, 144 pp., 18 pls, 4 col. Bournemouth, Richmond Hill.

Bristowe, W. S. (1958). *The world of spiders*, xvi, 304 pp., 4 col. pls, xxxii mono. pls, 116 text figs. (New Naturalist No. 38). London, Collins.

— (1967c) Fourteenth century butterfly nets. *Entomologist's Gaz.* **18**: 201, 1 pl.

Brown, J. (1812). Thomas Dandridge. *In* Nichols, J., *Illustrations of the literary history of the eighteenth century* **3**: 782. London.

Brown, T. (1832). *The book of butterflies, sphinxes and moths; &c.* **1**: xxxviii, 178 pp. [i.e., pp. 39–216], 60 col. pls. London, Whittaker, Treacher.

Buckler, W. (1886–1901). *The larvae of the British butterflies and moths*, 9 vols: 164 col. pls. London, Ray Society. (Ed. H. T. Stainton; and G. T. Porritt [after 1892]).

Burr, M. (1939). *The insect legion*, xiv, 321 pp., 16 pls. London, J. Nisbet.

Carson, R. (1963). *Silent Spring*, xxii, 304 pp. London, Hamish Hamilton.

Cater, W. F. (Ed.) (1980). *Love among the butterflies: travels and adventures of a Victorian lady, Margaret Fountaine*, 223 pp., 20 col. pls. London, Collins.

— (Ed.) (1986). *Butterflies and late loves: the further travels and adventures of a Victorian lady, Margaret Fountaine*, 141 pp., 1 pl. (frontis.). London, Collins.

Chalmers-Hunt, J. M. (1960–61). *The butterflies and moths of Kent* **1**, 144 pp. Arbroath, T. Buncle. (Originally published as supplements to *Entomologists's Rec. J. Var.* **72–73**.)

— (1976). Entomological Sales (pp. 3–14). In *Natural History Auctions 1700–1972. A register of sales in the British Isles*, xii, 189 pp. London, Sotheby Parke Bernet.

Chatfield, J. (1987). *F. W. Frohawk: his life and work*, 184 pp., incl. numerous b/w and col. illust. Marlborough, Crowood Press.

Coleman, W. S. (1860). *British butterflies*, vii, 179 pp., 16 pls. London, George Routledge. (Edn 2, 1895; edn 3, 1897; new edn, 1901)

Collier, R. V. (1986). The conservation of the chequered skipper in Britain, 16 pp. *Focus on Nature Conservation*, No. 16.

Collins, G. A. (1995). *Butterflies of Surrey*, [viii], 87 pp., 16 col. pls. Woking, Surrey Wildlife Trust.

Collins, N. M. & Thomas, J. A. (Eds) (1991). *The conservation of insects and their habitats*, xviii, 450 pp. (Symposium of the Royal Entomological Society of London, No.15). London, Academic Press.

Colvin, H. (1985). *Calke Abbey, Derbyshire. A hidden house revealed*, 128 pp., incl. 35 col. pls, 35 mono. pls. London, George Philip, for the National Trust.

Common, I. F. & Waterhouse, D. F. (1981). *Butterflies of Australia* (revd edn), 682 pp., 49 pls (incl. 28 col.), 25 text figs, 2 maps. London, Angus & Robertson.

Cox, N. (1686). *The gentleman's recreation* (edn 3), 445 pp., 4 pls as folded folios. London, Freeman Collins.

Curtis, J. (1824–39). *British entomology*, 16 vols (later bound as 8 vols), 770 col. pls. London, printed for the author.

— (1829). *A guide to an arrangement of British insects, &c.*, vi, 256 columns. London, Westley for the author.

[Curtis, W.] (1771). *Instructions for collecting and preserving insects, particularly moths and butterflies*, iv, 44 pp., 1 pl. London, printed for the author.

Dale, C. W. (1878). *The history of Glanville's Wootton in the County of Dorset, including its zoology and botany*, viii, 392 pp., 2 pls., London, Hatchards.

— (1886). *Lepidoptera of Dorsetshire*, [8], 76pp. Dorchester, Ling.

— (1890). *The history of our British butterflies*, xli, [iii], 232 pp. London, John Kempster. (Originally published as a supplement to *The Young Naturalist*, 1887.)

Dale, J. C. (1830). Notice of the capture of *Vanessa Huntera* for the first time in Britain, with a Catalogue of rare Insects captured. *Mag.nat.Hist.* **3**: 332–334.

Darwin, C. R. (1859). *On the origin of species by natural selection, or the preservation of favoured races in the struggle for life*, ix, 512 pp., 1 plan. London, Murray. (Very many later editions.)

Darwin, F. (Ed.) (1887). *The life and letters of Charles Darwin, including an autobiographical chapter*, 3 vols, illust. London, John Murray.

— (Ed.) (1929). *Autobiography of Charles Darwin*, iv, 154 pp. London, Watts.

de Beer, G. (Ed.) (1974). *Charles Darwin, Thomas Henry Huxley: autobiographies*, xxvi, 123 pp., 12 pls. London, Oxford University Press.

[Denis, M. & Schiffermüller, I.] (1775). *Ankündung eines systematischen Werkes von den Schmetterlingen der Wienergegend*, 324 pp., 3 pls. Vienna.

Dennis, R. L. H. (1977). *The British butterflies, their origin and establishment*, xviii, 318 pp., text figs, maps. Faringdon, Classey.

Desmond, R. (1994). *Dictionary of British and Irish botanists and horticulturalists, including plant collectors, flower painters and garden designers* (new, updated edn), xl, 825 pp. London, Taylor & Francis.

Dickens, C. (1838). *Memoirs of Joseph Grimaldi.* (Edited by 'Boz'), 2 vols. London. (Revd version, 1968. London, MacGibbon & Kee.)

Donovan, E. (1792–1813). *The natural history of British insects, &c.*, 16 vols: 569 col. pls, 7 plain pls. London, Rivington.

—– (1794). *Instructions for collecting and preserving various subjects of natural history*, ii, 86 pp., 2 pls. London, Rivington. (Edn 2, 1805.)

—– (1833). *Memorial.* London. Printed privately for the author.

Doubleday, E. & Westwood, J. O. (1846–52). *The genera of diurnal Lepidoptera*, 2 vols: 534 pp., 85 col. pls, 1 plain pl. London, Longmans.

Doubleday, H. (1847–50). *A synonymic list of British Lepidoptera*, 27 pp. London, Van Voorst.

Douglas, J. W. (1856). *The world of insects: a guide to its wonders*, x, 244 pp. London, Van Voorst.

--- & Scott, J. (1865). *The British Hemiptera* **1**: 1. *Hemiptera-Heteroptera*, xii, 628 pp., 21 pls. London, The Ray Society (all published).

Drury, D. (1770–82). *Illustrations of natural history, &c.*, 3 vols [mainly illus. by M. Harris]. London, White.

--- (*c.*1800). *Directions for collecting insects in foreign parts*, 3 folios.

Duddington, J. & Johnson, R. (1983). *The butterflies and larger moths of Lincolnshire and South Humberside*, 299 pp. Lincoln, Lincolnshire Naturalists' Union.

Dunbar, D. (1993). *Saving butterflies, a practical guide to the conservation of butterflies*, 80 pp., incl. many col. illust. Dedham, Butterfly Conservation.

Duncan, J. (1835). *The natural history of British butterflies*, xv, 268 pp., 31 pls. Edinburgh, Lizars.

— (1836). *The natural history of British moths, sphinxes, &c.*, xv, 268 pp., 30 col. pls, 1 portr. Edinburgh, Lizars.

— (1840). *Introduction to entomology*, xv, 17–331 pp., 36 col. pls, 1 portr. Edinburgh, Lizars.

Dutfield, J. (1748–49). *A new and complete natural history of English moths and butterflies, &c.*, [17] pp., 12 col. pls. London, M. Payne. (Originally issued in six fascicles, unfinished.)

Edwards, E. (1870). *Lives of the founders of the British Museum, with notices of its chief augmentors and other benefactors, 1570–1870*, x, 780 pp., 1 pl., 5 plans (col.), illust. in text. London, British Museum.

Edwards, G. (1743–51). *A natural history of uncommon birds, &c.*, (4 parts), 210 col. pls with descr. text. London, printed for the author.

— (1758–64). *Gleanings of natural history, exhibiting figures of quadrupeds, birds, insects, plants, &c.* (3 parts), xxxv, vii, 347 pp., 152 col. pls. London, printed for the author.

— (1770). *Essays upon natural history, and other miscellaneous subjects, &c.*, viii, 231 pp., incl. portr. London, Robson. (Facsimile reprint, 1972. Chicheley, Paul Minet.)

Emerson, R. W. (1841). *The works of Ralph Waldo Emerson, comprising his essays, lectures, poems and ovations*, 3 vols complete. (English edn, 1886). London, Geo. Bell.

Emmet, A. M. & Heath, J. (Eds) (1989). *The moths and butterflies of Great Britain and Ireland* **7**(1), ix, 370 pp., incl. 24 col. pls, 22 text figs, 74 maps. Colchester, Harley Books.

Esper, E. J. C. (1777–80). *Die Schmetterlinge: Tagschmetterlinge* **1**: 338 pp., 50 pls; **2**: 190 pp., 43 pls. Erlangen, Walther.

Fabricius J. (1775). *Systema Entomologiae sistens Insectorum classes, ordines, genera, species, adjectis synonymis, locis, descriptionibus, observationibus*, [xxviii], 832 pp. Flensburg & Leipzig, Korte.

— (1787). *Mantissa insectorum, &c.* **1**: 348 pp.; **2**: 382 pp. Hafnia [Copenhagen], Prost.

— (1792–94). *Entomologia systematica emendata et aucta, &c.*, 4 vols, **1** (1792): xxi, 538 pp.; **2** (1793): viii, 519 pp.; **3**(1) (1793): iv, 487 pp.; **3**(2) (1794): 349 pp.; **4** (1794): vi, 472 pp. Hafnia [Copenhagen], Prost.

— (1807). Systema glossatorum. *Magazin.Insektenk. (Illiger)* **6**: 277–296.

Firmin, J. (1992). Historical Notes, pp. 8–16. *In* Goodey, B. & Firmin, J. *Lepidoptera of North East Essex*. Colchester, Colchester Natural History Society.

—, Buck, F. D., Dewick, A. J., Down, D. G., Huggins, H. C., Pyman, G. A. & Williams, E. F. (1975). *A guide to the butterflies and larger moths of Essex*, 152 pp., 4 pls, 1 map. Colchester, Essex Naturalists' Trust.

Fitton, M. & Gilbert, P. (1994). Insect Collections, pp. 112–122. *In* MacGregor, A. (Ed.), *Sir Hans Sloane*, 292 pp., numerous illust. London, British Museum Press.

Ford, E. B. (1945). *Butterflies*. xiv, 368 pp., 48 col. pls, 24 b/w pls, 9 text figs, 8 tables, 32 maps. (New Naturalist No. 1). London, Collins.

— (1952). *Moths*, xix, 266 pp., 32 col. pls, 24 b/w pls, 7 text figs, 12 maps. (New Naturalist No. 30). London, Collins.

Ford, R. L. E. (1963). *Practical entomology*, ix, 198 pp., 12 b/w pls, 36 text figs. London, Warne.

— (1973). *Studying insects*, xii, 150 pp., 16 col. & b/w pls., 47 text figs. London, Warne.

Forster, J. George A. (1777). *A voyage round the world, in His Britannic Majesty's sloop, Resolution, commanded by Capt. James Cook, during the years 1772, 3, 4 and* 5, 2 vols. London, White, Robson, Elmsly & Robinson.

Forster, J. R. (1770). *A catalogue of British insects*, 16 pp. Warrington, William Eyres.

— (1771). *Novae species insectorum centuria 1* [all published], viii, 100 pp. London.

— (1778). *Observations made during a voyage round the world, on physical geography, natural history and ethnic philosophy, especially on 1. The earth and its strata, 2. Water and the ocean, 3. The atmosphere, 4. The changes of the globe, 5. Organic 'bodies', and 6. The human speech*, iv, iv, 649 pp. map, table. London, Robinson.

— (1781). *Indische zoologische, &c.* [*Zoologica indica, &c.*], [ii], iv, 42 pp., 15 col. pls. Halle, Gebauer.

Foster, W. (1924). *The East India House: its history and associations*, x, 250 pp., frontis., 37 b/w pls. London, John Lane.

Fowles, J. (1971). *The Blinded Eye*. London, Little, Brown.

Freeman, J. (1852). *Life of the Rev. W. Kirby, &c.* [with a chapter by W. Spence on *The Introduction to Entomology*], xii, 506 pp., 1 table, 2 pls; portr. London, Longman, Brown, Green & Longman.

Friday, L. E. (Ed.) (1997). *Wicken Fen, the making of a wetland nature reserve*, xvi, 306 pp., 16 col. pls, 82 text figs, 2 maps. Colchester, Harley Books.

Frohawk, F. W. (1923). *Carcharodus alceae* in Surrey. *Entomologist* **56**: 267–268.

— (1924). *Natural history of British butterflies* **1**: xiii, 207 pp., 34 col. pls; **2**: viii, 206, [4], 26 col. pls, 3 mono. pls. London, Hutchinson.

— (1934). *The complete book of British butterflies*, 384 pp., 32 col. pls, numerous text drawings. London, Ward Lock.

— (1938a). *Varieties of British butterflies.* 200 pp., 48 col. pls, London, Ward Lock.

Füessly, J. C. (1775). *Verzeichniss der ihm bekannten Schweizerischen Insecten*, [12], 62 pp., 1 pl. Zürich, Verfasser.

Furneaux, W. (1894). *Butterflies and moths*, 358 pp., 12 col. pls, 241 text figs. London, Longmans. (Edn 2, 1911.)

Gage, A. T. & Stearn, W. T. (1998). *A bicentenary history of the Linnean Society of London*, 242 pp., col. frontis., 20 pls. London, Academic Press.

Gardiner, B. O. C., Emmet, A. M., Hart, C., Skinner, B. & Waring, P. (compilers) (1995). An index to the modern names for use with J. W. Tutt's practical hints for the field lepidopterist. *The Amateur Entomologist*, Vol. 23A, iii, 58 pp. Colchester, AES Publications.

Gaskell, E. (1848). *Mary Barton*, London, Smith, Elder.

Geoffroy, E. L. (1785). *In* Fourcroy, A. F. *Entomologia parisiensis, sive catalogus insectorum, quae in agro parisiensi reperiuntur, &c.*, **1**: 231 pp. Paris.

Gesner, C. (1551–87). *Historiae animalium*, 5 vols, illust. Zürich.

Gilbert, P. (1977). *A compendium of the biographical literature on deceased entomologists*, [xiv], 455 pp., incl. frontis. and 27 portrs. London, British Museum (Natural History).

— (1998). *John Abbot. Birds, Butterflies and other wonders*, 128 pp., incl. 62 col. and 29 b/w illusts. London, Merrell Holberton and The Natural History Museum.

Goater, B. (1974). *The Butterflies and moths of Hampshire and the Isle of Wight*, xiv, 439 pp. Faringdon, Classey.

— (1992). *The butterflies and moths of Hampshire and the Isle of Wight: additions and corrections*, vi, 266 pp. (UK Nature Conservation publication No. 7.) Peterborough, Joint Nature Conservation Committee.

Gosse, P. H. (1851). *A naturalist's sojourn in Jamaica*, xxiv, 508 pp., 8 pls. London.

Gosse, E. (1907). *Father and son*, vi, 374 pp. London, Heinemann.

Greene, J. (1863). *The insect hunter's companion, being instructions for collecting and preserving butterflies and moths, and comprising an essay on pupa digging*, 164 pp., text figs. London, Van Voorst. (Edns 2, 1870; 3, 1880; 4, 1892; edn 5, revd and extended by A. B. Farn, 1907.)

Hagen, H. A. (1862–63). *Bibliotheca Entomologica*, **1**: 566 pp.; **2**: 512 pp. Leipzig, Engelmann. (Facsimile reprint in 1 vol., 1960.)

Harris, M. (1766). *The Aurelian or natural history of English insects; namely, moths and butterflies. Together with the plants on which they feed, &c.*, 92 pp., 44 col. pls. London, printed for the author.

— (1773). *The Aurelian, &c.* with Table of terms, Index, 4 additional col. pls. (1st edn, 2nd issue.)

— (1775a). *The Aurelian, or a natural history of insects and plants, &c.*, (edn 2), xvii, 90 pp., 49 col. pls. London, Robson.

— (1775b). *The English Lepidoptera: or, the Aurelian's pocket companion*, xv, 66 pp., 1 col. pl. London, Robson.

— (1776[–80]). *An exposition of English insects, &c.*, 166, [4] pp., 53 col. pls. London, printed for the author.

Harvey, J. M. V., Gilbert, P. & Martin, K. (1996). *A catalogue of manuscripts in the Entomology Library of the Natural History Museum, London*, xvi, 251 pp., 4 col pls. London, The Natural History Museum.

Harwood, W. (1903). *Lepidoptera. In* Doubleday, H. A. & Page, W. (Eds), *The Victoria History of the County of Essex* **1**: 136–177.

[Haworth, A. H.] (1802). *Prodromus Lepidopterorum Britannicorum*, vii, 39, (*Addenda*) 6 pp. Holt, Hurst.

Haworth, A. H. (1803–28). *Lepidoptera Britannica*, 4 vols: xxxvi, 609 pp. London, John Murray.

— (1807). Review of the Rise and Progress of the Science of Entomology in Great Britain; chronologically digested. *Trans.ent.Soc.Lond.* **1**: 1–69.

Hawes, F. W. (1890). *Hesperia lineola* Ochsenheimer: an addition to the list of British butterflies. *Entomologist* **23**: 3–4.

Heath, J. (1981). *Threatened Rhopalocera (butterflies) in Europe*, vi, 157 pp. (Nature and Environment Series, No. 23). Strasbourg, Council of Europe.

— & Emmet, A. M. (Eds) (1976–). *The Moths and Butterflies of Great Britain and Ireland*, 7 published vols. (Vol 1, ed. Heath; Vol. 2, 9 & 10, ed. Heath & Emmet; Vols 7(1) & 7(2), ed. Emmet & Heath; Vol. 3, ed. Emmet). Vols 1, 9: London, Curwen; Vols 2, 3, 7(1), 7(2), 9 & 10: Colchester, Harley.

Heslop, I. R. P. (1959). A new label list of British macrolepidoptera. *Entomologist's Gaz.* **10**: 179–181.

— (1964). *Revised indexed check-list of the British Lepidoptera, with the English name of each of the 2,404 species*, vi, 145 pp. Library edn. [Classey.]

—, Hyde, G. E. & Stockley, R. E. (1964). *Notes and views on the Purple Emperor*, 245 pp., 6 col. pls. Brighton, Southern Publishing.

Higgins, L. G. (1975). *The classification of European butterflies*, 320 pp., 402 figs. London, Collins.

— & Riley, N. D. (1970). *A field guide to the butterflies of Britain and Europe*, 380 pp., 60 col. pls, text figs, 384 maps. London, Collins.

Hoare, M. E. (1976). *The tactless philosopher: Johann Reinhold Forster (1729–1798)*, x, 419 pp., 13 pls. Melbourne, Hawthorne Press.

Holland, W. J. (1898). *The Butterfly Book*, xx, 382 pp., 48 col. pls, 183 figs. New York, Doubleday & McClure.

— (1916). *The Butterfly Guide, a popular guide to a knowledge of the butterflies of North America*, 237 pp., 295 col. figs. New York, Doubleday, Page & Co.

Hope, F. W. (1845). The auto-biography of John Christian Fabricius (translated from the Danish with additional notes and observations). *Trans.ent.Soc.London* **4**: i–xvi.

Hough, R. (1994). *Captain James Cook*, xviii, 398 pp., 25 pls. London, Hodder & Stoughton.

Howarth, T. G. (1973). *South's British butterflies*, xiii, 210 pp., 48 col. pls, 57 maps. London, Warne.

Hudson, W. H. (1903). *Hampshire days.* London, Longmans, Green.

Huggins, H.C. (1960). A Naturalist in the Kingdom of Kerry. *Proc.Trans.S.Lond.ent. nat.Hist.Soc.* **1959**: 176–183.

Humphreys, H. N. & Westwood, J. O. (1841). *British butterflies and their transformations. Arranged and illustrated by H. N. Humphreys ... with characters and descriptions by J. O. Westwood*, xii, 138 pp., 42 col. pls. London, Smith.

— & — (1843–45). *British moths and their transformations, &c.*, 2 vols, 124 col. pls. London, Smith.

Ingpen, A. (1827). *Instructions for collecting, rearing and preserving British and foreign insects; also for collecting and preserving Crustacea and shells; &c.*, xv, 92 pp., 1 col. pl. London, Bulcock.

James, M. J. (1973). *The new Aurelians, a centenary history, &c., with an account of the collections by A. E. Gardner*, 80 pp., 4 pls. London, British Entomological & Natural History Society.

[Jermyn, L.] ([1824]). *The butterfly collector's vade mecum or a synoptical table of English butterflies*, 68 pp., 2 col. pls. Ipswich, Raw. (Edn 2, 1827.)

Jessop, L. (1989). Notes on insects, 1692 & 1695, by Charles duBois. *Bull.Br.Mus.nat.Hist. (Historical Series)* **17**(1): 1–165.

Johns, B. G. (1891). *Among the butterflies*, 194 pp., 12 pls, text figs. London, Isbister.

Kane, W. F. de V. (1885). *European butterflies*, xxxi, 184 pp., 15 pls. London, Macmillan.

— (1893). A catalogue of the Lepidoptera of Ireland. *Entomologist* **26**: 69–73, 117–121, 157–159, 187–190, 212–215, 240–244, 269–273.

Kettlewell, H. B. D. (1973). *The evolution of melanism*, xv, 423 pp., 39 pls (3 col.). Oxford, Oxford University Press.

Kingsley, C. (1855). *Glaucus; or the wonders of the shore*, vi, 165 pp., 1 pl. Cambridge, Macmillan.

Kirby, W. (1802). *Monographia apum Angliae, &c.*, 2 vols in 1: xxii, 646 pp., 18 pls. Ipswich, Raw for the author.

— & Spence, W. (1815–26). *An introduction to entomology: or elements of the natural history of insects, &c.*, 4 vols: **1**, 1815 (edn 2, 1816; edn 3, 1818), xxiv, 519 pp., 3 col. pls); **2**, 1817 (edn 2, 1818), [ii], 530 pp., 2 col. pls; **3**, 1826, 722 pp., 20 b/w pls; **4**, 1826, iv, 634 pp., 4 b/w pls. London, Longman, Rees, Hurst, Orme, & Brown.

Kirby, W. E. ([1906]). *Butterflies and moths of the United Kingdom*, lii, 468 pp., 70 pls. London, George Routledge.

Kirby, W. F. (1862). *A manual of European butterflies on the plan of Stainton's Manual of British butterflies and moths*, xii, 153 pp., 1 pl. London, Williams & Norgate.

— (1871–77). *A synonymic catalogue of diurnal Lepidoptera*, supplement – appendix, v, 883 pp. London, British Museum.

— (1882). *European butterflies and moths ... based upon Berge's 'Schmetterlingsbuch'* [Stuttgart, 1842], lvi, 427 pp., 62 col. pls. London, Cassell. (Later edns to 1898.)

— (1885). *Elementary text book of entomology*, viii, 240 pp., 87 pls. London, Sonnenschein.

— (1896). *A hand-book to the order Lepidoptera 1–3: Butterflies*, 261 pp., 37 col. pls. London, Lloyd's Natural History.

Knaggs, H. G. (1869a). *The lepidopterist's guide*, 122 pp. London, Van Voorst.

— (1870). *A list of macro-lepidoptera occurring in the neighbourhood of Folkestone*, 24 pp. Folkestone, Folkestone Natural History Society.

Knoch, A. W. (1781–83). *Beiträge zur Insectengeschnichte* (3 parts) **1** (1781): 10, 98, 3 pp., 6 col. pls; **2** (1782): 8, 102, 2 pp., 7 col. pls; **3** (1783): 2, 138, 2 pp., 6 col. pls. Leipzig, Schwickert.

Lankester, E. (Ed.) (1848). *The Correspondence of John Ray: consisting of selections from the philosophical letters published by Dr. Derham, and original letters of John Ray, in the collections of the British Museum*, xvi, 503 pp., 1 pl., 1 portr. London, The Ray Society.

Lewin, W. (1795). *The Papilios of Great Britain*, 97 pp., 46 col. pls. London, J. Johnson.

Linnaeus, C. (1746). *Fauna Suecica, &c.*, xxvi, 411 pp., 2 pls. Stockholm; edn 2, (1761), [50], 578 pp., 2 pls. Stockholm, Laurentii Salvii.

— (1753). *Species plantarum, &c.* **1**: xii, 560 pp.; **2**: iv, 622, [31] pp. Stockholm Laurentii Salvii.

— (1758). *Systema Naturae* **1**: *Regnum animale* (edn 10), [iv], 824 pp. Stockholm, Laurentii Salvii.

— (1767). *Systema Naturae* **1**, *Pars 2* (edn 12): Lepidoptera, pp. 774–900. Stockholm, Laurentii Salvii.

Lisney, A. A. (1960). *A bibliography of the British Lepidoptera, 1608–1799*, xviii, 315 pp., many portrs, text figs. London, Eyre & Spottiswoode.

Llewellyn, J. T. D. (1876). A foreign visitor (*Danais Archippus*). *Entomologist's mon.Mag.* **13**: 107–108.

Longstaff, G. B. (1912). *Butterfly-hunting in many lands, &c.*, xx, 729 pp., 16 col. pls, text figs. London, Longmans, Green.

MacGregor, A. (1994). The life, character and career of Sir Hans Sloane, pp. 11–44 (incl. Appendix 3: Sloane's museum at Chelsea, as described by Per Kalm, pp. 31–34). *In* MacGregor, A. (Ed.), *Sir Hans Sloane*, 292 pp., numerous illust. London, British Museum Press.

McLean, R. & McLean, A. (1988). *Benjamin Fawcett, engraver and colour printer, 1808–1893, with a list of his books and plates*, 196pp., incl. col. & b/w pls. Aldershot, Scolar Press.

Marren, P. (1995). *The new naturalists*, 304 pp., 16 col. pls, numerous text illust. (New Naturalist No. 82). London, Collins.

— (1998). A short history of butterfly-collecting in Britain. *Brit. Wildlife* **9**: 362–370.

Martin, G. (1978). *Hieronymus Bosch.* London, Thames & Hudson.

Martyn, W. F. (1785). *A new dictionary of natural history: or, compleat universal display of animated nature, &c.*, 2 vols, illust, 100 col. pls [some by Moses Harris]. London, Harrison.

Masó. A. & Pijoan, M. (1997). *Observar mariposas*, 317 pp., numerous col. illust. Barcelona, Planeta.

Mays, R. [H.] (1978). *Henry Doubleday. The Epping naturalist*, 118 pp., 4 pls. Marlow, Precision Press.

— (Ed.) (1986). *The Aurelian, &c., by Moses Harris* (facsimile edn), 104 pp., 44 col. pls. London, Newnes Country Life Books.

Merrett, C. (1666). *Pinax rerum naturalium Britannicarum, continens vegetabilia, animalia et fossilia, in hac insula reperta inchoatus*, xxviii, 221 pp. London, Pulleyn.

Meyrick, E. (1895). *A handbook of British Lepidoptera*, vi, 843 pp., text figs. London, Macmillan. (Revd edn, [1928].)

Miller, S. H. & Skertchly, S. B. J. (1878). *The fenland: past and present*, xxxii, 649 pp., numerous illust. London, Longmans Green.

M[offet], T. (1599). *The silkwormes and their flies.* London

— (1634). *Insectorum sive minimorum animalium theatrum*, xx, 326 [4] pp., text illust. London. (Edn 2, 1658, in English: see Topsell, E.)

Moore, J. (1954). *The season of the year.* London, Collins.

Morris, F. O. (1853). *A history of British butterflies*, viii, 168 pp., 71 col. pls, 2 plain pls. London, Groombridge. (Edns 2, 1857; 3, 1870; 4, 1876; 5, 1879; 6, 1890; 7, 1893; 8, 1895; 9, 1904; 10, 1908; 8 additional pls after edn 4.)

— ([1859–]70). *A natural history of British moths*, 4 vols: 132 col. pls. London, Knox.

Müller, O. F. (1764). *Fauna insectorum Fridrichsdalina sive methodica descriptio insectorum agri Fridrichsdalensis, &c.* 8, xxiv, 96 pp. Hafnia [Copenhagen], Leipzig, Gleditsch.

Neal, E. (1994). *The badger man, memoirs of a biologist*, x, 274 pp. Ely, Providence Press.

Neave, S. A. (1933). *The centenary history of the Entomological Society of London, 1833–1933*, xlvi, 224 pp., 8 pls. London, Entomological Society of London.

Newman, E. (1835). *The grammar of entomology*, xvi, 304 pp., 4 pls (2 col.). London, Westley & Davis.

— (1841). *A familiar introduction to the history of insects* [new edn of *The grammar of entomology*], xiv, 288 pp., illust. London, Van Voorst.

[—] ([1857]). *The insect hunters; or, entomology in verse*, viii, 86pp. London, Newman.

— (1860). *A natural history of all the British butterflies, &c.*, 24 pp., text illust. London, Judd & Glass. (Originally published in *Young England.*)

— (1869). *An illustrated natural history of British moths, &c.*, viii, 486 pp., many text illust. London, Tweedie.

— (1871). *The illustrated natural history of British butterflies*, xvi, 176 pp., many text figs. London, Tweedie.

Newman, L. H. (1967). *Living with butterflies*, ix, 228 pp., 63 pls. London, John Baker.

Newman, L. W. & Leeds, H. A. (1913). *Text book of British butterflies and moths*, [iv], 216, [i] pp. St Albans, Gibbs & Bamforth.

[Newman, T. P.] (1876). *Memoir of the life and works of Edward Newman*, 32 pp., portr. London, Van Voorst.

Ochsenheimer, F. (1808). Falter oder Tagschmetterlinge, p. 230. *In* Ochsenheimer, F. & Treitschke, F. (1807–1834). *Die Schmetterlinge von Europa* **1**(2): 240, 30 pp. Leipzig, Er. Fleischer. (Treitschke completed the 10-volume work after the death of Ochsenheimer in 1822.)

Paget, C. J. & Paget, J. (1834). *Sketch of the natural history of Yarmouth and its neighbourhood containing catalogues of the species of animals ... and plants at present known*, xxxii, 88 pp. London & Yarmouth, Skill.

Pallas, P. S. (1771). *Reise durch verschiedene Provinzen des Russischen Reiches in den Jahren 1768–1774*, **1**: 471. Petersburg, Akadem. Buchhandl.

Palmer, M. G. (1946). Butterflies and moths, pp. 67–112. *In.* Palmer, M.G. (Ed.) *The fauna and flora of the Ilfracombe district of North Devon*, xii, 266 pp., 8 pls, 1 map. Exeter, Ilfracombe Field Club.

Petiver, J. (1695–1703). *Musei Petiveriani centuria prima–decima, rariora naturae continens, &c.*, 93 pp., 2 pls. London, printed for the author.

--- (*c.*1700). *Brief directions for the easie making and preserving collections of all natural curiosities*, 1-sided folio. London, printed for the author.

--- (1702–06). *Gazophylacii naturae & artis: decas prima–decas decima*, 12 pp., 100 pls. London, printed for the author.

--- (1717). *Papilionum Britanniae icones*, 2 pp., 6 pls (hand-col. in some copies). London, printed for the author.

--- (1767). *J. Petiveri opera historiam naturalem spectantia; or Gazophylacium, &c. to which are now added seventeen curious tracts, ... which completes all he ever wrote upon natural history*, 2 vols. London, Millan.

Pickard, H. A. (1860). Singular, considering the Season. *Entomologist's Wkly Intell.* **8**: 171–172.

— & members of the Entomological Societies of Oxford and Cambridge (1858). *An accentuated list of the British Lepidoptera with hints on the derivation of the names*, xliv, [iv], 118 pp. London, Van Voorst.

Pickard-Cambridge, A. W. (1918). *Memoir of the Reverend Octavius Pickard-Cambridge by his son*, 96 pp., 9 pls. Oxford, published privately.

Pickard-Cambridge, O. (1885). *Lycaena argiades* Pallas – A butterfly new to the British fauna. *Entomologist* **18**: 249–252.

Poda, N. (1761). *Insecta musei Graecensis, &c.*, [vii], 127, [12] pp., 2 pls. Graecii.

Postlethwayt, Malachy (1751–55). *Universal dictionary of trade and commerce* (2 vols). London, Knapton.

Poulton, E. B. (1890). *The colours of animals, their meaning and use, especially considered in the case of insects, &c.*, xiii, 360 pp., 1 col. pl., 1 table, text figs. London, Kegan Paul.

Pratt, C. R. (1981). *A history of the butterflies and moths of Sussex*, 356 pp., 8 col. pls, text figs, maps. Brighton, Booth Museum of Natural History.

Quarrell, W. H. & Mare, M. (Eds.) (1934). *London in 1710: from the Travels of Zacharias Conrad Von Uffenbach.* London.

Raven, C. E. (1942). *John Ray, naturalist: his life and works,* xx, 506 pp., 1 pl. Cambridge, Cambridge University Press. (edn 2, 1950)

Ray, J. (1660). *Catalogus plantarum circa Cantabrigiam nascentium, &c.* [Flora of Cambridge]. Cambridge, Field.

— (1704). *Methodus insectorum,* [iv], 16 pp. London, Royal Society.

— (1710). *Historia Insectorum ... opus posthumum, &c.* [includes *Appendix de scarabaeis Britannicis* by Martin Lister], xv, 400 pp. London, Royal Society.

— (1718). (Ed. W. Derham). *Philosophical letters between the late learned Mr. Ray and several of his Ingenious correspondents, &c,* [vi], 376, [10] pp. London, Royal Society.

Reaumur, R. A. F. de (1734–42). *Mémoires pour servir à l'histoire des insectes,* 6 vols, illust. Paris, Imprim. Royale.

Rennie, J. (1830). *Insect architecture,* xii, 420 pp., text figs. London, Knight. (Several later edns.)

— (1830). *Insect transformations,* xii, 420 pp., text figs. London, Knight.

— (1832). *A conspectus of the butterflies and moths found in Britain,* xxxvii, [iv], 287 pp. London, Orr.

Riley, N. D. (1964a). *The Department of Entomology of the British Museum (Natural History) 1904–1964. A brief historical sketch,* 48 pp. XII International Congress of Entomology, London.

— (1975). *A field guide to the butterflies of the West Indies,* 224 pp., illust. London, Collins.

Rothschild, M. (1979). *Nathaniel Charles Rothschild 1877–1923.* Cambridge, published privately.

— (1983). *Dear Lord Rothschild,* xx, [iv], 398 pp., 12 col. pls, 92 b/w pls, map. London, Hutchinson.

— & Marren, P. (1997). *Rothschild's reserves: time and fragile nature,* xvi, 242 pp., incl. 232 figs (some in col.). Rehovot, Balaban; Colchester, Harley Books.

Rothschild, W. & Jordan, K. (1895). A revision of the *Papilios* of the eastern hemisphere, exclusive of Africa. *Novit.zool.* **2**: 167–463, pl. 6.

— & — (1903). A revision of the lepidopterous family Sphingidae. *Novit.zool.* **9**, Suppl.: cxxv, 972 pp., 67 pls.

Rottemburg, S. A. von (1775). Anmerkungen zu den Hufnagelschen Tabellen der Schmetterlinge. *Naturforscher, Halle* **1775** (6): 1–34 (Diurna).

Rye, A. (1970). *Gilbert White & his Selborne,* 272 pp. London, Kimber.

St John, J. S. (1885). *Lycaena argiades* Pallas in Somerset. *Entomologist* **18**: 292–293.

Sabbagh, K. (1999). *A Rum affair. How botany's 'Piltdown Man' was unmasked,* 224 pp., 11 mono. pls. Harmondsworth, Allen Lane.

Samouelle, G. (1819). *The entomologist's useful compendium; or an introduction to the knowledge of British insects,* 496 pp., 12 pls. London, Thomas Boys.

— (1826). *General directions for collecting and preserving exotic insects and crustacea, &c.,* 70 pp., 4 col. pls, col. frontis. London.

Sassoon, S. (1938). *The old century and seven more years.* London, Faber and Faber.

Sawford, B. (1987). *The butterflies of Hertfordshire,* xii, 195 pp., 22 col. pls, 16 figs, 4 tables, maps. Ware, Castlemead.

Scopoli, G. A. (1763). *Entomologia Carniolica exhibens insecta Carnioliae indigena, &c., ... methodo Linneana,* [xxxvi], 420 pp., 43 pls. Austria, Trattner.

Sitwell, E. (1933). *English Eccentrics.* London, Faber and Faber.

Sloane, H. (1707–25). *A Voyage to the Islands Madera, Barbados, Nieves, S. Christophers and Jamaica, with the natural history of the herbs and trees, four-footed beasts, fishes, birds, insects, reptiles, &c., of the last of those islands, &c.,* 2 vols. illust. London.

Smiles, S. (1877). *Life of a Scotch naturalist: Thomas Edward. Appendix – selections from the Fauna of Banffshire,* xix, 438 pp., 6 pls, portr., text figs. London, John Murray. (New edn, 1891.)

Smith, A. Z. (1986). *A history of the Hope entomological collections in the University Museum, Oxford,* [xiv], 172 pp., 17 pls. Oxford, Oxford University Press.

Smith, C. H. (1842). Memoir of Dru Drury. In *Introduction to Mammalia, chiefly with reference to the principal families not described in the former volumes,* 313 pp., 31 col. pls, 1 portr. (Jardine's Naturalist's Library, vol. xxv). Edinburgh, Lizars.

Smith, F. H. N. (1997). *The moths and butterflies of Cornwall and the Isles of Scilly,* xiv, 434 pp., 32 col. pls. Wallingford, Gem Publishing.

Smith, J. E. (1791). Introductory discourse on the rise and progress of natural history. *Trans.Linn.Soc.Lond.* **1**: 1–56.

South, R. (1884). *'The Entomologist' synonymic list of the British Lepidoptera, &c.,* vi, 40 pp. London.

—— (1906). *The butterflies of the British Isles,* x, 204 pp., 127 col. pls, 27 text figs. London, Warne. (Several later edns.)

—— (1907–08). *The moths of the British Isles,* 2 vols: **1**, vi, 343 pp., 159 col. & b/w pls; **2**, vi, 376 pp., 159 col. & b/w pls. London, Warne. (Several later edns.)

Sowerby, J. ([1804]–06). *The British miscellany: or coloured figures of new, rare, or little known animal subjects; many not before ascertained to be inhabitants of the British Isles; and chiefly in the possession of the author,* 2 vols, 76 col. pls with text. London.

Stainton, H. T. (1852). *The entomologist's companion; being a guide to the collection of micro-lepidoptera,* iv, 75 pp. London, Van Voorst. (Revd enlarged edn, 1954.)

—— (1854). *Insecta Britannica. Lepidoptera: Tineina,* viii, 313 pp., 10 pls. London, Reeve.

—— (1855–73). *The natural history of the Tineina,* 13 vols, 104 col. pls. London, Van Voorst.

—— (1857–59). *A manual of British butterflies and moths* **1**: xii, 338 pp., b/w illust.; **2**: xii, 480 pp., b/w illust. London, Van Voorst.

—— (1867). *British butterflies and moths, an introduction to the study of our native Lepidoptera,* xii, 292 pp., 16 col. pls. London, Reeve.

Stearn, W. T. (1965). Adrian Hardy Haworth 1768–1833. Biographical introduction to Haworth's *Complete Works on Succulent Plants,* 80 pp. (facsimile edn). London, Gregg Press.

—— (1981). *The Natural History Museum at South Kensington. A history of the British Museum (Natural History) 1753–1980,* xxiv, 414 pp., 88 b/w pls, 31 text figs. London, Heinemann.

Stephens, J. F. ([1827]–28). *Illustrations of British entomology – (Haustellata)* **1**: 152 pp., 12 col. pls. London, Baldwin & Cradock.

—— (1829). *A systematic catalogue of British insects* (2 pts in 1 vol.), xxxiv, [ii], 416 + 388 pp. London, Baldwin & Cradock.

—— (1856). *List of specimens of British animals in the collection of the British Museum. 5: Lepidoptera,* xii, 353pp. London, British Museum.

Stewart, A. M. (1912). *British butterflies,* viii, 88pp., 16 pls (incl 8 col. pls). London, Black. (Reissued many times.)

Swainson, W. (1822). *The naturalist's guide for collecting and preserving all subjects of natural history and botany,* viii, 72 pp., 2 pls. London, Baldwin & Cradock.

—— (1834). *A preliminary discourse on the study of natural history,* viii, 462 pp. London, Longman, Rees, Orme, Brown, Green & Longman.

—— (1840). *Taxidermy; with the biography of zoologists and notices of their works,* 392 pp., text figs. London, Longman.

Thomas, J. [A.] & Lewington, R. (1991). *The butterflies of Britain and Ireland,* 224 pp., numerous col. figs. London, Dorling Kindersley.

— & Webb, N. (1984). *Butterflies of Dorset*, 128 pp., 16 col. pls, 8 text figs, 3 tables, maps. Dorchester, Friary Press.

Thompson, E. P. (1845). *The note-book of a naturalist*, xii, 275 pp. London, Smith Elder.

Thomson, G. (1980). *Butterflies of Scotland*, xvii, 267 pp., 39 pls. (incl. 8 col.), 97 text figs (incl. maps). London, Croom Helm.

Tolman, T. (1997). *Butterflies of Britain and Europe*, 320 pp., 106 double-page col. pls (A, B, 1–104). London, Collins.

Topsell, E. (1658). *The history of four-footed beasts and serpents ... collected out of the writings of C. Gesner and other authors by E. Topsel. Whereunto is now added The Theater of Insects by T. Muffet &c.*, (2 vols in English), **1**: [xiv], 818, [6] pp.; **2**: [xii], 889–1130 [6] pp., numerous text figs. London. (Facsimile in 3 vols (1967). New York, Da Capo Press.)

Trimen, H. & Dyer, W. T. Thiselton (1869). *Flora of Middlesex: a topographical and historical account of the plants found in the county, with sketches of its physical geography and climate, &c.*, xli, [iii], 428 pp., 1 map. London, Hardwicke.

Turner, D. (Ed.) (1835). *Extracts from the literary and scientific correspondence of Richard Richardson ... illustrative of the state and progress of botany*, lxvi, 451 pp. Yarmouth.

Turpyn, R. (1540). *Chronicles of Calais in the reigns of Henry VII and Henry VIII to the year 1540.* MSS in the British Museum (Reprinted by the Camden Society, 1846).

Tutt, J. W. (1891–92). *The British Noctuae and their varieties*, 4 vols. London, Sonnenschein.

— (1896a). *British butterflies*, 469 pp., 10 pls, text figs. London, George Gill.

— (1896b). *British moths*, xii, 368 pp., 12 col. pls. London, Routledge.

— (1899–1914). *A natural history of the British Lepidoptera*, 9 vols (Vols 6 & 7 not published): **1** (1899), [iv], 560 pp.; **2** (1900), viii, 584 pp., 7 pls; **3** (1902), xi, 558 pp.; **4** (1904), xvii, 535 pp., 2 pls; **5** (1906), xiii, 558 pp., 4 pls; **8** (1905–06), iii, 479 pp., 20 pls; **9** (1907–08), x, 494 pp., 28 pls; **10** (1908–09), viii, 410 pp., 53 pls; **11** (1910–14) (Wheeler, G., Ed.), v, 373 pp., 45 pls. London.

— (1901–05). *Practical hints for the field lepidopterist*, Parts 1–3: **1** (1901), 106 pp.; **2** (1902), 143 pp.; **3** (1905), 166 pp., 7 pls, Index. London, Eliot Stock.

— (1905–14). *A natural history of the British butterflies*, Vols **1– 4** (reissue of Vols **8–11** of *A natural history of the British Lepidoptera*). Vols **1–3** London, Swan Sonnenschein; Vol. **4** (Wheeler, G., Ed.) London, Elliot Stock.

Vane-Wright, R. I. & Ackery, P. R. (Eds) (1984). *The biology of butterflies*, xxiv, 429 pp. (Symposium of the Royal Entomological Society of London, No. 11). London, Academic Press.

Wallace, A. R. (1869). *The Malay Archipelago: the land of the orang-utan and the bird of paradise*, 2 vols. London, Macmillan.

Watkins, W. (1884). *Directions for collecting, rearing and preserving British and foreign butterflies and moths*, 36 pp., text figs. Croydon, Crystal Palace Insectarium.

Westwood, J. O. (1838). *The entomologist's text-book; an introduction to the natural history, structure, physiology and classification of insects, including the Crustacea and Arachnida*, x, 432 pp., 5 col. pls, text figs. London, Orr.

— (1839–40). *An introduction to the modern classification of insects; founded on the natural habits and corresponding organisation of the different families*, 2 vols [Vol. 2 contains synopsis of the genera of British insects], 1 col. pl., text figs. London, Longman.

— (1855). *The Butterflies of Great Britain, with transformations delineated and described*, xl, 140 pp., 21 col. pls. London, Orr.

White, F. B. W. (1871). Fauna Perthensis, Pt. 1. Lepidoptera. *Trans.Proc.Perthsh.Soc.nat.Sci.* **1871**: 1–31.

White, G. (1789). *The natural history and antiquities of Selborne in the county of Southampton*,

2 vols: v, 468 pp., 9 pls. London.
Wilkes, B. (1742a). *Twelve new designs of English butterflies*, dedication pl., 12 col. pls. London. (Facsimile edn, 1982. *Benjamin Wilkes: the British Aurelian*, with introduction by R. W. Wilkinson, 11 pp. Faringdon, Classey.)
— (1742b). 'Supplement' to *Twelve new designs* [Directions for making a collection of Moths and Butterflies], large folio broadside. (Included in facsimile edn of *Twelve new designs* in *Benjamin Wilkes: the British Aurelian*, 1982. Faringdon, Classey.)
— (1747–49). *The English moths and butterflies, &c.*, xxvi, 8, [22], 64, [4] pp., 120 col. pls. London, printed for the author.
— (1773). *One hundred and twenty copper-plates of English moths and butterflies, &c.*, xxiv, 63, 8 pp., 120 col. pls. London, printed for the author.
Wilkinson, R. S. (1966a). William Vernon, entomologist and botanist. *Entomologist's Rec.J.Var.* **78**: 115–121.
— (1966d). English Entomological Methods in the Seventeenth and Eighteenth centuries. Part 2. Wilkes and Dutfield. *Ibid.* **78**: 285–292.
— (1977a). Evidence concerning the death of Eleazar Albin. *Ibid.* **89**: 101.
Williams, C. B. (1930). *The migration of butterflies*, xi, 473 pp., 71 figs, tables. Edinburgh, Oliver & Boyd.
— (1958). *Insect migration*, xiii, 235 pp., 8 col. pls, 16 mono. pls, 49 figs, tables. (New Naturalist No. 36). London, Collins.
Wollaston, T. V. (1854). *Insecta Maderensia, being an account of the insects of the islands of the Madeira group*, xliii, 634 pp., 13 col. pls. London, Van Voorst.
— (1856). *On the variation of species, with especial reference to the Insecta, followed by an inquiry into the nature of genera*, vi, [ii], 206 pp. London, Van Voorst.
— (1868). *Lyra Devoniensis*. London, Macmillan.
Wood, J. G. (1857). *Common objects of the country*, iv, 132 pp., 12 col. pls. London, Routledge.
— (1871). *Insects at home, being a popular account of British insects, their structure, habits and transformations*, xx, 670 pp., 1 col. pl., numerous plain pls, text figs. London, Longmans, Green. (Several edns to 1892.)
Wood, W. (1839); *Index entomologicus; or a complete illustrated catalogue ... of the lepidopterous insects of Great Britain*, xii, 266 pp., 54 col. pls (1612 col. figs). London, Ward. (New edn, 1852 (Ed. J. O. Westwood), viii, ii, 298, 21 pp., 59 col. pls. (1934 col. figs); Edn 3, 1854).
Young, M. R. (1997). *The natural history of moths*, xiv, 271 pp., 16 col. pls. London, Poyser.

JOURNAL TITLES AND THEIR ABBREVIATIONS

Journals cited in the Footnotes and Bibliography are listed below, with their full titles, dates of publication, and abbreviations, and with the titles that have preceded and superseded them, often merely a change of name. The abbreviations are from *Serial Publications in the British Museum (Natural History) Library* (3 vols), edn 3, 1980. For publications not in that work, abbreviations are based on the same principles. For details of other provincial and regional journals, see Appendix 2.

Abstracts of Proceedings of the Royal Entomological Society of London (1935–36) (*Abstr.Proc.R.ent.Soc.*) (continued as *Proc.R.ent.Soc.*)

Abstracts of Proceedings of the South London Entomological & Natural History Society (1885–96) (*Abstr.Proc.S.Lond.ent.nat.Hist.Soc.*) (formerly *Rep.S.Lond.ent.nat.Hist.Soc.*; continued as *Proc.S.Lond.ent.nat.Hist.Soc.*)

Antenna (1977>) (formerly *Proc.R.ent.Soc.Lond.*, Series C)
Archives of Disease in Childhood (*Archs Dis.Childhood*)
Archives of Natural History (1981>) (*Archs nat.Hist.*) (formerly *J.Soc.Biblphy nat.Hist.*)
Atropos (1996>) (*Atropos*)
Biographical Memoirs of the Fellows of the Royal Society (1955>) (*Biogr.Mem.Fellows R.Soc.*)
British Journal of Entomology & Natural History (1988>) (*Br.J.ent.nat.Hist.*) (formerly *Proc.Trans.Br.ent.nat.Hist.Soc.*)
British Medical Journal (1908>) (*Br.med.J.*)
British Naturalist (1891–94) (*Br.Nat.*) (formerly *Young Nat.*)
Bulletin of the Amateur Entomologist's Society (1939>) (*Bull.amat.Ent.Soc.*) (formerly *Entomologists' Bulletin*)
Bulletin of Zoological Nomenclature (1943>) (*Bull.zool.Nom.*)
Butterfly Conservation News (1991>) (*Butterfly Cons.News*) (formerly *Br.Butterfly Cons.Soc.News*)
Canadian Entomologist (1869>) (*Can.Ent.*)
Ecological Entomology (1976>) (*Ecol.Ent.*) (formerly *Trans.R.ent.Soc.Lond.*)
Entomological Magazine (London) (1833–38) (*Ent.Mag.*)
Entomologist (1840–42; not issued 1843–63;1864–1973 [incorporated *J.Soc.Br.Ent.*, 1960–73)]; not issued 1973–77; revived 1978–97 (*Entomologist*)
Entomologist's Annual (1855–74) (*Entomologist's Annu.*)
Entomologists' Bulletin (1937–38) (*Entomologists' Bull.*) (continued as *Bull.amat.Ent.Soc.*)
Entomologist's Gazette (1950>) (*Entomologist's Gaz.*)
Entomologist's Monthly Magazine (1864>) (*Entomologist's mon.Mag.*)
Entomologist's Record & Journal of Variation (1890>) (*Entomologist's Rec.J.Var.*)
Entomologist's Weekly Intelligencer (1856–61) (*Entomologist's Wkly Intell.*)
Essex Moth Group Newsletter (1996>)
Essex Naturalist (1887>) (*Essex Nat.*) (formerly *Trans.Essex Fld Club*)
Gentleman's Magazine (1731–1833) (*Gentleman's Mag.*)
Geological Curator (1992>) (*Geol.Curator*)
Journal of Entomology, Series A & B (Royal Entomological Society of London), (1971–76) (formerly *Proc.R.ent.Soc.Lond*, Series A & B; continued as *Physiol.Ent.* [A] and *Syst.Ent.* [B])
Journal of the British Entomological & Natural History Society (*J.Br.ent.nat.Hist.Soc.*) (1988>) (formerly *Proc.Trans.Br.ent.nat.Hist.Soc.*)
Journal of the Entomological Society of the South of England. Southampton (1932–33) (*J.ent.Soc.S.Engl.*) (formerly *Trans.ent.Soc.S.Engl.*; continued as *J.Soc.Br.Ent.* and *Trans.Soc.Br.Ent.*)
Journal of the Linnean Society of London (Zool.) (1867–68) (formerly *J.Proc.Linn.Soc.London.*) (continued as *Zool.J.Linn.Soc.*)
Journal of the Proceedings of the Linnean Society. London (1855–65) (*J.Proc.Linn.Soc.Lond.*) (continued as *J.Linn.Soc.Zool.*)
Journal of the Society for the Bibliography of Natural History (1936–80) (*J.Soc.Biblphy nat.Hist.*) (continued as *Archs nat.Hist*)
Journal of the Society for British Entomology (Southampton) (1934–59) (*J.Soc.Br.Ent.*, incorporated in *The Entomologist* (1960–73)) (formerly *J.ent.Soc.S.Engl.*)
Lambillionea. Revue Mensuelle de l'Union des Entomologistes Belges. Bruxelles. (1927>) (*Lambillionea*)
Magazine of Natural History (1828–40) (*Mag.nat.Hist.*)

* N.B. From 1968–80, official abbreviation of journal omitted '*Trans.*', though journal title remained the same throughout.

Natural History Review (Dublin, 1853–65; London, 1861–65) (*Nat.Hist.Rev.*)

Naturalist (Hull, London) (1864/65; 1866/67; N.S. 1875>) (*Naturalist, Hull*)

Naturalist (Morris), London (1851–58) (*Naturalist (Morris)*)

New Naturalist, a Journal of British Natural History (1948–49) (*New Nat.*)

Nota Lepidopterologica (1977>) (*Nota lepid.*)

Physiological Entomology (1976>) (*Physiol.Ent.*) (formerly *J.Ent.*, Series A. Royal Ent.Soc. of Lond.)

Proceedings and Transactions of the British Entomological & Natural History Society (1968–87) (*Proc.Trans.Br.ent.nat.Hist.Soc.*)* (formerly *Proc.Trans.S.Lond.ent.nat.Hist.Soc.*; continued as *J.Br.ent.nat.Hist.Soc.*)

Proceedings and Transactions of the South London Entomological & Natural History Society (1933–67) (*Proc.Trans.S.Lond.ent.nat.Hist.Soc.*)* (formerly *Trans.Proc.S.Lond.ent.nat.Hist.Soc.*; continued as *Proc.Trans.S.Lond.ent.nat.Hist.Soc.*)

Proceedings of the Entomological Society of London (1926–33) (*Proc.ent.Soc.Lond.*)

Proceedings of the Royal Entomological Society of London (1933–36) (*Proc.R.ent.Soc.Lond.*)

Proceedings of the Royal Entomological Society of London, Series A (General Entomology) (1936–70) (*Proc.R.ent.Soc.Lond.*, Series A) (formerly *Proc.R.ent.Soc.Lond.*; continued as *Physiol.Ent.*)

Proceedings of the Royal Entomological Society of London, Series B (Taxonomy) (1936–70) (*Proc.R.ent.Soc.Lond.*, Series B) (formerly *Stylops*; continued as *Syst.Ent.*)

Proceedings of the Royal Entomological Society of London, Series C (Meetings) 1936–76) (*Proc.R.ent.Soc.Lond.*, Series C) (formerly *Abstr.Proc.R.ent.Soc.*; continued as *Antenna*)

Proceedings of the South London Entomological & Natural History Society (1897–1932) (*Proc.S.Lond.ent.nat.Hist.Soc.*) (formerly *Abstr.Proc.S.Lond.ent.nat.Hist.Soc.*; continued as *Trans.Proc.S.Lond.ent.nat.Hist.Soc.*)

Report of the South London Entomological & Natural History Society (1879–84) (*Rep.S.Lond.ent.nat.Hist.Soc.*) (continued as *Abstr.Proc.S.Lond.ent.nat.Hist.Soc.*)

Scottish Naturalist (1871–1952, with some gaps) (*Scott.Nat.*)

Stylops. Journal of Taxonomic Entomology (1932–35) (*Stylops*) (continued as *Proc.R.ent.Soc.*, Series B)

Substitute, London (1856–57) (*Substitute*)

Symposia of the Royal Entomological Society of London (1961>) (*Symp.R.ent.Soc.Lond.*)

Systematic Entomology (1976>) (*Syst.Ent.*) (formerly *J.Ent.*, Series B. Royal Ent.Soc.Lond.)

Transactions and Proceedings of the South London Entomological & Natural History Society (1932–33) (*Trans.Proc.S.Lond.ent.nat.Hist.Soc.*) (formerly *Proc.S.Lond.ent.nat.Hist.Soc.*; continued as *Proc.Trans.S.Lond.ent.nat.Hist.Soc.*)

Transactions of the Entomological Society of London (1807–12; 1834–1933) (*Trans.ent.Soc.Lond.*) (continued as *Trans.R.ent.Soc.Lond.*)

Transactions of the Entomological Society of [Hampshire and] the South of England. Southampton. (1929–32) (*Trans.ent.Soc.S.Engl.*) (formerly *Trans.Hamps.ent.Soc.*; continued as *Trans.Soc.Br.Ent.*)

Transactions of the Hampshire Entomological Society (1924–28) (*Trans.Hamps.ent.Soc.*) (continued as *Trans.ent.Soc.S.Engl.*)

Transactions of the Lincolnshire Naturalists' Union (1893–94; 1905>) (*Trans.Lincs.Nat.Un.*)

Transactions of the Linnean Society (1791–1875; Zool. 2nd series, 1875–1926; 3rd series, 1939–55) (*Trans.Linn.Soc.Lond.*)

Transactions of the Norfolk & Norwich Naturalists' Society (1869>) (*Trans.Norfolk Norwich Nat.Soc.*)

* N.B. From 1952–67, official abbreviation of journal omitted '*Trans.*', though journal title remained the same throughout.

Transactions of the Royal Entomological Society of London (1933–75) (*Trans.R.ent.Soc.Lond.*) (formerly *Trans.ent.Soc.Lond.*; replaced by *Ecol.Ent.*)

Transactions of the Society for British Entomology (1934–71) (*Trans.Soc.Br.Ent.*) (formerly *J.ent.Soc.S.Engl.*)

Transactions of the Tyneside Naturalists' Field Club (1846–64) (*Trans.Tyneside Nat.Fld Cl.*)

Transactions of the Woolhope Naturalists' Field Club. Hereford. (1852>) (*Trans.Woolhope Nat.Fld.Club*)

Young Naturalist (1879–1890) (*Young Nat.*) (continued as *Br.Nat.*)

Zoological Journal of the Linnean Society of London (1969>) (*Zool.J.Linn.Soc.*) (formerly *J.Linn.Soc.Zool.*)

Zoologist (1843–1916) (*Zoologist*)

INDEX

The index covers Chapters 1 to 5 and the appendices, though past common names of butterflies as listed in Appendix 1 are included here only where they have been mentioned independently within the main text. Page references in bold type under the names of individuals and under present common names of butterflies indicate the main entries from Chapter 4 and Appendix 1, respectively. Figure and plate numbers are shown in parentheses.

Abbot, Charles **124**, 139, 241, 243, 256; *Flora Bedfordiensis* 124
Abbot, John **123**(fig.**48**); *The Natural History of the Rarer Lepidopterous Insects of Georgia* 123
Abbot's Fritillary 335
aberrations 175, 202–203, 204, 211, 215, 221, 230, 265(pl.19), 267, 345, 369, 370
Acentria (Pyralidae) 169
acis, Papilio 297
acis, Polyommatus 50, 297, 299(pl.27)
acteon, Papilio 245
acteon, Thymelicus 245–9(pl.13), 378
adippe, Argynnis 194, 338, 390–91
Adkin, Robert 46(fig.10), **186–7**(fig.**90**), 189, 209
Admiral 58, 109, 112
Adonis Blue 131, 365, 386
aegeria, Pararge 392–3(fig.157)
aegon, Lycaena 293
AES (Amateur Entomologists' Society) 232, 233, 372, 406, 408; Conservation Committee 371
aethiops, Erebia 295, 345, 393
agestis, Aricia 295, 297, 385
aglaja, Argynnis 60, 194, 391(fig.156)
Albin, Eleazar 29, 32, 35, 60, 80, 105, **109–10**(fig.**40**), 268, 309–311, 318; *The Natural History of English Insects* 109–10; *The Natural History of Spiders and other Curious Insects* 106, 109(fig.40), 110
Albin, Elizabeth 109
Albin's Hampstead Eye 104, 110, 155, **309–311**(fig.**137**, pl.**30**), 327(pl.36), 387
albus minor, Papilio 379
alceae, Carcharodus 249(pl.14), 378
alceae, Papilio 249
alciphron, Lycaena 290
Alderman Butterfly 112
Aldrovandi, Ulysse, *De Animalibus Insectis* 98
Alexandra, Princess 206
alfacariensis, Colias 260–63(pl.18), 380
Allan, Philip Bertram Murray 124, 126, 134, 147, 162, 166, 177, 180, 203, 204, **218–19**(fig.**113**), 257, 284, 289, 298, 300, 327, 329–31, 336, 339, 345, 368; *A Moth-Hunter's Gossip* 218, 232; *Talking of Moths* 218; *Moths and Memories* 218, 232; *Leaves from a Moth-Hunter's Notebook* 218, 219
Allen, D. E. 55, 74, 84, 397, 398
alpinus, Papilio 249
amaryllis, Papilio 343
Amateur Entomologists' Society (AES) 232, 233, 372, 406, 408; Conservation Committee 371
American Painted Lady 114, 227, **316–17**(pl.**32**), 327(pl.36), 388
anachoreta, Clostera 168
Annals and Magazine of Natural History 400
antiopa, Nymphalis 321–7(fig.139, pls 34–5), 334, 389
antiopa, Papilio 321–3(fig.139, pl.34)
antiopa, Vanessa 324
'*antiopa* years' 324
Antrobus, Robert 271
ants 195, 207–208(fig.105), 305, 308, 309
aphrodite, Argynnis 155
Apollo 95, 148, **249–52**(pl.**15**), 327(pl.36), 379(fig.146)
apollo, Papilio 249
apollo, Parnassius 100, 249–52(pl.15), 379(fig.146)
apothecaries 55–6
Appleby, William 174
archippus, Danaus 346–51(pls 40–41)
arctica, Somatochlora 134
argiades, Everes 170, 293–5(fig.135), 385
argiades, Lycaena 293–5(fig.135)
argiades, Papilio 293
argiolus, Celastrina 386
argus, Plebejus 385
arion, Lycaena 306, 307
arion, Maculinea 300–309(pls 28–9), 386
arion, Papilio 300–301(pl.28, fig.136)
Armitage, Edward 367
Arran Argus 344
Arran Brown 251(pl.15), **343–6**(pl.**39**), 393–4
Arran, Isle of 85–6, 343
artaxerxes, Aricia 120, 295–7(pl.26), 385–6
artaxerxes, Hesperia 295–7(pl.26)
artists and illustrators 29, 32, 66, 109, 110–11, 112, 115–17, 120, 123, 130, 131, 136–7, 138–9, 157–9, 193, 278–9, 369; methods 110, 113; *see also* illustrations
Ash Pug moth 182
Ashmolean Natural History Society 44, 399
ashworthii, Xestia 152
Ashworth's Rustic moth 152
assembling 55
'At Home' invitations 39–41(fig.8), 85, 140, 142, 162, 163
atalanta, Vanessa 387–8(fig.151)
athalia, Mellicta 392
Atropos magazine 408
auctions 367–8(figs 144–5)
Aurelian Cabinet 35
'Aurelian Macaroni' 33(fig.3)
Aurelian Societies 29–35(fig.6), 56, 105, 111, 114, 117, 127–8, 280, 398(fig.159), 399; *Transactions* 27, 35–6
aurinia, Eurodryas 342, 392
Austin, G. 250, 254
Austin, Stephen 32, 398
australis, Colias 260–62(pl.18)
authenticity 50, 117, 124, 133–4, 144, 153, 216, 257, 265, 267, 276–7, 282, 288, 329–31, 332–4, 336–8, 343–6; *see also* fraud
avis, Callophrys 177
Azure Blue 299(pl.27)
Babington, C. C. 44
Backhouse, William 324
Bacon, Francis 27
Baird, A. 400
Baker, Henry 29, 32, 77, 110, 398
Banks, Sir Joseph 119, 127, 397
Barrett, C. G. 64, 74, **171–2**(fig.**78**), 245, 277–8, 288, 297, 308, 314, 317, 331, 370; *Lepidoptera of the British Islands* 171, 172, 254
Barrett, James Platt 94, **173–4**(fig.**80**)
Barrett's Marbled Coronet moth 171
Barrington, Daines 99
Barrington, Rupert 372
Barron, Charles 63
Batchelor, T. 330
Bates, H. W. 351
Bath White 28, 29, 108, 124, 241, **271–4**(fig.**133**, pl.**21**), 314, 382
Bause, I. F. 118(fig.44)
Baynes, E. S. A. 307
Beaufort, Duchess of 29, 109
Beautiful Painted Lady 316
Bedford Blue 299(pl.27)
beetles 80, 153, 197
Beirne, B. P. 367
Bell, Ephraim 398
Bella donna, Papilio 311, 312(fig.138)
bellargus, Lysandra 386
Bennett, A. W. 338–9
Bentinck, Lady Margaret Cavendish, later Duchess of Portland 29, 114
Berger, Lucien 260
Berger's Clouded Yellow **260–63**(pl.**18**), 295, 380
Berkenhout, John **121–2**(fig.**47**), 263, 286, 322; *Outlines of the Natural History of Great Britain and Ireland* 121, 256
Berridge, W. S. 366–7
Berwickshire Naturalists' Club 43–4, 400
Bethnal Green Museum 82
betulae, Thecla 282, 383
betularia, Biston 227–8
Biddle, Robert 281
Bignell beating tray 166
Bignell, George Carter **166–7**(fig.**74**), 305–306
bignellii, Apanteles (Hymenoptera) 166
bignellii, Cotesia (Hymenoptera) 166
bignellii, Mesoleius (Hymenoptera) 166
Birchall, Edwin 50
bird-watching 241, 348
birdwing butterfly 26
Black Hairstreak 138, **274–7**(pl.**22**), 384
Black Rustic moth 128
Black-veined Brown 346, 347(pl.40)
Black-veined White 58, 95, 109, 115, 210, **268–70**(fig.**132**), 380–81
Blackburn, T. 43, 402
Blackburn, T. and J. B. 402
Blackett, Walter 32, 398
blandina, Erebia 343, 344
Blathwayt, C. S. H. 274
Blathwayte, F. L. 273
Blomer, Charles 139–40, 317
blomeri, Discoloxia 138
Blomer's Rivulet moth 138
Bloomfield, E. N. 331
Bloxworth Blue 170, **293–5**(fig.**135**), 385
Bloxworth Snout moth 170
BM(NH) (British Museum (Natural History)) 183, 185, 186, 186, 207, 213, 219; *see also* British Museum; Natural History Museum, London
boeticus, Lampides 291–2, 384
boeticus, Papilio 291–2
Bolingbroke, Valezina, Lady 195, 196–7(fig.95, pl.10)
Bond, Frederick **155–6**(fig.**64**), 168, 171, 338
bondii, Chortodes 168
Bond's Wainscot moth 156
Bonhote, J. L. and Rothschild, N. C., *Harrow Butterflies and Moths* 213
Bosch, Hieronymus, *The Garden of Worldly Delights* 69

Botanical Magazine 117
Bouck, Baron J. 308, 365
Bowden and Altrincham Entomological Society 41, 402
Bowman, Mabel Jane Hart 195, 196(fig.95)
Bradley, J. D. 406
Branson, Andrew 407
brassicae, Pieris 381
Bree, C. R. 318
Bree, William T. 87, 133, 277, 284, 288, 298, 302, 329
Bretherton, Russell 226, **227(fig.120)**, 253, 261, 266, 281, 297, 317, 327, 345, 346, 368
Bright, Percy M. and Leeds, H. A., *Monograph of the British Aberrations of the Chalk-hill Blue* 211
Bright, Percy May **202(fig.99)**, 211, 368
Brighton Argus 291
brimstone 61, 63
Brimstone butterfly 28, 58, 59(pl.4), 66, 95, 101, 103, 109, 194, 222, 266–7(pl.20), 318, 380(fig.147)
Bristowe, W. S. 105, 106, 108, 110
British Association for the Advancement of Science 163, 207
British Butterfly Conservation Society 407
British Entomological and Natural History Society 220, 224, 227, 234, 345, 403, 407, 408; *see also* South London Entomological and Natural History Society
British Entomological Society *see* British Entomological and Natural History Society
British Journal of Entomology and Natural History 407
British Museum 28, 66, 76, 106, 137, 138, 141, 142, 338, 400; *A List of Lepidopterous Insects in the British Museum* 152; *see also* British Museum (Natural History) (BM(NH)); Natural History Museum, London
British Museum (Natural History) (BM(NH)) 183, 185, 186, 186, 207, 213, 219; *see also* British Museum; Natural History Museum, London
British Naturalist 167, 404
British Trust for Entomology Ltd 406
British Wildlife 407–408
'Brittish Eye-winged Butterflies' 309, 310(fig.137), 311
Brodie, N. S. 331
Brown Argus 295, 296(pl.26), 297, 385
Brown Blue 296(pl.26)
Brown Hairstreak 28, 58(pl.3), 274–5(pl.22), 282, 383
Brown, James 106
Brown, Thomas 302
Brown Whitespot 120, 295, 296(pl.26)
Brownsword, Elias 32, 398
bucephalus, Hesperia 124
Buckler, William **157–9(fig.66)**; *The Larvae of the British Butterflies and Moths* 158, 167, 307–308, 401
Bucks, Berks & Oxon Naturalists' Trust 224
Buckton, G. Bowdler 63
Buddle, Adam 29, 104, 107, 397; herbarium 58, 59(pl.4), 331
Bug Club 406
Bulletin of Zoological Nomenclature 222–3
Burr, Malcolm 189
Burrell, John 399
Burtt, E. T. 216–17
Bute, Isle of 343, 344
Butt Ekins, T. 328
Butterfly Conservation 244, 328, 371, 407, 408
Butterfly Conservation News 407
butterfly farms 187, 202, 209–210(figs 107–108), 231, 346, 403; *see also* rearing

'C. R.' *see* Russell, Sidney George Castle
Cabbage White 28, 109
cabinets 78–80
cadgers 89
caesia mananii, Hadena 159
Caius, John 95
c-album, Polygonia 160–61, 389(fig.155)
c-album, Vanessa 160–61
calida, Colias 260–61
Calke Abbey 183
Camberwell Beauty 26, 111, 131, 174, 252, 320, **321–7(fig.139, pls 34–5)**, 373, 389
Cambridge, University of 95, 108
camilla, Ladoga 194, 387
Campbell, John Lorne 216
Canterbury Journal 240
Capper, Samuel J. 45, 403
cardamines, Anthocaris 382–3
Cardinal, The **338–9**, 391
cardui, Cynthia 263, 311–16(fig.138, pl.31), 388(fig.152)
cardui, Papilio 311–16(fig.138, pl.31)
cardui virginiensis, Nymphalis 114, 316–17(pl.32)
Carpenter, G. D. Hale 188
Cary, Lucius, Viscount Falkland 25
cassiope, Hipparchia 134, 135(pl.7)
Castle Eden Argus 297
cataloguing 77, 82, 104, 139, 141, 164, 210
Cater, W. F.: *Butterflies and Late Loves* 201; *Love among the Butterflies* 201
Catesby, M. 397
Celsius, Anders 397
Chalk Hill Blue 87, 202, 204, 211, 212(pl.11), 386
Chalmers-Hunt, J. Michael 61, 191, 227, 252
Chamberlain, Neville 25
Chapman, Thomas Algernon 46(fig.10), **176–7(fig.82)**, 189, 189, 300, 308
Chapman's Blue 177
Chapman's Green Hairstreak 177
Charles II, King 397
Charlesworth, Edward 400
Charleton, William *see* Courten, William
Charlton, William *see* Courten, William
Charlton's Brimstone Butterfly 66(pl.6)
Chequered Skipper 124, 130, 195, **241–4(pl.12)**, 256, 368, 372, 377
Chequered Skipper, The (inn) 88
Chequered White 271
Chequers, The (inn) 88
Children, J. G. 37, 400
Christie, L. 404
chryseis, Lycaena 288
chryseis, Papilio 290–91
Churchill, Winston 25, 210
cimon, Papilio 297–8
Cinnabar moth 210
cinxia, Melitaea 339–42(pl.38), 392
cinxia, Papilio 339
City of London Entomological and Natural History Society 404–405
Clark, J. A. 404
Clarke, Sir Cyril 228, 229, 370
Classey, Eric 229, 232, 404, 406(fig.162)
Classey, Ivy 406, 407
Classey, Peter 407
classification 27–8, 35, 100, 116, 119, 121, 122, 142, 152, 164, 186; Copper butterflies 278, 286–8; Hairstreak butterflies 274–6
Cleopatra 64, **266–8(pl.20)**, 380
cleopatra, Gonepteryx 64, 266–8(pl.20), 380
cleopatra, Papilio 266
clergy 86
Clifden Blue 131
Clifden Nonpareil moth 111
Clouded Drab moth 166
Clouded Orange 264
Clouded Sulphur 258
Clouded Yellow 95, **263–6(fig.131, pl.19)**, 380
Clouded Yellows 95, 128, **257–66(pls 17–19, figs 130–31)**, 327; *see also* individual entries
Cockayne, Edward Alfred 204, **214–15(fig.111)**, 228, 325, 370
Code for Insect Collecting (Joint Committee for the Conservation of Insects) 371
Codling moth 111
coenosa, Laelia 140, 285
Coffee Houses 31(pl.2), 33, 108
Colchester Natural History Society 222
Coleman W. S. 74, 245, 250, 258, 264, 266, 267, 288, 291, 321, 322, 329; *British Butterflies* 251(pl.15)
Colenutt, George W. 405
collecting: bans 372; boxes 67; clothing 60–61, 201; methods 57–8(fig.16), 84, 111, 126, 143, 155, 168, 297; overseas 57–8, 66, 114–15, 117–18, 200, 206, 373, 403; trends 367–8, 371; trips 51–3, 55–6(figs 14, 15), 60, 83–6, 365; *see also* over-collecting
Collier, A. E. 277
Collins, G. A. 277, 327–8
Collins, W. W. 271–3
Collinson, Peter 32, 102, 398
coloration 164, 191, 228, 244; *see also* discoloration
Comma 95, 101, 102(fig.38), 148, 160–61, 175, 389(fig.155)
comma, Hesperia 378
commercialism 132, 284, 285, 350, 367–8, 372
Common Blue 95, 386(fig.150)
conservation 95, 172, 214, 232, 234, 241, 244, 285, 328, 366–7, 371–3, 407
Conservation Bulletin see *Invertebrate Conservation News*
Conservation of Wild Creatures and Wild Plants Act 371
Constable, Philip, junior 32, 398
Continental Swallowtail 373
Cook, James 117, 118, 119
Coombe Wood, Surrey 111
Coppers **277–91(pls 23–5)**; *see also* individual entries
Corbin, C. G. 337
coridon, Lysandra 386
cork 67, 75, 76, 80
Cossonidae (Coleoptera) 161
Cotteswold Naturalists' Field Club 44, 45, 401
Courten, William 66
Cousin German moth 134
Cox, William, *The Gentleman's Recreation* 69
Crallan, Thomas 348
craneflies 141
crataegi, Aporia 268–70(fig.132), 380–81
crataegi, Papilio 268–9
Crewe, Henry Harpur 166, **181–4(fig.87)**
Crewe, Henry Robert 181
Crewe, Sir George, 8th baronet of Calke 181
Crewe, Sir Vauncey Harpur, 10th baronet of Calke 181, **182–4(fig.88, pl.9)**; *Natural History of Calke and Warslow* 183
Cribb, Peter William **232–3(fig.125)**, 342
Cribb, Phillip 233, 331
croceus, Colias 260, 263–6(fig.131, pl.19), 380
croceus, Papilio 258, 263–4
croesus, Ornithoptera 26
Cross, L. B. 188
Crotch, W. D. 64, 76, 282, 284, 365
Cruickshank, George 131
Crystal Palace Insectarium 403
Curtis, C. M. 134
Curtis, John 44, 110, 111, 112, **138–9(fig.54)**, 142, 250, 267, 268, 276–7, 288, 290, 399, 400; *British Entomology* 138,

139, 246, 247(pl.13), 255(pl.16), 257, 274; *Guide to an Arrangement of British Insects* 139
Curtis, W. Parkinson 405
Curtis, William 117
Curtis's Botanical Magazine 182

da Costa, Emanuel Mendes 105, 106, 280, 397, 398
Daily Advertiser 33
Dale, Charles William 155, 281, 285, 306; *Lepidoptera of Dorsetshire* 141, 266, 270
Dale Collection 134, 139–41, 271
Dale, James Charles 25, 83, 108, 124, 138, **139–41(fig.55, pl.8)**, 171, 241, 245–6, 250, 253, 256, 257, 295, 317, 342
Dale, Samuel 100, 298, 339, 397
Daley, T. J. 345
Dandridge, Joseph 29, 32, 76, 83, 100, 104, **105–106**, 107, 109, 110, 339, 369, 398
Dandridge's midling Black Fritillary 105
daplidice, Papilio 271–2(fig.133)
daplidice, Pontia 271–4(fig.133, pl.21), 334, 337, 382
Darenth, Kent 87–8(fig.30)
Dark Blue 122, 298
Dark Green Fritillary 60, 95, 148, 194, 373, 391(fig.156)
Dartford Blue 131–2
Darwin, Charles 25, 26, 95, 162, 164, 171, 364; *The Origin of Species* 86, 149, 187
Davis, A. H. 146
Davy, Sir Humphry 60
Dawson, J. F. 342
de Grey, Thomas, Lord Walsingham 25, **177–8(fig.83)**
de Grise, Jehan, *The Romance of Alexander* 68(fig.17)
de Latour, A. 291
de Mayerne, Sir Theodore 98(fig.35)
de Worms, Baron Charles George Maurice 211, **226–7(fig.119)**
dealers 124, 144, 153, 155, 218, 288, 290–91, 308, 329, 332–4, 335, 336–8, 345, 365, 366
Demuth, R. P. 228
Dennis, R. L. H. 289, 292
Denny, Henry 399
dentata, Drypta (Coleoptera) 153
dia, Argynnis 251(pl.15), 330
dia, Boloria 133–4, 155, 251(pl.15), 329–31(fig.140), 390
dia, Melitaea 251(pl.15), 329–31(fig.140)
dia, Papilio 329–31(fig.140)
Diana, Princess 407
Dickens, Charles, *Memoirs of Joseph Grimaldi* 131–2
dimorphism 264
Dingy Skipper 222, 286–7(pl.25), 378
discoloration 64; *see also* coloration
dispar batavus, Lycaena 281, 285
dispar, Chrysophanus 281, 298
dispar, Lycaena 278–85(pl.23–4), 288, 298, 300, 384
dispar, Lymantria 285
dispar, Papilio 278
dispar rutilus, Lycaena 281, 285
Disraeli, Benjamin 364
distribution maps 233
Doncaster, Arthur 403–404
Donovan, Edward 63, 115, 123, **129–31**, 263, 264, 271, 288, 295, 298, 302, 322, 331, 343; *Instructions for Collecting* 130; '*Memorial Respecting My Works on Natural History*' 129; *Natural History of British Insects* 129–30, 242, 243(pl.12), 249
Dorset Natural History and Antiquarian Society 403
Doubleday, Edward 84, 146, **150–52(fig.61)**, 298
Doubleday, Edward and Westwood, John Obadiah, *On the Genera of Diurnal Lepidoptera* 152
Doubleday, Henry 37, 50, 82, 147, **150–53(fig.60)**, 291, 318, 332–4, 336–7, 402; *Synonymic List of the British Lepidoptera* 152, 288, 291
doubledayana, Olethreutes 152
Douglas, J. W. 40, 42, **156–7(fig.65)**, 163, 246, 366(fig.144), 401; 'The Reigate Gathering' 51–3; *The World of Insects* 156
Douglas, J. W. and Scott, J., *The British Hemiptera* 156
Drake, Sir Francis 98
Drewitt, Frederick 121
Drewitt, John 121
Druce, G. C., *Flora of Buckinghamshire* 182
Drury, Dru 34, **114–15(fig.42)**, 120–21, 268, 286, 317, 398; *Directions for Collecting Insects in Foreign Countries* 114; *Illustrations of Natural History* 114
duBois, Charles **101–103**, 311, 312, 397
Ducie, Earl of 45
Duke of Burgundy Fritillary 108, 242, 387
Duke of York Fritillary 242
Dunbar, David 328
Duncan, James **147–8**, 340–42; *British Moths* 148; *Introduction to Entomology* 148; *The Natural History of British Butterflies* 148
Dunning, J. W. 154, 155, **168–9(fig.76)**
D'Urban, W. S. M. 284
Durham, Bishop of 400
Dutfield, James 32, **112**, 340

East London Entomological Club 46
Eastbourne Natural History Society 187
ecclipsis, Papilio 66(pl.6)
Edelstein, H. M. 216
edusa, Papilio 263
'Edusa Years' 266
Edward VII, King 403
Edward, Thomas 289
Edwards, George **112–13(fig.41)**; *Gleanings of Natural History* 113; *Natural History of Uncommon Birds* 113
Edwards, Peter 373
Edwards, W. 365
Edwards, William 344, 345
Elliott, Sir William 148
Ellis, Hon. C. A. 330
Ellison, R. E. 249
Elm Butterfly 318
Elwes, Henry John **181(fig.86)**
Emerson, Ralph Waldo 37
E.M.M. see *Entomologist's Monthly Magazine*
Emperor Moth 111
Encyclopaedia Britannica 148
Enfield Eyes 103
Entomological Club 48, 137–8, 146, 147, 185, 399, 404; collection 152
Entomological Correspondence Club 399
Entomological Exchange and Correspondence Club 406
Entomological Magazine 41, 84, 85, 138, 139, 146, 151, 152, 324, 399, 400
Entomological Society of Great Britain 36, 399
Entomological Society of Hampshire 405
Entomological Society of London 25, 35, 36–7, 48–9, 125, 128, 133, 142, 143, 144, 146, 149, 153, 156, 157, 162, 165, 168–9, 178, 180, 328, 399, 400, 401, 403, 404; *see also* Royal Entomological Society of London
Entomologist 41, 43, 49, 84, 146, 156, 167, 179, 185, 186, 187, 194, 216, 217, 219–20, 223, 244, 245, 248, 250, 258, 260, 284, 293(fig.135), 328, 331, 332–4, 336–8, 400–401, 406, 407
Entomologist's Annual 38, 39, 163, 267, 324–5, 340
Entomologist's Bulletin 406
Entomologist's Gazette 232, 280, 406, 408
Entomologist's Monthly Magazine 43, 143, 157, 168, 169, 172, 179, 181, 270, 307, 402, 408
Entomologist's Record 84, 190, 200, 203, 210, 218, 228, 405, 408
Entomologist's Weekly Intelligencer 39, 41, 47, 50(fig.13), 51, 64, 84, 145, 163, 179, 182, 267, 330, 365(fig.143), 401–402
epiphron, Erebia 134, 135(pl.7), 138, 344, 393
Epping Forest and County of Essex Naturalist's Field Club 404
equipment 32, 34, 37, 43(fig.9), 55, 61, 65, 166; Heath trap 233; nets 67–74(figs 18–9, 21–3), 81(fig.28), 111, 143, 155
Esper, E. J. C. 286–7
Esper's Marbled White 213
Essex Field Club 44, 45, 404
Essex Naturalist 404
Essex Skipper **244–5**, 377–8
euphrosyne, Boloria 331, 390
Eupithecia moths 128, 182
European Committee for the Conservation of Nature and Natural Resources of the Council of Europe 234
European Invertebrate Survey 234
European Magazine 121(fig.47), 122
European Map butterfly 176, **327–8(pl.36)**, 390
europome, Colias 258
Evans, C. Ethel 244
Evelyn, John 397
evolution 95, 162, 165, 171, 187, 188, 224, 225, 228
E. W. Classey Ltd 406, 407, 408
Exeter Naturalist's Club 284
exhibitions 47, 403
Exotic Microlepidoptera 192
extensaria occidua, Eupithecia 171
extinction 172, 285, 364, 366, 368–9, 371; Black-veined White 218, 270, 368; Chequered Skipper 242, 244; Comma 160; Glanville Fritillary 342; Large Blue 302–306, 308–309, 366; Large Copper 121, 218, 277, 284–5, 368; Large Tortoiseshell 318, 368; Lulworth Skipper 366; Mazarine Blue 176–7, 218, 297, 300, 368; Northern Brown Argus 365; Purple-edged Copper 277–8, 291; Scarce Copper 277–8; Scotch Argus 365

Faber, John, Portrait of Sir Hans Sloane 77(fig.25)
Fabricius, Johann Christian 34(fig.5), 35, 66, 119, 120, 122, 263–4, 295, 316–17, 343
Farn, Albert Brydges **175–6(fig.81)**, 202, 222, 261, 328, 368
Farren, W. 50
Fawcett, Benjamin 311
fens, drainage 284–5
Field, The 146, 193, 194
field guides 220
field trips *see* collecting trips
Folkes, Martin 35
Folkestone Natural History & Microscopical Society 402
Fontaine, Maurice 260–61
Ford, Edmund Brisco 95, 108, **224–6(fig.118)**, 228, 271, 290, 298, 338–9, 350, 370; *Butterflies* 224, 289, 295, 328; *Moths* 224
Ford, Harry Dodsworth 224
Ford, James 137
Ford, L. T. 231
Ford, Richard Lawrence Edward 78, 79(fig.26), **231–2(fig.124)**, 404, 406; *The Observer's Book of Larger Moths* 232; *Practical Entomology* 232; *Studying Insects* 232
Ford, Robin 406
foreign introductions 155, 176, 270, 328, 346
Forestry Commission 372
forgeries *see* fraud
Forster, George 118, 119;

Voyage round the World 120
Forster, Johann (John) Reinhold 75, **117–20(fig.44)**, 122, 288, 289, 290; *A Catalogue of British Insects* 119(fig.45), 285–6; *Novae Species Insectorum* 119; *Observations made during a Voyage round the World, on Physical Geography, Natural History and Ethnic Philosophy* 120; *Zoologica Indica* 120
Fountaine, John 200
Fountaine, Margaret Elizabeth **199–201(figs 97–8)**
Fowles, John, *The Collector* 366
Frankland, Thomas 407
fraud 66, 124, 134, 143–4, 155, 216, 218, 267, 288, 290–91, 332–4, 336–8; *see also* authenticity
fraxinata, Eupithecia 182
French, D. J. 83
fritillaries 87; *see also* individual entries
Frohawk, F. W. 61, 95, 175, 189, **193–7(fig.94–5)**, 208(fig.105), 249, 257, 264, 267, 271–3, 297, 306, 317, 328, 366, 370; *The Complete Book of British Butterflies* 193, 194; *Natural History of British Butterflies* 193, 194, 346; *Varieties of British Butterflies* 193, 194–5
Fry, W. T. 125, 132
Fullerton, John 267

Gainsford, A. P. 316
Gairdner, Douglas 214
galathea, Melanargia 394
Gamlingay Wood Nature Reserve 332
Gardener's Chronicle 139, 149
Gaskell, Mrs 46
Gatekeeper 148, 394–5
Gay, John 397
Geldart, E. M. 402
Genesis, Book of 364
genetics 224–5, 228, 370
George III, King 26
Gerrard, Mr 338
Gesner, Conrad 96(fig.32), 97; *Historiae Animalium* 97
Gibbs, A. E. 344
Gibbs, Julian 407
Gillespie, A. B. and J. W. 344
Gillespie, A. C. 344
Gillett, J. D. 188
Glanville Fritillary 107–108, 112, **339–42(pl.38)**, 373, 392
Glanville, Eleanor 28, 80, **106–108**
Glanville, Richard 107
Glanville's Wootton 139(pl.8), 140, 246, 253, 298
globulariae, Adscita 164
glomeratus, Apanteles (Hymenoptera) 270
God 162, 364
Golden brown double Streak 58(pl.3)
Golden Copper 291
Goodden, Robert 406, 407
Goodden, Rosemary 407
Goodricke, Richard and Muriel 106
Goodricke, Sir Henry 107
Goss, Herbert 246–8, 277
Gosse, Philip Henry 37
Grace, Thomas 32, 398
Grand National Entomological Exhibition 47, 403
Grand Surprize, The 321(fig.139)
Grant, Margaret 195
Gray, E. W. 66
Gray, G. R. 400
Gray, J. E. 400
Gray, Mr, junior 333
Gray, Thomas 25
Grayling 394
Great Cornhill Fire 33, 398
Great Fire of London 99
Great Tortoiseshell 318, 320
Greater Silver Spotted Fritillary 58, 60
Green Chequered White 271
Green Hairstreak 383
Green-veined White 95, 107, 382
Greene, Joseph **165–6(fig.73)**; *The Insect Hunter's Companion* 165, 175
Greenwich Natural History Club 44
Gregson, Charles Stuart **159–60(fig.68)**
Gregson's Dart moth 159
Grey moth 159
Grimaldi, Joseph **131–2(fig.51)**
Grizzled Skipper 105, 122, 148, 378–9
grossulariata, Abraxas 166
Guenée, A. 152
Gulliver, Henry William 88(fig.29)
Gurney, Thomas 78
Gypsy Moth 285

habitat, loss of 284–5, 300, 309, 368–9, 371–2
Haggerstone Entomological Society 46, 402, 404–405
Half-mourner 28, 103, 58, 271–3(fig.133)
Hall, F. 257
Hambrough, W. 338
Hampshire Field Club 404
Hampson, Sir George 219
Hampstead, The 311
hampstediensis, Papilio 309–11(fig.137, pl.30)
Harding, H. J. 284
Harley, Edward, 2nd Earl of Oxford 29
Harmer, A. S. 263
Harris, Moses, junior 26, 29, 32, 60, 61, 72, 75, 80, 107, 110, 114, **115–17(fig.43)**, 123, 252–3, 264, 268, 331, 339, 369, 398; *The Aurelian* 32(fig.2), 34(fig.4), 35, 67, 69–70(fig.18), 75, 106, 115–16, 117, 263, 274, 278–9(fig.134b), 286–7(pl.25), 300, 301(pl.28), 312, 313(pl.31), 321, 322, 398(fig.159); *The English Lepidoptera, or, the Aurelian's Pocket Companion* 116–17, 258; *An Exposition of English Insects* 116 (fig.43)
Harris, Moses, senior 32, 115, 398
Harrison, D. Percy 264–6
Harrison, John William Heslop **216–17(fig.112)**, 307
Hartert, Ernst 197, 205
Hartley, Samuel 32, 398
Harwood, Bernard 175
Harwood, Philip 175
Harwood, William Henry **174–5**
Hawes, F. W. 244–5
hawk moths 174
Haworth, Adrian Hardy 35, 36, 37, 120, 121, 124, **127–9(fig.50)**, 242, 249, 256, 264, 271, 278, 280, 288, 290, 298, 302, 322, 325, 332, 365, 399; *Lepidoptera Britannica* 124, 129, 258; *Prodromus Lepidopterorum Britannicorum* 128
haworthii, Celaena 128
Haworth's Minor moth 128
Hawthorn, The 268
Head, H. W. 210
Healy, C. 402
Heath Fritillary 111, 174, 372, 373, 392
Heath, J. and Emmet, A. M., *The Moths and Butterflies of Great Britain and Ireland* 227, 231, 233–4, 263
Heath, Joan 234
Heath, John **233–4(fig.126)**, 280; *Threatened Rhopalocera (butterflies) in Europe* 234
Hellins, John 158–9(fig.67), 307–308
Hemming, Arthur Francis **222–3(fig.116)**, 261
Henslow, J. S. 44
hero, Coenonympha 144, 155
Heslop, I. R. P. 74(fig.23), 370
Heslop-Harrison, John 217
Hesperiidae 181, 186, 377–9
Higgins, L. G. and Riley, N. D., *Field Guide to the Butterflies of Britain and Europe* 220, 221
Higgins, Lionel George **220–21(fig.115)**; *The Classification of European Butterflies* 220
Higgins, Nesta 221
Higgs, Sarah 112
High Brown Fritillary 60, 107, 194, 338, 372, 390–91
Hill, John 397
hippothoe, Lycaena 277–9, 286, 289, 290–91
hippothoe, Papilio 122, 290–91
hoaxes *see* fraud
hogs 103(fig.39)
Holdgate, Martin 364
Holland, W. J. 316–17
Holland, William **180–81**
Holly Blue 101, 299(pl.27), 386
Holmesdale Natural History Club 401
Hooker, Sir William 250
Hope Chair of Zoology, Oxford University 145, 149, 187–8
Hope Department of Zoology 95, 108, 144, 180, 271, 149, 187
Hope Entomological Collections 29, 121, 129, 139, 130, 144, 188, 267
Hope, E. H. M., Mrs 400
Hope, Frederick William **144–5(fig.57)**, 149, 256–7, 400
Hopkins, G. H. E. 214
Hopley, E. W. J.: 'Bagged' 56(fig.15); 'The Chase' 56(fig.14)
hornets 56
Horniman Museum 174
Horsfield, T. 400
Horticultural Magazine 143
Hough, Richard 117
Howarth, T. G. 289, 328, 339
Hoyle, W. E. 405
Hudd, A. E. 300
Hudson, W. H. 88; *Hampshire Days* 366
Huggins, Henry Charles **221–2**, 307; *A Guide to the Butterflies and Larger Moths of Essex* 222; 'The Old Days at Chattenden Roughs' 89
Hughes, A. W. 328
humidalis, Hypenodes 152
Humphreys, H. N. 149, 151, 274, 275, 299
Humphreys, H. N. and Westwood, Obadiah J., *British Butterflies and their Transformations* 149, 151, 274, 275(pl.22), 299(pl.27)
huntera, Papilio 316–17(pl.32)
Hunter's Butterfly 316
Hutchinson, Emma **160–61(fig.69)**
hutchinsoni, Polygonia c-album f. 160(fig.69)
hyale, Colias 89–90, 195, 258–60(pl.17), 379–80
hyale, Papilio 258–9(pl.17), 263
Hyde, G. E. 370
hyperantus, Aphantopus 395

I Zingari Entomological Society 46, 403
icarus, Lycaena 294
icarus, Polyommatus 386(fig.150)
ICZN (International Commission on Zoological Nomenclature) 220, 222, 223
identification *see* authenticity
illustrations 32, 98, 109, 111, 122, 194, 220, 343; inaccuracies 111, 129, 138, 142, 340; *see also* artists and illustrators; photography
imperator, Pulex 148, 150
infestations 80–81, 107
Ingpen, Abel 55, 61, 65, 67, 76, 84, 86, **143–4**; *Instructions for Collecting, Rearing and Preserving British Insects* 143
Inquisitor (Newman) 334
Inquisitor (Stainton) 47, 50
Insectline 348
insigniata, Eupithecia 160
International Commission on Zoological Nomenclature (ICZN) 220, 222, 223
Invertebrate Conservation News 371
io, Inachis 389(fig.154)
Ipswich Museum 331, 338
iris, Apatura 194, 387

irregularis, Hadena 174
Isle of Wight 340, 342
Isle of Wight Natural History and Archaeological Society 405

Jasione Pug moth 182
jasioneata, Eupithecia 182
Jenkinson, Francis 176
Jermyn, George 136
Jermyn, George Bitton 400
Jermyn, Lætitia 65, 81, **136–7**, 281, 290, 311, 322, 331, 369–70; *The Butterfly Collector's Vade Mecum* 70, 136(fig.53), 254, 288
Johns, B. G. 373
Johnson, Maurice 397
Johnston, George 401
Joint Committee for the Conservation of Insects 328, 371
Jones, Jezreel 70
Jones, William 66, 115(fig.46), **120–21(fig.46)**, 127, 295, 296
Jordan, Heinrich Ernst Karl **197–9(fig.96)**, 205, 207, 213, 219
Jordan, R. C. R. 348
jurtina, Maniola 395

Kalm, Pehr 77
'Kentish Buccaneers' 218, 333), 345
Kentish Glory moth 111
Kershaw, G. B. 175
Kershaw, S. H. 203, 204, 366
Kettlewell, Henry Bernard Davis 204, 215, **227–9(fig.121)**, 273–4, 370; *The Evolution of Melanism* 228
Keys, J. H. 167
Kidd, John 144
killing agents *see* killing methods
killing bottles 55, 63
killing methods 61–5, 115, 126, 142, 155
Kingsley, Charles 37, 38
kirbii, Stylops 127
Kirby, W. and Spence, W. 70, 72; *Introduction to Entomology* 94, 124–5, 126, 137, 139
Kirby, W. E. 179
Kirby, William 25, 36, 37, 38, 44, 61, 65, 67–8, 76, 80, 114, 115, **124–7(fig.49)**, 136, 141, 149, 400; *Monographia Apum Angliae* 126
Kirby, William Forsell **179–80(fig.85)**, 250, 254, 277–8, 291; *Elementary Text-Book of Entomology* 180; *European Butterflies and Moths* 170, 180, 288; *Evolution and Natural Theology* 180; *List of British Rhopalocera* 179; *Manual of European Butterflies* 179; *Synonymic Catalogue of Diurnal Lepidoptera* 179–80
Kirchner, John 336
Knaggs, Henry Guard 38, 43, 63, 64, 70, 72, 74, 84, 156, **167–8(fig.75)**, 324, 330, 402; *A List of the Macro-Lepidoptera of Folkestone* 168; *The Lepidopterist's Guide* 168
Knaggs, Valentine Guard 168
Kneller, Sir Godfrey 77
Knoch, A. W. 276
Knowlton, Thomas 32, 398
Koenig, Charles 77–8
Krieg, David 331

labelling 82, 134, 138, 153, 211, 212(pl.11), 344
Lambillionea 260(fig.130)
Lancashire and Cheshire Entomological Society 45, 403
Langley, A. F. 300
Langport 282, 305
Lang's Short-tailed Blue 295
laodice, Argynnis 213
Large Blue 87, 122, 123, 124, 155, 177, 195, 207–208, 216, 220, 222, 298, **300–309(pls 28–9)**, 371, 386
Large Copper 45, 121, 122, 128, 154, 207, 209, **278–85(pls 23–4)**, 289, 295, 298–300, 384, 368, 397
Large Heath 122, 148, 395
Large Skipper 103(fig.39), 116, 378
Large-tailed Blue 291
Large Tortoiseshell 148, 174, 193, 194, **318–21(pl.33)**, 389
Large White 95, 240, 381
lathonia, Argynnis 331–5(pl.37), 390
lathonia, Papilio 331
'*lathonia* years' 334
Leach, W. E. 137, 138, 290
Lee, James 398
Lee, Samuel 32, 398
Leech, J. H. 186
Leeds, Henry Attfield **211–12**, 277
Legge, Thomas 95
Leman (*or* Lemon), James 32, 106, 398
Lesser Belle moth 171
Lesser Silver-spotted Fritillary 331
'Letter of Rusticus' 146
leucomelanus, Papilio 382
levana, Araschnia 176, 327–8(pl.36), 390
levana, Papilio 327(pl.36)
Lewin, William **122–3**, 258, 264, 271, 278, 286, 288, 295, 296, 298, 312, 318, 322, 331; *The Papilios of Great Britain* 256(pl.17), 259(pl.17), 300, 302
Lewington, Richard 370
Lewis, W. Arnold 330–31
ligea, Erebia 343–6(pl.39), 393–4
ligea, Papilio 343
Lime Hawkmoth 210, 221
Lincolnshire Fritillary 107, 139
linea, Hesperia 245(fig.127)
lineola, Papilio 244–5
lineola, Thymelicus 244–5, 377–8
Lingwood, R. M. 401
Linnaean classification 27–8, 35, 60, 100, 107, 116, 119, 121, 122, 128, 142, 258, 263, 274, 278, 285, 312, 327, 343
Linnaeus 27–8, 32, 35, 60, 66, 99, 114, 119, 121, 252, 263, 286; *Systema Naturae* 28(fig.1), 35, 60, 66, 95, 116, 279
Linné, Carl von *see* Linnaeus
Linné, Fru von 36
Linnean Society 33, 36, 37, 66, 121, 241, 398, 401, 403; Zoological Club 36, 142, 399
Linton, W. R.: *Flora of Derbyshire* 182
Lipscomb, C. G. 317
Lister, Martin 33, 100, 252, 397
Lizars, W. H. 100, 114
Llewellyn, J. T. D. 348
Lobster Moth 141
Logan, R. F. 297, 365
London Magazine 33
London Natural History Society 215, 405
Long-tailed Blue 251(pl.15), **291–2**, 314, 384
loopers 107
Lorkin, John 95
Loudon, J. C. 86, 133, 317, 329, 342, 400
Lubbock, John 26
lucina, Hamearis 387
Lullingstone Silk Farm 407
Lulworth Skipper 138, 140, **245–9(pl.13)**, 378
luteago barrettii, Hadena 171
Lycaenidae 176–7, 186, 211, 383–7(fig.150)
Lyell, M. A. C. 295
Lyell, Sir Charles 401

machaon, Papilio 83, 252–4(fig.128), 379
Mackworth-Praed, C. W. 244
Magazine of Natural History 72, 86, 133, 146, 317, 329, 342, 400
Magpie Moth 166
Mallow Skipper **249(pl.14)**, 327(pl.36), 378
malvae, Pyrgus 378–9
Malvern College 344, 345
Malvern Naturalists' Field Club 401
Manchester Argus 122
Manchester Club 45
Manchester Entomological Society 405
Map butterfly *see* European Map butterfly
maps, distribution 233
Marbled White 89, 103, 175, 368, 373, 394
Marcon, J. A. 204
margariellus, Crambus 86
Marsh Fritillary 105, 342, 372, 392
Marsh Moth 171
Marsh Oblique-barred moth 152
Marshal, Daniel 32, 398
Marsham, Thomas 125
Martyn, Thomas 123
Martyn, William, *New Dictionary of Natural History* 117
Mays, Robert 112, 153, 227(fig.120)
Mazarine Blue 28, 50, 85, 101, 122, 123, 140, 176–7, 195, **297–300(pl.27)**, 386
McArthur, N. 291
McLachlan, Robert 43, 150, 402
McRae, W. 248
Meadow Brown 62(pl.5), 63, 69, 395
meconium 56
Mediterranean Fritillary **338–9**, 391
Meek, E. G. 265(pl.19)
megera, Lasiommata 393(fig.158)
melanism, industrial 191, 227–8
Meldola, Ralph 404
Melville R. V. 223
Meredith, Ellen 144
Merian, Maria Sibylla 123
Merrett, Christopher, **98–9**, 286, 290, 397; *Pinax rerum naturalium Britannicarum* 98, 268
Merrin, Joseph 281–2, 302–304
Mette, Frederick 404
Meyrick, Edward **191–2(fig.93)**; *Handbook of British Lepidoptera* 192; *Exotic Microlepidoptera* 192
mica 29, 58, 75, 77, 78
microlepidoptera 163–4, 171–2
microscopes 37, 142
Middle Copper *see* Scarce Copper
migration 227, 240–41, 348; Bath White 271, 274; Camberwell Beauty 324–5; Clouded Yellows 257–60, 262–3, 266; Large Tortoiseshell 320–21; Long-tailed Blue 291–2; Monarch 346; Painted Lady 312–16
Milkweed Butterfly *see* Monarch
Millan, John 104, 116
Milton, Lord 242
minimus, Cupido 298, 385
Minsmere RSPB Reserve 334–5
Moffat, Thomas *see* Moffet, Thomas
Moffet, Thomas 26, 28, 33, 35, 60, 67, **95–8(figs 31–2)**, 249, 252, 258, 268, 297–8; *Insectorum Theatrum* 56, 95–7(figs 31–2, 35), 256(fig.129), 263(fig.131), 379–82(figs 146–9), 386–9(figs 150–55), 391–3(figs 156–8); *The Silkwormes and their Flies* 98
Monarch 320, **346–51(pls 40–41)**, 396
Moore, John 373
Moorland Clouded Yellow 258, 327(pl.36)
Morris, Beverley 155
Morris, Francis Orpen 61, 63, 86–7, 131, **154–5(fig.63)**, 311, 314, 317, 318–20, 344; *A History of British Butterflies* 43(fig.9), 65, 70, 73(fig.22), 154, 155; *A History of British Moths* 154
Morris, Henry 154
Morris, Richard 69
morrisii, Chortodes 168
morrisii, Photedes 155
Morrow, George 49
Mosley, S. L. 404, 405
moths 111, 128, 134, 138, 140, 164, 169, 170–72, 173, 182, 210, 227–8, 369
Mouffet, Thomas *see* Moffet, Thomas
Mountsfield 40–41, 164
Mr Dandridge's Marsh Fritillary

105
Mr Ray's Alpine Butterfly 100, 249
Mr Ray's Purple Streak 100
Mr Vernon's half-Mourner 108, 271–3(fig.133)
Mr Vernon's small Fritillary 108
Muffet, Thomas *see* Moffet, Thomas
Mull, Isle of 345
Müller, J. S. 113
Murchison, Sir Roderick 401
Museum of Natural History, Birmingham 134–6
Museum of the Royal Dublin Society 180
mutation 217, 267

napi, Pieris 382
National Museum, Dublin 180, 307
National Trust 184, 372
Natural History Museum 28, 105, 130, 152, 167, 172, 173, 175, 176, 178, 181, 186, 192, 194, 215, 216, 222, 224, 228, 230, 231, 234, 331, 343; Botany Department 29, 58, 60; Department of Entomology 29, 60, 339; Entomology Library 61; *see also* British Museum; British Museum (Natural History) (BM(NH))
Natural History Society of Northumberland, Durham and Newcastle-upon-Tyne 400
Natural History Society of Northumbria 400
Naturalist 44, 155
Naturalists' Journal 405
Nature Conservancy 233
Nature Conservancy Council 372
Nature Study 405
Neal, Ernest 261, 304–305
Neimy, Khalil 199–200
Netley, Shropshire 256–7
nets 67–74(figs 18–19, 21–3), 81(fig.28), 111, 143, 155
New Clouded Yellow 258, 261
Newbury District Field Club 224
Newman, Edward 37, 38, 41, 43, 48, 50, 65, 70, 74, 78, 80, 84, 89–90, 138, 142, **145–7(fig.58)**, 152, 153, 163, 172, 270, 274, 291, 295–7, 298, 305, 314, 322, 332, 342, 344, 367(fig.144), 399, 400–402; *A History of British Ferns* 146; *History of Insects* 43(fig.9a); *Illustrated Natural History of British Butterflies* 146–7, 166, 250, 276–7, 336(fig.141); *Illustrated Natural History of British Moths* 146, 336(fig.141); 'The Song of Bugfliwatha' 157
Newman, L. Hugh 202, 209, 266, 324, 335; *Living with Butterflies* 210
Newman, L. W. and Leeds, H. A., *Text-book of British Butterflies and Moths* 210, 211
Newman, Leonard Woods 187, **209–210(figs 106–108)**, 366–7
Newton, Sir Isaac 27, 35, 397
Nicholas, Bishop of Carlisle 105
Nicholls, Sutton 103
nigra, Aporophyla 128
Niobe Fritillary 327(pl.36), **335–8(fig.141)**, 390
niobe, Argynnis 335–8(fig.141), 390
niobe, Papilio 335
nomenclature 28, 103, 116, 119, 122, 142, 152, 211, 212, 222–3, 261, 263, 274–6, 286, 342
Norfolk and Norwich Naturalists' Society 44, 46, 402
North London Entomological Society 46, 403
North Staffordshire Naturalists' Field Club and Archaeological Society 402
Northern Arches moth 134
Northern Brown Argus 120, **295–7(pl.26)**, 385–6
Northern Emerald dragonfly 134
Northern Footman moth 159
Northern Naturalists' Union 217
Norwich Entomological Society 44, 399, 402
Norwich Museum 199, 288
Nymphalidae 387–96(figs 151–8)

Oak Eggar moth 210
obsitalis, Hypena 170
oileus, Pyrgus 124
Oliver, G. B. 328
ononaria, Aplasta 173
Orange Argus of Elloe 280(pl.23)
Orange Tip 95, 101, 112, 174, 271, 382–3
ornata, Ctenophora (Diptera) 141
Overall, Edmund 32, 398
over-collecting 284, 297, 308–309, 364–6, 368–9; *see also* collecting
Oxford, University of 144–5
Oxford University Entomological Society 402

Painted Lady 58, 101, 103, 263, **311–16(fig.138, pl.31)**, 317, 373, 388(fig.152)
Palaearctic butterflies 221
palaemon, Carterocephalus 241–4(pl.12), 377
palaemon, Papilio 241
palaeno, Colias 258
Pale Clouded Yellow 89–90, 195, **258–60(pl.17)**, 261, 379–80
Pallas's Fritillary 213
pallustris, Athetis 171
paludum, Buckleria 171
paludum, Trichoptilus 170–71
pamphilus, Coenonympha 395
pandora, Argynnis 338–9, 391
pandora, Papilio 338–9
paniscus, Papilio 130, 241–3(pl.12)
paphia, Argynnis 193, 338, 391
Papilionidae 379–83(figs 146–9)
Parry, George 332, 336, 337
Parry, Thomas 298, 330
patronage, societies' 48, 397
Peach Blossom moth 111
Peacock 28, 62(pl.5), 63, 64, 95, 318, 389(fig.154)
Peacock, A. D. 217
Peacock's Eye 58
Pea-pod Argus 291
Pearl-bordered Fritillary 224, 331, 390
Peck, William Dandridge 63, 76
Pedley, Esther 114
Pegg, Agathos 183–4
Pelham-Clinton, Edward Charles, 10th Duke of Newcastle 147, **230–31(fig.123)**, 344
Pennant, Thomas 397
Penny, Thomas 6, 96(fig.32), 97(fig.34)
Peppered Moth 227–8
Pepys, Samuel 397
Perceval, M. J. 286
Perry, John 72, 73
Perthshire Society of Natural Sciences 178, 179
Petiver, James 26, 27, 28, 29, 35, 57–60, 66, 67, 68, 69, 83, 100, 102, **103–105**, 107, 108, 114, 124, 252, 253, 258, 263, 265(pl.19), 274, 286, 309, 311, 317, 318, 331, 339–40, 397; *Brief Directions for the Easie Making and Preserving Collections of all Natural Curiosities* 57(fig.16); collections 58, 77, 80; *Gazophylacium* 101, 103, 107, 339; *J. Petiveri Opera* 104; *Musei Petiveriani Centuria* 58, 271, 272(fig.133), 273; *Papilionum Britanniae Icones* 103(fig.39), 104, 268, 269(fig.132)
phlaeas, Lycaena 384
phlaeas, Papilio 286
photography 185, 351, 371; *see also* illustrations
Pickard, H. Adair 267, 402
Pickard-Cambridge, Octavius 148, 150, **169–71(fig.77)**, 293–4
Pickles, A. J. & C. T. 263
Pigmy Footman moth 152
pill-boxes 43(fig.9)
pincushions 65
Pine Processionary moth 337
Pinion-spotted Pug moth 160
pinning 66–7, 75, 76, 77
pins 65, 67, 76, 82, 115, 181
pirithous, Leptotes 295
pityocampa, Cnethocampa 337
plagiarism 142
Plantain Fritillary 340
Plastazote© 80
Plastead 124, 144, 288, 290–91, 335
Platten, E. W. 331
plexippus, Anosia 347(pl.40)
plexippus, Danaus 317, 346–51(pls 40–41), 396
plexippus, Papilio 346
Plukenet, Leonard 28–9, 58, **101**, 102, 397; collection 60, 62(pl.5), 63
plume moths 170–71
Plymouth Borough Museum 167
pocket guides 185
pockets 61
podalirius, Iphiclides 144, 254–7(pl.16), 337, 379
podalirius, Papilio 254–7(pl.16)
polychloros, Nymphalis 194, 318–21(pl.33), 389
polychloros, Papilio 318–19(pl.33)
Pope, Alexander 28, 397; *The Rape of the Lock* 29
Portland, Duchess of 29, 114
Poulton, Sir Edward Bagnall 115, 180, **187–9(fig.91)**; *The Colours of Animals* 188
Power, John Arthur 44, **153–4(fig.62)**, 400
Pratt, C. R. 270
prejudice 364, 371
preservation methods 58, 77–8
pressing 58
Pretor, A. 248
Prince of Orange Butterfly 112, 340–41(pl.38)
Princess, The 331
pruni, Papilio 274
pruni, Satyrium 274–7(pl.22), 384
pruni, Thecla 276
pug moths 128, 182
Punch 49(fig.12), 190
pupa digging 165–6
Purefoy, Edward Bagwell 195, **207–209(figs 103–5)**, 265, 285, 308, 407
Purple-edged Copper 99, 153, 251(pl.15), 277, 278, 288, 289, **290–91**, 384
Purple Emperor 74, 84, 86–7, 123, 174, 194, 370, 387
Purple Hairstreak 100, 383
Purple-shot Copper 290
pygmaeola, Eilema 152
pygmina, Photedes 128

Queen of Spain Fritillary 28, 60, 108, 124, 153, **331–5(pl.37)**, 373, 390
Quekett, E. J. and J. T. 282
quercus, Quercusia 383

Raffles, Sir Stamford 36, 399
rail travel 51, 83
Rainbow Coffee House 31
Rait-Smith, William **211–13(fig.109)**, 267
rapae, Pieris 381–2
Raven, C. E. 99, 298
Raven, John 216
Raw, John 136
Ray Society 156, 157, 163, 187, 401
Ray, John 27–8, 29, 32, 35, 55, 56, 60, 67, 68, 76, **99–101(fig.36)**, 102, 108, 122, 249, 252, 253, 256, 263, 265, 268, 271, 274, 312, 318, 339, 397, 401; *Catalogus plantarum circa Cantabrigium nascentium* 56, 99; *Historia Insectorum* 95, 98, 99, 103, 240, 252, 253(fig.128), 258, 286, 297–8; *Methodus Insectorum* 100
Read Collection 257
Read, W. H. Rudston 255(pl.16), 257

Reading Museum 180, 181, 227
rearing 107, 109, 160, 174, 194, 208–209, 215, 228, 232, 257, 290, 307, 346; *see also* butterfly farms
Réaumur, René 61
record-keeping 82, 134, 137, 230
Red Admiral 28, 100–101, 112, 240, 387–8(fig.151)
Redhorns 257
Redshaw, E. J. 280
Reed Tussock moth 140, 285
Reid, I. G. M. 268
relaxing methods 64, 76
Rennie, James **142–3**, 258, 271, 286, 288, 290–91, 295, 322, 331, 335; *Conspectus of the Butterflies and Moths found in Britain* 142; *Insect Architecture* 143; *Insect Transformations* 143
Rest Harrow moth 173
rhamni, Gonepteryx 66, 194, 266–7(pl.20), 380(fig.147)
Ribbe, C. 260
Richardson, Richard 80
Richmond and North Riding Field Club 45
Ricord, A. 63
Rider, T. 115
Riga Fritillary 60, 108, 331
rigor mortis 64, 75, 76
Riley, Norman Denbigh 180, 181, 184–5, 186, 195, 197, 198–9, 205, 216, **219–20(fig.114)**, 223, 295; *Butterflies of the West Indies* 220; *The Department of Entomology of the British Museum (Natural History) 1904–1964. A brief historical sketch* 220
Ringlet 395
Robson, John E. 404
Rocket, The 271
Roesel von Rosenhof, A. J., *Insecten-Belustigung* 322
Rogers, William 95, 96
Rothschild-Cockayne-Kettlewell collection 202, 207, 215, 228
Rothschild Museum, Tring 198, 204, 205, 215, 344
Rothschild, Hon. Nathaniel Charles 88, 175, 195, 198, **213–14(fig.110)**
Rothschild, Lionel Walter, 2nd Baron of Tring 89, 172, 175, 194, 197, 198, 202, **205–207(figs 101–102)**, 257, 267, 277, 403
Rothschild, M. and Marren, P., *Rothschild's Reserves* 214
Rothschild, Miriam 195, 197, 198, 199, 205, 206, 213–14, 222, 224, 225, 254, 399
Rothschild, W. and Jordan, K., *Revision of the Papilios of the Eastern Hemisphere* 197
Rowland-Brown, H. 328
Royal Agricultural Society 139
Royal Botanic Gardens, Kew 183, 217, 233, 346
Royal College of Physicians 99, 112–13
Royal Entomological Society of London 48(fig.11), 49, 128, 169, 186–7, 188, 207, 214, 215, 223, 339, 367, 400, 404, 405, 407, 408; emblem 127; *see also* Entomological Society of London
Royal Scottish Museum 227, 230, 231, 344
Royal Society 27, 28, 35, 48, 100, 117, 120, 123, 144, 149, 397
Royal William, 103, 252
Royal Zoological Society Gardens 403
rubi, Callophrys 383
Rudd, G. 400
Rum, Isle of 216, 307
Russell, Sidney George Castle **202–204(fig.100)**, 289
russiae, Melanargia 213
Rye, E. C. 43, 365, 402

Sabbagh, K. 216
Sabine collection 175
sabuleti, Myrmica 308, 309
Saffron Butterfly 263, 265(pl.19)
salicalis, Colobochyla 171
salmacis, Polyommatus 295
Salvage, J. 333
Salwey, Posthumus 66
sambucaria, Ourapteryx 160
Samouelle, George 48, 82, **137–8**, 242, 295, 399; *Entomologist's Useful Compendium* 48, 137; *General Directions for collecting and Preserving Exotic Insects and Crustacea* 137; *The Entomological Cabinet* 137
Sampson, C. P. J. 339
Sassoon, Siegfried 26, 325–6
Satyridae 211
Saundby, Sir Robert Henry Magnus Spencer **223–4(fig.117)**, 231
Saunders, William Wilson 401
scabrinodis, Myrmica 308
Scallop-winged Fritillary 331
Scarce Chocolate-tip moth 168
Scarce Copper 122, 153, 277, **285–90(pl.25)**, 300, 384
Scarce Forester moth 164
Scarce Heath 144, 155
Scarce Painted Lady 316
Scarce Pug moth 171
Scarce Scotch Argus 344
Scarce Skipper 242
Scarce Spotted Skipper 122
Scarce Swallowtail 124, 144, 251(pl.15), **254–7(pl.16, fig.129)**, 327(pl.36), 379
Schofield, James 209, 285
scoliaeformis, Synanthedon 159
Scotch Argus 295, 343, 344, 365, 393
Scotch Brown Argus 297
Scotin, J. 109
Scott, John 330
Scott, P. 252
Scott, Sir Peter 407
Scottish Naturalist 178, 179
Seaman 153, 288, 291
Sedgwick, A. 401
SEL (Societas Europaea Lepidopterologica) 234
Selby, P. J. 147–8, 151
selene, Boloria 329, 390
selene, Melitaea 133
semele, Hipparchia 394
semiargus, Cyaniris 85, 177, 297–300(pl.27), 386
semiargus, Papilio 297
sericea, Eilema 159
setting 75–6(fig.24), 111, 153, 164, 365
setting boards 75(fig.24), 76, 111
Seymer, Henry 278–9(fig.134a), 286, 300
Shakespeare, William, *As You Like It* 369
Sharp, David 81(fig.28)
Shaw, George 28, 76
Sheldon, W. G. 186, 199
Short-tailed Blue 170, **293–5(fig.135)**, 385
Shropshire Natural History and Antiquarian Society 400
Silver-bordered Ringlet 155
Silver-spotted Skipper 195, 378
Silver-studded Blue 293, 296(pl.26), 298, 385
Silver-washed Fritillary 193, 195, 196(pl.10), 197, 286–7(pl.25), 338, 391
simbling *see* assembling
Simpson-Hayward Collection 344–5
Simpson-Hayward, G. H. 344
sinapis, Leptidea 379
Sitwell, Edith 38
Skertchly, S. B. J. 315
Skinner, A. 346
Skinner, Bernard 263
Skippers: exotic 124; Abyssinian 221; Oriental 181; *see also* individual entries
Skrimshire, Fenwick 280–81
Skrimshire, Thomas 399
Skrimshire, William 120, 280–81, 295
sleeving 210(fig.107)
Sloane, Sir Hans 58, 60, 77(fig.25), 100, 101, 102, 104, 105, 110, 112, 397; collections 28–9, 76–7, 100, 107
Small Blue 122, 296(pl.26), 298, 299(pl.27), 385
Small Copper 58, 59(pl.4), 101, 122–3, 286–7(pl.25), 384
Small Heath 395
Small Mountain Ringlet 36, 134, 135(pl.7), 138, 344, 393
Small Pearl-bordered Fritillary 133, 329, 390
Small Skipper 103(fig.39), 245, 246, 377
Small-tailed Blue 293
Small Tortoiseshell 62(pl.5), 63, 101, 318, 388–9(fig.152)
Small Wainscot moth 128
Small White 95, 101, 240, 381–2
Smeathman, Henry 115, 123, 398
Smetham, H. 189
Smith Collection 36
Smith, Audrey 188
Smith, B. 330
Smith, C. H., *Memoir of Dru Drury* 114(fig.42)
Smith, Sir James Edward 36, 121, 398
Smith, Wallace A. 331
sobrina, Paradiasia 134
Societas Europaea Lepidopterologica (SEL) 234
Society for British Entomology 405, 406
Society for the Promotion of Nature Reserves 214
Society of Antiquaries 113
Society of Apothecaries 55, 397
Society of Aurelians *see* Aurelian Societies
Society of Entomologists 34, 398
Solander, Daniel C. 398
Somerset County Museum 282, 288
Somerset, Mary, Duchess of Beaufort 29, 109
Somersetshire Archaeological & Natural History Society 401
Sothern, George 399
South Essex Natural History Society 222
South London Entomological and Natural History Society 46–7(fig.10), 94, 165, 172–3, 174, 186–7, 190, 195, 202, 211, 213, 215, 403, 404, 407; *see also* British Entomological and Natural History Society
South London Entomological Society 403, 404
South, Richard 46(fig.10), 49, **184–6(fig.89)**, 189, 219, 277, 293, 297, 322, 338–9, 345, 346; *The Butterflies of the British Isles* 185, 317; *Moths of the British Isles* 185; *Synonymic List of British Lepidoptera* 186
Southwood, Sir Richard 231
Sowerby, James 290; *British Miscellany* 343
Spalding Gentlemen's Society 27, 280, 397
Sparrman, André 67
Sparshall, Joseph 399; collection 288
Speckled Wood 95, 392–3(fig.157)
Spence, William 25, 38, **132–3(fig.52)**, 401
spiders 98
spinifera, Agrotis 159
Spooner, G. M. 309
Spotted Saffron Butterfly 263, 265(pl.19)
Spotted Skipper 242
Spotted Sulphur moth 169
Square and Compass, The (inn) 89
St John, J. S. 294
Stainton, Henry Tibbats 25, 26, 37–43, 47, 49–50, 73, 85, 142, 145, 150–51, **162–4(fig.71)**, 171, 267, 297, 308, 330, 340, 344, 365, 366, 367, 401, 402; 'At Home' invitations 39–41(fig.8), 85, 140, 142, 162, 163; *The Entomologist's Companion* 85; *Insecta Britannica* 164; *Manual of British Butterflies and Moths* 127, 158, 162, 163, 246, 291; *The Natural History of the Tineina* 156, 158, 163
Stainton, Isabel 164
Standish, Benjamin 284
Starrenburgh, John 68

Stearn, W. T. 180, 181
Step, Edward 46
Stephens, James Francis 37, 39, 44, 63, 112, 124, 128, 129, **141–2(fig.56)**, 143, 154, 242, 276, 302, 335, 340, 343, 344, 400; *Illustrations of British Entomology* 134, 135, 139, 141, 142, 149; *A Systematic Catalogue of British Insects* 141
Stevens, Samuel 25, 42, 368, 402
Stevens's Auction Rooms 368(fig.145)
Stewart, Alexander Morrison 147
Stockley, R. E. 370
Stothard, Thomas 138
Strepsiptera 127
Substitute; or Entomological Exchange Facilitator, and Entomologist's Fire-side Companion, The 401–402
Suffolk Naturalists' Society 338, 405
sugaring 150–51, 168, 189–90
sulphurata, Perispudea (Hymenoptera) 167
superstition 56–7
Surrey Skipper 249
Sutton, R. G. 282
Swaffham Prior Natural History Society 44, 400
Swainson, J. J. 271, 278
Swainson, William 61, 65
Swallow-tailed Moth 160
Swallowtail 45, 83, 95, 103, 115, 128, 136(fig.53), 137, 140, 197, **252–4(fig.128)**, 256, 372, 379; black 204, 209, 210, 300, 301(pl.28), 404
Swan Tavern 31, 32
Swedenborg, Emanuel 120
Swift Copper 291
Sylvester, E. T. 365
sylvestris, Thymelicus 245, 377
Symonds, W. G. 401

tages, Erynnis 378
Tailed Blue 291
Tattersall, J. W. 282
Taylor, H. Stuart 324
Temple Coffee-House Botanic Club 108, 397
Terasson, H. 268
Tesch, L. R. 406
Thatched House Tavern 37(fig.7)
thersites, Plebicula 177
Thomas, J. A. and Lewington, R., *The Butterflies of Britain and Ireland* 370
Thomas, Jeremy 208, 209, 245, 248–9, 308, 320, 328, 370, 371
Thomson, George 252, 289, 340–42; *The Butterflies of Scotland* 147
Tineina 163–4
tithonus, Pyronia 394–5
Tolman, Tom 221
Topsell, Edward, *History of Four-footed Beasts and Serpents* 98
trabealis, Emmelia 169
Tracey, H. E. 257
trading *see* commercialism
travel 83–4, 138; overseas 57–8, 66, 77, 114–15, 117–18, 200, 206
Tremewan, W. G. 406
Trevelyan, Sir Walter Calverley 401
Tring Museum 198, 204, 205, 215, 344
tullia, Coenonympha 395
Tunmore, Mark 408
Turner H. J. 46(fig.10), 173
Turner, William 56
Turpyn, Richard 240
Tutt, James W. 26, 37, 47–8, 49–50, 61, 70, 82, 86, 163, 165, 181, 186, **189–91(fig.92)**, 195, 265, 282, 288, 365–6, 368, 370, 405, 406; *British Butterflies* 191; 'British Lepidoptera' 190; *British Moths* 191; *The British Noctuae and their Varieties* 190; *Natural History of the British Lepidoptera* 190–91; *Practical Hints* 191; 'The Wonders of Insect Life' 190
twitchers 241, 348
Tyneside Naturalists' Field Club 400

Uffenbach, Z. von 104
Union of Naturalists' Societies in Scotland 179
urticae, Aglais 388–9(fig.152)

valesina, Argynnis paphia f. 195–7(pl.10), 286(pl.25)
Vallins, F. T., Dewick, A. J. and Harbottle, A. H. H. 261–2
venata, Ochlodes 378
venation, wings 117, 121
Venus Fritillary 155
Vernon, William 28, 100, 106, 107, **108**, 271, 332, 397
Verrall Suppers 48–9, 399, 404
Verrall, G. H. 48–9, 209, 285, 404
Vertue, George 99, 397
Victorian History of the County of Essex 175
Victoria, Queen 48
Vigors, N. A. 400
villida, Junonia 110, 309–311(fig.137, pl.30), 387
villida, Papilio 309–311(fig.137, pl.30)
Vinter, C. H. S. 405
Viper's Bugloss moth 174
virgaureae, Lycaena 277, 284, 285–90(pl.25), 384
virgaureae, Papilio 285–88(pl.25)
virginiensis, Cynthia 316–17(pl.32), 388
virginiensis, Vanessa 227
Volunteer Movement 41

Wagstaff, C. 134
w-album, Satyrium 383
w-album, Thecla 276
Walker, Francis 42, 146, 147
Walker, John 147, 343(fig.142)
Walker, Sir Patrick 343, 344
Wall butterfly 95, 103, 116, 393(fig.158)
Wallace, A. J. 134–6
Wallace, Alfred Russel 26, 164
Wallace, Rose 169
Wallis, H. M. 180
Walsingham, Lord *see* de Grey, Thomas, Lord Walsingham
wasps 56, 80
watercolours 110, 120
Waterhouse, E. A. 42, 306
Watkins & Doncaster 70, 71(fig.20), 78–9(figs 26–7), 231, 232, 403–404(fig.161), 406, 407, 408
Watkins, William 403–404(fig.160)
Watson, Robert W. **229–30(fig.122)**, 342
Weaver, Richard **133–6**, 138, 344; Museum of Natural History 134–6
Weaver's Fritillary 133–4, 155, 251(pl.15), **329–31(fig.140)**, 390
Webb, Sydney 160, 338
Weekly Entomologist 41, 402
Weir, John Jenner 147, **164–5(fig.72)**
Wellman, J. R. 403
Wells, William 32, 398
Welsh Clearwing moth 159
Werner, A. G. 148
Wernerian Natural History Society of Edinburgh 148
West London Entomological Society 46, 402
West, William 94, **172–3(fig.79)**
Westwood, Eliza 149
Westwood, John 149
Westwood, John Obadiah 37, 42, 44, 50, 127, 128–9, 139, 141, 145, **148–50(fig.59)**, 171, 187, 274, 288, 317, 400; *Entomologist's Text Book* 149; *Introduction to the Modern Classification of Insects* 149
Whalley, P. E. S. 110
Wheeler, F. D. 171
Wheeler, George 191
White Admiral 194, 387
White Border 322
White Clouded Yellow 264
White Dullidge Fritillary 340
White, Francis Buchanan **178–9(fig.84)**, 367(fig.144); *Fauna Perthensis* 179; *Insecta Scotica – Lepidoptera* 179
White, Gilbert 25, 99; *Natural History of Selborne* 145
White-letter Hairstreak 28, 274–6, 383
White-spot Argus 295
Whitehouse, Sir Beckwith 209, 368
Wigan, W. 336–7
Wildlife and Countryside Act, 1981 328, 371–2
Wildlife Trusts 214
Wilkes, Benjamin 26, 32, 35, 37, 60, 69, 70, 106, **110–12**, 123, 253, 398; '*Collecting Directions*' 75, 111; *Directions for making a Collection see* Wilkes, Benjamin, '*Collecting Directions*'; *English Moths and Butterflies* 29, 110, 111–12, 321, 322; *Twelve New Designs of English Butterflies* 29–31(pl.1), 111, 263
Wilkin, Simon 139
Wilkinson, Ronald S. 33, 106, 112, 219
Wilkinson, Walter 87, 254
Williams, C. B. 266, 271, 315
Willis, George 336
Willow Butterfly 111, 322, 323(pl.34)
Willughby, Francis **99–100(fig.37)**, 252, 397
wing-position, setting 75
wings: pattern variations 225; venation 117, 121
Wollaston, George 250
Wollaston, Henry 162
Wollaston, Thomas Vernon **161–2(fig.80)**; *Insecta Maderensia* 161; *On the Variation of Species* 162
Wollaston, William Hyde 161
Wollaston, William, *Religion of Nature* 161
women 25, 221, 225; entomologists 29, 60, 167, 136–7, 154, 160, 199–201
Wood Lady 112, 340–41(pl.38)
Wood White 379
Wood, Edward 45
Wood, J. G. 37, 320; *Common Objects of the Country* 37; *Insects at Home* 83
Wood, William, junior, 'Doubtful British Species' 327(pl.36)
Wood, William, senior 317; *Index Entomologicus* 249, 276, 327(pl.36)
Woodland, John 282
Woolhope Naturalists' Field Club 44, 401
Worldwide Butterflies 406–407, 408
Wotton, Edward 96(fig.32), 97(fig.33)
Wren, Sir Christopher 397
Wurzell, Brian 292

Yarrell, W. 400
Yorkshire Naturalists' Union 44
Young Naturalist 404, 405

Zeller, P. C. 297, 307
zeta assimilis, Apamea 134
Zoological Club of the Linnean Society of London 36, 142, 399
Zoological Record Association 163
Zoological Society of London 36–7, 144, 156, 163, 399; *Proceedings* 36; *Transactions* 36
Zoologist 41, 84, 133, 146, 153, 182, 192, 250, 344, 400–401

Printed in the United States
by Baker & Taylor Publisher Services